WOMEN IN THE HISTORY OF QUANTUM PHYSICS

Capturing the stories of sixteen women who made significant contributions to the development of quantum physics, this anthology highlights how, from the very beginning, women played a notable role in shaping one of the most fascinating and profound scientific fields of our time. Rigorously researched and written by historians, scientists, and philosophers of science, the findings in this interdisciplinary book transform traditional physics historiography. Entirely new sources are included alongside established sources that are examined from a fresh perspective. These concise biographies serve as a valuable counterweight to the prevailing narrative of male genius, and demonstrate that in the history of quantum physics, women of all backgrounds have been essential contributors all along. Accessible and engaging, this book is relevant for a wide audience including historians, scientists, science educators, gender theorists, and sociologists.

PATRICK CHARBONNEAU is Professor of Chemistry and Physics at Duke. His research focuses on theoretical aspects of soft matter and statistical physics. He also co-curated an exhibit on macromolecular visualization, leads an oral history project, and lectures on the history of chemistry.

MICHELLE FRANK is a 2024–2025 Public Scholar with the National Endowment for the Humanities. She was the 2023–2024 Sloan Fellow at the Leon Levy Center for Biography. She holds a JD from the University of Michigan and an MA from the CUNY Graduate Center. She is a former fellow with the Consortium for History of Science, Technology and Medicine.

MARGRIET VAN DER HEIJDEN is a particle physicist by training and a part time Professor of Science Communication in Physics at the Applied Physics Department of Eindhoven University of Technology. She is also a renowned science journalist and writer in the Netherlands, having published several nonfiction books on women's contributions to physics and mathematics.

DANIELA MONALDI is Assistant Professor in the Science, Technology and Society Department of York University, Canada. She teaches science and technology studies, the history of science, gender in STEM, and science, technology, and food. She was a postdoctoral fellow at the Max-Planck Institute for the History of Science in Berlin, Germany.

WOMEN IN THE HISTORY OF QUANTUM PHYSICS

Capturing the stories of sixteen women who made significant contributions to the development of quantum physics, this anthology highlights how, from the very beginning, women played a notable role in shaping one of the most fascinating and profound scientific fields of our time. Rigorously researched and written by historians, scientists, and philosophers of science, the findings in this interdisciplinary book transform traditional physics historiography. Entirely new sources are included alongside established sources that are examined from a fresh perspective. These concise biographies serve as a valuable counterweight to the prevailing narrative of male genius, and demonstrate that in the history of quantum physics, women of all backgrounds have been essential contributors all along. Accessible and engaging, this book is relevant for a wide audience including historians, scientists, science educators, gender theorists, and sociologists.

PATRICK CHARBONNEAU is Professor of Chemistry and Physics at Duke. His research focuses on theoretical aspects of soft matter and statistical physics. He also co-curated an exhibit on macromolecular visualization, leads an oral history project, and lectures on the history of chemistry.

MICHELLE FRANK is a 2024–2025 Public Scholar with the National Endowment for the Humanities. She was the 2023–2024 Sloan Fellow at the Leon Levy Center for Biography. She holds a JD from the University of Michigan and an MA from the CUNY Graduate Center. She is a former fellow with the Consortium for History of Science, Technology and Medicine.

MARGRIET VAN DER HEIJDEN is a particle physicist by training and a part time Professor of Science Communication in Physics at the Applied Physics Department of Eindhoven University of Technology. She is also a renowned science journalist and writer in the Netherlands, having published several nonfiction books on women's contributions to physics and mathematics.

DANIELA MONALDI is Assistant Professor in the Science, Technology and Society Department of York University, Canada. She teaches science and technology studies, the history of science, gender in STEM, and science, technology, and food. She was a postdoctoral fellow at the Max-Planck Institute for the History of Science in Berlin, Germany.

WOMEN IN THE HISTORY OF QUANTUM PHYSICS

Beyond *Knabenphysik*

Edited by

PATRICK CHARBONNEAU
Duke University

MICHELLE FRANK
2024–2025 Public Scholar, National Endowment for the Humanities

MARGRIET VAN DER HEIJDEN
Eindhoven University of Technology

DANIELA MONALDI
York University

CAMBRIDGE
UNIVERSITY PRESS

Shaftesbury Road, Cambridge CB2 8EA, United Kingdom

One Liberty Plaza, 20th Floor, New York, NY 10006, USA

477 Williamstown Road, Port Melbourne, VIC 3207, Australia

314–321, 3rd Floor, Plot 3, Splendor Forum, Jasola District Centre,
New Delhi – 110025, India

103 Penang Road, #05–06/07, Visioncrest Commercial, Singapore 238467

Cambridge University Press is part of Cambridge University Press & Assessment,
a department of the University of Cambridge.

We share the University's mission to contribute to society through the pursuit of
education, learning and research at the highest international levels of excellence.

www.cambridge.org
Information on this title: www.cambridge.org/9781009535830

DOI: 10.1017/9781009535816

First published 2025

A catalogue record for this publication is available from the British Library

A Cataloging-in-Publication data record for this book is available from the Library of Congress

ISBN 978-1-009-53583-0 Hardback

Cambridge University Press & Assessment has no responsibility for the persistence
or accuracy of URLs for external or third-party internet websites referred to in this
publication and does not guarantee that any content on such websites is, or will remain,
accurate or appropriate.

Contents

Contents

Contributors

Miriam Blaauboer
Department of Applied Sciences, Delft University of Technology, Delft, Netherlands

Patrick Charbonneau
Department of Chemistry and Department of Physics, Duke University, Durham, North Carolina, USA

Elise Crull
Department of Philosophy, The City College of New York & CUNY Graduate Center, New York, USA

Maria Paula Diogo
Centro Interuniversitário de História das Ciências e da Tecnologia (CIUHCT) and NOVA School of Science and Technology, Lisbon, Portugal

Michelle Frank
2024–2025 Public Scholar, National Endowment for the Humanities. 2023–2024 Sloan Fellow, Leon Levy Center for Biography at the CUNY Graduate Center, New York, USA

Bretislav Friedrich
Fritz-Haber-Institut der Max-Planck-Gesellschaft, Berlin, Germany

Stefano Furlan
Max Planck Institut für WissenschaftGeschichte (MPIWG), Berlin, Germany
Université de Genève, Geneva, Switzerland

Ivã Gurgel
Department of Nuclear Physics, Physics Institute, University of São Paulo, São Paulo, Brazil

Margriet van der Heijden
Department of Applied Physics, Eindhoven University of Technology, Eindhoven, Netherlands

Michel Janssen
School of Physics and Astronomy, University of Minnesota, Minneapolis, Minnesota, USA

Charnell Chasten Long
College of Education, North Carolina A&T State University, Greensboro, North Carolina, USA

Maria McEachern
Center for Astrophysics | Harvard & Smithsonian, Cambridge, Massachusetts, USA

Barbra Miguele
Interunit Postgraduate Program in Science Teaching, Physics Institute, University of São Paulo, São Paulo, Brazil

Adriana Minor
Centro de Estudios Históricos, El Colegio de México, Tlalpan, Ciudad de México, México

Daniela Monaldi
Science, Technology and Society Department, York University, Toronto, Ontario, Canada

Gernot Münster
Department of Physics, Universität Münster, Münster, Germany

Andrea Reichenberger
Department of Science, Technology and Society, TUM School of Social Sciences and Technology, Technical University of Munich, Munich, Germany

Mar Rivera Colomer
Independent Researcher and member of the Societat Catalana d'Història de la Ciència i la Tècnica, Catalonia, Spain

Jens Salomon
Wolfgang-Borchert-Gesamtschule, Recklinghausen, Germany

Ana Simões
Centro Interuniversitário de História das Ciências e da Tecnologia (CIUHCT), Faculdade de Ciências, Universidade de Lisboa, Lisbon, Portugal

Marta Jordi Taltavull
Institut Menorquí d'Estudis, Maó, Illes Balears, Spain

Foreword

Since I was awarded the Nobel Prize in Physics in 2023, one of the most frequent questions that people ask me is: Why aren't there more women in physics? Why aren't there more female role models for students to look up to? A few stand out: Marie Skłodowska Curie and Maria Goeppert Mayer, the only two women who were awarded the Nobel Prize in Physics during the twentieth century. Lise Meitner is another primary figure who is well-recognized. When I was asked to write the foreword of this book, I accepted, hoping that maybe I could find some of the answers to these questions in the field of quantum physics.

As I write these lines, I am also preparing a talk for the opening of the International Year of Quantum Science and Technology. Looking back in time at the history of early quantum mechanics, it is striking how much it has been dominated by male scientists, for example, Max Planck, Niels Bohr, Erwin Schrödinger, Werner Heisenberg, Paul Dirac, and Max Born, to name a few. Early quantum physics even received the nickname "boys' physics" (*Knabenphysik*), highlighting that some of these scientists were also very young.

The book *Women in the History of Quantum Physics: Beyond Knabenphysik*, edited by Patrick Charbonneau, Michelle Frank, Margriet van der Heijden, and Daniela Monaldi, tells the life stories and scientific contributions of 16 outstanding women in quantum science from different countries and at different times. It also attempts to explain why the contributions of these women have almost all been forgotten. We learn that there is no single explanation but several: motherhood, career changes, lack of mentorship, lack of support, discrimination, collaborating couples, and isolation. To give a flavor of the stories presented in this book, I summarize below the fate of four of these women.

Williamina Paton Fleming, employed by the Harvard College Observatory in the US, identified the spectrum of ionized helium from stellar spectra, which was essential for showing the validity of Bohr's atom model beyond neutral hydrogen.

Her discovery was called the "Pickering series" after the observatory director, Edward Charles Pickering. Her own name did not make it into physics textbooks.

Johanna van Leeuwen was a theoretical physicist in the Netherlands. She showed theoretically that magnetism could not be reconciled with a classical physics description. Her work was carried out at around the same time as Bohr's doctoral thesis and reached the same conclusion. This result was named the Bohr–Van Leeuwen's theorem. She is one of the few women described in this book that remained in academia. She suffered, however, from discrimination, becoming a reader at the Technical University of Delft only at the age of 59.

Elisabeth Monroe Boggs was a bright computational quantum chemist in the US. During World War II, she participated in the Manhattan Project. She left academia after giving birth to a disabled child in 1945 to take care of him. She became a high-level defender of the disability rights movement in the US.

Chien-Shiung Wu, who emigrated from China to the US in 1936, is a counter-example: In spite of discrimination both as a woman and as a Chinese immigrant in the US, she made a bright career in physics, being the first person to receive the prestigious Wolf Prize in 1978. She performed seminal experiments in particle physics and conducted early studies regarding quantum entanglement with photons.

Reading the life stories and career paths of the 16 women presented in this book filled me with admiration. The hurdles that they had to overcome to fulfill their dreams were incredibly high. So high that some of them had to give up at an early stage or accept being overshadowed by a supportive partner. I was born, on average, 50 years after these women. My career path has been much easier than theirs. Was I luckier? Maybe. In any case, fortunately, I believe that things have become easier for women in physics during the last 50 years.

This book shows that quantum physics was developed not only by male physicists, but also by the contributions of many talented women. The impressive historical and scientific research summarized in the different chapters will help us remember these women and what they contributed to science.

Lund, January 2025
Anne L'Huillier
Professor of Physics at Lund University, Sweden
Nobel Laureate in Physics, 2023

Introduction

PATRICK CHARBONNEAU, MICHELLE FRANK, MARGRIET VAN DER HEIJDEN,
AND DANIELA MONALDI

The year 2025 marks the centennial of quantum mechanics, a theory that has revolutionized scientific thinking about matter, energy, causality, and information. That theory continues to amaze researchers and the general public not only because of its internal intricacies, such as entanglement and superposition, but also because of the phenomena it has helped reveal, such as the quantum Hall effect and Bose–Einstein condensation, and not least because of the technologies it has brought forward, from transistors and lasers to the putative building blocks for future quantum computers. These advances underlie the rationale for making the 2025 centennial the International Year of Quantum Science and Technologies (IYQ, 2025).

Over a Century of Quantum Physics

Although the birth of quantum mechanics is generally acknowledged as a major turning point in the development of modern physics, it is of course part of a larger story. Quantum mechanics – like any other theory – did not simply fall from the sky in 1925. It had taken roughly a quarter of a century of instrumental, experimental, and theoretical work seeking to understand phenomena at the smallest accessible scales to shape the ideas that, starting in 1925, were cast in coherent – albeit difficult to interpret – mathematical formulas. The quantum-mechanical formalism, in turn,

Sincere appreciation is due to Michel Janssen and the APS Forum on History and Philosophy of Physics for launching the Women in the History of Quantum Physics (WiHQP) working group and for their ongoing advice over the years. Special thanks are due to Elise Crull for convening and leading WiHQP during 2021–2022. Her co-organization – with Guido Bacciagaluppi and Margriet van der Heijden – of the July 2022 Hermann and Friends workshop at the University of Utrecht, where many of the chapters in this volume were initially presented and discussed, was key to anchoring the group. The co-editors also thank Daniela Monaldi for subsequently chairing WiHQP, notably throughout this volume's production. We also thank the wonderfully diverse set of contributors to this anthology and all other WiHQP participants for their sustained dedication and engagement. Patrick Charbonneau finally thanks the Chimera group of Sapienza for hosting his sabbatical year during which this volume was edited.

proved to be a powerful theoretical basis for the growth of diverse and fruitful theoretical, experimental, and applied research programs.[1]

Interestingly, quantum mechanics is distinct from some of the other well-known and revolutionary theories in physics, such as Isaac Newton's mechanics and Albert Einstein's theory of relativity. These theories were mainly the result of strenuous efforts by single individuals, whose names remain attached to their work. By contrast, the formulation of quantum mechanics was a collective endeavor right from its beginning. It resulted not only from the development of theoretical physics as a professional specialization at the beginning of the twentieth century and the ongoing conversation between experimentalists and theorists, but also from the interplay between theoretical physicists from different schools. Such exchanges were facilitated by the emergence of centers of theoretical physics in various European countries in the early twentieth century. In the aftermath of World War I, interactions between these centers could and indeed did grow markedly (Seth, 2013; Schirrmacher, 2019; Kojevnikov, 2020).

Conventional and popular narratives of the quantum revolution tend to highlight a sequence of theoretical (and sometimes experimental) breakthroughs, each linked to one or two of the big names of physics history. In a typical version, the buildup begins in 1900 with Max Planck's proposal to mathematically quantize the energy to derive an empirically accurate formula for the entropy of thermal radiation. It is furthered in 1905 by Einstein's interpretation of the "light quanta" as physical entities (though this proposal would not be widely accepted for nearly two decades) and by his 1907 application of energy quantization to the specific heats of solids. The quantum hypothesis then gains additional buoyancy in 1913, when Niels Bohr successfully applies it to a solar-system model of the hydrogen atom. In the following decade, Arnold Sommerfeld, Bohr, and others extend and refine the quantum model of the atom. These studies raise a wealth of new questions about the behavior of electrons in heavier atoms, as well as questions about the interaction between such electrons and external electric and magnetic fields, and the underlying role of quantization. As theorists and experimentalists address the newest scientific questions, they pave the way toward the mathematically sophisticated and radical theory of quantum mechanics, which is successfully formulated in 1925–1926 by Werner Heisenberg, Max Born, Pascual Jordan, Paul Dirac, and (by a different route that passes through Louis de Broglie's ideas about matter waves) Erwin Schrödinger.

---

[1] It is impossible to do justice to the vast and multifaceted literature on the history of the quantum revolution. Among relevant and relatively recent works, we note Beller (1999), Darrigol (2002), Carson et al. (2011), Katzir et al. (2017), Schirrmacher (2019), Kojevnikov (2020), Duncan and Janssen (2019, 2023), and Freire (2022). A more general volume that places the development of quantum physics in the broader context of twentieth-century physics is Kragh (1999). A work that integrates quantum physics in an even broader perspective on the history of science in the twentieth century and beyond is Agar (2012). Other important works explore different facets and spotlight specific historical actors, events, and sites; they are too numerous to mention here. Helpful reviews of the historiography include Staley (2013), Badino (2016), and Joas and Hartz (2019).

By as early as 1968, however, science historian John Heilbron pointed out the nexus between social and scientific developments, writing, "The development of quantum physics was intimately linked to that of other branches of physical science, particularly statistical mechanics, the study of radioactivity, spectral analysis, and the theory of atomic structure. So, of course, it was connected somehow with the general cultural and social milieu in which it grew" (Heilbron, 1968, p. 90). Historians of physics have heeded the call. Since the 1970s, employing a rich methodological toolbox, historians have moved beyond the conventional story line by exploring the diversity of research streams, local contexts, and cultural, social, and institutional factors that fed into the canonical set of milestones and branched out from them. Nevertheless, as the philosopher and historian of science Massimiliano Badino noted, "Where HQP [history of quantum physics] lags shamefully behind other kinds of histories – and philosophy of science as well – is in the incorporation of geographical and gender perspectives. There has been a reassuring increase of studies on quantum physics in European and World peripheries, ... but much work still needs to be done. Analogously, the narratives remain as male-dominated as the discipline as a whole" (Badino, 2016, p. 334).[2]

By pointing to the persistent underrepresentation of women in the field of physics, Badino rightly underscores that gendered narratives and gender participation in the discipline go hand in hand. Especially at the higher rungs of the career ladder and in gate-keeping positions, women are still severely underrepresented. At the top levels of professional recognition, imbalances are stark: of 226 Nobel laureates in physics, only five are women, and three of them – Donna Strickland, Andrea Ghez, and Anne L'Huillier – were awarded the prize only in the last seven years. Progress has unquestionably been made, but disparities persist despite several decades of studies, evidence-based recommendations, and equity-oriented policies.[3] The persistent gender gap is a complex issue, attributable to subtle dynamics of institutional, social, cultural, and individual factors. Current analyses, however, widely agree that deepseated Western stereotypes attributing a masculine character to the hard sciences and associating a masculine identity with scientists play a large role in producing biases, professional segregation, and unfavorable conditions for the recruitment and

[2] A gender perspective should include all genders and not be limited to a focus on women. Focusing on women's contributions to science, however, does oftentimes bring up a broader consideration of gender. In this sense, the substantial subset of the literature on women in science that is dedicated to women in physics in the twentieth century, including biographies, biographical dictionaries, prosopographies, and gender analyses of research cultures, does offer a gender perspective on the field. An English-language sample of works specifically relevant to the history of women in quantum physics includes Kevles (1977, pp. 202–207), Traweek (1988), Schiebinger (1999), Rossiter (2002), Byers and Williams (2006), Götschel (2011), Rentetzi (2007), and Howes and Herzenberg (1999, 2015).

[3] The landscape of gender representation in science, technology, engineering, and mathematics (STEM) is complex and ever changing. There are significant regional and national differences, as well as differences between STEM disciplines and sub-disciplines, and between groups of different class, race, and ethnicity. In Western Europe, UK, US, Canada, and Australia, physics shares the problem of obdurate gender imbalances with other "hard" sciences, such as mathematics, engineering, computer science and technology, and (in a lesser measure) chemistry. Theoretical physics is one of the most male-dominated sub-specializations (Porter and Ivie, 2019; Nature Reviews, 2019; STEP UP, 2020; Roy et al., 2020; Schneegans et al., 2021).

retention of women in these disciplines; see, for example, Hill et al. (2010), Sekuła et al. (2018), and Thébaud and Charles (2018). By spotlighting a handful of male "geniuses," conventional narratives of the quantum revolution somehow throw a longer shadow on women than on men. In part this is because the historical lens, polished by tales of heroic genius, at times tacitly reinforces the stubborn stereotypes that portray women as insufficiently interested in or not brilliant enough to excel in the field of physics. New historical narratives that complement the conventional all-male story line, by shining light on women's participation in the enterprise and the structural obstacles they faced, may therefore help to dismantle the tenacious gender stereotypes that stand in the way of diversity, equity, and inclusion.

The scientists who created quantum mechanics famously formed a youthful group. In his book, *The Copenhagen Network: The Birth of Quantum Mechanics from a Postdoctoral Perspective* (2020), Alexei Kojevnikov notes that: "Over 80 authors took part in that brainstorming effort: The majority of them were under 30 years of age and they authored almost 70% of all publications. Some were still working on their dissertations, but more commonly, they were recent PhDs, having obtained their degrees after 1920, and would have been considered postdoctoral students by today's standards" (Kojevnikov, 2020, p. 3). The young age of several of the most prominent contributors to quantum mechanics prompted bittersweet jokes at the time, and quantum mechanics itself was sometimes colloquially referred to as *Knabenphysik*, or "boys' physics" (Weyl, 1946, p. 216).[4] In the public imagination, therefore, the genesis of quantum mechanics is tied to the idea of a special innate ability, the "raw talent" of a select group of young men, who were united not only by many personal and social characteristics but also bound together and to their mentors (such as Bohr and Born) by homosocial relationships (or, more colloquially, a "boys' club" or a "boys' network"): collaboration, competition, friendship, and mentorship. Although quantum mechanics was not the creation of a solitary hero of science, the rhetoric of virile heroism is not absent from its narrative. The evocation of *Knabenphysik* reconciles two images that may other-wise seem mutually exclusive: that of the rebellious and creative "solitary genius," which dominated older narratives, and that of science as a collective enterprise. According to the *Knabenphysik* trope, quantum mechanics was created by an all-male team of scientific heroes.[5] Certainly, the *Knabenphysik* characterization did

---

[4] Paul Ehrenfest, one of the mentors to the generation of theoretical physicists that came of age around 1925, felt overwhelmed by the storm of publications that heralded the new theory. In a letter to his brother Arthur, he wrote that he felt like an asthmatic dachshund chasing a tram full of young physicists (Ehrenfest to Arthur Ehrenfest, August 28, 1928; cited in Van der Heijden (2021, p. 290)).

[5] Kojevnikov attributes the subversive creativity of the quantum mechanics pioneers not (or not only) to their raw talent but to the dynamics of junior status in the academic hierarchies of the time, their professional insecurity, and their mobility in the historical context of the interwar period in Europe (Kojevnikov, 2020). The heroic overtones of the *Knabenphysik* trope still resonate in the genre of historical science popularizations, as illustrated by two recent bestsellers, Segrè (2007) and Rovelli (2022).

not intend to refer to the gender of the protagonists, only to their age and independent spirit. *Knabenphysik*'s gender connotation has long gone unnoticed and unquestioned due to the prevailing stereotype that physicists would be men and scientific genius a masculine attribute.

Women in the History of Quantum Physics

The Women in the History of Quantum Physics (WiHQP) working group first convened in early 2021 in direct response to the challenge of broadening the gender perspective on the field. This international and interdisciplinary team of physicists, historians, philosophers, and writers, including renowned academics and early-career researchers, seeks to deconstruct the myth that women somehow lacked enthusiasm, talent, or character to participate in quantum developments. Our working group does so here by shedding light on the contributions of 16 women scientists. Through this new lens on quantum developments, we aim to reach beyond *Knabenphysik* and to add a new dimension to the prevailing narrative that suggests quantum physics resulted from the efforts of small group of brilliant men. Our working group seeks to ensure that women are discussed as part of the rich history of quantum physics, throughout the IYQ and beyond.

For this volume, the WiHQP working group has opted *not* to focus on the more well-known heroines of physics, Marie Skłodowska Curie, Lise Meitner, and Maria Goeppert Mayer. (For their stories, we refer interested readers to existing scholarship, including, e.g., Boudia (2001), Emling (2012), Goldsmith (2011), Sime (1996b), Wuensch (2013), and Masters (2017).) They have by now become legendary figures. But by perpetuating a mythology of uniqueness, these women also seem to have become inimitable by definition, as historian of gender Julie Des Jardins pointed out in *The Madame Curie Complex: The Hidden History of Women in Science* (Des Jardins, 2010). Moreover, their high visibility may – inadvertently – reinforce the common idea that fewer than a handful exceptional women made rare contributions to one of the most fruitful intellectual revolutions of the twentieth century. As a counterweight, our anthology purposefully highlights scientists who are less well known or have hitherto remained in the shadows. Far more women have contributed to the progress of quantum physics than just the celebrities. Shifting the focus to them rebalances the emphasis on the exceptional and helps to "show us more about everyday science and the opportunities open and closed to most women," as historian Margaret Rossiter points out (Rossiter, 2002, p. 59).[6] And as historian of science Ruth Lewin Sime mentions, in so doing, it further contributes "to an expanded, more

---

[6] See also Rossiter's trilogy, *Women Scientists in America* (Rossiter, 1982, 1995, 2012).

nuanced understanding of social institutions, scientific practice, the personal lives of scientists, and science itself" (Sime, 1996a).

The WiHQP working group has also elected to use a broad definition of quantum physics, including the old quantum theory and the experiments that supported it, the birth of quantum mechanics and the philosophical enigmas associated with it, as well as quantum field theory and nuclear and particle physics, reaching beyond the traditional emphasis on leading centers of physics in Europe and the US. The table of contents of the resulting anthology is ordered chronologically in terms of key contributions to quantum science, starting with Williamina Fleming's (Chapter 1) discovery of spectroscopic lines that would prove crucial for validating Bohr's model, and ending with the present day, highlighting the activism for international scientific cooperation and the IYQ proposal by Mexican physicist Ana María Cetto (Chapter 16). The 16 narratives not only highlight women's contributions to quantum developments, but also illustrate how individual women scientists struggled with social conventions, scientific culture, and the – often unconscious or internalized – prejudices they confronted. Taken together, the chapters suggest several overarching mechanisms and possible explanations for why so many women scientists fell into obscurity. We offer these observations with the caveat that our contributors are not specifically trained as gender theory or women's studies scholars. This volume should nevertheless contribute to a broad and inter-disciplinary conversation on the theme of inclusion.

We are acutely aware that our book does not present the full breadth of women's contributions to quantum physics. A much larger group of women scientists still remains hidden in the shadows. Too often, the missing voices are those of women of color, and women from countries and regions that are often lumped together under the umbrellas of "the peripheries" and "the global south." In some cases, despite intense and lengthy recruitment efforts, we could not secure authors for scientists we hoped to include; in other cases, the scarcity of historical sources – a problem that especially plagues the archival collections pertaining to women and to people of color – proved discouraging to potential authors before they had a chance to begin.[7] We find the missing voices deeply troubling, not only because the absence of stories and images of scientists from diverse genders and backgrounds erases their historical contributions, but also because their invisibility has a particularly negative impact on women and other underrepresented groups in the present day. We hope that this volume will not perpetuate long-standing omissions, but instead will stimulate and inspire further scholarship with an increasingly wide lens.

---

[7] We hesitate to include names for women scientists we were unable to feature, as mentioning some implies omitting many others. Willie Hobbes Moore, Purnima Sinha, Bibha Chowdhuri, and Shirley Ann Jackson are but a few examples.

Emerging Themes

The chapters of this book encompass a diversity of time periods, contexts, and individual experiences. Each of them provides a detailed description and analysis of a scientist's unique trajectory in its specific context, to be appreciated in all its complexity and depth. In addition, we see several themes related to women's experiences in science emerging across the chronological timeline of the chapters. They notably include: isolation and invisibility; preconceptions about raw talent and the culture of competition; interrupted careers; hidden labor in science; intersectionality; and the role of collaborative couples.

Isolation and Invisibility

Women's erasure and omission from historical records may be one of the few constants of history. History of science is no different. In 1897, the French mathematician Alphonse Rebière published a volume about 650(!) women in the sciences who had previously been overlooked (Rebière, 1897). He was neither the first nor the last to do so (Boucard, 2020). Fifty years after Rebière, a US Department of Labor report titled *The Outlook for Women in Physics and Astronomy* bemoaned the "paucity of published information on women in science" despite the authors having consulted "more than 800 books, articles and pamphlets" (Zapoleon et al., 1948, p. 6-III). The Sources for History of Quantum Physics project, conducted before the emergence of the field of studies on gender and science, was no exception (Chapter 16). The general dearth of records about women remains a genuine obstacle, as several leading archives and archivists publicly acknowledge (Zanish-Belcher and Voss, 2017).

Women's erasure – documentary or otherwise – is difficult to remedy. The tendency to shape scientific discoveries into heroic tales has perhaps compounded the problem. As historian of science Naomi Oreskes points out, "Emphasizing activities that might be considered irresponsible if undertaken by a woman, the heroic ideology relegates women's work to the realm of the inconsequential. The marginalization of women in science is a predictable consequence of heroic rhetoric . . ." (Oreskes, 1996, p. 111). But it would be an oversimplification, as Oreskes also notes, to conclude that women have remained largely invisible in the historiography and popular narratives of quantum physics, exclusively *because* so few were able to attain the rank of "heroines of science." Instead, it is worth asking what results from "heroic ideology" with its concomitant impulse to neglect collaboration and collective effort.

Another crucial question is why women scientists have less frequently been found in the highest ranks of academia. Here, too, it helps to shift our gaze from the individual to the broader structure of the institution. Too often, a scientist found herself the sole woman in a laboratory or lecture room and was not readily admitted to the old (or young) boys' networks. For instance, Jo van Leeuwen (Chapter 2) was always a bit of an outsider among the young theoretical physicists in Leiden. Her supervisor, Hendrik Antoon Lorentz, had retired from the university and lived out of town. His successor, Paul Ehrenfest, gathered a circle of promising students, but Van Leeuwen was not truly part of this close-knit group, who would visit Ehrenfest at home, attend colloquia in the study of Ehrenfest's large house, and help Tatiana Ehrenfest-Afanassjewa in the garden. Moreover, although Van Leeuwen had been part of a small wave of women who enrolled in Leiden University to study physics, she went on to become the only woman in her department when she landed a position at Delft Technical University. She would remain in that solitary situation for the rest of her career.

The solitary and subordinate position of women scientists often had the effect of making them dependent on goodwill from select mentors who dared to break with conventions that otherwise interfered with women's full participation. Too often, such goodwill and support were partial and incomplete. When Jane Dewey (Chapter 5), for example, carried out an experiment on the Stark effect in helium while on a fellowship at Bohr's institute in 1926, Bohr prioritized the publication of the parallel experimental results of another visitor, J. Stuart Foster, with whom he had a relation of friendship and informal mentorship. At Princeton, where Dewey later landed as the first female postdoc thanks to the support of William Francis Magie, she was nicknamed "Magie's Folly" by colleagues who saw no room for women in physics (see also Kevles (1977), p. 207). Foster, in the meantime, befriended and collaborated with Heisenberg, who later would effusively recall Foster's work on the Stark effect in his written and oral memoirs about the birth of quantum mechanics. Heisenberg completely ignored Dewey's parallel work, a further illustration of how isolation (from the boys' networks) and invisibility (through blind spots of bias and prejudice) can combine to push women toward the margins. Even Foster's PhD student, Laura Chalk (Chapter 6), who actually produced the initial data validating the first new prediction of quantum mechanics while measuring the Stark effect in hydrogen, was left out of Heisenberg's narrative.

Whereas Heisenberg and Foster, as well as many other male scientists, celebrated one another's work with reciprocal gestures that were a steady element of their social and professional networks, Chalk, Dewey, and Van Leeuwen were kept out of these networks by gender norms, and worked in the absence of female colleagues who would propagate and advertise their work. Women scientists persisted in often isolated and dependent positions, rendering it nearly impossible to reach out to one another to obtain career advice from *girls'* networks until much later in the twentieth century (Rentetzi and Kohlstedt, 2009). A prime example of someone who has made a distinct commitment to improving networking opportunities for women is the Mexican quantum physicist Ana María Cetto (Chapter 16), who sees scientific collaboration as one of the pillars of international diplomacy. When, in 1989, the Third World Organization for Women in Science (TWOWS, now the Organization for Women in Science for the Developing World, OWSD) was established, Cetto became one of its inaugural vice presidents, making the elimination of gender bias in science one of her priorities.

Throughout history there also have been men who successfully supported women in physics research. Still, the gender-normative exclusion of women from male socio-professional circles had an overpowering dual impact. First, it disadvantaged women in scientific productivity and career advancement. Second, it contributed to *invisibilizing* women in scientific narratives, especially when those narratives were cast under the guise of virile heroism. In the male-dominated world of twentieth-century physics, what biophysicist Ellen Weaver wrote about female scientists who, like her, participated in the Manhattan Project, was especially poignant:

When I read the personal reminiscences of the men who were pioneers in the nuclear field, I am struck by the importance they attach to their friends and colleagues, and to the intense interaction often present among them, which could lead to important insights in both the theory and practice of science. And I am a little jealous. In general, women did not participate in that exchange of ideas (Weaver, 1999, p. viii).

Raw Talent Preconception and Culture of Competition

A recent psychological analysis highlights that even today, after half a century of equality policies and efforts to eliminate gender gaps in academia, "women are stereotyped to possess less [raw intellectual talent] than men" (Meyer et al., 2015, p. 1). Noting that these stereotypes impact the fields' gatekeepers as well as other participants, Meyer et al. find that the disciplines with the widest and most persistent gender gaps – such as physics and mathematics – are often those in which success is mainly attributed to brilliance and raw talent rather than collaborative, diligent, and persistent work. According to these social scientists, one key mechanism through which this underrepresentation is produced is the existence in such fields of "masculinity-contest cultures," organizational environments in which ruthless competition discourages women's participation. For Vial et al., women express less interest and a lesser sense of belonging in fields of study and professions whose image emphasizes brilliance and competitiveness (Vial et al., 2022).

Lucy Mensing (Chapter 4), a forgotten pioneer of quantum mechanics who obtained her doctorate in 1925 in Hamburg with Wilhelm Lenz and Wolfgang

Pauli and was a postdoctoral scholar in Göttingen during the key years of the birth of quantum mechanics, may be a prominent example of this effect. In 1928, she indeed left physics ostensibly to marry and start a family, but also likely as a result of the fiercely competitive (and destructive) climate of the physics research environment she had experienced during her subsequent appointment in Tübingen.

A bit more speculatively, one can also wonder if Katharine Way's (Chapter 8) choice of a backstage role as creator and curator of the foundational Nuclear Data Project reflected a desire to remain involved in nuclear physics while keeping clear of the fierce competition in the field. Way's dissertation (with John A. Wheeler, on the instability of a rotating heavy atomic nucleus) and subsequent roles in the construction of the first nuclear reactors for the Manhattan Project could certainly have served as stepping stones for an altogether different, and perhaps more prominent, career in theoretical physics, had she chosen to pursue it.

Cetto (Chapter 16) has critiqued the culture of competition that she encountered as a graduate student at Harvard University. As Mar Rivera Colomer explains in that chapter, Cetto described the setting at Harvard as "characterized by competition and a notable absence of solidarity." In contrast, she found the Mexican scientific field to be "more open, flexible, and accommodating," which she tentatively attributed to it having a "relatively lower maturity level in scientific production."

Interrupted Careers

Gender norms, stereotypes, and biases, and the related social pressure to conform to such norms can help explain some of the interrupted careers of women in twentieth-century physics. As early as 1965, the American sociologist Alice Rossi pondered, "Women in science: why so few?" (Rossi, 1965). Her answer – along with that of many others – focused on the different social roles played by men and women, including the priority that society placed (and often still places) on women committing to marriage and motherhood. These factors certainly have presented barriers to women's full participation in scientific research as well as in academic and professional life, but they are not the only considerations. The concept of the *leaky pipeline* aims to shine light on how structural flaws in workplaces, and in broader society, disproportionately place the burden upon women to balance personal and professional responsibilities, and how those same structural flaws simultaneously blame women for exiting the professional realm. The leaky pipeline metaphor can help point out where fields like physics suffer from a loss of available talent. That metaphor, however, is not without limitations. While it conveys losses from the perspective of the talent pool for scientific research it also raises what are perhaps inappropriate concerns about the balance between educational investments and professional output. In so doing, it risks minimizing valuable contributions to social progress that many women have made in other capacities after they leave the scientific pipeline.

Without a doubt, the stories of several women in the present volume illustrate how difficult it was to reconcile a career in physics with the gender normative roles of wife, caregiver, and mother. But many also point out that even in cases where systemic, political, or personal obstacles prompted women to leave quantum physics entirely, it was not uncommon for those same women to subsequently make major contributions in other fields. After the fall of the Nazi regime, Grete Hermann (Chapter 11) left the philosophy of quantum mechanics to pursue political and educational reform in West Germany. When faced with institutional obstacles at Southern Illinois University, Maria Lluïsa Canut (Chapter 15) turned her attention from quantum physics to ethical and societal issues, in her case second-wave feminism in the US. Frieda Friedman Salzman (Chapter 14) would valiantly fight against gender discrimination after anti-nepotism policies deprived her of a secure faculty role at the University of Massachusetts Boston. Elizabeth Monroe Boggs (Chapter 7), who trained as a computational quantum chemist at the University of Cambridge and later worked on the Manhattan Project, left the scientific workforce after giving birth to a child with disabilities, and pivoted to a remarkable life of public advocacy.

Some career interruptions, however, were altogether unavoidable. A particularly dramatic one affected Sonja Ashauer (Chapter 9), the first Brazilian woman to obtain a doctorate in physics. She died of bronchopneumonia at the age of 25, six months after defending her thesis on the nonphysical consequences of the equation for the point electron in quantum theory at the University of Cambridge. She was one of Paul Dirac's few doctoral students, and the only woman he mentored.

Perhaps also for reasons of ill health, Carolyn Parker withdrew suddenly from her PhD program in 1954 and then again 1955. Little more than a decade later, she passed away at the young age of 49. Archival silences, however, make it difficult to determine whether these interruptions in Parker's later career were related to illness or to other hardship.

Hidden Variables

In 1989, the historian of science Steven Shapin described how the experiments of the famous seventeenth-century chemist and natural philosopher Robert Boyle were in fact conducted by the invisible hands of a fleet of (male) operators and assistants. Shapin's paper, "The invisible technician," makes it clear that others designed and built instruments, collected data, and sometimes even drafted the publications, rather than Boyle himself. "The predominant biases in the Western academic world," Shapin wrote, "have traditionally portrayed science as a traditional and wholly rational enterprise carried out by reflective individual thinkers." It led him to conclude that: "People who are really present but invisible are those whose roles are considered to be unimportant" (Shapin, 1989, p. 563). This observation is not only true for technicians, but also for lower-rank physicists more generally. Rather than standing on the shoulders of giants, the scientists who have achieved celebrity status often stood on the backs of a great number of "hidden figures," both men and women (Star, 1991; Shapin, 1989, 1994; Bangham et al., 2022; Shetterly, 2016). In this sense, the low-visibility work of many women physicists becomes hard to distinguish from that of the majority of (male) physicists in subordinate positions or behind-the-scenes roles. For women, however, the issue is exacerbated by the larger proportion of them remaining at the lower ranks of professional hierarchies.

Williamina Fleming (Chapter 1) is a salient example of such erasure. She was one of the women *computers* hired on grossly unequal terms compared with men who were employed by the Harvard College Observatory. Later, as the curator of the Astronomical Photographic Glass Plate Collection, she discovered a peculiar pattern of spectroscopic lines. Her discovery, which later played a singular role on the path to quantum mechanics by serving as proving ground for Bohr's model of the atom, nevertheless became known as the "Pickering series" after the observatory director, Edward Charles Pickering.

Similarly, Hertha Sponer (Chapter 3), who ran James Franck's spectroscopy laboratory in Göttingen in the 1910s, designed and executed ground-breaking experimental work that applied quantization rules to molecular spectroscopy, for which Franck received significant credit. During her year-long Rockefeller Fellowship at the University of California, Berkeley, she also worked with prominent American spectrographer Raymond Birge; Sponer taught the group, including Edward Condon, how to apply quantization methods to radiation. She convinced Birge to jump on the quantum physics bandwagon early, and it paved the way for Condon's famous achievement (albeit without Sponer): the Franck–Condon principle. Meanwhile, Sponer's name slipped quietly from public and scientific consciousness.

Way (Chapter 8), as mentioned above, was the leading force behind the Nuclear Data Project, which quickly became an indispensable reference for the experimental, theoretical, engineering, and even biomedical communities as well as an opportunity to establish common standards among them. For Way's backstage work, Wheeler and others nominated her for the 1978 APS Tom W. Bonner Prize in Nuclear Physics. But, despite the strong supporting party, the prize was not awarded to her, neither that year nor later.

Intersectionality

For Chien-Shiung Wu (Chapter 10) underrepresentation was manifold. She immigrated to the US from China and became one of relatively few women who studied physics *and* who had a successful career in experimental research. Still, she was described as "a decorative addition to any laboratory" and compared to a "lotus" as a young woman. Later, colleagues referred to Wu with derogatory ethnic stereotypes, even at the peak of her career. She conducted groundbreaking experiments that did not receive the recognition they deserved. Despite her pivotal role, Wu was passed over for the 1957 Nobel Prize that celebrated the "penetrating investigation of the so-called parity laws," and many years later, when "experiments with entangled photons" led to the 2022 Nobel Prize, her 1949 experiments in this field also seemed to have been glossed over.

Carolyn Parker's (Chapter 13) life and her physics career took shape in the midst of racial oppression in the Jim Crow South and in the grip of enduring northern US racism. Her story illustrates the intersectional barriers that a young Black woman aspiring to a nontraditional field encountered in the mid-century US. The intricate patchwork of traditional archival and Black counter-archival sources supporting this chapter also shows how scholars of Black history often must contend with silences and gaps in historical records (Hartman, 2008). The work notably reveals how circumstances forced Parker to detour at seemingly every step along the way. Between obtaining her degree in physics from Fisk University in 1938, contributing to applied research for the US military during World War II, and then obtaining a master's degree in physics at MIT in 1953, she also taught mathematics and physics in segregated high schools and in historically Black colleges and universities – detours which were arguably beneficial to the broader Black community, while perhaps also slowing her scientific trajectory.

Collaborative Couples

For women in this volume, participation in research as part of a collaborative scientific couple emerges as a two-sided coin. In the foreword to the anthology *For Better or For Worse? Collaborative Couples in the Sciences*, science historian Sally Gregory Kohlstedt remarked that "Viewed collectively, the results seem to be most consistently 'better,' [for women working in collaborative couples] especially if one of the measures is the science produced" (Lykknes et al., 2012, p. viii).[8] Marital and scientific partnership likely opened doors for women, making research opportunities more readily available than would otherwise have been possible; couplehood facilitated women's collaboration when they worked in partnership

---

[8] See also Pycior et al. (1996).

with a supportive spouse. At the same time, anti-nepotism policies often blocked married women who wished to continue their scientific and academic careers.

Being part of a collaborative couple certainly was a double-edged sword for Maria Lluïsa Canut (Chapter 15). She built her career in the hostile environment of Francoist Spain, where gendered societal roles were promoted through segregated boys' and girls' education and curricular differences. In doing so, Canut, from an elite family in Menorca, worked in tandem with her husband and collaborator José Luis Amorós. He initially acted as a supportive mentor because he held higher positions. Eventually, however, he would overshadow her. The same fate befell Cetto (Chapter 16), who then diversified her interests, partly as a strategy to uphold her own scientific identity.

Freda Friedman Salzman (Chapter 14) and George Salzman had always striven to stay together despite the "two-body problem" that often plagues scientific couples looking jointly for a new position. After they both obtained a professor-ship at the Boston campus of the University of Massachusetts, it was Friedman Salzman, and not her husband, who faced (and fought) the threat of exclusion under anti-nepotism policies – policies that disproportionately impacted univer-sity women.

Finally, the narrative of the Portuguese Lídia Salgueiro (Chapter 12) and her partner and collaborator José Francisco Gomes Ferreira illustrates how gendered perceptions of the outside world influence gender roles within the laboratory. The two supported one another when forced to reinvent their research agenda in response to national political pressures. Salgueiro was a behind-the-scenes researcher, whose scientific guidance and relevance was core to the group, but in contrast to Gomes Ferreira she seemed almost invisible to the outside world. The couple thus enacted the expected gendered roles of female self-effacement and male visibility, a mechanism that seems even more subtle than the tendency of the outside world to attribute a woman's work to a male colleague, the so-called Matilda effect (Rossiter, 1993).[9]

All things being equal, in these chapters it is thus women's visibility that suffers.

Epilogue

In 1906, when Van Leeuwen began to study physics in Leiden and 10 years after Fleming measured her peculiar pattern of spectroscopic lines, the Indian political activist, poet, and feminist Sarojini Naidu discussed the education of women in a speech to the Indian National Congress. "In the matter of education you cannot

---

[9] Margaret Rossiter named the effect in 1993 in commemoration of suffragist Matilda Joslyn Gage, who in an 1893 essay protested against the idea that a "woman ... possesses no inventive or mechanical genius."

say *thus far and no further*," she said. "Neither can you say to the winds of Heaven 'Blow not where ye list,' nor forbid waves to cross their boundaries, nor yet the human soul to soar beyond the bounds of arbitrary limitations" (Sarojini, 1906).

Yet, obstacles like the ones outlined above would continue to hinder women socially and institutionally throughout the twentieth century, raising questions about how substantial women's influence on quantum developments otherwise might have been. Questions of this type are not new. As early as 1738 French scientist Marquise Émilie du Châtelet wrote (du Châtelet, 1735, par. 24)[10]:

> Were I king, I would like to try this physical experiment. I would redress an abuse which cuts back, as it were, one half of humankind. I would have women participate in all human rights, especially those of the mind.

The result of such an experiment is obviously unknowable. But by shining a bright light on women in the history of quantum physics on the occasion of the IYQ, we hope that this volume contributes toward redressing the field's unbalanced history, and that it can be a sure step toward achieving a more inclusive world of physics, of science, and beyond, within our lifetime.

References

Agar, Jon. 2012. *Science in the Twentieth Century and Beyond*. Cambridge: Polity Press.

Badino, Massimiliano. 2016. What have the historians of quantum physics ever done for us? *Centaurus*, 58(4), 327–346. https://doi.org/10.1111/1600-0498.12127

Bangham, Jenny, Chacko, Xan, and Kaplan, Judith (eds.). 2022. *Invisible Labour in Modern Science*. Lanham, MD: Rowman & Littlefield Publishers.

Beller, Mara. 1999. *Quantum Dialogue: The Making of a Revolution*. Chicago: University of Chicago Press.

Boucard, Jenny. 2020. Arithmetic and memorial practices by and around Sophie Germain in the 19th century. In Kaufholz-Soldat, Eva and Oswald, Nicola M. R. (eds.), *Against All Odds: Women's Ways to Mathematical Research Since 1800*. Cham: Springer. https://doi.org/10.1007/978-3-030-47610-6_7

Boudia, Soraya. 2001. *Marie Curie et son laboratoire: sciences et industrie de la radioactivité en France*. Paris: Editions des archives contemporaines.

Byers, Nina and Williams, Gary (eds.). 2006. *Out of the Shadows: Contributions of Twentieth Century Women to Physics*. Cambridge: Cambridge University Press.

Carson, Catryn, Kojevnikov, Alexei, and Trischler, Helmut (eds.). 2011. *Weimar Culture and Quantum Mechanics: Selected Papers by Paul Forman and Contemporary Perspectives on the Forman Thesis*. London: Imperial College Press and World Scientific.

Darrigol, Olivier. 2002. Quantum theory and atomic structure, 1900–1927. In Nye, Mary Jo (ed.), *The Cambridge History of Science*. Cambridge: Cambridge University Press, pp. 329–349.

---

[10] "Pour moy j'avoüe que si j'etois roy, je voudrois faire cette experience de physique. Je reformerois un abus qui retranche, pour ainsi dire[,] la moitié du genre humain. Je ferois participer les femmes à tous les droits de l'humanité, et sur tout à ceux de l'esprit."

Des Jardins, Julie. 2010. *The Madame Curie Complex: The Hidden History of Women in Science*. New York: Feminist Press at CUNY.

du Châtelet, Émilie. 1735. Préface du traducteur. In *L'adaptation de La Fable des abeilles*. Center for the History of Women Philosophers and Scientists, University of Paderborn. https://historyofwomenphilosophers.org/stp/documents/view/mandeville

Duncan, Anthony and Janssen, Michel. 2019. *Constructing Quantum Mechanics: Volume 1: The Scaffold: 1900–1923*. Oxford: Oxford University Press.

Duncan, Anthony and Janssen, Michel. 2023. *Constructing Quantum Mechanics Volume 2: The Arch, 1923–1927*. Oxford: Oxford University Press.

Emling, Shelley. 2012. *Marie Curie and Her Daughters: The Private Lives of Science's First Family*. New York: St. Martin's Publishing Group.

Freire, Olival, Jr. (ed.). 2022. *The Oxford Handbook of Quantum Interpretations*. Oxford: Oxford University Press.

Goldsmith, Barbara. 2011. *Obsessive Genius: The Inner World of Marie Curie*. New York: W. W. Norton.

Götschel, Helene. 2011. The entanglement of gender and physics: Human actors, workplace cultures, and knowledge production. *Science Studies*, 24(1), 66–80. https://doi.org/10.23987/sts.55270

Hartman, Saidiya. 2008. Venus in two acts. *Small Axe: A Caribbean Journal of Criticism*, 12(2), 1–14. https://doi.org/10.1215/-12-2-1

Heilbron, John L. 1968. Quantum historiography and the archive for history of quantum physics. *History of Science*, 7(1), 90–111. https://doi.org/10.1177/007327536800700103

Hill, Catherine, Corbett, Christianne, and St. Rose, Andresse. 2010. *Why So Few? Women in Science, Technology, Engineering, and Mathematics*. Washington: American Association of University Women.

Howes, Ruth H. and Herzenberg, Caroline L. 1999. *Their Day in the Sun: Women of the Manhattan Project*. Philadelphia, PA: Temple University Press.

Howes, Ruth H. and Herzenberg, Caroline L. 2015. *After the War: Women in Physics in the United States*. San Rafael, CA: Morgan and Claypool.

IYQ. 2025. *International Year of Quantum Science and Technology*. https://quantum2025.org

Joas, Christian and Hartz, Thiago. 2019. Quantum cultures: Historical perspectives on the practices of quantum physicists. *Berichte zur Wissenchaftsgeschichte*, 42(4), 286–289. https://doi.org/10.1002/bewi.201970043

Katzir, Shaul, Lehner, Christoph, and Renn, Jürgen. 2017. *Traditions and Transformations in the History of Quantum Physics*. Berlin: Edition Open Access.

Kevles, Daniel J. 1977. *The Physicists: The History of a Scientific Community in Modern America*. New York: Alfred A. Knopf.

Kojevnikov, Alexei. 2020. *The Copenhagen Network: The Birth of Quantum Mechanics from a Postdoctoral Perspective*. Cham: Springer.

Kragh, Helge. 1999. *Quantum Generations: A History of Physics in the Twentieth Century*. Princeton, NJ: Princeton University Press.

Lykknes, Annette, Opitz, Donald L., and van Tiggelen, Brigitte (eds.). 2012. *For Better or For Worse? Collaborative Couples in the Sciences*. Heidelberg: Springer.

Masters, Barry. 2017. The origins of Maria Göppert's dissertation on two-photon quantum transitions at Göttingen's Institutes of Physics 1920–1933. In Katzir, Shaul, Lehner, Christoph, and Renn, Jürgen (eds.), *Traditions and Transformations in the History of Quantum Physics*. Berlin: Edition Open Access, pp. 209–230.

Meyer, Meredith, Cimpian, Andrei, and Leslie, Sarah-Jane. 2015. Women are underrepresented in fields where success is believed to require brilliance. *Frontiers in Psychology*, 6, 235. http://dx.doi.org/10.3389/fpsyg.2015.00235

Nature Reviews. 2019. Data on women in physics. *Nature Reviews Physics*, 1(5), 297. https://doi.org/10.1038/s42254-019-0061-3

Oreskes, Naomi. 1996. Objectivity or heroism? On the invisibility of women in science. *Osiris*, 11(1), 87–113. https://doi.org/10.1086/368756

Porter, Anne Marie and Ivie, Rachel. 2019. *Women in Physics and Astronomy, 2019*. College Park, MD: American Institute of Physics. https://ww2.aip.org/statistics/women-in-physics-and-astronomy-2019

Pycior, Helena, Slack, Nancy, and Abir-Am, Pnina G. (eds.). 1996. *Creative Couples in the Sciences*. New Brunswick, NJ: Rutgers University Press.

Rebière, Alphonse. 1897. *Les femmes dans la science, 2nd ed*. Paris: Nony & cie.

Rentetzi, Maria. 2007. *Trafficking Materials and Gendered Experimental Practices: Radium Research in Early 20th Century Vienna*. New York: Columbia University Press.

Rentetzi, Maria and Kohlstedt, Sally G. 2009. Introduction: Gender and networking in twentieth-century physical sciences. *Centaurus*, 51(1), 5–11. https://doi.org/10.1111/j.1600-0498.2008.00133.x

Rossi, Alice S. 1965. Women in science: Why so few? Social and psychological influences restrict women's choice and pursuit of careers in science. *Science*, 148(3674), 1196–1202. https://doi.org/10.1126/science.148.3674.1196

Rossiter, Margaret W. 1982. *Women Scientists in America: Struggles and Strategies to 1940*. Baltimore, MD: Johns Hopkins University Press.

Rossiter, Margaret W. 1993. The ~~Matthew~~ Matilda effect in science. *Social Studies of Science*, 23(2). https://doi.org/10.1177/030631293023002004

Rossiter, Margaret W. 1995. *Women Scientists in America: Before Affirmative Action, 1940–1972*. Baltimore, MD: Johns Hopkins University Press.

Rossiter, Margaret W. 2002. A twisted tale: Women in the physical sciences in the nineteenth and twentieth centuries. In Nye, Mary J. (ed.), *The Cambridge History of Science, Volume 5, The Modern Physical and Mathematical Sciences*. Cambridge: Cambridge University Press, pp. 54–71. https://doi.org/10.1017/CHOL9780521571999.005

Rossiter, Margaret W. 2012. *Women Scientists in America: Forging a New World since 1972*. Baltimore, MD: Johns Hopkins University Press.

Rovelli, Carlo. 2022. *Helgoland: The Strange and Beautiful Story of Quantum Physics*. London: Penguin Books.

Roy, Marie-Françoise, Guillopé, Colette, Cesa, Mark, et al. 2020. *A Global Approach to the Gender Gap in Mathematical, Computing, and Natural Sciences: How to Measure It, How to Reduce It?* Genève: Zenodo. https://doi.org/10.5281/zenodo.3882609

Sarojini, Naidu. 1906. *Education of Indian Women. The Calcutta Congress & Conferences*. Calcutta: G. A. Natesan & Company, pp. 176–177.

Schiebinger, Londa. 1999. Physics and Math. *Has Feminism Changed Science?* Cambridge, MA: Harvard University Press, pp. 159–179.

Schirrmacher, Arne. 2019. *Establishing Quantum Physics in Göttingen: David Hilbert, Max Born, and Peter Debye in Context, 1900–1926*. New York: Springer International Publishing.

Schneegans, Susan, Straza, Tiffany, and Lewis, Jake (eds.). 2021. *UNESCO Science Report: The Race Against Time for Smarter Development*. Paris: UNESCO Publishing. https://unesdoc.unesco.org/ark:/48223/pf0000377433

Segrè, Gino. 2007. *Faust In Copenhagen: A Struggle for the Soul of Physics and the Birth of the Nuclear Age*. New York: Viking.

Sekuła, Paulina, Struzik, Justyna, Krzaklewska, Ewa, and Ciaputa, Ewelina. 2018. *Gender Dimensions of Physics: A Qualitative Study from the European Research Area*. Krakow: GENERA Network. https://www.genera-network.eu/gip:generainterviews

Seth, Suman. 2013. Quantum physics. In Buchwald, Jed Z. and Fox, Robert (eds.), *The Oxford Handbook of the History of Physics*. Oxford: Oxford University Press, pp. 814–859. https://doi.org/10.1093/oxfordhb/9780199696253.013.28

Shapin, Steven. 1989. The invisible technician. *American Scientist*, 77(6), 554–563. https://www.jstor.org/stable/27856006

Shapin, Steven. 1994. *A Social History of Truth: Civility and Science in Seventeenth-Century England*. Chicago, IL: University of Chicago Press.

Shetterly, Margot L. 2016. *Hidden Figures: The American Dream and the Untold Story of the Black Women Who Helped Win the Space Race*. New York: William Morrow.

Sime, Ruth L. 1996a. Partnerships. *Science*, 273(5273), 316. https://doi.org/10.1126/science.273.5273.316-a

Sime, Ruth L. 1996b. *Lise Meitner: A Life in Physics*. Berkeley, CA: University of California Press.

Staley, Richard. 2013. Trajectories in the history and historiography of physics in the twentieth century. *History of Science*, 51(2), 151–177. https://doi.org/10.1177/007327531305100202

Star, Susan L. 1991. The sociology of the invisible: The primacy of work in the writings of Anselm Strauss. In Strauss, Anselm L. and Maines, David R. (eds.), *Social Organization and Social Process: Essays in Honor of Anselm Strauss*. Hawthorne, NY: Aldine de Gruyter, pp. 265–283.

STEP UP. 2020. *Women in Physics International Factsheet*. APS STEP UP Physics Together. https://higherlogicdownload.s3.amazonaws.com/APS/2c0c9f07-6428-4f8e-b9aa-a76098a80cd0/UploadedImages/WiP_InternationalFacts_2020.pdf

Thébaud, Sarah and Charles, Marie. 2018. Segregation, stereotypes, and STEM. *Social Sciences*, 7(7), 11. https://doi.org/10.3390/socsci7070111

Traweek, Sharon. 1988. *Beamtimes and Lifetimes: The World of High Energy Physics*. Cambridge, MA: Harvard University Press.

Van der Heijden, Margriet. 2021. *Denken is verrukkelijk, het leven van Tatiana Afanassjewa en Paul Ehrenfest*. Amsterdam: Uitgeverij Prometheus.

Vial, Andrea C., Muradoglu, Melis, Newman, George E., and Cimpian, Andrei. 2022. An emphasis on brilliance fosters masculinity-contest cultures. *Psychological Science*, 33(4), 595–612. https://doi.org/10.1177/09567976211044133

Weaver, Ellen C. 1999. Foreword. In *Howes, Ruth H. and Herzenberg, Caroline L. Their Day in the Sun: Women of the Manhattan Project*. Philadelphia: Temple University Press, pp. vii–viii.

Weyl, Hermann. 1946. Encomium. *Science*, 103(2669), 216–218. https://doi.org/10.1126/science.103.2669.216

Wuensch, Daniela. 2013. *Der letzte Physiknobelpreis für eine Frau? Maria Goeppert Mayer: eine Göttingerin erobert die Atomkerne: Nobelpreis 1963: zum 50. Jubiläum*. Göttingen: Termessos.

Zanish-Belcher, Tanya and Voss, Anke. 2017. *Perspectives on Women's Archives*. Chicago, IL: American Library Association.

Zapoleon, Marguerite W, Goodman, Elsie K., and H., Brilla Mary. 1948. *The Outlook for Women in Physics and Astronomy*. Washington, DC: US Department of Labor Women's Bureau. Bulletin No. 223–6.

1

The Spectrum of He$^+$ as a Proving Ground for Bohr's Model of the Atom: A Legacy of Williamina Fleming's Astrophysical Discovery

MARIA McEACHERN AND BRETISLAV FRIEDRICH

1.1 Prologue

The Bohr model of the atom was conceived at a time characterized by Abraham Pais (1918–2000)[1] as "the epoch of belief, . . . the epoch of incredulity" (Pais, 1986, p. 211). The model entailed the "fourth coming" of Planck's constant, following upon its previous appearances in Planck's black body radiation law, Einstein's light quantum hypothesis, and Einstein's quantum treatment of the heat capacity of solids. See, for example, Kragh (2002).

Although the Bohr model had its precursors in the work of Arthur Erich Haas (1884–1941) (Haas, 1910; Wiescher, 2021) and John William Nicholson (1881–1955) (Nicholson, 1911), it was Bohr's 1913 postulates concerning the electron dynamics in a planetary atom that led to the explanation of the "Balmer series" in terms of transitions between stationary electronic states and an interpretation of the Rydberg constant of atomic hydrogen in terms of fundamental constants (Bohr, 1913a). However, the model remained tentative until the advent of quantum mechanics in 1925–1926 and was regarded as ad hoc by Bohr's contemporaries as it was tailored for atomic hydrogen. Then, Bohr thought of the "Pickering series," as the spectrum of the helium atomic ion (He$^+$) was known, and recognized that his model should be as applicable to He$^+$ as it was to atomic hydrogen. But was it?

For assistance with images, we wish to thank Thomas Burns, Curator of Astronomical Photographs, Harvard College Observatory along with the staff of the Harvard College Observatory Plate Stacks. We also express our appreciation for help with searching for plates to Lisa Bravata, Curatorial Assistant, Harvard College Observatory Plate Stacks. We would also like to express our thanks to Sara J. Schechner, the David P. Wheatland Curator of the Collection of Historical Instruments, Harvard University, and to Sara Frankel, Collections Manager, Collection of Historical Instruments, Harvard University. We also thank the staff of the John G. Wolbach Library at the Center for Astrophysics | Harvard & Smithsonian for their support. BF thanks Georgene and Dudley Herschbach for discussions related to this chapter, and John Doyle and Hossein Sadeghpour for their hospitality during his stay in 2021–2023 at Harvard Physics and at the Institute for Theoretical Atomic, Molecular, and Optical Physics (ITAMP) of the Center for Astrophysics | Harvard & Smithsonian.
[1] Pais borrowed these words from *A Tale of Two Cities* by Charles Dickens.

In this chapter, we describe the history of the discovery of the spectrum of He^+ and show how it served as a proving ground for Bohr's model of the one-electron atom. In the process, we demonstrate that the Pickering series had in fact been discovered by one of the women astronomers who worked as a "computer" at the Astronomical Observatory of Harvard College (or Harvard College Observatory or simply the Observatory, and abbreviated as HCO) that was headed at the time by Edward Charles Pickering. This astronomer and stellar spectroscopist, who worked under Pickering, was Williamina Paton Fleming.

Although the postulates on which Bohr built his atomic model remained incomprehensible until the discovery of quantum mechanics, the model's credibility was boosted by its ability to explain – with a five-digit accuracy – the Pickering series.

This chapter is structured as follows: In Section 1.2, we provide brief biographies of the chapter's protagonists, Edward Pickering and Williamina Fleming. Section 1.3 introduces the workings of the HCO during the Pickering era (1877–1919). Subsections 1.3.1 and 1.3.2 detail the project of mapping the starry skies by means of astrophotography and its spectrally resolved variant. Subsections 1.3.3 and 1.3.4 introduce stellar classifications and variable stars. In Section 1.4, we explore the holdings of the HCO, including its collection of photographic glass plates, and find evidence that the Pickering series was actually discovered by Williamina Fleming as a "peculiar spectrum" in the spectrally resolved photographs of the southern variable star ζ Puppis (Zeta Puppis). Section 1.5 describes how the spectrum of He^+ became a proving ground for Bohr's model of the one-electron atom and what impression the model's extended applicability to a species for which it had not been tailored left on the community. Finally, in Section 1.6, we comment on the significance of He^+ for astronomy itself as well as for astrochemistry.

1.2 The Protagonists

1.2.1 Edward Charles Pickering

Edward Charles Pickering (Figure 1.1) was born in Boston on July 19, 1846 into a patrician family. Upon graduating from the Lawrence Scientific School of Harvard University[2] at the age of 19, and a stint as an Instructor of Mathematics at his alma mater, he was hired in 1867 by the Massachusetts Institute of Technology as a professor of physics. During the following decade, Pickering developed MIT's Physical Laboratory, the first such establishment in the US (Pickering, 1869). The way the laboratory was set up and administered reflected Pickering's credo: "There are no secrets in Science" (Cannon, 1919). The laboratory was open not only to students and teachers of physics but also to the lay public

---

[2] A forerunner of what is today Harvard's School of Engineering and Applied Sciences.

Figure 1.1 Edward Charles Pickering (1846–1919). Photo taken between 1891 and 1896 at the Pach Brothers studio in Cambridge, Massachusetts. Courtesy of the Harvard College Observatory.

wishing to carry out their research. In 1876, Pickering was appointed by Harvard President Charles Eliot (1834–1926) as the fourth Director of the HCO.

Pickering entered his duties at the HCO in 1877 with a plan to investigate the brightness of stars by photometry. In 1882, he extended traditional photometric techniques by launching a program in *photographic astronomy*. This innovation was based on dry photographic plates (i.e., glass plates coated with a gelatin emulsion of silver bromide that could be stored before exposure and developed at leisure after exposure). In 1886, he launched the Henry Draper Memorial project of spectrally resolved photographic mapping of both the northern and southern skies. The telescopes used entailed a dispersive element (an ocular prism) in order to

spectrally resolve the star images recorded as structured smudges on the photographic glass plates. Pickering's predilection for spectroscopy originated in his
training as a physicist. The spectrally resolved astrophotography enabled an unprecedented accumulation of empirical data intended to prepare the soil for a better
understanding of the nature of stars. Moreover, given that the astrophotographs
were taken at known times (the exposures lasted minutes to tens of minutes), a set of
subsequent astrophotographs contained information about the variation in time of
the stars' brightness and spectra. As a result, the study of *variable stars* had become
a major preoccupation of the astrophysical community. Enhanced sensitivity to
time variation was achieved by the technique of superposing a negative and
a positive taken at different times, leading to the discovery of almost 3,500
variables at the HCO during Pickering's tenure.

Pickering cultivated relations with potential sponsors of the HCO such as Mary
Anna Draper (1839–1914), Catherine Bruce (1816–1900), and Uriah Boyden
(1804–1879). Mrs. Draper would fund the Henry Draper Memorial catalog of
stellar spectra, so named after her late husband, the physician and amateur astronomer Henry Draper (1837–1882), who pioneered astrophotography and, assisted by
his wife, recorded stellar spectra as early as 1872. The Boyden fund made it
possible to build the HCO's permanent station in Arequipa, Peru, to map out the
southern sky. It was at Arequipa where the spectrum of ζ Puppis was recorded.
Miss Bruce's gift covered the cost of a double-telescope (Bruce Telescope) consisting of a tracking telescope that would be locked to a guiding star (to compensate for
the star's apparent motion due to the Earth's rotation) and a telescope through
which the astrophotograph of a given star was taken. It was the Bruce Telescope
that was installed at Arequipa. If need be, Pickering would also make financial
contributions to the HCO from his own pocket (Bailey, 1932).

In 1891, William Henry Pickering (1858–1938), the brother of Director
Pickering, who had been hired as an assistant, was put in charge of the southern
outpost in Arequipa. Much to Director Pickering's dismay, William was submitting
his observations of the planet Mars via telegraph to the *New York Herald* in
anticipation of the 1892 opposition of the Red Planet. These missives helped to
stir the frenzy of the general public eager to receive any news regarding the Martian
channels interpreted as canals created, perhaps, by the Martian public works. There
was great general interest in the potential for communication with the inhabitants of
Mars as the upcoming opposition approached. It was, as journalistically described,
"the time of the great Mars boom, when public imbecility and journalistic enterprise
combined to flood the papers and society with 'news from Mars,' and queries
concerning Mars, most exasperating to grave thinkers and workers in science"
(Clerke, 1896; Shindell, 2023). Edward Pickering was furious and, given William's
great liberties taken in ignoring the work which he had actually been assigned to do

and his continually mounting extravagances, he recalled his brother with the approval of the Harvard Corporation.

As noted by Annie Jump Cannon (1863–1941) (Cannon, 1919):

In the early days of photographic astronomy, Professor Pickering foresaw a great opportunity for woman's work, and gradually, from 1884, the staff of woman assistants was instituted [at the HCO], first for simple computing and examination of the photographs, then for posts of greater responsibility, until independent investigations were made by them.

Fleming, Cannon, and Henrietta Swan Leavitt (1868–1921) were among the most prominent beneficiaries of Pickering's foresight. The exemplary partnership of Mary Anna and Henry Draper was likely suggestive of the possibilities of involving women in research.

As a man of the new Industrial Age, Pickering believed in the specialization of labor. The women astronomical computers, known as the Harvard Computers, were chosen "to work, not to think" (Payne-Gaposchkin, 1979a, p. 54). Their work was to analyze and not to interpret the data derived from the astrophotographic glass plates. Thus, thanks to the growth of astronomical spectroscopy which enabled astronomical work in daylight (Lankford, 1997, p. 310), the HCO came to embody the "factory observatory" (Lankford, 1997, p. 309). In an address to the Harvard branch of Phi Beta Kappa, Pickering observed that (Johnson, 2005, p. 18):

A great observatory should be as carefully organized and administered as a railroad. Every expenditure should be watched, every real improvement introduced, advice from experts welcomed, and if good, followed, and every care taken to secure the greatest possible output for every dollar expenditure. A great savings may be effectuated by employing unskilled and therefore inexpensive labor, of course under careful supervision.

Inexpensive labor indeed: the women's pay was typically 25 cents per hour (Sobel, 2016, p. 113), see also Section 1.2.2.

Pickering was the recipient of the highest honors from both US and European institutions; these included membership in the American Academy of Arts and Sciences (1867), the National Academy of Sciences (1873), as well as national societies of England, Germany, Ireland, Italy, Mexico, Russia, and Sweden. Moreover, he served as President of the American Astronomical Society (1905–1919) as well as of the American Association of Variable Star Observers (AAVSO), founded in 1911, which coordinated observations and their analysis by mainly amateur astronomers (Cannon, 1919). He was also the founder and first President of the Appalachian Mountain Club. Edward Pickering died in Cambridge, Massachusetts, on February 3, 1919. The International Astronomical Union (IAU) named in 1935 a lunar crater (IAU, 2010b) and in 1973 a Martian crater (DPedia, 2023) after him. In addition, Asteroid 784, discovered in 1914, was named Pickeringia (Schmadel, 2007).

1.2.2 Williamina Fleming

Williamina Paton Stevens Fleming (Figure 1.2) was born in Dundee, Scotland, on May 15, 1857. She was one of several surviving children in her family at a time when child mortality accounted for nearly half of all deaths in Scotland (Anderson and Roughley, 2018). Her father, a craftsman, was an early enthusiast of the recently discovered daguerreotype imaging. Williamina thus already as a child became familiar with this early photographic technique and its utility for recording images.

Williamina became a teaching apprentice (pupil–teacher) at the age of fourteen and would continue in this occupation for five more years. In 1877, she married James Orr Fleming and, in 1878, the couple followed other members of her family in leaving Scotland with the hope of starting a better life in the US (Massachusetts Naturalization Records, 1900). Within a year, Williamina Fleming found herself pregnant and abandoned by her husband. Left to fend for herself and her unborn child, she found employment in the household of Edward Charles Pickering, where she was hired to provide household help along with copying assistance. Director Pickering and his wife, Lizzie Sparks Pickering (1849–1906), may have been impressed with her intelligence and facility with numbers in managing household accounts. Eventually, and quite possibly at the urging of Mrs. Pickering, Fleming began to work part-time at the HCO as a copyist and computer.

Figure 1.2 Williamina Paton Stevens Fleming (1857–1911). Courtesy of the Harvard College Observatory.

Williamina, or Mina, as she was familiarly known to friends and colleagues, returned home to Scotland to be with her family for the birth of her son, whom she named Edward Charles Pickering Fleming in gratitude for Pickering's kindness. In 1881, on her return to Cambridge, Massachusetts, Fleming began full-time employment at the HCO. She became a member of the regular staff, one of several who comprised the corps of women astronomical computers. Included among her duties was to oversee and edit several periodicals and other publications issued by the HCO.

During her early career, she worked on the Harvard Photometry project, in which all of the stars visible from the HCO were photographed and assigned a magnitude (a degree of brightness), relative to Polaris, the North Star, which stood as a standard.

Once the Henry Draper Memorial project was launched in 1886, Williamina Fleming became a full-fledged participant. She was involved in the investigation and cataloging of stellar spectra as recorded on photographic glass plates that held images of the night sky in both the northern and southern hemispheres. Under her charge labored the corps of women astronomical computers who, besides herself, inspected and analyzed the recorded images. Fleming organized the workflow and oversaw every phase of producing data gleaned from the plates. The data were published in multiple volumes of the *Annals of the Astronomical Observatory of Harvard College*. Observations of objects of special note might be hurried to press in one of the *Harvard College Observatory Circulars*. Each issue of these titles was edited, proofread, prepared for publication and, following printing, inspected by Fleming. During the first three directorships of the HCO, eight volumes of the *Annals* had been published while, under the Pickering administration, this number rose to nearly 100.

In 1899, additional responsibility, and a corresponding job title, were bestowed upon Fleming. On Monday morning, January 16, 1899, *The Boston Herald* carried the front-page headline "Harvard honors women" while a bold-print subheading conveyed the news, "Mrs. Fleming appointed curator of astronomical photographs, charged with their care" and, in less bold print, "She is the first of her sex to have her name placed with the list of the officers" (Boston Herald, 1899, p. 1). The article went on to say (Boston Herald, 1899, p. 2):

Mrs. Mina Fleming, the recently appointed Curator of Astronomical Photographs, has a worldwide reputation at once as a painstaking and patient investigator and as a brilliant discoverer in the field that is covered by the Henry Draper Memorial. The names of Mrs. Draper and Mrs. Fleming will go down in astronomical history in honorable conjunction with those of Caroline Herschel, Mary Somerville, and Maria Mitchell.

It is entirely appropriate that Fleming should have been named as the first Curator of Astronomical Photographs at the Harvard College Observatory.

Besides her numerous discoveries made via the astrophotographic glass plates, thanks to her familiarity with daguerreotype images since childhood, she knew and thoroughly trusted the photographic process. So much so that, in 1893, she defended this new method of astronomical investigation against its detractors writing that (Fleming, 1893, p. 687):

One must not always cling to the earliest method of accomplishing anything and assume that because it was the earliest and has held sway for centuries, it must consequently be the best, and also the only way.

Cannon would later write in regard to Fleming's staunch defense of the use of photography in astronomical observation (Cannon, 1911):

In the early days, when celestial photographs were rare, and some of these discoveries were attributed by skeptics to defects on the film, she never doubted the validity of the photographic evidence. Her industry was combined with great courage and independence.

During that year, the Columbian Exposition was held in Chicago, and the HCO was expected to participate and provide exhibition material, which would be prepared by Fleming. Besides describing, in depth, the work at the HCO, the many discoveries made by her colleagues (to whom she was not the least bit stingy in giving credit) and herself, she discussed, at some length, the great benefit of the use of photography in astronomy and its utility in studying the true elemental makeup of stars from their spectra. Fleming wrote (Fleming, 1893, pp. 685–686) that:

[A]stronomical photographs must be considered more reliable for, in the case of a visual observer, you have simply his statement of how the object appeared at a given time as seen by him alone, while here you have a photograph in which every star speaks for itself, and which can at any time, now or in the years to come, be compared with any other photographs of the same part of the sky.

Fleming would also write a paper on women's participation in astronomical study titled "A field for woman's work in astronomy" to be read at the Exposition and, subsequently, published in the journal *Astronomy and Astrophysics* (Fleming, 1893). Given her numerous responsibilities and what must have been, at least at times, repetitive work, it is little wonder that, in her diary for March 1900, which was kept as part of Harvard University's celebration of the new century and would be placed into a time capsule, Fleming noted on March 8 at 8:00 p.m., "Miss Christopherson came to massage [my] hand and arm" (Fleming, 1900, p. 18). She mused that, "If one could only go on and on with original work, looking for new stars, variables, classifying spectra and studying their peculiarities and changes, life would be a most beautiful dream" (Fleming, 1900, p. 9); however, she added, "but you come down to its realities when you have to put all that is interesting aside in order to use most of your available time preparing the work of others for publication" (Fleming, 1900, pp. 9–10). Still, she appreciated her role at the HCO, "contented to have such excellent opportunities for work … and proud to be considered of any assistance to such a thoroughly capable Scientific man as our Director [Pickering]" (Fleming, 1900, p. 10).

On the subject of pay, Fleming felt the unfairness of her situation. In her 1900 journal (Fleming, 1900, pp. 18–19), see also Figure 1.3, she noted on March 12:

I had some conversations with the Director regarding women's salaries. He seems to think that no work is too much or too hard for me, no matter what the responsibility or how long the hours are. But let me raise the question of salary and I am immediately told that I receive an excellent salary as women's salaries stand. … Does he ever think that I have a home to keep and a family to take care of as well as the men? But I suppose a woman has no claim to such comforts. And this is considered an enlightened age! I cannot make my salary meet my present expenses with Edward [Fleming's son] in the Institute [MIT] and still another year

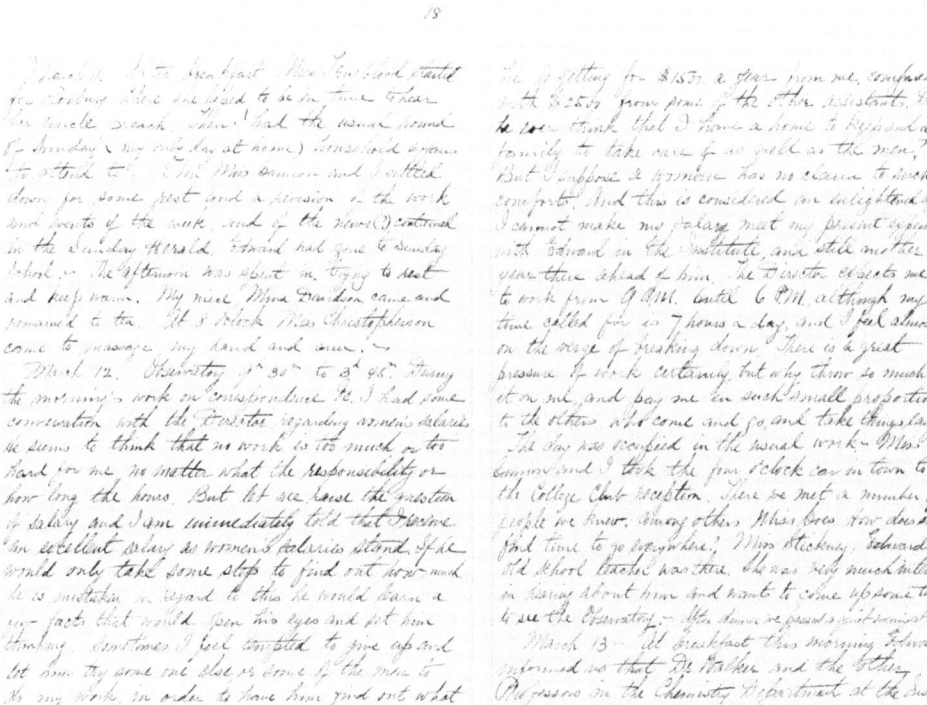

Figure 1.3 Pages 18 and 19 from Fleming's 1900 Journal (Fleming, 1900). Fleming's annual salary of $1,500 as Curator of Astronomical Photographs would correspond to about 50 cents per hour. For male assistants, the annual salary was $2,500 according to this record. Courtesy of Harvard University Archives.

there ahead of time.[3] The Director expects me to work from 9 a.m. to 6 p.m., although my time called for is 7 hours a day, and I feel almost on the verge of breaking down. There is a great pressure of work certainly, but why throw so much of it on me, and pay me in such small proportion to the others, who come and go, and take things easy?

During her many years of active research, Fleming classified 10,351 stellar spectra and discovered more than 300 variable stars and 10 novae by recognizing the particular type of bright lines in their spectra. She also identified 59 new gaseous nebulae, including that which came to be known as the Horsehead Nebula, familiar to many with even a passing interest in astronomy. She discovered 94 Wolf–Rayet stars and, along with Pickering and Henry Norris Russell, Fleming is credited with the discovery of white dwarf-type stars. Additionally, she found "ninety-one stars of the fifth type, Class O, and sixty-nine stars of the Orion type having bright hydrogen lines" (Cannon, 1911). In 1906 – in recognition of her contributions to the study of astronomy – Fleming became the first American woman to be named an Honorary Member of the Royal Astronomical Society of London. She was also appointed as an Honorary Fellow in Astronomy at Wellesley College. Also in 1906, Fleming was elected as an Honorary Member of the Sociedad Astronómica de Mexico and received its Guadelupe Almendaro Gold Medal for the discovery of new stars.

Early in May 1911, Fleming was feeling "not so well as usual" and entered a hospital to rest (Cannon, 1911). It turned out that her extreme fatigue was due to a severe case of pneumonia which proved to be fatal. She passed away on May 21, 1911. In the Sixty-Sixth Annual Report of the Director of the Astronomical Observatory of Harvard College, the first published following Fleming's death, Pickering wrote of the HCO's "severe loss" (Pickering, 1912):

Mrs. Fleming's record as a discoverer of new stars, of stars of the fifth type, and of other objects having peculiar spectra, was unequalled. Her gifts as an administrative officer, especially in the preparation of the *Annals*, although seriously interfering with her scientific work, were of the greatest value to the Observatory.

On Monday morning, May 22, 1911, *The Boston Herald* featured the passing of Fleming on its front page with the headline "Leading woman of science dies in a hospital." Of Fleming and her work a subheadline stated that "In 20 years of work at the little brick building on Observatory Hill in North Cambridge ... she made more discoveries than all other astronomers had made in 200 years" (Boston Herald, 1911).

---

[3] After Fleming's departure for the US in 1881, her son was being cared for in Scotland by her mother. The boy joined Fleming in Cambridge, Massachusetts, when he was eight, accompanied by a female caretaker. He graduated from MIT in 1901 (in mining engineering) and was naturalized as a US citizen in 1904. He died in California in 1962 (Hughes, 2012, pp. 691–692).

Writing on behalf of the Royal Astronomical Society in its February 1912 issue of the *Monthly Notices*, its editor Herbert H. Turner paid the following tribute to Fleming (Turner, 1912):

As an astronomer Mrs. Fleming was somewhat exceptional in being a woman; and in putting her work alongside that of others, it would be unjust not to remember that she left her heavy daily labours at the observatory to undertake on her return home those household cares of which a man usually expects to be relieved. She was fully equal to the double task, as those who have had the good fortune to be her guests can testify; and it is perhaps worthy of record, as indicating how lightly the double burden sat on her, that she yielded to none in her enjoyment of a football match, especially a match between Harvard and Yale.

As the years have passed, Fleming's name has been associated with the grouping of the HCO's women employees and she is often referred to as one of a number of computers. But, in her many years at the HCO in which she drove forward the work of the Draper Memorial and discovered, analyzed, and classified innumerable stars of various types, she considered herself to be an astronomer. On May 10, 1907, Fleming submitted to the Massachusetts District Court her petition for naturalization requesting "to be admitted a citizen of the United States" which was signed and witnessed by Director Pickering and the Harvard astronomy professor Solon Irving Bailey (1854–1931). On the document, Fleming listed her occupation as "Astronomer" (Massachusetts Naturalization Records, 1900).

For the final word on the preferred nomenclature for Fleming's occupation, we may look to her gravestone in the Mount Auburn Cemetery in Cambridge, Massachusetts, which reads, simply: Williamina Paton Fleming, Astronomer.

In 1970, the International Astronomical Union named a lunar crater (IAU, 2010a) and in 1991 Asteroid 5747 (BIAU, 2022) after her.

1.3 Stellar Spectroscopy at the Harvard College Observatory

1.3.1 Stellar Photometry

Early in his directorship of the HCO, which began in 1877, Pickering took note of research programs at other institutions and found the field of photometry to be lacking as a major undertaking. On October 25, 1879, he launched a project in which the brightness of stars as recorded on photographic glass plates would be accurately measured (Bailey, 1931, p. 188). He hoped to create a uniform scale of stellar magnitude and, with the assistance of the well-known Cambridgeport, Massachusetts, astronomical optics firm Alvan Clark & Sons, he would go about creating the specialized instrumentation required to do this work (Plotkin, 1990). Of Pickering's many photometric inventions, the Meridian Photometer is one of the best known and became the project's most relied-upon instrument. It made use of

Harvard's Great Refractor[4] and dual-objective optics that enabled comparing the brightness of a given star with that of Polaris chosen as a standard (Lewis, 1885). In 1882, the HCO published its first work devoted to these photometric efforts, known as the Harvard Photometry, which appeared in Volume XIV of the *Annals of the Astronomical Observatory of Harvard College*. The work included analysis of the magnitudes of 4,260 stars brighter than 6th magnitude and received a stellar review which appeared in 1885 in Volume 8 of the publication *Observatory*, a monthly review of astronomy (Lewis, 1885):

To say that the volume recently issued from the Observatory of Harvard College is, in this branch of astronomy, epoch making, is to do no more than justice to a work which must henceforth be regarded as at once the foundation and treasury of scientific stellar photometry.

Over the course of three years, Pickering and the HCO staff members who had worked on the Harvard Photometry had managed to assign a designation of magnitude to every star visible from Cambridge, Massachusetts (Sobel, 2016, p. 12).

During this time, besides fulfilling her copying duties, Fleming learned how to measure the brightness of stars on photographic plates and to establish their magnitudes. She also became familiar with the appropriate methods of handling these photographs on glass and placing them in the special frame, backlit by a mirror, which allowed her to view the images. The process also involved the inspection of the glass plates in order to determine the quality of each image. The preservation of these fragile images, eventually numbering 200,000 during her tenure, along with their proper cataloging and overall care, fell directly on Fleming's shoulders as, in 1899, she would be granted the title Curator of Astronomical Photographs by the Harvard Corporation, thus becoming the first woman to hold a professional position at Harvard.

1.3.2 Mapping Out Stars According to Their Spectra: The Henry Draper Memorial Catalog

An astronomer and pioneering astrophotographer by avocation if not profession, Henry Draper had passed away merely days after a dinner party at which the HCO Director was in attendance along with many members of the National Academy of Sciences. Draper, thanks in part to the largesse and enthusiasm of his wife, Mary Anna Draper, had followed in the path of his own father, John William Draper (1811–1882),[5] in the newly developed pursuit of astrophotography. Supported by family wealth, Henry Draper was able to retire from his position as Professor of

---

[4] The acquisition of this 15-inch telescope by Harvard in 1847 was funded by the Boston citizenry, whose interest in astronomy was aroused by the Great Comet of 1843 (Bailey, 1931, p. 84).
[5] John William Draper was the first president of the American Chemical Society.

Physiology at New York University in 1882 – abandoning academia in favor of pursuing his freelancing interests in astronomy.

It was not only photography of the stars and other astronomical objects that had inspired Henry Draper's imagination, but also the much more specialized and challenging process of imaging their spectra that captivated him. Capturing the spectrum of a particular star was more technically difficult than merely imaging the object itself and the process of trial and error was heavily involved in this undertaking.[6] By the time of his untimely death at the age of 45, Draper had taken more than 100 spectral images.

In early 1883, Pickering sent a letter of condolence to Anna Draper, in which he suggested that the HCO might, with her blessing, continue the spectroscopic work of her late husband. Pickering, although aware that astronomical "firsts" were especially notable, knew that if others achieved success in spectral astrophotography, the efforts of Henry Draper would become a mere footnote. Mrs. Draper was pleased to hear from Pickering and, after some deliberation, decided to fund his idea which, she perceived, would be a suitable tribute to her late husband. Originally, she intended to establish an observatory in her husband's name and to run it on her own, but that plan never came to fruition. Thus, she agreed with Pickering that February 14, 1886 would stand as the official date of the establishment of the great undertaking by the HCO, named the Henry Draper Memorial (Jones and Boyd, 1971, p. 228). And not only stars of the northern hemisphere were to be included in the spectroscopic survey but also those of the southern sky. Fortunately for Pickering and the fate of the enterprise, the will of engineer Uriah A. Boyden was finally settled in 1887 following five years of legal deliberations, which resulted in a sum of money being left to Harvard for the sole purpose of establishing a high-altitude observatory. This funding helped to create the Boyden Department, as the HCO's southern hemisphere site came to be known (Jones and Boyd, 1971, pp. 246–252).

With a steady source of financial backing, thanks mainly to Mrs. Draper's generosity, Pickering was able to provide both the northern and southern departments of the HCO with the equipment needed. Mrs. Draper transferred her husband's 11-inch-refractor to Pickering on loan for however long it would take to complete the required work. Another telescope was purchased in 1885 with money granted by Alexander Dallas Bache, a physicist and the former Superintendent of the United States Coast Survey. The Bache fund was administered by the National Academy of Sciences. The 8-inch Bache telescope was eventually dismantled and sent from Cambridge, Massachusetts, to Arequipa, Peru. The instrument had been very productive at the HCO in Cambridge, so Mrs. Draper ordered a new 8-inch

---

[6] His first success occurred in 1872 when he photographed the spectrum of the star Vega, or α-Lyrae. Among his many triumphs that followed was the imaging of the spectra of α-Aquila, Arcturus, Capella, the Moon, Venus, Mars, and Jupiter (Barker, 1888) and that of the Great Nebula of Orion (New York Times, 1888).

telescope that would be used there instead of the Bache telescope (Hoffleit, 1991). Several other instruments from the Drapers' collection, along with other assorted apparatus, including lenses and various prisms, were given over to the work of the Draper Memorial or purchased as needed.

In 1888, Pickering issued a further appeal for donations with the aim of implementing his astrophotographic agenda for the inventory and mapping of the heavens. This resulted in a gift which, as Pickering had hoped, provided a larger telescope and lens for the project. Catherine Wolfe Bruce, a philanthropist who happened to be a devotee of astronomical science, had received a sizable inheritance from her father. She agreed to provide $50,000 toward building a 24-inch photographic telescope. Miss Bruce was so enthusiastic about funding astronomical endeavors, and Pickering was so persuasive in convincing her of the great need for support of astronomical institutions in general, that she agreed to participate in worthy efforts at observatories worldwide. The two accepted applications for funding and chose "five scientists in the United States to receive her support" along with 10 other grants to astronomers working in England, Norway, Russia, India, and Africa (Sobel, 2016, p. 48). Following installation and testing in Cambridge beginning in 1893, the Bruce telescope was conveyed to Arequipa in 1896. Pickering himself donated $100,000 of his own money anonymously to further assist the project. Additional income was derived from the bequest of Robert Treat Paine, grandson of one of the original signers of the Declaration of Independence, who was a lawyer and an amateur astronomer and who left $250,000 to the HCO.

Late in 1888, HCO staff member Solon Bailey departed for Peru on a steamship and railway journey, finally reaching Lima in March, 1889. From there, he traveled to the town of Chosica to explore the local environs and inspected various mountain peaks in search of a suitable site to establish Harvard's southern hemisphere observatory. A nameless peak was finally chosen high above the town of Chosica and named "Mount Harvard" (Jones and Boyd, 1971, p. 291). Bailey's first impulse had been to name the peak "Mount Pickering" but, when informed by mail, the HCO Director declined the honor. The first necessity was to build a trail by which to deliver the nearly 100 boxes along with the 8-inch Bache telescope up the mountain. In the interest of hurrying the process along, Bailey and his family helped to provide labor. Of the area in general, Bailey noted (Bailey, 1895):

From the time of our first visit to Mount Harvard on March 11 throughout the rest of the year the region seemed extremely dry and barren. The ravine, however, through which our path to the mountain led, showed unmistakable evidence of tremendous rainfall sometime in the past.

Once the observatory was finally established and in working order, the process of imaging the southern sky and recording its splendors on astrophotographic glass plates commenced. The first boxes arrived back at the HCO in August 1889, though it was found that some of the plates had suffered slight damage during shipment. Photographic work continued at Mount Harvard through the Peruvian autumn. Meanwhile, other potentially desirable sites for a more permanent establishment were inspected, including the Atacama Desert plateau in Chile. Eventually, in late 1889 and early 1890, fog and mists enveloped Mount Harvard and were followed by heavy rains. Bailey had reported to Pickering that an alternative site might exist at Arequipa, Peru. As the year progressed, the weather remained unstable through the summer with more heavy rain. The buildings at the observatory site, with walls constructed of building paper, began to sag and collapse. Given the discouraging reports from Peru, Pickering had decided to move the Boyden Station to Arequipa as its permanent site, and work ceased at Mount Harvard. The original site was dismantled and the instrumentation packed for shipment. Bailey and his party moved on to Arequipa, where the observational and photographic work continued until May 9, 1891. Over the course of two years, Bailey had managed to establish the magnitudes of approximately 8,000 bright stars thus completing the Harvard Photometry (Jones and Boyd, 1971, p. 291). He had also obtained around 2,500 photographic plates of the spectra of the southern stars with the aid of his brother Marshall, a professional photographer, and of his Peruvian assistant Elias Vieyra. On May 15, 1891, Solon Bailey and his family departed Arequipa. After an unfortunate interregnum at Arequipa (see Section 1.2.1), Bailey returned there on February 25, 1893, and remained in charge of the outpost until 1919.

In Fleming, Pickering had found the most capable person possible to keep the work of his observatory on track. Besides her scientific work – the reduction of the data contained in the astrophotographic images – Fleming, with her great diligence, determination, and attention to detail, became the foreperson who was put in charge of the women astronomical computers, their hiring, workflow, and output. She also oversaw the entire process of publishing all HCO-related materials. Figure 1.4 shows Harvard Computers during a visit by Anna Draper.

By 1893, there were a total of 40 assistants working at the HCO, 17 of them women (Sobel, 2016, p. 53). Figure 1.5 shows a 1913 photo of 13 female members of Pickering's team.

In a satire, "HCO Pinafore," written in 1879 by Winslow Upton, then a telescope assistant and later Professor of Astronomy at Brown University (Bailey, 1931), to the tune of Gilbert & Sullivan's 1878 comic opera "HMS Pinafore," the chorus of women computers sang (Upton, 2001, p. 161):

> We work from morn 'till night,
> For computing is our duty;
> We're faithful and polite,

Figure 1.4 Mary Anna Draper (1838–1914), seated at right, during her visit to the HCO in 1891. She sponsored the Henry Draper Catalog named in honor of her late husband, the physician and amateur astronomer – and pioneer of astrophotography – Henry Draper (1837–1882). Standing next to her at the table is Fleming, the Curator of the Astronomical Photographic Glass Plate Collection and supervisor of the women astronomers at the HCO. HUV 1210 (9-5), olvwork289692. Harvard University Archives, 1891.

> And our record book's a beauty;
> With Crelle and Gauss, Chauvenet and Peirce,
> We labor hard all day;
> We add, subtract, multiply and divide,
> And we never have time to play.

This telling satire was performed for the first time only in 1929, at the residence of the then HCO Director Harlow Shapley (1885–1972) for about 100 members of the American Astronomical Society. We note that when the HCO Pinafore was written, there were four women and several men computers employed by the HCO (LaFortune, 2001). The satire's sting had apparently a more general aim than the HCO, with the Naval Laboratory and other government services being among its

Figure 1.5 Female members of Pickering's team, also known as "Pickering's harem" (Sobel, 2016, pp. 105–122). At the far left is Margaret Harwood (AB Radcliffe, 1907; MA University of California, 1916), who was later appointed director of the Maria Mitchell Observatory. Beside her in the back row is Mollie O'Reilly, a computer from 1906 to 1918. Next to Pickering is Edith Gill, a computer since 1889. Then Annie Jump Cannon (BA Wellesley, 1884), who at this time was about halfway through classifying stellar spectra for the Draper Catalog. Behind Cannon is Evelyn Leland, a computer from 1889 to 1925. Next is Florence Cushman, a computer since 1888. Behind Cushman is Marion Whyte, who worked for Cannon as a recorder from 1911 to 1913. At the far right of this row is Grace Brooks, a computer from 1906 to 1920. Ahead of Harwood in the front row is Arville Walker (AB Radcliffe, 1906), who served as assistant from 1906 until 1922 and later secretary to Pickering's successor Harlow Shapley. The next woman may be Johanna Mackie, an assistant from 1903 to 1920. In front of Pickering is Alta Carpenter, a computer from 1906 to 1920. Next is Mabel Gill, a computer since 1892. And finally, Ida Woods (BA Wellesley, 1893), who joined the corps of women computers just after graduation. In 1920, she received the first AAVSO nova medal; by 1927, she had seven bars on it for her discoveries of novae on photographs of the Milky Way (Welther, 1982). HUPSF Observatory (14), olvwork360662. Harvard University Archives, c. 1911.

targets as well (Grier, 2013, pp. 83–84). The piece also spoofed, or perhaps not (Lewis, 1885; Fiss, 2023), Pickering for his choice of Polaris, a variable, as a luminosity standard (for whose variation the photometric measurements had to be corrected):

> Pole Star, to thee I sing
> Bright pivot of the heavens,
> Why are all our magnitudes
> Either at sixes or at sevens?
> I have lived hitherto
> Free from the breath of slander.
> Beloved all my crew
> A really popular commander.
> But now my prisms all rebel
> And ruin the photometer,
> And damage also, sad to tell,
> My fame as an astronomer.
> Pole Star, to thee I sing
> Bright pivot of the heavens,
> Why are all our magnitudes
> Either at sixes or at sevens?

1.3.3 Harvard Stellar Classification System

In 1882, Pickering laid out a plan suggesting that interested parties might search for five types of stars with varying magnitudes (Pickering, 1882a, p. 5):

I. temporary stars, or those that shine out suddenly, sometimes with great brilliancy, and gradually fade away, otherwise known as novae; II. long-period variables, or those undergoing great variations of light; III. stars undergoing slight changes according to laws as yet unknown; IV. short period variables, or stars whose light is continually varying, but the changes are repeated with great regularity in a period not exceeding a few days; V. Algol stars, or stars which ... every few days suffer a remarkable diminution in light for a few hours.

Toward the conclusion of his tract, Pickering provided motivation for undertaking the study of variable stars in case any reader should need convincing (Pickering, 1882a, p. 14–15):

Apart from the value of the results attained it is believed that many amateurs will find it a benefit to accustom themselves to work in a systematic manner, and that they will thus receive a training in their work not otherwise easily obtained outside of a large observatory. The lesson should be taught that time spent at a telescope is nearly wasted unless results are secured worthy of publication and having a permanent value.

Heeding Pickering's clarion call, many amateurs followed through with the plan of observing variable stars and reported the resulting data to the HCO. Hence,

Fleming scrutinized not only astrophotographic plates taken in Cambridge, Massachusetts, and those shipped from Arequipa, but she also sought to verify claims by the amateurs of variability in regard to particular stars. In her thorough investigations of these stars, she discovered that a preponderance of those categorized as variables were Type III stars according to Angelo Secchi's system based on spectral appearance. Initially, Secchi's system contained two classes, one of which consisted of red and yellow stars like the Sun while the other contained blue stars such as Sirius (Holberg, 2007, p. 87). As he was exposed to more and more spectra, Secchi changed his classifications to allow for five different types: Class I: blue stars such as Sirius, which display a small number of hydrogen absorption lines; Class II: stars similar to the Sun, with spectra rich in narrow dark lines; Class III: reddish stars such as Betelgeuse, which display broad dark bands; Class IV: stars with strong carbon lines (Holberg, 2007, p. 87). In 1877, Secchi added a fifth class of emission-line stars. Thus a natural part of Fleming's job was devising a new classification system (Cannon, 1915):

The divisions into five types made by Secchi proved altogether inadequate to represent the numerous differences seen on the photographs. A new system had to be adopted which would permit the reader to understand the various aspects of the spectra as shown by the photographs. ... This classification is purely empirical, being based wholly on the external appearances, without any idea of expressing differences in temperature or stages of evolution.

Fleming elaborated on Secchi's spectral classes and added a number of subdivisions, each of which received a letter designation: A, B, C, D, E, F, G, H, I, K, L, M, and N. Class Q would be added later for stars with spectra that didn't really fit into any other category. In Fleming's system (Cannon, 1915):

Stars of Type A and B were bright blue stars like Sirius and those in Orion . . . while stars of Type F and G had spectra similar to that of the sun, and K and M stars were characterized by reddish spectra containing strong dark bands, in addition to the narrow metallic lines seen in other stars.

The letter "O" was added to the classification to indicate those stars which Fleming herself considered to be "the most striking class of stellar spectra, in fact the only one that is at once detected visually with a small telescope" (Fleming, 1912, p. 177). Included in this class would be Wolf–Rayet-type stars along with 21 stars of this spectral type that had been discovered in the Large Magellanic Cloud. A number of others were found to "occur near the central line of the Milky Way." This group of stars would become known as being of the fifth type, of which Fleming discovered many.

The Second Conference of Astronomers and Astrophysicists was held at Harvard in August 1898, where it was resolved that the American Society of Astrophysicists, now known as the American Astronomical Society, would come into existence. Fleming submitted a paper to the conference titled, "Stars of the fifth type in the Magellanic Clouds," which was read by Pickering. Up to that time, according to the paper, 92 such stars exhibiting the particular bright line spectra had been discovered. At the end of his reading of the paper, Pickering took the liberty of informing the crowd that, of those 92 stars, 78 had been discovered by Fleming (who had not mentioned this fact in the paper) and who was in attendance. A spontaneous burst of applause impelled her to come forward whereupon she responded to questions that had been generated by her paper from the crowd of astronomers in attendance (Donaghe, 1898).

Cannon simplified Fleming's system, combining certain classes and doing away with others, such as little-used C, D, E, H, I, L, N, and Q. She noted that (Cannon, 1915, pp. 209–210):

A modification of the system of letters used for the Draper Catalogue was adopted [in 1897] ... The stellar sequence was found to be in some respects less complex than was at first supposed. The appearances for which some of the letters, such as C, D and E, had been assigned, were not confirmed by later and better photographs. Therefore, the letters were dropped from the sequence.

Cannon reordered the spectral types to the now-familiar types O, B, A, F, G, K, and M. The new ordering system allowed for differentiations in spectra to be expressed within the various types either by a supplementary letter such as Oa ("A broad bright band whose centre is at wavelength 4633 is the most conspicuous feature of this spectrum" (Cannon, 1912, p. 66)) or by the addition of a digit such as F2 ("This spectrum resembles Class F, except that there is a slight appearance of continuity in band G" (Cannon, 1912, p. 68)). Cannon's reordering aligns with the abscissa of the Hertzsprung–Russell diagram of stellar luminosity versus temperature.

In 1910, Edward Pickering attended the fourth meeting of the International Union for Cooperation in Solar Research, which was held on the summit of Mount Wilson, California, at the solar observatory of the Carnegie Institution. At this time, international astronomy did not abide by a single stellar classification system. Pickering saw an opportunity for the widespread adoption of the Harvard Pickering/Fleming/Cannon system. Recording his thoughts in a diary on the long railroad trip to California, he wrote that "the Harvard system will probably be the system of the world. I am repaid for my journey of 2,000 miles, had I done nothing else ... My part in this will be regarded as one of the most important things I have ever done." To his great relief and delight the Union gave it "the strongest endorsement I could have desired" (Plotkin, 1990, p. 57).

1.3.4 Variable Stars

Formerly, verification of a variable – a star with a time-dependent luminosity – required a tedious process of following the star over a series of plates, taken at various times, and determining whether its magnitude actually did vary and, if so, whether there was a regular, discernible pattern of brightening and darkening. In 1881, Edward Pickering forwarded an announcement to the journal *Astronomische Nachrichten* titled "New variable star in Puppis" (Pickering, 1881a). The star, in Pickering's opinion, was a Mira-type star as it resembled Omicron Ceti, an old, pulsating red giant known as Mira or "The Wonderful" in Latin. Mira's variability had been recorded by German pastor and amateur astronomer David Fabricius (1564–1617) beginning on August 3, 1596, but evidence of its capricious nature is hinted at in earlier Chinese, Babylonian, and Greek records (Wilk, 1996). In his account, Pickering went on to write that:

Its variability was first suspected from the character of its spectrum, which resembled that of some other variable stars of long period. A method of detecting such objects is thus indicated, far more rapid than the usual one of repeatedly observing the magnitude of the same star.

Fleming proved the HCO Director's statement right as she, herself, went on to discover 125 Mira-type variables spectroscopically (Hoffleit, 1997). In 1907, she looked back upon this newer, more convenient method of determining variables (Fleming, 1907):

The examination of photographs of the Henry Draper Memorial has led to the discovery of a large number of variable stars of long period. The greater portion of these has been found from the characteristic spectrum, Md, which indicates the third type traversed by bright hydrogen lines. While many variable stars of long period have spectra of the fourth type, and a few have spectra of the third type in which the hydrogen lines are not bright in the photographs, no case has yet been found in which a star having the spectrum Md, described above, is not a variable.

   Among the HCO staff members – north and south – there was a competitive spirit involved in the finding of variables. For instance, Bailey was interested in investigating variables found in globular clusters. Unfortunately for Fleming, her finding that particular formations of spectra were indicators of a star's variability provided a key to their discovery. One no longer needed to consecutively view the star itself in order to determine whether it was variable but had only to see its spectrum as viewed on a glass plate. The assistants at the Boyden Station who were responsible for imaging the southern sky, with the encouragement of Bailey, would often inspect the plates following exposure and look for the variables thus partaking in the thrill of discovery and a modicum of credit for the achievement. As a result, Fleming would receive plates back in Cambridge that had already been marked, and she was quite vexed. She felt that, since the plates were her responsibility, the process of ferreting out the variables should be hers as well. Very much displeased, she made her feelings known to Pickering. A message was dispatched to Bailey explaining that Fleming's work was being duplicated and that the Peruvian assistants would, in the current situation, receive credit for the discoveries that they made while it was she who was still involved in doing the greater part of the work. As Pickering wrote on September 29, 1897 (Jones and Boyd, 1971, pp. 354–345):

She is obliged to measure the positions, the variations in brightness, if any, and to identify the individual lines, classify the object and see if it is a catalogue star. She also has to reexamine the plates since the fainter objects, including about half of the peculiar objects, and as many more having slight peculiarities are omitted. All of this is part of her regular routine work and has been for the past ten years, and much of it could not be done in Peru … The delay might also prevent the early discovery of a new star or other object of special interest.

Pickering hoped that Bailey might be able to provide some means of mitigating this "source of friction" but, instead, Bailey came to the defense of his assistants. He didn't think that it was very realistic, or very fair, to prevent the employees at Arequipa from looking over the photographic results and marking any unusual objects. He added (Jones and Boyd, 1971, p. 355):

Personally, I think that the ability to make first class plates is greater than that required in the mere picking up of new objects by certain well-known characteristics, as it has been done generally by the assistants here. But as the latter has been publicly recognized and the former seldom or never, it is perhaps not strange that an ambitious assistant should desire to try also the latter … Mrs. Fleming is not the only one who has felt vexed at times.

Despite the occasional difficulties, personal and otherwise, and misunderstandings that sometimes arose from the establishment and maintenance of the Arequipa station, the photographic results produced there were more than suitable, as Pickering would later express with great satisfaction (Pickering, 1897a):

A distinguishing feature of the climate of Arequipa is the great steadiness of the air. The value of this location as an observing station is largely due to this fact. Good definition, under high powers, is obtained there on many more nights than in Europe, or in the United States, where nine-tenths of the observatories of the world are presently located.

Already in 1881, Pickering submitted an advisory from Cambridge, Massachusetts, to *Astronomische Nachrichten*, which began (Pickering, 1881b): "Herewith I send you a list of objects discovered here, and having some peculiarity either of color or of the distribution of light in their spectra." The list consisted of 39 stars, many of which were provided with a corresponding remark such as "Perhaps variable; red," "Bands distinct; fine specimen," or the name of its discoverer. This list was continued in the *Astronomische Nachrichten* the following year as "Stars with peculiar spectra, discovered at the Astronomical Observatory of Harvard College" (Pickering, 1882b). This series of announcements of the discovery of stars with peculiar spectra at the HCO would continue on for several decades and see publication in *Harvard College Observatory Circulars*, the journal *Sidereal Messenger* (renamed in 1892 *Astronomy and Astro-Physics*) and, also, in 1917, the finding of a "peculiar asteroid" was announced in an issue of the *Harvard College Observatory Bulletin*. Through the years, quite often, a single announcement would appear in several different titles simultaneously – verbatim except for acknowledgment of the particular publication's title.

On September 13, 1890, another notice appeared in the *Sidereal Messenger*; while the notice had been "Communicated by Edward C. Pickering, Director of Harvard College Observatory," the submission was credited to "M. Fleming" (Pickering and Fleming, 1890). In the notice related to plates taken at Chosica, Peru, Fleming described the spectral peculiarities of stars and the suspicion of their variability due to the bright hydrogen lines exhibited photographically. Her suspicions were ultimately proven correct. From this publication onward, notices of spectrally peculiar stars would either be submitted by Fleming or, when authored by Director Pickering, they would mainly feature her findings. A year later, in 1891, another submission to *Astronomische Nachrichten* by Fleming, and, once again, "communicated" by Pickering, announced the finding of a star showing "a spectrum consisting mainly of bright lines, and similar to that of the Wolf and Rayet stars in Cygnus" (Pickering and Fleming, 1891). Fleming would go on to discover 94 of the 107 known stars of the Wolf–Rayet type.

In 1912, Pickering wrote (Pickering, 1915):

On one plate more than a thousand spectra were classified. The late Williamina P. Fleming, Curator of Astronomical Photographs, from an examination of these plates, discovered several thousand objects having peculiar spectra. In fact, probably few bright objects of this class escaped her. Of the nineteen new stars, known to have appeared during the progress of this work, she discovered ten, and five others were found by other observers here. In this work alone also, the number of stars of the peculiar class known as fifth type, has been increased from seventeen to one hundred and eight.

Any star detected by Fleming that did not fit into Harvard's established classes was considered to be *peculiar*. The reason might be due to some variation in the spectrum such as an unexpectedly thick or thin spectral band or an unusually bright, dark, or curiously placed spectral line. Many such stars were of a type that had not yet been closely studied or were undergoing hitherto unseen processes in their evolution. Certain elements, the existence of which altered the appearance of particular spectra were, as of yet, undetected when Fleming was classifying her multitudes of stars. Neither were other processes obvious such as that of ionization, which strips away electrons due to elemental superheating and changes the elements' spectra.

Even Henry Norris Russell, the acknowledged expert on stellar composition, opined (Russell, 1914):

The first great problem of stellar spectroscopy is the identification of this predominant cause of the spectral differences. The hypothesis which suggested itself immediately upon the first studies of stellar spectra was that the differences arose from variations in the chemical compositions of the stars.

It would not be until further work was carried out by Cecilia Payne-Gaposchkin (Payne, 1925), that an understanding of stellar nucleosynthesis would begin to take hold. Payne-Gaposchkin measured the output of the energies from stars of the various stellar classes and correlated their spectra with their elemental makeup. She observed and measured the abundances of the various elements related to each class and took into consideration the effects of ionization, employing astrophysicist Meghnad Saha's 1920 theory (Hearnshaw, 2014, pp. 138–140). As summarized by William Sheehan (Sheehan, 2015, p. 162):

The spectral sequence is thus a result almost entirely of temperature progression in the atmosphere of stars . . . The higher the temperature, the faster the atoms are moving, and the more electrons will be stripped away. Since the number of atoms in one state of ionization versus another depends on competition between collisions of all kinds and radiation causing ionization and rates of electron-ion collisions producing recombination, temperature – not chemical composition – determines which absorption lines appear in a star's spectrum.

1.4 Discovery of the Peculiar Spectrum of ζ Puppis

In one of Williamina Fleming's many logbooks, the notes dated October 29, 1896, stand out as particularly striking (Fleming, 1896). On pages 146–147 in Sequence No. 9 of her logbook series, see Figure 1.6, in which she examined photographic plates for the presence of variable stars, appears the note "Meas. of lines in Spectrum of ζ Puppis," which records her documentation of the wavelengths found in the image borne on plate X6257, see Figure 1.7.

The plate series X denoted spectral plates with images obtained at Arequipa. With astounding speed, Fleming's documentation appeared swiftly in print in the form of *Harvard College Observatory Circular* No. 12 on November 2, 1896 (Pickering, 1896a). The same announcement, though slightly tailored to the individual journals, also appeared in the December 1896 issues of both *Astronomische Nachrichten* (Pickering, 1896b) and *Astrophysical Journal* (Pickering, 1896c) under Pickering's name. In these announcements, Pickering wrote:

A list of stars having peculiar spectra and found by Mrs. Fleming in her regular examination of the Draper Memorial photographs are given in the annexed table.

Then, he goes on to say that the first star in the list provided, ζ Puppis, exhibits a spectrum:

which is very remarkable and unlike any other yet obtained. The continuous spectrum is traversed by three systems of lines, First, the hydrogen lines and the line K, which are dark, as in stars of the first type. Second, two bright bands or lines whose approximate wave lengths are 4652 and 4698, which may be identical with the adjacent lines in spectra of the fifth type. Third, a series of lines whose approximate wave lengths are 3814, 3857, 3923; 4028, 4203, and 4505, the last line being very faint. *These six lines form a rhythmical series*

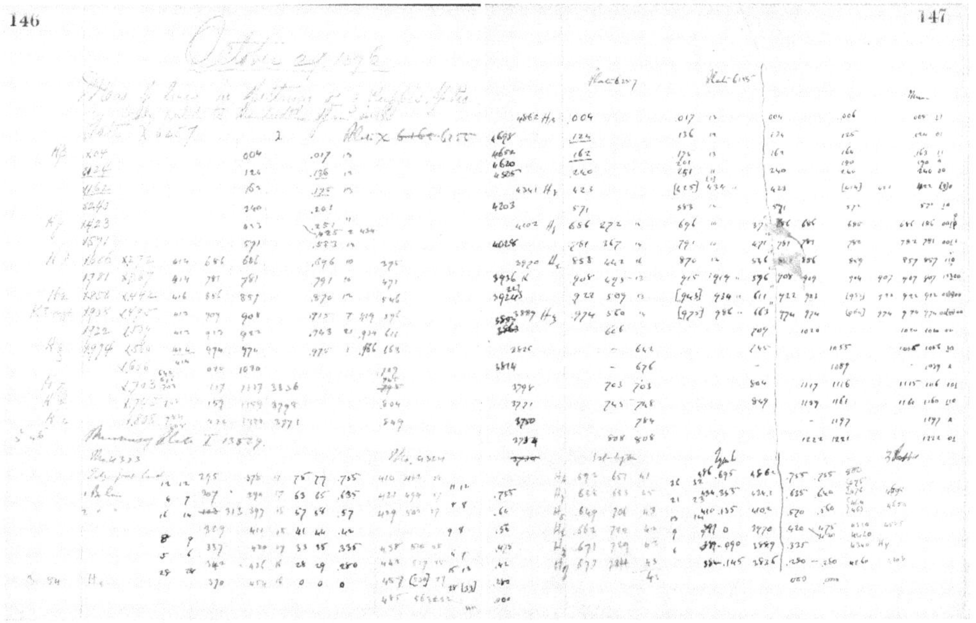

Figure 1.6 Pages 146 and 147 from Williamina Fleming's logbook of October 29, 1896 (Fleming, 1896), pertaining to the glass plate number X6257 shown in Figure 1.7. Note that on the left margin of page 147 are listed the wavelengths of the emission lines of the Pickering series of He$^+$: 4505, 4203, 4028, 3923 (average of 3922 and 3924), 3857, and 3814 Å. Courtesy of Harvard College Observatory.

like that of hydrogen and apparently are due to some element not yet found in other stars or on earth (emphasis added).

He then concluded that "The formula of Balmer will not represent this series" – known since as the *Pickering series* – but went on to show that, when suitably modified, a formula involving squares of an integer pertaining to the individual lines fit the spectrum, see Figure 1.8,

$$\lambda = 4650\frac{m^2}{m^2 - 4} - 1032 \tag{1.1}$$

with λ the wavelength in Å and m an integer. These wavelengths were featured along with the Balmer-series-like fit, cf. Eq. (1.1), in Pickering (1896b) and Pickering (1896c), whose manuscripts were expedited to *Astronomische Nachrichten* and *Astrophysical Journal* just four days later and published in December 1896. As noted, the first announcement of the discovery of the Pickering series appeared in *Harvard College Observatory Circular* No. 12 on November 2, 1896 (Pickering, 1896a), the submission date of the paper.

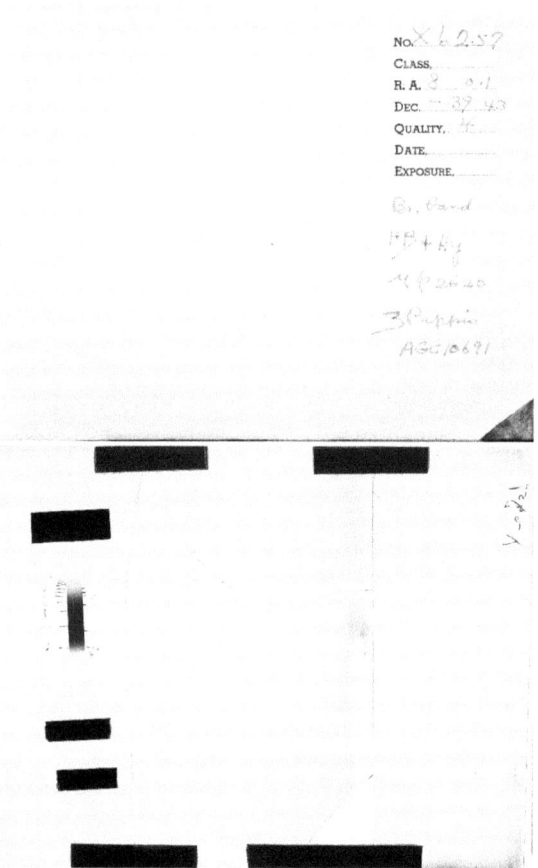

Figure 1.7 Photographic glass plate number X6257 with an image of the spectrum of ζ Puppis taken with the 13-inch Boyden refractor on December 17, 1894 at HCO's Boyden Station in Arequipa, Peru. Also shown is a remake of the original plate jacket. Courtesy of Harvard College Observatory.

The formula for the Balmer series of atomic hydrogen[7] was found empirically in 1885 by the Swiss high-school teacher Johann Jakob Balmer (1825–1898) and generalized in 1890 by Johannes Rydberg (1854–1919) for spectral series of other atoms (especially those of Groups 1–3) (Rydberg, 1890).

In a follow-up paper "The spectrum of ζ Puppis" published in 1897 in *Harvard College Observatory Circular* No. 16 (Pickering, 1897b), in *Astronomische Nachrichten* (Pickering, 1897c), and in the *Astrophysical Journal* (Pickering, 1897d), Pickering confirmed the wavelengths of the lines of the Pickering series but changed tack and ascribed them to atomic hydrogen. He

---

[7] Balmer wrote his formula in the form $\lambda = 3645 \frac{m^2}{m^2-4}$ with the wavelength λ in Å for $m = 5 - 10$, see Balmer (1885).

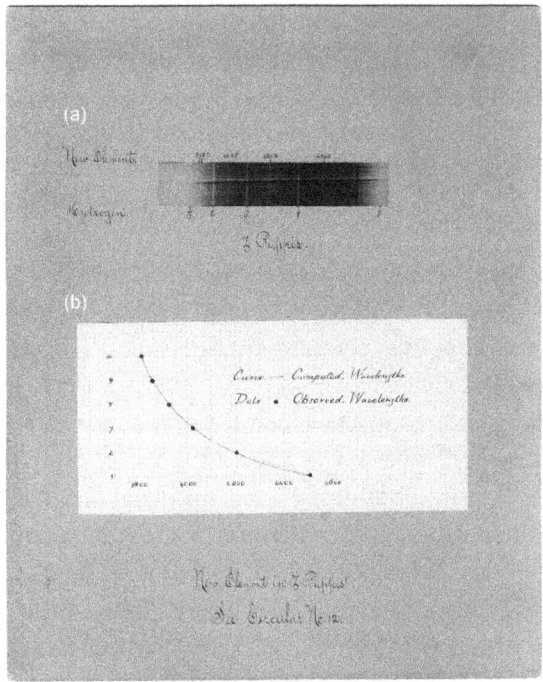

Figure 1.8 (a) Spectrum of ζ Puppis photographed at HCO's Boyden Station in Arequipa, Peru, and evaluated by Williamina Fleming on October 29, 1896, cf. Figures 1.6 and 1.7. The wavelengths shown along the upper abscissa pertain to the spectrum of the "New Element," that is, to the Pickering series of He$^+$, while the absorption lines of the Balmer series (β through ζ) of atomic hydrogen are shown on the lower abscissa and serve as wavelength standards. (b) A numerical fit, $m = 2\sqrt{\frac{\lambda + 1032}{\lambda - 3618}}$, of the wavelengths λ [Å] of the Pickering series of He$^+$ for $m = 5 - 10$. This fit was obtained by inverting the empirical Balmer-series-like fit, Eq. (1.1), featured in Pickering (1896d). Note that the handwriting in the panels is likely Williamina Fleming's. Published with permission from the Collection of Historical Scientific Instruments, Harvard University.

reported a unifying empirical formula involving squares of an integer that reproduced the Balmer series and Pickering series for the integer's even and odd values, respectively. This was very much in the spirit of the times, when physicists were groping for the conceptual foundations of spectroscopy (Robotti, 1983).

In the 1897 follow-up (Pickering, 1897d), Pickering noted: "Miss A. J. Cannon has found that the same series of lines [i.e., the Pickering series] occurs in the star 29 Canis Majoris." A related side note appears in the 1979 work by Cecilia

Payne-Gaposchkin (1900–1979), "Stars and clusters," in regard to both 29 Canis Majoris and ζ Puppis (Payne-Gaposchkin, 1979b, p. 43):

It is not without significance that the known stars of very high mass are luminous blue stars: once they have left the main sequence, such stars cannot last long. The giant eclipsing variable UW Canis Majoris (29 Canis Majoris, another star of many names) and the second magnitude ζ Puppis, one of the hottest known stars, are of very high mass, at least thirty times the sun's. We do not know what the future holds for them.

Today, we know that ζ Puppis is a blue supergiant of type O4, one of the hottest and most luminous stars of the Milky Way with a surface temperature of 4.2×10^4 K (Schilbach, 2008) where the "element not yet found" is especially abundant.

Harvard College Observatory Circular No. 55 (Pickering, 1901), added further updates to the findings related to the spectral significance of ζ Puppis as more and better plates could provide sharper images. Of course, the phenomenon of these mysterious lines of "cosmic" hydrogen, or "proto-hydrogen," as labeled by Sir Norman Lockyer, attracted scientific attention as those interested tried to determine the precise cause of these fascinating spectral features (Nature, 1912).

We note that in Fleming's final published work, "Stars having peculiar spectra," which appeared in the 1912 Volume LVI of the *Annals of the Astrophysical Observatory of Harvard College* (Fleming, 1912), the discoveries of the peculiar spectra of both ζ Puppis and 29 Canis Majoris are credited to Pickering. The "reference" given for ζ Puppis in this work is the 1897 *Harvard College Observatory Circular* No. 16, "The spectrum of ζ Puppis" (Pickering, 1897b), while that for 29 Canis Majoris is the 1897 *Harvard College Observatory Circular* No. 17, "Stars having peculiar spectra" (Pickering, 1897e), which were both written by Pickering. Fleming's "Stars having peculiar spectra" featured the "Pickering series" and was published, with a preface by Pickering dated October 21, 1912, more than a year after Fleming's death. By the time she had authored and overseen the production of her final publication, the notion of a "Pickering series" may have gained such notoriety in scientific circles that she did not wish to detract from the renown of the Director and left things the way they were.

1.5 The Spectrum of He$^+$ as a Proving Ground for Bohr's Model of the Atom

In the first sequel of his 1913 trilogy on the atomic model, Bohr noted (Bohr, 1913a):

We shall ... see that, [with the] help of the above theory [Bohr's model of the hydrogen atom], we can account naturally for these series of lines [i.e.] the series first observed by Pickering in the spectrum of the star ζ-Puppis, and the set of series recently found by Fowler in experiments with vacuum tubes containing a mixture of hydrogen and helium ... if we ascribe them to helium [with one electron missing].

According to Bohr's early version of his atomic model, the wavelength, λ, of an emission line from state n_2 to state n_1 (where $n_2 > n_1$) of a hydrogenic (one-electron) atom was given by (in modern notation but in the cgs units used by Bohr):

$$\lambda^{-1} = R_Z \left(\frac{1}{n_1^2} - \frac{1}{n_2^2} \right) \tag{1.2}$$

where R_Z is the Rydberg constant pertaining to a one-electron atom with atomic number Z,

$$R_z = \frac{2\pi^2 m_e Z^2 e^4}{h^3 c} \tag{1.3}$$

and m_e is the electron mass, e is the electron charge, h is Planck's constant, and c is the speed of light. Upon reading Bohr's paper (Bohr, 1913a), Alfred Fowler pointed out to Bohr (Fowler, 1913a) a discrepancy between Bohr's treatment of the Pickering series and the positions of the spectral lines of the series as measured in his own laboratory (Fowler, 1913b) and confirmed by Evan Jenkins Evans at Rutherford's laboratory (Evans, 1913). Indeed, the ratio of the Rydberg constants given by Bohr's formula, Eq. (1.3), for He$^+$ and for atomic hydrogen would be

$$\frac{R_{He^+}}{R_H} = \frac{Z_{He^+}^2}{Z_H^2} = 4 \tag{1.4}$$

whereas for the Pickering series and the Balmer series (Ames, 1890), this ratio came out experimentally as

$$\frac{R_{He^+}}{R_H} \approx 4.0016 \tag{1.5}$$

As a result, Bohr went back to the drawing board and replaced in his expression, Eq. (1.3), for the Rydberg constant the electron mass with the reduced mass, $m_e \mapsto \mu_Z \equiv m_e M_Z/(m_e + M_Z)$, of the nucleus–electron system, as he should have done to begin with (Bohr, 1913b):

$$R_z = \frac{2\pi^2 \mu_z Z^2 e^4}{h^3 c} \tag{1.6}$$

With this amendment, the ratio of the Rydberg constants for He$^+$ and H became

$$\frac{R_{He^+}}{R_H} = \frac{Z_{He^+}^2 \mu_{He^+}}{Z_H^2 \mu_H} = 4.00163 \tag{1.7}$$

in excellent agreement with the accurate laboratory data, as subsequently acknowledged by Fowler (Fowler, 1913b).

Had the wavelengths of the Pickering series published by Pickering in 1897 been
used to determine R_{He^+}, the ratio $\frac{R_{He^+}}{R_H}$ would have come out as 4.00101 using the
data from the left ζ column of the table in Pickering (1897d)[8] and 4.00011 using the
data from the right ζ column of the table in Pickering (1897d).[9] Interestingly
enough, the blue $\lambda = 4686$ Å line corresponding to the transition between the
$n_1 = 3$ and $n_2 = 4$ levels was not reported in communications from the HCO in
the 1890s. This line typifies the Pickering series as observed in laboratory
discharge-tube spectra (Fowler, 1912).

In 1915, Bohr provided further comments on the differences between the Balmer
series and the Pickering series and strengthened his case by referring to 1914
experimental data on the ionization potentials of hydrogen and helium (Bohr,
1915).

The explanation of the Pickering series was a major triumph for Bohr's theory.
Thereby, the theory proved capable of not just reproducing what was well known
and well established, such as the Balmer series or the Rydberg constant for
hydrogen, but it also demonstrated its predictive and explanatory power. In particular, the accurate rendition of the ratio, Eq. (1.7), of the Rydberg constants for
He$^+$ and H made a deep impression. When Albert Einstein (1879–1955) heard
about it at a 1913 conference, he exclaimed (Stachel, 2001, pp. 369–370): "Then the
frequency of the light does not depend at all on the frequency of the electron . . . this
is an enormous achievement. The theory of Bohr must then be right."

In his 1951 reflections on the advent of quantum theory, Einstein took the
opportunity to pay tribute to Bohr's achievement once more (Einstein, 1951):

[W]ithout having a substitute for classical mechanics, I could nevertheless see to what kind
of consequences this law of temperature–radiation [black body radiation law] leads for the
photoelectric effect and for other related phenomena of the transformation of radiation–
energy, as well as for the specific heat of (especially) solid bodies. All my attempts,
however, to adapt the theoretical foundation of physics to this (new type of) knowledge
failed completely. It was as if the ground had been pulled out from under one, with no firm
foundation to be seen anywhere, upon which one could have built. That this insecure and
contradictory foundation was sufficient to enable a man of Bohr's unique instinct and tact to
discover the major laws of the spectral lines and of the electron-shells of the atoms, together

---

[8] This is for a value of R_{He^+} determined as an average of the values $[R_{He^+}]^{-1} = \lambda(n_1^{-2} - n_2^{-2})$ for
$\lambda(n_1 = 4; n_2 = 11) = 4199.2$ Å, $\lambda(n_1 = 4; n_2 = 13) = 4027.1$ Å, $\lambda(n_1 = 4; n_2 = 15) = 3924.6$ Å,
$\lambda(n_1 = 4; n_2 = 17) = 3858.7$ Å, and $\lambda(n_1 = 4; n_2 = 19) = 3814.7$ Å listed in the left ζ column of the table in
Pickering (1897d).

[9] This is for a value of R_{He^+} determined as an average of the values $[R_{He^+}]^{-1} = \lambda(n_1^{-2} - n_2^{-2})$ for
$\lambda(n_1 = 4; n_2 = 11) = 4201.6$ Å, $\lambda(n_1 = 4; n_2 = 13) = 4026.5$ Å, $\lambda(n_1 = 4; n_2 = 15) = 3924.9$ Å,
$\lambda(n_1 = 4; n_2 = 17) = 3858.6$ Å, and $\lambda(n_1 = 4; n_2 = 19) = 3817.2$ Å listed in the right ζ column of the table in
Pickering (1897d).

with their significance for chemistry, appeared to me like a miracle – and appears to me as a miracle even today. This is the highest form of musicality in the sphere of thought.

Bohr's model of the hydrogenic atom was amended in 1916 by Arnold Sommerfeld (1868–1951), who made use of relativistic old quantum theory to explain the fine structure of the H and He^+ spectra and to provide the means to meet the challenge of unriddling the anomalous Zeeman effect (Sommerfeld, 1916; Paschen, 1916; Schmidt-Böcking et al., 2022). According to Walther Gerlach (1889–1979), it was the fine structure of the spectrum of He^+ that convinced his teacher Friedrich Paschen (1865–1947) – a skeptical experimentalist – that the theory of relativity was correct: "Since today, there's relativity" (Gerlach, 1963, p. 9). Sommerfeld and, independently, Peter Debye (1884–1966) (Debye, 1916), concluded that not just the magnitudes of the electronic orbital angular momenta but also the spatial orientations of the electronic orbits with respect to an external magnetic field are quantized. The Bohr–Sommerfeld–Debye model was subjected in 1922 to a nonspectroscopic test, the Stern–Gerlach experiment (Gerlach and Stern, 1922), which confirmed the existence of space quantization and thus ruled unequivocally in favor of quantum theory as epitomized by the Bohr–Sommerfeld–Debye model (Friedrich, 2023).

Bohr's model of the atom thus continued playing the role of a touchstone on the path to quantum mechanics until its advent in 1925–1927. We add that a definitive treatment of the fine structure of hydrogenic atoms as due to spin–orbit coupling and relativistic effects was provided by Sommerfeld in 1940 (Sommerfeld, 1940) and reviewed by numerous authors, including Condon and Shortley (Condon and Shortley, 1951).

1.6 Epilogue

Cecilia Payne-Gaposchkin, in her 1925 PhD thesis written at the HCO, provided this assessment of the significance of the test of Bohr's model afforded by the measurement of the Pickering series (Payne, 1925, p. 14):

The detection and resolution of the alternate components of [the Pickering] series, which fall very near to the Balmer lines of hydrogen in the spectra of the hottest stars, and the consequent derivation of the Rydberg constant for helium [plus] ... represents an astrophysical contribution to pure physics which is of the highest importance.

However, the spectral properties of He^+ proved to be of great consequence also for astronomy itself, following the discovery of the enabling role of He^+ in the pulsation of the Cepheids and of other variable stars populating chiefly the instability strip of the Herzsprung–Russell diagram (Eddington, 1917; Vitense, 1951; Zhevakin, 1953; Cox, 1963): In the dimmest (and hottest) phase of the Cepheid cycle, the radiation emanating from the stellar interior is scattered and absorbed by He^+ in the stellar envelope, leading to the envelope's expansion and, thus, cooling. At the coolest point of the cycle, the passage of the thermal radiation from the stellar interior through the envelope is the least impeded and so the star shines the brightest. The workings of variable stars could thus be likened to a Carnot cycle with He^+ playing the role of the intake/exhaust valve.[10]

Moreover, in the 1990s, the chemical properties of the He^+ ion would be recognized as key to astrochemistry, through the ion's role in the kinetically rather than thermodynamically controlled chemistry of the interstellar medium. In the 1970s, laboratory experiments conducted in conjunction with quantum-chemical calculations (Mahan, 1975) showed that while He^+ does not react with the most abundant interstellar molecule, hydrogen (H_2), it does react more than willingly with the second most abundant molecule, carbon monoxide (CO),

$$He^+ + CO \rightarrow He + C^+ + O \qquad (1.8)$$

yielding C^+ in essentially every $He^+ + CO$ collision (Klemperer, 1995, 2006). This enhances the concentration of the C^+ ion by the He/CO abundance ratio, that is, by about a factor of a thousand. The abundant C^+ ion reacts only reluctantly with the prevalent H_2 molecules and so is spared for its avid reactions with methane and acetylene, giving rise to a reaction sequence responsible for the formation of many organic compounds. As Dudley Herschbach put it (Herschbach, 1999):

The paradoxical irony is that the mutual distaste of the simplest inorganic species, He^+ and H_2, [for one another] gives rise to the proliferation of complex organic molecules in the cold interstellar clouds.

Edward Pickering's approach of collecting hosts of quality data in the face of ignorance resulted in many unexpected payoffs. Williamina Fleming's sifting through the data uncovered precious nuggets that keep dazzling us to this day.

References

Ames, Josef S. 1890. V. On some gaseous spectra: hydrogen, nitrogen. *The London, Edinburgh, and Dublin Philosophical Magazine and Journal of Science*, 30, 48–58. https://doi.org/10.1080/14786449008619987

Anderson, Michael and Roughley, Corinne. 2018. Causes of death. In Anderson, Michael (ed.), *Scotland's Populations from the 1850s to Today*, Oxford: Oxford University Press, pp. 348–373. https://doi.org/10.1093/oso/9780198805830.001.0001

---

[10] Henrietta Leavitt (1868–1921) discovered in 1908 at the HCO the nearly linear relationship between the maximum luminosity and the period of the Cepheids (Leavitt, 1908), which would become instrumental for establishing Hubble's law from observational data in 1929 (Hubble, 1929).

Bailey, Solon I. 1895. Photometric observations of southern stars. Chapter 1: History of the expedition. *Annals of the Astronomical Observatory of Harvard College*, 34, 22.

Bailey, Solon I. 1931. *The History and Work of Harvard Observatory, 1839–1927*. New York and London: McGraw-Hill Book Company.

Bailey, Solon I. 1932. Biographical memoir of Edward Charles Pickering 1846–1919. *National Academy of Sciences Biographical Memoirs*, 15, 169–189. https://www.nasonline.org/wp-content/uploads/2024/06/pickering-edward.pdf

Balmer, Johann. 1885. Notiz über die Spektrallinien des Wasserstoffs. *Verhandlungen der Naturforscher Gesellschaft zu Basel*, 7, p. 548. https://onlinelibrary.wiley.com/doi/10.1002/andp.18852610506.

Barker, George F. 1888. Biographical memoir of Henry Draper 1837–1882. *National Academy of Sciences Biographical Memoirs*, 81–139. https://www.nasonline.org/wp-content/uploads/2024/06/draper-henry.pdf

BIAU. 2022. *Bulletin of the International Astronomical Union*, 2, 5. https://www.iau.org/static/publications/wgsbn-bulletins/wgsbn-bulletin-2201.pdf

Bohr, Niels, 1913a. I. On the constitution of atoms and molecules. *The London, Edinburgh, and Dublin Philosophical Magazine and Journal of Science*, 26, 1–25. https://www.tandfonline.com/doi/abs/10.1080/14786441308634955

Bohr, Niels. 1913b. The spectra of helium and hydrogen. *Nature*, 92, 231–232. https://www.nature.com/articles/092231d0

Bohr, Niels. 1915. The spectra of hydrogen and helium. *Nature*, 95, 6–7. https://www.nature.com/articles/095006a0

Boston Herald. 1899. Harvard honors women. *Boston Herald, January* 16, 1–2. https://www.newsbank.com/schools/solutions/core4/americas-historical-newspapers

Boston Herald. 1911. Leading woman in science dies in a hospital: Mrs. Fleming, Harvard's famous astronomer, 30 years at observatory; honored by Royal Society; discovered more stars than all others in profession in 200 years; original in her methods; first to detect approach of Halley's comet only two years ago. *Boston Herald*, May 22, 1, 8. www.newsbank.com/schools/solutions/core4/americas-historical-newspapers

Cannon, Annie J. 1911. Williamina Paton Fleming. *Astrophysical Journal*, 34, 314–316. https://doi.org/10.1086/141894

Cannon, Annie J. 1912. Classification of 1,477 stars by means of their photographic spectra. *Annals of the Astronomical Observatory of Harvard College*, 56, 66.

Cannon, Annie J. 1915. The Henry Draper Memorial. *Journal of the Royal Astronomical Society of Canada*, 9, 205–206.

Cannon, Annie J. 1919. Edward Charles Pickering. *Popular Astronomy*, 27, 177–182.

Clerke, Mary A. 1896. New views about Mars. *Edinburgh Review*, CLXXXIV, 368–385. https://babel.hathitrust.org/cgi/pt?id=uc1.b2973350&seq=382

Condon, Edward U. and Shortley, George H. 1951. *The Theory of Atomic Spectra*. Cambridge: Cambridge University Press.

Cox, John P. 1963. On second helium ionization as a cause of pulsational instability in stars. *Astrophysical Journal*, 138, 487. https://doi.org/10.1086/147661

Debye, Peter 1916. Quantenhypothese und Zeeman-Effekt. *Physikalische Zeitschrift*, 17, 507–512.

Donaghe, Harriet R. 1898. Photographic flashes from Harvard Observatory. *Popular Astronomy*, 6, 483.

DPedia. 2023. Pickering (Martian Crater). DBpedia. https://planetarynames.wr.usgs.gov/Feature/4723

Eddington, Arthur S. 1917. The pulsation theory of Cepheid variables. *The Observatory*, 40, 290–293.

Einstein, Albert. 1951. The advent of the quantum theory. *Science*, 113, 82–84. https://www.science.org/doi/10.1126/science.113.2926.82

Evans, Evan J. 1913. The spectra of helium and hydrogen. *Nature*, 92, 5. https://www.nature.com/articles/092005a0

Fiss, Andrew 2023. For computing is our duty. In Ames, Morgan G. and Mazzotti, Massimo (eds.), *Algorithmic Modernity*, Oxford: Oxford University Press. https://doi.org/10.1093/oso/9780197502426.003.0008

Fleming, Williamina. 1893. A field for woman's work in astronomy. *Astronomy and Astrophysics*, 12, 683–689. https://archive.org/details/sim_astronomy-and-astrophysics_1893-10_12_8/page/682/mode/2up

Fleming, Williamina. 1896. Variable stars. *phaedra0809*, Cambridge, MA: John G. Wolbach Library, Harvard College Observatory, Project PHAEDRA. Harvard College Observatory observations, logs, instrument readings, and calculations, 9, 146–147, 1896/1897. https://articles.adsabs.harvard.edu/pdf/1896phae.proj..809F

Fleming, Williamina P. S. 1900. Journal of Williamina Paton Fleming, 1900 Mar. 1–Apr. 18; Curator of Astronomical Photographs, Harvard College Observatory. Harvard College Observatory. Chest of 1900. https://iiif.lib.harvard.edu/manifests/view/drs:3007384$1i

Fleming, Williamina P. 1907. Photographic study of variable stars. *Annals of the Astronomical Observatory of Harvard College*, 47, 2. https://articles.adsabs.harvard.edu/pdf/1907AnHar..47....1F

Fleming, Williamina P. 1912. Stars having peculiar spectra. *Annals of the Astronomical Observatory of Harvard College*, 56, 178. https://articles.adsabs.harvard.edu/pdf/1912AnHar..56..165F

Fowler, Alfred. 1912. Observations of the principal and other series of lines in the spectrum of hydrogen (plates 2–4). *Monthly Notices of the Royal Astronomical Society*, 73, 62–72. https://doi.org/10.1093/mnras/73.2.62

Fowler, Alfred. 1913a. The spectra of helium and hydrogen. *Nature*, 92, 95–96. https://www.nature.com/articles/092095b0

Fowler, Alfred. 1913b. Letter to the editor. *Nature*, 92, 232–233. https://www.nature.com/articles/092232a0

Friedrich, Bretislav. 2023. A century ago the Stern–Gerlach experiment ruled unequivocally in favor of quantum mechanics. *Israel Journal of Chemistry*, 63, e2023000. https://doi.org/10.1002/ijch.202300047

Gerlach, Walther. 1963. Interview with Thomas Kuhn, February 16, 1963. *Sources for History of Quantum Physics*. College Park, MD: American Institute of Physics. https://repository.aip.org/islandora/object/nbla:266549#page/1/mode/2up

Gerlach, Walther and Stern, Otto. 1922. Der experimentelle Nachweis der Richtungsquantelung im Magnetfeld. *Zeitschrift für Physik*, 9, 349–352. https://link.springer.com/article/10.1007/BF01326983

Grier, David A. 2013. *When Computers Were Human*. Princeton: Princeton University Press. https://doi.org/10.1515/9781400849369

Haas, Arthur E. 1910. Über eine neue theoretische Bestimmung des elektrischen Elementarquantums und des Halbmessers des Wasserstoffatoms. *Physikalische Zeitschrift*, 11, 537. https://babel.hathitrust.org/cgi/pt?id=mdp.39015035421174&seq=651

Hearnshaw, John B. 2014. *Analysis of Star Light: Two Centuries of Astronomical Spectroscopy*, 2nd edition. Cambridge: Cambridge University Press. https://www.cambridge.org/core/books/analysis-of-starlight/FCAEF94416B01AB09DA100506ED62FF9

Herschbach, Dudley R. 1999. Chemical physics: Molecular clouds, clusters, and corrals. In Bederson, Benjamin (ed.), *More Things in Heaven and Earth. A Celebration of Physics at the Millennium, Volume II.* Heidelberg: Springer, pp. 693–705. https://doi.org/10.1007/978-1-4612-1512-7_45

Hoffleit, Dorrit. 1991. Evolution of the Draper Memorial. *Vistas in Astronomy*, 34, 118. https://www.sciencedirect.com/science/article/pii/008366569190022K

Hoffleit, Dorrit. 1997. History of the discovery of Mira stars. *Journal of the American Association of Variable Star Observers*, 25, 132. https://adsabs.harvard.edu/full/1997JAVSO..25..115

Holberg, Jay. 2007. *Sirius: The Brightest Diamond in the Night Sky.* Berlin, New York, Chichester: Springer. http://doi.org/10.1007/978-0-387-48942-1

Hubble, Edward. 1929. A relation between distance and radial velocity among extra-galactic nebulae. *Proceedings of the National Academy of Sciences*, 15, 168–173. https://doi.org/10.1073/pnas.15.3.168

Hughes, Stefan. 2012. *Catchers of the Light. A History of Astrophotography.* ArtDeCiel Publishing.

IAU. 2010a. Lunar crater Fleming. *Gazetteer of Planetary Nomenclature.* International Astronomical Union. https://planetarynames.wr.usgs.gov/Feature/1979

IAU. 2010b. Lunar crater Pickering. *Gazetteer of Planetary Nomenclature.* International Astronomical Union. https://planetarynames.wr.usgs.gov/Feature/4722

Johnson, George. 2005. *Miss Leavitt's Stars. The Untold Story of the Woman Who Discovered How to Measure the Universe.* New York: Norton.

Jones, Bessie Z. and Boyd, Lyle G. 1971. *Harvard College Observatory. The First Four Directorships.* Cambridge, MA: Harvard University Press. https://www.degruyter.com/document/doi/10.4159/harvard.9780674418806/html

Klemperer, William A. 1995. Some spectroscopic reminiscences. *Annual Review of Physical Chemistry*, 46(1), 1–28. https://doi.org/10.1146/annurev.pc.46.100195.000245

Klemperer, William A. 2006. Interstellar chemistry. *Proceedings of the National Academy of Sciences*, 103(33), 12232–12234. https://doi.org/10.1073/pnas.0605352103

Kragh, Helge. 2002. *Quantum Generations: A History of Physics in the Twentieth Century.* Heidelberg: Princeton University Press, 2002.

LaFortune, Keith. 2001. *Women at the Harvard College Observatory, 1877–1919: "Women's Work," the "New" Sociality of Astronomy and Scientific Labor.* MA Thesis. Notre Dame, IN: University of Notre Dame.

Lankford, John. 1997. *American Astronomy: Community, Careers, and Power.* Chicago: University of Chicago Press.

Leavitt, Henrietta S. 1908. 1777 variables in the Magellanic Clouds. *Annals of Harvard College Observatory*, 60, 87–108.

Lewis, T. 1885. Harvard Photometry. *Observatory*, 8, 49.

Mahan, Bruce H. 1975. Electronic structure and chemical dynamics. *Accounts of Chemical Research*, 8, 55–61. https://doi.org/10.1021/ar50086a002

Massachusetts Naturalization Records. 1900. Massachusetts Naturalization Records – originals, 1906–1929 for Williamina Fleming. www.ancestry.com

Nature. 1912. New hydrogen spectra. *Nature*, 90, 466–467. https://www.nature.com/articles/090466a0

New York Times. 1888. Nebula in Orion: Prof. Henry Draper's photographs of the spectrum. *The New York Times*, April 2, 9. https://www.nytimes.com/1882/04/02/archives/the-nebula-in-orion-prof-henry-drapers-photographs-of-the-spectrum.html

Nicholson, John W. 1911. A structural theory of the chemical elements. *Philosophical Magazine*, 22, 864. https://doi.org/10.1080/14786441208637185

Pais, Abraham. 1986. *Inward Bound*. Oxford: Oxford University Press.

Paschen, Friedrich. 1916. Bohr's Heliumlinien. *Annalen der Physik*, 355(16), 901–940. https://doi.org/10.1002/andp.19163551603

Payne-Gaposchkin, Cecilia 1979a. *Dyer's Hand: An Autobiography*. Privately printed, 1979b.

Payne-Gaposchkin, Cecilia 1979b. *Stars and Clusters*. Cambridge, MA: Harvard University Press.

Payne, Cecilia H. 1925. *Stellar Atmospheres: A Contribution to the Observational Study of High Temperature in the Reversing Layers of Stars*. PhD Thesis, The Observatory, Cambridge, MA.

Pickering, Edward C. 1869. *Plan of the Physical Laboratory of the Massachusetts Institute of Technology*. Boston, MA: Press of A.A. Kingman, Museum of the Boston Society of Natural History.

Pickering, Edward C. 1881a. New variable star in Puppis. *Astronomische Nachrichten*, 100, 13–14. https://articles.adsabs.harvard.edu/cgi-bin/nph-journal_query?volume=100&plate_select=NO&page=13&plate=&cover=&journal=AN

Pickering, Edward C. 1881b. Schreiben des Herrn Professor Edw. C. Pickering an den Herausgeber. *Astronomische Nachrichten*, 99, 375–378.

Pickering, Edward C. 1882a. *Plan for Securing Observations of the Variable Stars. Harvard College Observatory Papers*, Vol. 1, 1882.

Pickering, Edward C. 1882b. Stars with peculiar spectra discovered at the Astronomical Observatory of Harvard College. *Astronomische Nachrichten*, 101, 73–74.

Pickering, Edward C. 1896a. Stars having peculiar spectra. New variable stars in Crux and Cygnus. *Harvard College Observatory Circular*, 12, 1–2.

Pickering, Edward C. 1896b. Stars having peculiar spectra. New variable stars in Crux and Cygnus. *Astronomische Nachrichten*, 142, 87–90.

Pickering, Edward C. 1896c. Stars having peculiar spectra. New variable stars in Crux and Cygnus. *Astrophysical Journal*, 4, 369–370. https://articles.adsabs.harvard.edu/pdf/1897AN....142...87P

Pickering, Edward C. 1896d. Stars having peculiar spectra. New variable stars in Crux and Cygnus. *Astronomische Nachrichten*, 142(6), 87–90.

Pickering, Edward C. 1897a. Southern double stars. *Harvard College Observatory Circular*, 18, 1.

Pickering, Edward C. 1897b. Spectrum of ζ Puppis. *Harvard College Observatory Circular*, 16, 1–2.

Pickering, Edward C. 1897c. Spectrum of ζ Puppis. *Astronomische Nachrichten*, 142, 399–402.

Pickering, Edward C. 1897d. Spectrum of ζ Puppis. *Astrophysical Journal*, 5, 92–94.

Pickering, Edward C. 1897e. Stars having peculiar spectra. *Harvard College Observatory Circular*, 17, 1–2.

Pickering, Edward C. 1901. Spectrum of ζ **Puppis**. *Harvard College Observatory Circular*, 55, 1–3.

Pickering, Edward C. 1912. *Sixty-Sixth Annual Report of the Director of The Astronomical Observatory of Harvard College for the Year Ending September 30, 1911*. Cambridge, MA: The University, 1912.

Pickering, Edward C. 1915. Objective prism. *Popular Astronomy*, 23, 489.

Pickering, Edward C. and Fleming, Mina. 1890. Stars having peculiar spectra. *Sidereal Messenger*, 379, 379–380.

Pickering, Edward C. and Fleming, Mina. 1891. Stars having peculiar spectra. new variable stars in Aquarius, Delphinus and Camelopardalus. *Astronomische Nachrichten*, 127, 5.

Plotkin, Howard. 1990. Edward Charles Pickering. *Journal for the History of Astronomy*, xxi, 48. https://journals.sagepub.com/doi/10.1177/002182869002100106

Robotti, Nadia. 1983. The spectrum of Puppis and the historical evolution of empirical data. *Historical Studies in the Physical Sciences*, 14(1), 123–145. https://online.ucpress.edu/hsns/article-abstract/14/1/123/47630/The-Spectrum-of-Puppis-and-the-Historical?redirectedFrom=fulltext

Russell, Henry N. 1914. Relations between the spectra and other characteristics of the stars, *Popular Astronomy*, 22, 276.

Rydberg, Johannes R. 1890. On the structure of the line-spectra of the chemical elements. *The London, Edinburgh, and Dublin Philosophical Magazine and Journal of Science*, 29(179), 331–337. https://doi.org/10.1080/14786449008619945

Schilbach, Elena and Röser, Siegfried. 2008. On the origin of field O-type stars. *Astronomy and Astrophysics*, 489(1), 105–114. https://doi.org/10.1051/0004-6361:200809936

Schmadel, Lutz D. 2007. (784) Pickeringia. In *Dictionary of Minor Planet Names*, Berlin, Heidelberg: Springer, pp. 74–74. https://doi.org/10.1007/978-3-540-29925-7

Schmidt-Böcking, Horst and Gruber, Gernot and Friedrich, Bretislav. 2022. One hundred years ago Alfred Landé unriddled the anomalous Zeeman effect and presaged electron spin. *Physica Scripta*, 98, 014005.

Sheehan, William. 2015. *Galactic Encounters: Our Majestic and Evolving Star System, from the Big Bang to Time's End*. New York: Springer. https://doi.org/10.1007/978-0-387-85347-5

Shindell, Matthew. 2023. *For the Love of Mars*. Chicago: Chicago University Press. https://doi.org/10.7208/chicago/9780226821900.001.0001

Sobel, Dava. 2016. *The Glass Universe. How the Ladies of the Harvard Observatory Took the Measure of the Stars*. New York: Viking.

Sommerfeld, Arnold. 1916. Zur Quantentheorie der Spektrallinien, *Annalen der Physik*, 51, 125–167. https://doi.org/10.1002/andp.19163561702

Sommerfeld, Arnold. 1940. Zur Feinstruktur der Wasserstofflinien. Geschichte und gegenwärtiger Stand der Theorie. *Naturwissenschaften*, 28, 417–423. https://link.springer.com/article/10.1007/BF01490583

Stachel, John. 2001. *Einstein from 'B' to 'Z.'* Boston, MA: Birkhäuser.

Turner, Herbert H. 1912. Report of the Council to the Ninety-Second Annual General Meeting: Mrs. Fleming. *Monthly Notices of the Royal Astronomical Society*, 72, 261–264. https://doi.org/10.1093/mnras/72.4.261

Upton, Winslow. 2001. HCO Pinafore. In Davis Philip, A. G. and Koopmann, Rebecca A. (eds.), *The Starry Universe. The Cecilia Payne-Gaposchkin Centenary*, Schenectady, NY: L. Davis Press, 2001. See also https://planet4589.org/astro/cfa/play/play.html

Vitense, Erika. 1951. Der Aufbau der Sternatmosphären IV. Teil, Kontinuierliche Absorption und Streuung als Funktion von Druck und Temperatur. *Zeitschrift für Astrophysik*, 28, 81–112.

Welther, Barbara L. 1982. Pickering's harem. *Isis*, 73, 94. https://doi.org/10.1086/352911

Wiescher, Michael. 2021. *Arthur E. Haas: The Hidden Pioneer of Quantum Mechanics. A Biography*. Cham: Springer Nature. https://doi.org/10.1007/978-3-030-80606-4_1

Wilk, Stephen R. 1996. Mythological evidence for ancient observations of variable stars. *Journal of the American Association of Variable Stars*, 24, 129–133. https://apps.aavso.org/jaavso/article/1559/

Zhevakin, Sergei A. 1953. On the theory of the Cepheids. *Russian Astronomical Journal*, 30, 161.

2

H. Johanna van Leeuwen, the Other Scientist behind the Bohr–Van Leeuwen Theorem

MIRIAM BLAAUBOER AND MARGRIET VAN DER HEIJDEN

2.1 Two Sisters in Leiden

In the early twentieth century, Hendrika Johanna ("Jo"; 1887–1974) and Cornelia ("Nel"; 1889–1974) van Leeuwen, two sisters from The Hague in the Netherlands, both travelled to Leiden to study physics. The two sisters were not the first women in the Netherlands to do so, but their choice was not so common either. Just like in other European countries, Dutch universities had only slowly begun to open their doors to women during the second half of the nineteenth century. The very first woman to complete any studies in a Dutch university was Aletta Jacobs, who subsequently obtained a doctorate in medicine at Groningen University in 1879. Her path had not been straightforward; she had to seek permission from the Dutch minister of education, Johan Thorbecke, and her case sparked extensive debates in newspapers of the time (Bosch, 2005). It is not surprising that women – following Jacobs' footsteps – only slowly trickled into the universities; by the end of the nineteenth century just about one in every 17 Dutch students was a woman (Kirejczyk, 1993).[1]

One of the hurdles on their path to higher education – as in most other European countries – was the lack of preparatory education. Traditionally, the so-called gymnasia prepared the sons of the Dutch elite to study law, medicine, or the humanities, and henceforth for a career in the judiciary, the government, or the military. In 1863, the same minister Thorbecke who would later grant Jacobs permission to study, established an alternative school type, the *Hogere Burger School* (Higher Civic

We would like to thank Eduard Pannenborg and Loek Kraamer for sharing personal memories; Hans Wilschut for sharing the letter from Dirk Coster's archives in Groningen; Anne Kox for mentioning Van Leeuwen's letters in the Noord-Hollands Archief and for sharing the materials mentioned in footnote 18, and Heidi Kristjankroon for sharing personal correspondence and photographs. One of us, Margriet van der Heijden, would also like to express her gratitude for the stimulating and inspiring collaboration with co-editors Patrick Charbonneau, Daniela Monaldi, and Michelle Frank. The present chapter is a substantially extended and adapted version of an earlier article in Dutch that was published in the *Dutch Journal of Physics* (Blaauboer, 2015).
[1] The total number of students in the Netherlands in 1900 was roughly 2,800 according to Statistics Netherlands (Centraal Bureau voor de Statistiek (CBS)), which began to register these numbers in 1900. CBS, "Leerlingen en studenten; onderwijssoort, vanaf 1900": https://www.cbs.nl/nl-nl/cijfers/detail/37220

School, or HBS for short), to modernize the educational system. The HBS aimed to prepare boys for careers in the rapidly expanding commerce and industry sectors. Being more affordable than the traditional gymnasia, it drew sons from families of lower socio economic status, who, along with the professors' and reverends' sons in their classrooms, shared an interest in the natural sciences. The HBS's emphasis on mathematics, physics, and chemistry soon also led to a new pathway to university: Talented boys with an HBS diploma who passed the state exams in Latin and Greek as well, could now pursue university studies, a privilege previously reserved for those with a gymnasium diploma (Boekholt and De Booy, 1987, pp. 179–190; Van Steen, 2003, pp. 89–93; Baneke and Maas, 2018).[2]

For girls, circumstances were different. The law did not formally forbid them from attending school. But while Thorbecke's 1863 law on higher education mandated that every town with over 10,000 inhabitants should establish an HBS for boys, the provision of schooling for girls was left at the discretion of local city councils. Even if an HBS for girls was established, its curriculum usually diverged from that of the "traditional" HBS. English, French, German, geography, history, and biology were taught in roughly the same manner, but math, physics, and chemistry were combined into a single "natural sciences" course. The hours freed as a result were instead devoted to courses in needle work, dietetics, health, and sometimes music (Van Steen, 2003, pp. 85–86 and pp. 100–102). Thus, for girls with an interest in the natural sciences, the only option was to enrol in an HBS for boys, but this option came with hurdles too: the minister of education had to grant them an exemption and the school had to permit girls to attend (Bosch, 1993).

Clearly, such a step also required parents or other family members who were receptive to the idea of an education that did not solely prepare girls for future roles as wives, mothers, and caregivers, *and* who were able and willing to finance such an education. Pieter van Leeuwen and Maria van Leeuwen-Schepman, both trained as elementary school teachers, were such parents. They must have been closely observing the developments in The Hague, for in 1901, as soon as one of the city's HBS schools for boys began admitting girls, they enrolled their daughters: Nel in the first grade and Jo, who had previously spent a year at another school, in the second. Later, they also allowed their daughters to take the state exam in Greek and Latin and subsequently to study physics in Leiden, with Jo starting in 1906[3] and Nel in 1907[4] (Kroon and Blok, 2021).

---

[2] A famous example of someone who took this route is Nobel laureate Hendrik Lorentz, the son of a horticulturist (Capponi and Frenken, 2021; Kox and Schatz, 2021).

[3] Jo spent this year at the *Hoogere Burgerschool met driejarigen cursus voor jongens* (Higher Civic School with a three-year course for boys), which offered a reduced HBS curriculum and was open to girls from 1899 on. Archief van de Dalton hbs, 1883–1963, Haags Gemeentearchief, Den Haag.

[4] In 1901 the city council of The Hague decided that the *Hoogere Burgerschool met vijfjarigen cursus* (Higher Civic School with a five-year course) could open its doors to girls. Thorbecke Scholengemeenschap, Haags

In this chapter, we discuss how circumstances, chance, societal expectations, and (internalized) gender stereotypes shaped the lives of the two sisters in vastly different fashion. We focus especially on the career as (quantum) physicist of the elder sister, Jo van Leeuwen.

2.2 Flourishing Physics in Leiden

In retrospect and given the limited number of primary sources about the Van Leeuwen sisters, we can only guess what it must have been like for them to take the train to Leiden and to walk to the red brick physics building – het Natuurkundig Laboratorium (Physics Laboratory) – that quietly faced one of Leiden's canals. Leiden University, founded in 1575, prided itself on being the oldest university of the Netherlands. In the seventeenth and eighteenth centuries, it had been the working place of scientists like physicist Pieter van Musschenbroek (1692–1761), co-inventor of the well-known Leyden jar, and famed physician and botanist Herman Boerhaave (1668–1738). Admittedly, during the first half of the nineteenth century, academic life in this city surrounded by low polders had become a bit sleepy and dull (Kox and Schatz, 2021, p. 77). But when the two sisters arrived in the early twentieth century, it was booming again, especially in the natural sciences, bringing forth several Nobel prizes in quick succession.

The steep rise in academic activity and recognition started about 10 years before the Van Leeuwen sisters began their studies, with the 1896 discovery of the splitting of spectral lines in a magnetic field by experimental physicist Pieter Zeeman. Theoretical physicist and Leiden professor Hendrik Lorentz would soon explain this unexpected "Zeeman effect" with his so-called electron theory, and in 1902 both men shared the Nobel Prize in Physics. Roughly a decade later, Heike Kamerlingh Onnes (1853–1926) would liquefy helium to study metals at extremely low temperatures, leading him to serendipitously discover superconductivity. He too would receive the Nobel Prize in Physics for his discovery, in 1913. The prestige of Leiden University was further boosted by two other Dutch Nobel laureates, Johannes van der Waals (Nobel Prize in Physics, 1910) and Jacobus van 't Hoff (Nobel Prize in Chemistry, 1901), who had studied there.

Historians of science often attribute the "mild rain" of Nobel prizes to the introduction of the HBS (Baneke and Maas, 2018; Kox and Schatz, 2021, pp. 84–86). It was even more remarkable given that around 1900 the combined physics departments of all Dutch universities employed only 18 staff members, six of whom (Lorentz, Kamerlingh Onnes and four assistants) worked in Leiden (Kox and Schatz, 2021, p. 78). As a result,

---

Gemeentearchief, Den Haag. Until 1906, girls still had to obtain permission from the minister of education as well (Bosch, 1993).

from the onset of their studies, Leiden physics students were taught by the Nobel laureates themselves. Whether or not these lectures were inspiring is another question. As a former student, Dirk Jan Struik, later noted: "We had to attend once a week a lecture by Kamerlingh Onnes, who hated to leave his lab and demonstrated it by reading his notes in as dull a way as possible" (Van Delft, 2005, p. 336). By contrast, Lorentz's lectures were often qualified as marvellous, flawless, and crystal clear, though perhaps even too smooth: "you could just copy it [the blackboard] and print it" (Gorter, 1962). Much later, Jo van Leeuwen would recall how much Lorentz meant to students outside of the lecture room as well (Van Leeuwen, 1925):

It was not a coincidence that the first thing I was told by one of the elder students that guided me a bit after having arrived in Leiden, was this: "If we realize that one of us needs a bit of extra encouragement, we always ask if professor Lorentz wants to talk to him [sic] and he will then help him overcome his temporary discouragement."[5]

Perhaps Jo and Nel may have felt that Lorentz was in favor of women attending the university. His own daughter Berta (full name: Geertruida Luberta) was among the first women to study physics in Leiden and would in 1912 obtain a PhD in physics under her father's supervision. Around 1905, Lorentz had even supplied the newly founded Vereeniging van Vrouwelijke Studenten Leiden (Association of Female Students Leiden, VVSL) with feminist literature (Kox and Schatz, 2021, p. 49). This was in line with his long-running stance on women's rights. Around the turn of the century he had, for example, accompanied his wife Aletta Lorentz-Kaiser to meetings of the Nederlandse Vereeniging voor Vrouwenkiesrecht (Dutch Association for Women's Suffrage), established at the initiative of Jacobs. Later he had also taken part in meetings and dinners of its more moderate demerger, the Nederlandsche Bond voor Vrouwenkiesrecht (Dutch Union for Women's Suffrage) in which Lorentz-Kaiser played a leading role. And even though Lorentz himself had always expressed his views with moderation – conforming to his character – these activities nevertheless made him more progressive than many of his colleagues, some of whom even called him a "feminist à outrance [extremist]" (Kox and Schatz, 2021, p. 45).

What Nel and Jo van Leeuwen thought of this, if anything, cannot be inferred from the scarce sources available. We do know that this early generation of women students struggled to find their place. Interestingly, many of them (up to 50%) opted for joining the faculty of *wis- en natuurwetenschappen* (mathematics and natural sciences), most choosing pharmacology as their specialty (Kirejczyk, 1993).

---

[5] Unless otherwise specified, all translations are from the authors. Original: "Toeval was het dus niet, dat onder het eerste wat mij verteld werd door een van de oudere studenten in de wis- en natuurkunde, die mij bij aankomst in Leiden wat wegwijs maakte, het volgende was, 'Als we weten, dat er een van ons «en riem onder het hart gestoken moet worden, vragen we of professor Lorentz eens met hem praten wil en hij helpt hem over de tijdelijke ontmoediging in zijn studie heen'."

Yearbooks of the associations of female students in Delft and Leiden reveal that they were afraid to be judged as overly serious, on the one hand, and too feminist, on the other. "Don't be a prudent little miss, nor a shrew with a 'do-not-come-near-me-attitude,' but especially avoid turning into a male-like person with offhand manners and coarse language," was the advice still given to new members of the Association of Female Students in Delft in 1930 for example.[6] In sum, the Van Leeuwen sisters, finding their way in the Leiden physics buildings where all lecturers and supporting staff were male and where few women had ventured into before, must have felt like pioneers exploring new possibilities for leading a woman's life.

2.3 A Graduate Student of Hendrik Lorentz

Did the two Van Leeuwen sisters support one another during their studies in Leiden? Did they perhaps share notes and ideas? Once more, the scarcity of sources makes it impossible to answer such questions. What we do know is that Nel, the younger of the two, began to work on a doctoral thesis under the supervision of Willem Keesom in 1916. In that same year, she and her supervisor wrote an article on the second virial coefficient of hard quadrupole molecules that appeared in the *Proceedings of the Royal Dutch Academy of Sciences* (Keesom and Van Leeuwen, 1916). However, Nel's research came to a halt after the summer of 1917, when she married the Finnish theoretical physicist Gunnar Nordström, an expert in the field of gravity who had been working in Leiden with Paul Ehrenfest for two years.[7] By autumn of 1918 Nel gave birth to a son,[8] and shortly afterwards she and her family moved to Finland where Nordström had been appointed a professor of physics at Helsinki University of Technology. Nel's path thus followed the conventions of traditional Dutch society which dictated that wives and mothers stayed at home, and she left the field of physics – although it remains unclear whether she did so with regret.

Her elder sister Jo would follow an altogether different path. In 1914, under the supervision of Lorentz, she began to investigate the microscopic origin of magnetic phenomena in solids from a classical-mechanical point of view. Lorentz, in his sixties and formally retired from the university since 1912, had by then even more prestige than shortly after receiving the Nobel Prize (which was not as glamorous

---

[6] Homans, L. N. S. (1930). Toespraak tot de nieuwe leden. In Almanak der Delftsche Vrouwelijke Studenten Vereniging. Delft: J. Waltman. https://erfgoed.tudelft.nl/en/objects/trt-784. Original: "Wees geen preutsch juffertje, geen kat met een raak-me-niet-aan-houding tegenover haar mede-studenten, maar vooral geen ver-manlijkt wezen met onverschillige manieren en ruwe woorden."

[7] Marriage Certificate, August 14, 1917, 0335 – Huwelijksakten, Inventaris nr. 883, Haags Gemeentearchief, Den Haag. Note, Paul Ehrenfest was one of the witnesses.

[8] Birth Certificate, October 23, 1918, 0351 – Bevolkingsregister, Inventaris nr.1291, Haags Gemeentearchief, Den Haag.

then as it is today). His growing international standing was related to his role as chair of the famous Solvay Councils that allowed him to display both his theoretical physics knowledge as well as his language fluency (Berends, 2015). In the words of Einstein, written shortly after the very first such council in 1911 in Brussels: "H. A. Lorentz chaired the meeting with incomparable tact and unbelievable virtuosity. He speaks all three languages [English, French, and German] equally well and has a unique scientific acumen."[9]

Not only did he reap rewards from many decades of work and study, Lorentz also supervised four female graduate students between 1912 and 1919.[10] Other than his daughter Berta (see Section 2.2), who worked on the Brownian motion of electrons, thus making her one of the first scientists to look into the phenomenon of thermal noise (De Haas-Lorentz, 1912, p. 82),[11] they were Johanna Reudler, who worked on black-body radiation (Reudler, 1912); Eva Dina Bruins, who like Van Leeuwen studied topics from the field of magnetism (Bruins, 1918); and Van Leeuwen herself. Lorentz first mentioned her as one of his PhD students in a letter to his daughter Berta in early 1914, adding that she was still unsure about her dissertation topic.[12] Five years later, on January 20, 1919, Van Leeuwen received a doctorate for her thesis *Vraagstukken uit de electronentheorie van het magnetisme* (Problems from the electron theory of magnetism) (Van Leeuwen, 1919).

2.4 First Steps

When Van Leeuwen (see Figure 2.1) began her graduate work in 1914, quantum physics was still in its infancy. Not many years had passed since Max Planck, in 1900, had managed to properly describe the black-body radiation spectrum by introducing the fundamental constant, h, thus laying the foundation for the concept of energy quanta. And it was less than a year since Niels Bohr, in 1913, had incorporated this concept of energy quantization in his revolutionary model of the hydrogen atom, as well as in a model for the hydrogen molecule that Van Leeuwen would later build upon (Bohr, 1913a, b, c).

Building on the work of Planck and many others, Bohr's brand-new model described hydrogen atoms as consisting of a tiny, dense nucleus around which an electron revolved – the charged microscopic constituent of matter that according to Lorentz's electron theory was responsible for the interaction between light and matter.

---

[9] Einstein to Zangger, November 7, 1911: (Berends and Van Delft, 2021, p. 316).
[10] During his career Lorentz supervised 26 graduate students in total: https://www.lorentz.leidenuniv.nl/IL-publications/dissertations/lorentz.html
[11] This thermal noise is nowadays called Johnson–Nyquist noise after John Johnson, who first measured it at Bell Laboratories in 1926, and Harry Nyquist, who then was able to explain the results.
[12] Lorentz to De Haas-Lorentz, February 22, 1914, 364 – Prof. Dr. H. A. Lorentz Archive, Inventaris nr. 752-754, Noord-Hollands Archief, Haarlem.

Figure 2.1 Portrait photo of Jo van Leeuwen (date unknown, but surely before 1925). Image courtesy of the Van Leeuwen family.

Bohr realized that such an electron could only revolve without radiating (and thus losing) energy when moving in well-defined, stable orbits and he postulated that in such orbits the angular momentum of the electron was equal to an integer multiple of the Planck constant. In other words, he associated each orbit with a specific angular momentum and, consequently, with an energy level of the electron (and henceforth the atom). Finally, Bohr proposed that the electron could jump from one orbit (or energy level) to another by emitting or absorbing light of a specific frequency, and hence carrying a specific energy corresponding to the energy difference between the two levels (Bohr, 1913a, b). In one bold sweep his model thus illustrated how quantization was an intrinsic property of not only light but of matter as well, *and* that it played an essential role in the interaction between the two.[13]

---

[13] Bohr notably also built upon Ernest Rutherford's experimental results showing that the atom has a dense nucleus at its core, and on theoretical work especially by Einstein who could account for the photoelectric effect

Support for his groundbreaking idea came from calculations of the observed absorption and emission spectra of hydrogen that were made using Bohr's model of the atom (see Chapter 1). More specifically, the results of such calculations were in accordance with the empirically established Rydberg formula that recapitulated the emission spectrum of hydrogen. However, Van Leeuwen – during her first, explorative steps in research – left such calculations aside. Instead, she built on the third part of Bohr's 1913 trilogy on the structure of atoms and molecules, in which he proposed a model for the hydrogen molecule (Bohr, 1913c; for a further discussion of his trilogy, see: Kragh, 2012; Aaserud and Heilbron, 2013; Duncan and Janssen, 2019, pp.145–201).

Bohr described the hydrogen molecule as consisting of two nuclei with their two electrons revolving on a circular orbit, midway and perpendicular to the axis connecting the nuclei. Expanding on this work and refinements by Peter Debye (Debije), Van Leeuwen performed calculations to understand if oscillations might occur in such molecules in the absence of any external forces (Debije, 1915). By assuming, as Debye had done, that the nuclei themselves would to a good approximation stay in place, such "free oscillations" would consist of vibrations of the electron orbit expanding and shrinking at well-defined frequencies. Van Leeuwen then tried to ascertain if these vibrations could be triggered by the fluctuating electric field of light waves passing the hydrogen molecules, a bit comparable to triggering a tuning fork by hitting it on a surface (Van Leeuwen, 1916). In other words, she combined Bohr's quantum model of the molecule with the classical ideas about the interaction between light and matter from Lorentz's electron theory (see Section 2.5). She did not succeed, however.

Her work did nevertheless catch the attention of two young colleagues, who felt tempted to improve and expand upon it. The first was Jan Burgers, who would receive his doctorate from Ehrenfest in 1918 and who would subsequently become professor of fluid mechanics at the Technische Hoogeschool (Technical College) in Delft (TH Delft); the current Delft University of Technology. Much later he recalled: "Before I started to do anything myself, in 1916 Miss (van Leeuwen) [...] had worked out the free oscillations of the Debye model of the hydrogen molecule. She just assumed that the [electron] orbits were given there, and then calculated the free oscillations to see whether you could explain the absorption spectrum of hydrogen, and that she did not succeed in doing" (Burgers, 1962). For his own calculations Burgers decided to employ elements of Edmund T. Whittaker's *Analytical Dynamics*, which he had just been studying (Burgers,

on the assumption that light sometimes behaves as particles with quantized energies and, similarly, for the specific heats of solids at low temperature on the basis of a model of a solid consisting of harmonic oscillators with quantized energies (Duncan and Janssen, 2019, pp. 45–201).

1917). However, his work did not lead to significant improvements, as he later admitted: "I did not get much result either" (Burgers, 1962).

Something similar happened to Polish physicist Adalbert (Wojciech) Rubinowicz, who would become professor of theoretical physics at the University of Ljubljana in 1920. In 1917, while in Salzburg and having some time to spare, "I read a paper by Miss Van Leeuwen. And so, I thought it would perhaps be interesting to generalise the thing, to see how the nuclei move," he later recalled (Rubinowicz, 1963). Once his calculations were published in *Physikalische Zeitschrift* (Rubinowicz, 1917), Rubinowicz sent an offprint to Van Leeuwen, who politely thanked "Dear Dr. Rubinowicz" for his "paper on the natural oscillations of the Bohr–Debije H_2 molecule."[14] Yet, she also seemed to subtly suggest that his work merely confirmed hers: "[Y]ou have had to do a lot of calculating before you came to the insight that there are indeed no new unstable modes of vibration to add to the already known ones."[15] Much later, Rubinowicz himself seemed to agree with that judgement when he described his own paper, laughingly, as "surely of no importance" (Rubinowicz, 1963).

2.5 Magnetism as a Quantum Phenomenon

The largest and by far most important part of Van Leeuwen's doctoral work built on the centerpiece of Lorentz's long career: his electron theory (*electronentheorie*). Her work would culminate in the demonstration that magnetism is an intrinsically quantum phenomenon, but to understand how this conclusion came about, we must first briefly describe Lorentz's theory of the electron.

Beginning with his 1875 dissertation, Lorentz had built on Maxwell's theory of electromagnetism[16] that so beautifully describes magnetic and electric phenomena, defines the relationships between electricity and magnetism, and shows that light consists of self-propagating, alternating, electric and magnetic waves, that is, electromagnetic waves. That theory, however, left open the details of the microscopic interactions between light and matter that underpinned macroscopic phenomena such as light reflecting from a surface, or the refraction of light that passes from one material to another.

By 1878, Lorentz had proposed that tiny, electrically charged constituents of matter were responsible for these interactions at the microscopic scale. He initially

---

[14] Van Leeuwen to Rubinowicz, June 26, 1917. Archive for the History of Quantum Physics, APS. Original: "Sehr geehrter Herr Dr. Rubinowicz, Für den mir zugeschickten Sonderdruck Ihrer Arbeit über die Eigenschwingungen des Bohr–Debijeschen H_2 Molekül danke ich bestens (. . .). Hochachtungsvoll, H. J. v. Leeuwen."

[15] Original: "Sie haben eine grosze Rechenarbeit gehabt, bevor sie wussten dass sich in der Tat keine neuen instabilen Schwingungsweisen zu den alten fügten."

[16] For his thesis work and in the first years thereafter he used Hermann von Helmholtz' formulation of Maxwell's theory (Kox and Schatz, 2021, pp. 56–65)

called these "ions," but by 1899 he had adopted the name "electrons."[17] Such charged particles would begin to oscillate when "hit" by a light wave and, since oscillation entails acceleration, the oscillating charged particles would consequently emit radiation themselves. Crucial to Lorentz's final electron theory was that it also helped translate electric and magnetic interactions on microscopic scales (where the electric and magnetic field strengths vary rapidly) to phenomena observed at a macroscopic scale by calculating average field values for groups of many atoms and molecules (Kox and Schatz, 2021, pp. 56–65). Because Maxwell's theory gives that an electric current (or, for Lorentz, a moving electron) induces a magnetic field, one could then try to predict if and how moving electrons inside matter (i.e., at the microscopic scale) would give rise to magnetism (at the macroscopic scale).

Building on Lorentz's classical-mechanical work, Van Leeuwen investigated a range of magnetic phenomena. The first chapters of her thesis (see Figure 2.2) explore whether and how specific types of magnetic behavior may arise in hypothetical gases consisting of neutral atoms and molecules that interact via elastic collisions. Van Leeuwen studies cases in which these molecules have an electric dipole moment (i.e., an asymmetric charge distribution causing the molecule to have a positive and a negative pole) or some other internal charge distributions and compares her results with earlier (semi-classical) work by French physicist Pierre Langevin (Langevin, 1905). She also examines the magnetic behavior of free electrons in metals, elaborating on two slightly different proofs that these do not produce a net magnetic moment given by Lorentz during a lecture in the academic year 1910–1911, and comparing and contrasting these results with earlier work by Waldemar Voigt, Erwin Schrödinger, and Joseph Thomson (Van Leeuwen, 1919).[18]

The most relevant aspect of Van Leeuwen's thesis from the perspective of quantum physics is her showing that any dynamical classical-mechanical system in a magnetic field in thermal equilibrium (i.e., with no net energy flowing into or out of the system) has no net magnetic dipole moment, and thus cannot give rise to

---

[17] Lorentz called these constituents ions at first but adopted the name "electrons" after the German physicist Emil Wiechert and the British physicist Joseph Thomson in their so-called "cathode-ray experiments" discovered these negatively charged particles outside atoms as well. The name "electron" had already been introduced in 1891 by the Irish physicist George Johnstone Stoney for hypothetical units of electrical charge (Kragh, 2023, pp. 75–76).

[18] Indeed, on page 66 of her thesis, Van Leeuwen refers to two lectures by Lorentz in which he suggested that the free electrons in a metal do not in fact give rise to a net magnetic moment. His argumentation was succinct and somewhat sketchy in Göttingen in 1913, where according to the proceedings he concluded "that one cannot attribute the diamagnetic properties of a metal to the curvature of the electron orbits caused by the magnetic field, as J.J. Thomson tried to do" ("Wie mir scheint, kann man hieraus schliessen, dass man nicht, wie J. J. Thomson versucht hat, die diamagnetische Eigenschaften eines Metalles auf die durch das Magnetfeld verursachte Krümmung der Elektronenbahnen zurückführen kann." See Lorentz (1914)). A more detailed proof was given in Lorentz's 1910–1911 "Capita Selecta" lecture series at the University of Leiden (see: 364 – Prof. Dr. H. A. Lorentz Archive, Inventaris nr. 265, pp. 47–51, Noord-Hollands Archief, Haarlem).

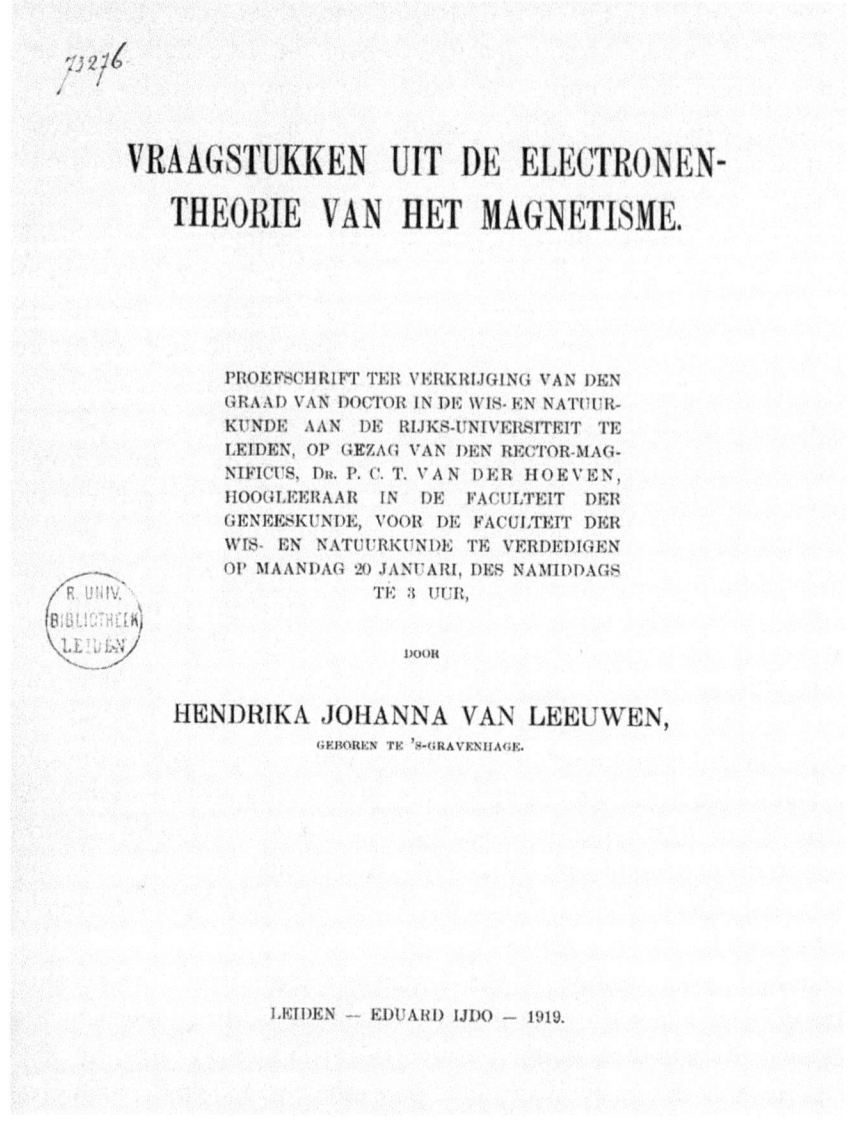

Figure 2.2 Title page of the dissertation of Hendrika Johanna van Leeuwen.

magnetism. In other words, using statistical methods she proves that the thermal mean of the magnetization of such a system is identically zero and that the net magnetic susceptibility (i.e., the degree to which a material can be magnetized in an applied magnetic field) vanishes completely (Van Leeuwen, 1919, p. 34 and p. 49).

Her finding directly implies that classical-mechanical models cannot explain the origin of the various types of magnetism such as paramagnetism, diamagnetism, and ferromagnetism. Or, as the American physicist and Nobel laureate John Van Vleck (1899–1980) wrote in his seminal 1932 book on the topic (Van Vleck, 1932, pp. 94–95):

[I]n 1919 Miss van Leeuwen demonstrated the remarkable and rather disconcerting fact that when classical Boltzmann statistics are applied completely to any dynamical system, the [magnetic] susceptibility is zero. We shall refer to these results as "Miss van Leeuwen's theorem".[19]

However, whereas Van Leeuwen's work might be viewed as "disconcerting" from a classical-mechanical perspective, it could also be interpreted quite positively from a quantum perspective: magnetism must be a quantum property. Van Leeuwen herself, who seemed as moderately tempered as Lorentz, did not explicitly make such a bold assertion, even though she did refer to the quanta a few times in her clearly, carefully, and cautiously written dissertation. That she was indeed thinking of magnetism as a quantum phenomenon follows only indirectly from one of the propositions at the end of her thesis: "The explanation for paramagnetism, as given by Kroo with the help of quantum assumptions, is unsatisfactory (J. N. Kroo, *Verh. d. Deutsch. Phys. Ges.* 1915, p. 452)" (Van Leeuwen, 1919, p. 118).[20] A true quantum understanding of magnetism would of course only become possible after the work of Ehrenfest's students George Uhlenbeck and Sam Goudsmit on electron spin, Wolfgang Pauli's formulation of the exclusion principle, and the development of a complete theory of quantum mechanics several years later. In 1919, Van Leeuwen's finding nevertheless did seem to provide yet another piece of evidence in support of the idea that nature must be quantized at the smallest scales.

2.6 Scooped

What Van Leeuwen and Lorentz apparently did not know was that the very same result – at least for the case of free electrons – had also been obtained already in 1911 by none other than Niels Bohr during his PhD research (Bohr, 1911). Expanding on Lorentz's electron theory, Bohr had concluded that "the presence of free electrons in a metal will give rise neither to diamagnetic nor to paramagnetic properties." That Lorentz and Van Leeuwen seemed unaware of this is understandable, given that Bohr's thesis was written in Danish and filed in the university archives, and that his results were not published in academic journals.[21] Or, as Van Vleck remarks in a footnote: "Many of Miss van Leeuwen's results were previously

---

[19] Van Vleck phrased the theorem as: "At any finite temperature, and in all finite applied electrical or magnetic fields, the net magnetization of a collection of electrons in thermal equilibrium vanishes identically."

[20] Original: "De verklaring, door Kroo met behulp van quanta-onderstellingen gegeven van het paramagnetisme van vaste lichamen, is onbevredigend (J. N. Kroo, *Verh. d. deutsch. phys. Ges.* 1915, blz. 452)."

[21] The English translation was only published in 1972. By then, the tradition of publishing theses in languages other than English had almost disappeared in many European countries.

obtained in Bohr's dissertation (Copenhagen, 1911), but this unfortunately is probably rather inaccessible to most readers" (Van Vleck, 1932, p. 94).

Interestingly, the roles of Bohr and Van Leeuwen were completely reversed in Van Vleck's 1977 Nobel lecture. That time, Van Vleck only mentioned Bohr's thesis work, and Van Leeuwen's work ended up in a footnote. The change is closely related to the shift in perspective that was already visible in the title of his lecture, "Quantum mechanics: The key to understanding magnetism." In 1932, Van Vleck praised the completeness of Van Leeuwen's critical analysis of the phenomenon of magnetism by asserting that, even though "other investigators had previously predicted zero magnetic susceptibilities under certain conditions, it remained for Miss van Leeuwen to review critically the whole subject of susceptibilities in classical theory" (Van Vleck, 1932, p. 95). In 1977, instead, Van Vleck focuses on the role of Bohr's thesis work ("the perhaps most deflationary publication of all time in physics")[22] in quantum developments, and suggests that this work may well have been "one reason why Bohr broke with tradition and came forth with his remarkable theory of the hydrogen spectrum in 1913" (Van Vleck, 1992, p. 354).

In fact, Van Vleck's shift in perspective already emerges from his 1963 interview with Thomas Kuhn: "One always thinks of its [quantum mechanics'] effect and successes in connection with spectroscopy, but I remember Niels Bohr saying that one of the great arguments for quantum mechanics was its success in these non-spectroscopic things such as magnetic and electric susceptibilities" (Van Vleck, 1963). In his 1977 Nobel lecture Van Vleck discussed this observation in more detail (Van Vleck, 1992, p. 354):

That year [1913] can be regarded as the debut of what is called the old quantum theory of atomic structure, which utilized classical mechanics supplemented by quantum conditions. In particular it quantized angular momentum [through the well-defined quantized electron orbits] and hence the magnetic moment of the atom [induced by the negatively charged electron revolving around the atomic nucleus] as was verified experimentally in the molecular beam experiments of Stern and Gerlach [Stern and Gerlach, 1922]. Hence there was no longer the statistical continuous distribution of [positive and negative] values of the dipole moment which was essential to the proof of zero magnetism in classical theory.

Van Vleck continued by briefly mentioning the work of Langevin who had in 1905 concluded that para- and diamagnetism could arise in classical-mechanical systems in thermal equilibrium, a result seemingly contradicting the work of Bohr and Van Leeuwen. The caveat here was that Langevin had unknowingly added a form of quantization to the system, Van Vleck explained. In his own words: "When Langevin assumed that the magnetic moment of the atom or molecule had a fixed value μ, he was quantizing the system without realizing it, just as in

---

[22] "Deflationary" in the sense that the work completely deflated the role of classical physics in explaining magnetism.

Moliere's [sic] Bourgeois Gentilhomme, Monsieur Jourdain had been writing prose all his life, without appreciating it, and was overjoyed to discover he had been doing anything so elevated." What Van Vleck did not mention here was that Van Leeuwen had already drawn this conclusion in her 1919 thesis.[23]

In retrospect it seems rather disappointing that while Van Leeuwen's finding seemed to provide an additional piece of evidence for the idea that nature is quantized at its smallest scales, the very similar finding by Bohr in 1911 may in fact have inspired Bohr to introduce his quantized model of the hydrogen atom in the first place. It also makes one wonder if Van Leeuwen, Bohr, and Lorentz ever discussed these matters. In the spring of 1919 Bohr did spend two weeks in Leiden at the invitation of Ehrenfest. Bohr, who had supervised a major part of Hendrik Kramers' thesis work during the three years prior, participated in Kramers' PhD defense (Dresden, 1987, pp. 110–113; Klein, 2010, pp. 308–311; Van der Heijden, 2021, pp. 210–211). But the two men also had lengthy discussions, took long walks, talked to Lorentz, and visited Kamerlingh Onnes' laboratory. Available evidence, however, does not clarify whether Van Leeuwen was involved in any of these activities. What is certain, though, is that her work is captured in what from the late 1930s onwards came to be known as the Bohr–Van Leeuwen theorem.[24]

2.7 Not One of the Boys; Not One of the Women

Van Leeuwen might have been scooped by Bohr, but at the same time, one might just as easily interpret this situation as a sign that her thesis work was on par with that of the well-known Bohr. From this perspective her work seemed like an excellent start for a promising career in physics. Yet, while Burger and Rubinowicz were soon promoted to full professorships, Van Leeuwen struggled to find a position.

She had always been a bit of an outsider among the young theoretical physicists in training in Leiden. Her PhD supervisor Lorentz lived in Haarlem, where he had become the director of the small physics laboratory of the Teylers Museum in 1909.[25] His successor in Leiden, Ehrenfest, had since his arrival in 1912 gathered

---

[23] A footnote in his 1932 book suggests that he was aware of her work: "She [Van Leeuwen] mentions and discusses at some length the fact that a susceptibility different from zero can be obtained if in statistical mechanics there is imposed some auxiliary condition (*Nebenbedingung*) which restricts to a definite numerical value some other function of the dynamical variables of the assembly of molecules besides its total energy. There is, however, no known justification for the imposition of such an extra condition in assemblies such as are encountered in the theory of magnetism" (Van Vleck, 1932, pp. 94–95).

[24] Van Vleck spoke of the Van Leeuwen theorem in 1932, but by 1927, Rice physicist Claude Heaps spoke of the Bohr–Lorentz–Van Leeuwen theory (Heaps, 1927) In 1937, geophysicist Edward Hulburt mentioned an argument by "Bohr (thesis) and by Van Leeuwen" (Hulburt, 1937). From then on, both names seem to have been attached to the work increasingly often, until this label became common use after World War II.

[25] Lorentz combined this post with his Leiden professorship until he formally retired from Leiden University in 1912. He did continue to give his 'Monday Lectures' and to act as a PhD supervisor until the early 1920s.

a circle of promising students around him such as Jan Burgers, Dirk Jan Struik (who would become a renowned mathematician in the US), Dirk Coster (who would become professor of physics in Groningen), and Kramers, who would later work with Bohr. But Van Leeuwen was not part of this close-knit group who would visit Ehrenfest at home, attend colloquia in the study of Ehrenfest's large house, and help Tatiana Ehrenfest-Afanassjewa in the garden. Her correspondence with the Ehrenfests is limited to a handful of postcards, and her name only rarely appears in memoirs or interviews with Ehrenfest's students (Burgers, 1962; Casimir, 2004, p. 94; Rowe, 2018).

An illustration of her relative outsider position can also be seen in Ehrenfest's correspondence from 1917. During the summer of that year and in a rather unusual move, Ehrenfest visited his student Jan Burgers' parents in their modest house in Arnhem. There, he also met Burgers' younger brother Willy, who was about to begin studying in Leiden, like his elder brother. In the aftermath of the visit an inspired Ehrenfest wrote a series of postcards to the two brothers expressing his hope that they would come live in a room in the Witte Rozenstraat, close to his own large house, where they could then work in his large study. In fact, before the new semester began, an overly enthusiastic Ehrenfest had already reserved that room and paid the landlady one month's rent in advance. It was the room where the Van Leeuwen sisters had lived until Nel's recent marriage. Apparently, Jo van Leeuwen planned to leave the apartment: her father had recently died and she intended to move back to The Hague to live with her mother, Ehrenfest explained briefly on one of the postcards to the brothers.[26] Almost casually his remark thus captures the losses that Jo was experiencing: her father dead, her younger sister Nel moving in with her newly-wed husband Nordström, and she herself having to depart from her room in Leiden.

Perhaps Ehrenfest kept his distance because he did not want to interfere with Lorentz, Van Leeuwen's supervisor, whom he deeply admired. It may also be that Ehrenfest – born as the youngest child in a family of five sons, and friends with famous physicists including Einstein, and later also Bohr, Schrödinger, and Pauli – was unconsciously inclined to see physicists in a gender stereotypical way. Such stereotypical views may have been exacerbated in the period after moving to Leiden when he, appointed as professor, and his wife Tatiana Ehrenfest-Afanassjewa, staying at home, no longer managed to live up to their ideal of working and studying together as two equals (Van der Heijden, 2024). It may also be that there was simply a bit of an *incompabilité d'humeur* (clash of personalities) between the extraverted, demanding, humorous, but also sharp-witted Ehrenfest and the somewhat reserved Van Leeuwen, who moreover was a couple of years older than most of Ehrenfest's PhD students.

---

[26] Ehrenfest to Burgers, August 27 and 29, 1917, Burgers Archive, Delft University of Technology.

In any case, it was Lorentz who supported Van Leeuwen most during her graduate studies, as is also clear from his correspondence. Just after World War I broke out in September 1914, in a note to Ehrenfest, Lorentz suggested that Van Leeuwen should become Ehrenfest's assistant. Such a temporary assistant position entailed giving lectures if necessary, writing lecture notes, assisting with a professor's correspondence, etc., in exchange for a small income. Lorentz wrote: "The assistant position may become temporarily vacant now that Droste has left. We could suggest [Adriaan] Fokker as his successor with the announcement that he has been called to arms and temporarily appoint Miss van Leeuwen as assistant. She would then resign when demobilization takes place."[27] Probably to avoid giving his successor the impression that he still wanted to manage university matters from Haarlem, Lorentz added that "it seems to me that you will have to choose an assistant from now on." Yet, when a few days later it turned out that temporary cutbacks prevented the university from employing new assistants, Lorentz immediately pressed Ehrenfest to find another solution for Van Leeuwen. "Can't the University Fund provide a small amount of money so Miss van Leeuwen can then take on the Reading Room? You can write to Kluyver about this," he wrote.[28]

Fortunately, Van Leeuwen got the assistant position and was able keep it until Fokker was demobilized in the autumn of 1917. In September 1917, she herself reminded Lorentz that she had promised to resign as assistant upon Fokker's return. Perhaps she also felt obliged to give up the position because unemployment rates were rising and the sentiment in Dutch society was that women should not "steal" jobs from men. Lorentz was of another opinion, however. He wrote to Ehrenfest: "Miss van Leeuwen asks me whether, now that Fokker has been granted study leave [from his mobilization], the time has come for her to resign as assistant. It seems to me that we can leave the situation as it is now for another six months. Fokker only has leave of absence during that time. Besides, he is wealthy enough and does not need the money and will want to devote himself to his studies first."[29]

Although previous authors have assumed that Fokker and Van Leeuwen both agreed with Lorentz's proposal for Van Leeuwen to keep the assistant position for a while longer (Berends and Van Delft, 2021), archival documents suggest otherwise. In October 1917, she moved to Deventer to teach physics and cosmography at the local HBS for three months, most likely at the recommendation of Lorentz's

---

[27] Lorentz to Ehrenfest, September 3, 1914, Ehrenfest Archive, Rijksmuseum Boerhaave, Leiden.
[28] Lorentz to Ehrenfest, September 8, 1914. Ehrenfest Archive, Rijksmuseum Boerhaave, Leiden.
[29] Lorentz to Ehrenfest, September 23, 1917, Ehrenfest Archive, Rijksmuseum Boerhaave, Leiden. Original: "Mej. van Leeuwen vraagt mij of, nu Fokker studieverlof heeft gekregen, niet de tijd is gekomen dat zij ontslag neemt als assistente. Mij dunkt dat we de toestand nog een half jaar zoo kunnen laten als hij nu is. Fokker heeft slechts voor die tijd verlof. Bovendien is hij vermogend genoeg en zal hij zich eerst aan zijn studie willen wijden."

son-in-law, Wander de Haas, who had taught there in prior years.[30] Afterwards, she indeed returned to The Hague, presumably to live with her mother, as Ehrenfest had written.[31] It is unclear if Van Leeuwen also worked as a teacher in 1918 and after her thesis defense on January 20, 1919. What can be verified is that she worked as a physics teacher in Delft, at the Gemeentelijk Gymnasium, from September 1919 until the summer of 1920.[32] Starting September 1, 1920, most likely with Lorentz's support, she became employed as assistant at the TH Delft, one of the first women to be hired by the institution.

That Van Leeuwen herself was not unaware of her unique position as one of the few women in her field is revealed by the last of the 13 propositions at the end of her dissertation. As is common in Dutch theses, this proposition does not pertain to a result from her thesis or to a physical phenomenon and it reads[33]: "On the basis of statistical material collected on the study of women (see G. Heijmans: *Die Psychologie der Frauen*, 1910, pp. 105–155, [Heidelberg: Winter]) no conclusion can be drawn, at least for the time being, as to the degree of suitability of women for scientific work (Van Leeuwen, 1919, p. 119)."

2.8 Assistant at the Technical College of Delft

In Delft, Van Leeuwen's main task was to supervise the physics laboratory courses in the Faculty of Electrical Engineering.[34] The job must, especially at first, have left her little time to pursue theoretical work on magnetism. Moreover, she was now truly a pioneer-woman. None of Lorentz's other female graduate students pursued an academic career with a formal position at a university or research institute. In the 1920s Johanna Reudler and Eva Bruins helped compile Lorentz's lecture notes and turn them into scholarly books that were translated to German by Lorentz's daughter Berta (Lorentz, 1921, 1927, 1928). Berta de Haas-Lorentz published a book herself as well, in Dutch, on thermodynamics (De Haas-Lorentz, 1938), and a paper on surface currents in superconducting materials that is still regularly cited (De Haas-Lorentz, 1925; Fossheim and Sudbø, 2005, pp. 141–143). But such

---

[30] Gedenkboek Deventer Hogere Burgerschool 1864–1939, Delpher.
[31] On December 22, 1917, she registered at Bentinckstraat 28, Den Haag, where she had previously lived with her parents; Haags Gemeentearchief, Den Haag. A postcard suggests that she still lived there with her mother in December 1921: Van Leeuwen to the Ehrenfests, December 27, 1921. Ehrenfest Archive, Rijksmuseum Boerhaave, Leiden.
[32] Annual report 1919–1920 of the curates of the Stedelijk Gymnasium Delft: "Curatoren (scholarchen) van de Latijnse School en Regenten van het Fraterhuis en het Stedelijk Gymnasium (sinds 1950 Grotius Gymnasium), 1610–1966," inventaris nr. 30, Stadsarchief Delft.
[33] Original: "Uit statistisch materiaal, verzameld over het studeeren der vrouwen (zie G. Heijmans: *Die Psychologie der Frauen*, 1910, blz. 105–55) kan althans voorloopig nog geen besluit worden getrokken over de mate van geschiktheid der vrouw voor wetenschappelijk werk."
[34] Applied physics did not become an independent study programme at TH Delft until 1928, although physics was part of all other engineering programmes.

projects were undertaken from home where Berta devoted much of her time to household duties and taking care of the children.[35]

This outcome was not surprising since traditional views on gender roles kept a strong foothold in the Netherlands. On the one hand, Dutch women had finally obtained the right to vote in 1919, thanks to the fighting spirit of Jacobs and her Association for Women's Suffrage (Feinberg, 1999; Bosch, 2005). On the other hand, 1924 brought a severe backlash for working women when a new law, confirmed by Royal Decree, stipulated that women were to be honorably discharged from public service and government jobs on the day of their marriage. The law effectively closed off roles as teacher, university lecturer, and researcher to married women, with most private companies following the government's example. Women thus had to choose between working and staying single or marrying and losing their job. As a result, slightly over 90% of the Dutch married women holding a PhD in physics were unemployed in the 1930s (Kirejczyk, 1993, pp. 240–241).

It would take until 1957 for this discriminatory law to be abolished, and for Dutch married women to – very slowly – return to the job market. For Jo van Leeuwen, this change came too late – if she had wanted to marry at all, that is. Little is known about how Van Leeuwen thought and felt about these matters though. For many of her contemporaries, she came across as someone who kept to herself, and they often described her as introverted and somewhat reserved. A former physics student from Delft, for example, recalls that Van Leeuwen was more distant than the other members of the scientific staff, stating that "the personal element was lacking in interactions with her."[36]

A former technician, who worked in the same "beautiful and cosy building with nooks and crannies" as Van Leeuwen, the Laboratory for Technical Physics at the Mijnbouwplein in Delft, remembers that she was also friendly:

As a boy of about nineteen, I started my work in the building in the recently established "Electronic Repair Shop" ... and went on to work in the "Calibration Room." ... It was there that ... I first met with Miss van Leeuwen, because the equipment of her laboratory class also had to be calibrated. It consisted of multimeters, a lot of tube voltmeters, low-frequency tone generators, oscilloscopes, stabilized power supplies and so on.... Miss van Leeuwen addressed everyone as "Sir," including me, the just twenty-year-old. Whenever I returned the calibrated equipment, she often stopped for a chat and asked if I had noted anything unusual related to her equipment. She almost always sat at a laboratory table that had been converted into a desk.[37]

Van Leeuwen did like manners though, he adds.

---

[35] In 1910, while still a graduate student, Berta Lorentz married Wander de Haas, who would become professor of experimental physics at Leiden University, and with whom she had two daughters and two sons.

[36] Dr. Ir. A.E. Pannenborg, personal communication by letter, dated May 31, 2014.

[37] Loek Kraamer, personal communication by email, February 29, 2016.

Nobody was afraid of her, not even the students I think, but you had to watch your step. I remember a master's thesis student explaining the apparatus he had designed while standing with his hands in his pockets. And yes, Miss van Leeuwen remarked: "Mr v.d. *** would you please take your hands out of your trouser pockets." It did not sound bitter or reproachful, rather more like a friendly request. Yet the face of the person addressed momentarily changed colour.

While thus supervising the physics laboratory, Van Leeuwen also remained in close contact with Lorentz and other members of the Leiden Physics Department. In 1925, together with Kramers and other Leiden colleagues, she organized a symposium commemorating the 50th anniversary of Lorentz's doctorate. It was attended by Einstein, Marie Skłodowska Curie, and quite a few other well-established physicists (see Figure 2.3). On this occasion she also wrote an article for the *Nieuwe Rotterdamsche Courant*, a Dutch national newspaper, about Lorentz's scientific work and mentorship. The article shows affection and admiration for her former advisor whom she describes as kind and modest. Van Leeuwen recalls, for example, how she and the other students slowly realized that those discoveries and findings that were discussed during his lectures without him mentioning their discoverer's name, as was Lorentz's custom, were made "by professor Lorentz himself." In short, for her and for the other students, Lorentz was "not just this great researcher, but also this kind, cheerful man with a strict devotion to duty, amiable simplicity and warm interest in all those around him" (Van Leeuwen, 1925).[38]

In addition, Van Leeuwen continued exchanging letters with Lorentz about theoretical problems in magnetism until at least 1927, one year before his death, when she suggested to him to include some of these problems in the notes of his Leiden "Monday Lectures" that he was compiling at the time with the help of Bruins, Reudler, and his daughter Berta.[39]

In Delft, meanwhile, it did not go unnoticed that Van Leeuwen, alongside her work on the laboratory courses, was still engaged in theoretical research on magnetism. In the early 1930s, for instance, Bram van Heel, professor of optics in Delft, composed a ballad for the annual "lab evening" of the Faculty of Applied Physics in which each of the more than 20 stanzas light-heartedly sang the praises of the work of a staff member.[40] Whether her colleagues valued what she was working on, though, cannot be deduced from the part for Van Leeuwen that only playfully refers to quantum theory and reads (translated):[41]

---

[38] Original: "Lorentz, voor zijn leerlingen, was niet alleen de groote onderzoeker, maar ook de vriendelijke, opgewekte mensch met strenge plichtsbetrachting, beminnelijken eenvoud en warme belangstelling voor allen om hem heen."

[39] Van Leeuwen to Lorentz, April 26, 1927, 364 – Prof. Dr. H. A. Lorentz Archive, Inventaris nr. 47, Noord-Hollands Archief, Haarlem.

[40] 433 – A. C. S. van Heel en H. G. van Heel-Meerburg, 1903–2001 Archive, Stadsarchief Delft, Delft.

[41] The original Dutch:

Figure 2.3 Group picture of the attendees at the celebration of Lorentz's golden doctorate on December 11, 1925. Jo van Leeuwen is in the third row, to the left of Einstein. Lorentz's daughter Berta is sitting next to Marie Skłodowska Curie, and Skłodowska Curie is sitting to the right of Lorentz. Image courtesy of the North Holland Archive/1100 – Beeldcollectie van de gemeente Haarlem, inv. nr. 21303.

> Miss van Leeuwen with her magnets
> Is perspi-sweating on the theory
> That is the the-ory
> That is the the-ory
> That is the the-ory . . .. of Heisenberg

2.9 Reader at Last

Van Leeuwen held the position of assistant for more than 20 years before becoming principal assistant in theoretical and applied physics in 1943. Even then she continued her research on magnetic phenomena and, among other things, proposed a model for the reduction of permeability in ferromagnetic metals that, in contrast to

"Juffrouw van Leeuwen met haar magneten zit op de theorie te transpizweeten Dat is de thė-orie Dat is de thė-orie Dat is de thė-orie . . . van Heisenberg"

previously proposed models, agreed well with experimentally measured data. She notably gave a lecture on the subject at a symposium on ferromagnetism in Delft in April 1947 (Van Leeuwen, 1947).

Shortly thereafter, at age 59, Van Leeuwen was promoted to reader in theoretical and applied physics. Her appointment not only came late by present-day standards but also in the eyes of some of her contemporaries, as can be deduced from a congratulatory note that Dirk Coster, professor of physical meteorology at the University of Groningen in the Netherlands, sent her. On October 25, 1948, Coster, who in the 1910s had been one of Ehrenfest's promising students, wrote: "Dear Miss van Leeuwen, I warmly congratulate you on your appointment at the TH [Delft]. Late, but luckily not too late, you have found at least partial recognition for your talents. Those who know you will have read the announcement with approval. I would also like to congratulate you on behalf of my wife."[42] The careful phrase "partial recognition" makes one wonder what career she could have pursued had she not been trapped in a position centred around guiding students towards a successful future – male students who would never encounter the hurdles with which she had been confronted.

Within two years two other women – Antonia Korvezee, who became a reader in physical chemistry in 1948, and Jentina Leene, who obtained a position as reader in fibre technology in 1949 – would receive a similar promotion at the TH Delft. The threefold elevation was perhaps less of a break with traditional views than a pragmatic measure. As science historian Frida de Jong noted: "Shortly after the war it was difficult for the TH to get academic staff. Wages were low compared with those in industry and laboratory facilities were less. At that time, shortly after each other, three women were appointed lecturer (. . .) These three ladies were all from an older generation, trained with a Ph.D. well before the war, and unmarried" (De Jong, 1997, p. 238).

For Van Leeuwen it meant that she was finally allowed to lecture. The 1952 "Programme of lectures of the Technical College Delft" informs us that Miss Dr H. J. van Leeuwen lectured on ferromagnetism and principles of special relativity theory. She also continued to run the advanced physics *practicum* (laboratory course) and many former colleagues later testified that the way in which she organized and ran this course for third-year students was highly appreciated by both her students and her colleagues. Some of these last ones had even attended "her" laboratory course as students themselves, before becoming staff members at the university. One of them, Jan Berend Westerdijk, addressed Van Leeuwen in his

---

[42] Coster to Van Leeuwen, October 25, 1948. Collectie-Coster, Groninger Archieven, Archief van de Senaat en de Faculteiten, Groningen. Original: "Zeer geachte Mejuffrouw van Leeuwen, Van harte wens ik U geluk met Uw benoeming aan de T.H. Het is wel laat geworden, maar gelukkig niet te laat dat U althans gedeeltelijk erkenning van Uw bekwaamheden hebt gevonden. Zij, die U kennen, zullen het bericht met instemming gelezen hebben. Ook namens mijn vrouw breng ik de gelukwensen over."

inaugural speech when he became professor of technical physics at the TH Delft in 1952, shortly before Van Leeuwen's retirement (Westerdijk, 1952, p. 20):

Distinguished Miss van Leeuwen, It makes me happy, before you bid us farewell, to be able to express to you my great admiration for the way in which, during so many years, you have trained our students to become independent learners and thinkers in your practical classes ... The often high-level graduation work in the department would never have been possible if the important "practical of Miss van Leeuwen" had not been constantly modernized and managed by you in such a dedicated manner.

2.10 Last Years

What had happened to her younger sister, Nel, during all those years? Not long after the young Nordström-Van Leeuwen family moved to Finland, two more children were born and soon thereafter, at the end of 1923, Nordström died from pernicious anaemia, perhaps induced by his experiments with radioactivity. Ironically, this meant that Nel, as a widow, would have been allowed to pursue a career in the Netherlands, even after the 1924 Royal Decree. Her distance from the field of physics had become insurmountable, however, and she decided to spend the rest of her life in Finland, where she worked as a translator and language teacher.

In 1923, Tatiana Ehrenfest-Afanassjewa mentioned Nel's activities in a letter to her husband, who often complained about the high costs of the large Ehrenfest-Afanassjewa household in Leiden. "I would like to have been able to do that; and most of all to win such assignments," Afanassjewa commented on Nel's jobs (Van der Heijden, 2021, p. 248). For Ehrenfest-Afanassjewa, forced into the conventional roles of a mother and professor's wife, it seemed a dream to earn a salary as Nel had done, let alone to secure a formal position at a Dutch Technical College as Jo had managed to do.

The two sisters themselves stayed in touch via visits, and through letters and postcards.[43] A few of these cards have been preserved, dating mostly from the last period of their lives, by which time Jo van Leeuwen had moved to Huyse van Sint Christoffel, an old-age home for ladies in the city centre of Delft (nowadays an apartment complex). She lived there from 1960 until her death in 1974 "in the midst of many kind ladies in rooms on two floors enclosing a beautiful and big garden."[44] There is no indication that she was still scientifically active during these years, but postcards and letters preserved in the City Archives of Delft suggest that she kept

---

[43] H. Kristjankroon, copies of postcards and personal communication.
[44] J. van Leeuwen to T. Ehrenfest-Afanassjewa, June 17, 1961, Ehrenfest Archive, Rijksmuseum Boerhaave, Leiden. Original: 'Ik leef hier temidden van vele aardige vrouwen in een kring van kamers in 2 lagen om een mooie en grote tuin.'

informed about and in some cases participated in important events in the lives of her former colleagues.

Her sister Nel survived Jo by only half a year and died in Finland in the late summer of 1974. Like much else, whether the two Van Leeuwen sisters – one single with a scientific career and the other widowed and self-supporting with children – ever discussed the different obstacles and hurdles they had encountered and how these had affected their own lives and careers, we will probably never know.

References

Aaserud, Finn and Heilbron, John L. 2013. *Love, Literature, and the Quantum Atom: Niels Bohr's 1913 Trilogy Revisited*. Oxford: Oxford University Press.

Baneke, David and Maas, Ad. 2018. De Hogere Burgerschool: onderwijs en emancipatie. *Studium* 10, 117–122. https://doi.org/10.18352/studium.10155

Berends, Frits A. 2015. Lorentz, the Solvay Councils and the Physics Institute. *European Physical Journal Special Topics*, 224, 2091–2111. https://doi.org/10.1140/epjst/e2015-02525-5

Blaauboer, Miriam. 2015. Hendrika J. van Leeuwen. *Nederlands Tijdschrift voor Natuurkunde*, 81(8), 4–6. https://ilorentz.org/IL-publications/dissertations/biographies/NTvN_vanLeeuwen.pdf

Berends, Frits and Van Delft, Dirk. 2021. *Lorentz. Gevierd fysicus, geboren verzoener*. Amsterdam: Prometheus.

Boekholt, P. Th. F. M. and De Booy, E. P. 1987. *Geschiedenis van de school in Nederland*. Assen-Maastricht: Van Gorcum.

Bohr, Niels. 1911. *Studier over Metallernes Elektrontheorie*. PhD Thesis. Copenhagen: J. Jørgens & co. Reprinted as Bohr, Niels. 1972. The doctor's dissertation (text and translation). In Rosenfeld, L. and Rud Nielsen, J. (eds.), *Niels Bohr Collected Works, Vol I*. Amsterdam: Elsevier, pp. 165–393.

Bohr, Niels. 1913a. On the constitution of atoms and molecules. *The London, Edinburgh, and Dublin Philosophical Magazine and Journal of Science*, 26(151), 1–25. https://doi.org/10.1080/14786441308634955

Bohr, Niels. 1913b. On the constitution of atoms and molecules. Part II Systems containing only a single nucleus. *The London, Edinburgh, and Dublin Philosophical Magazine and Journal of Science*, 26(153), 476–502. https://doi.org/10.1080/14786441308634993

Bohr, Niels. 1913c. On the constitution of atoms and molecules. Part III Systems containing several nuclei. *The London, Edinburgh, and Dublin Philosophical Magazine and Journal of Science*, 26(155), 857–875. https://doi.org/10.1080/14786441308635031

Bosch, Mineke. 1993. The gender of science: a historical analysis of the relation between women and higher education in the Netherlands, 1878–1948. *European Review of History – Revue Européenne d'Histoire*, 1, 161–164. https://doi.org/10.1080/13507486.1993.12035700

Bosch, Mineke. 2005. *Een onwrikbaar geloof in rechtvaardigheid. Aletta Jacobs, 1854–1929*. Amsterdam: Uitgeverij Balans.

Bruins, Eva D. 1918. *Vraagstukken uit de theorie van het magnetisme*. PhD Thesis. Leiden: E. J. Brill. https://www.lorentz.leidenuniv.nl/history/proefschriften/sources/Bruins_1918.pdf

Burgers, Jan M. 1917. Note on the model of hydrogen-molecule of Bohr and Debije. *Proceedings of the koninklijke Academie van wetenschappen te Amsterdam*, 19(I), 480–488.

Burgers, Jan M. 1962. Interview of Jan M. Burger by Thomas S. Kuhn and Martin Klein, June 9, 1962. Niels Bohr Library & Archives. https://www.aip.org/history-programs/niels-bohr-library/oral-histories/3092

Capponi, Giovanna and Frenken, Koen. 2021. On the sudden rise of Dutch science at the end of the nineteenth century: a core-periphery approach. *Industry and Innovation*, 28(9), 1175–1195. https://doi.org/10.1080/13662716.2021.1929867

Casimir, Hendrik B. G. 2004. *Het toeval van de werkelijkheid*. Amsterdam: Meulenhoff.

Debije, Peter. 1915. Die Konstitution des Wasserstoff-Moleküls. *Sitzungsberichte München* 1915:1. https://publikationen.badw.de/de/003395485

De Haas-Lorentz Geertruida L. 1912. *Over de theorie van de Brown'schen beweging en daarmede verwante verschijnselen*. PhD Thesis. Leiden: Eduard IJdo. https://www.lorentz.leidenuniv.nl/history/proefschriften/sources/deHaasLorentz_1912.pdf

De Haas-Lorentz, Geertruida L. 1925. Iets over het Mechanisme van Inductieverschijnselen. *Physica* 5, 384–388. https://www.lorentz.leidenuniv.nl/history/proefschriften/Physica/Physica_5_1925_05395.pdf

De Haas-Lorentz, Geertruida L. 1938. *De beide hoofdwetten der thermodynamica en hare voornaamste toepassingen*. 's-Gravenhage: Nijhof.

De Jong, Frida. 1997. Standhouden in Delft. *Gewina*, 20(4), 227–242. http://resolver.tudelft.nl/uuid:088b1e99-ccda-4c75-a512-f72188316d37

Dresden, Max. 1987. *H. A. Kramers. Between Tradition and Revolution*. New York: Springer-Verlag.

Duncan, Anthony and Janssen, Michel. 2019. *Constructing Quantum Mechanics, Volume 1: The Scaffold*. Oxford: Oxford University Press.

Feinberg, Harriet. 1999. *Aletta Henriette Jacobs. Shalvi/Hyman Encyclopedia of Jewish Women*; Jewish Women's Archive. http://jwa.org/encyclopedia/article/jacobs-aletta-henriette

Fossheim, Kristian and Sudbø, Asle. 2005. *Superconductivity: Physics and Applications*. Chichester: John Wiley & Sons.

Gerlach, Walther and Stern, Otto. 1922. Der experimentelle Nachweis der Richtungsquantelung im Magnetfeld. *Zeitschrift für Physik*, 9(1), 349. https://doi.org/10.1007/BF01326983

Gorter, Cornelis J. 1962. Interview of Cornelis J. Gorter by John L. Heilbron, November 15, 1962. Niels Bohr Library & Archives. https://www.aip.org/history-programs/niels-bohr-library/oral-histories/4639

Heaps, Claude W. 1927. The photomagnetic effect in silver chloride and in selenium. *Journal of the Optical Society of America*, 15, 190–193. https://doi.org/10.1364/JOSA.15.000190

Hulburt, Edward O. 1937. Terrestrial magnetic variations and aurorae. *Reviews of Modern Physics* 9, 44-68. https://doi.org/10.1103/RevModPhys.9.44

Keesom, Willem H. and van Leeuwen, Cornelia. 1916. On the second virial coefficient for rigid spherical molecules carrying quadruplets. *Proceedings of the koninklijke Academie van wetenschappen te Amsterdam*, 18(II), 1568. https://dwc.knaw.nl/DL/publications/PU00012625.pdf

Kirejczyk, Marta. 1993. Vrouwen kozen exact; studie en beroepsuitoefening rond de eeuwwisseling. *Gewina*, 16(3), 234–247. https://dspace.library.uu.nl/handle/1874/251237

Klein, Martin J. 2010. Paul Ehrenfest, Niels Bohr, and Albert Einstein: colleagues and friends. *Physics in Perspective*, 12, 307–337. https://doi.org/10.1007/s00016-010-0025-6

Kox, Anne and Schatz, Henriëtte. 2021. *A Living Work of Art: The Life and Science of Hendrik Antoon Lorentz*. Oxford: Oxford University Press.

Kragh, Helge. 2012. *Niels Bohr and the Quantum Atom*. Oxford: Oxford University Press.

Kragh, Helge. 2023. A terminological history of elementary particle physics. *Archive for History of Exact Sciences*, 77, 73–120. https://doi.org/10.1007/s00407-022-00299-2

Kroon, Just E. and Blok, Anthony J. 2021. *Album studiosorum Academiæ Lugduno-batavæ MDCCCLXXV–MCMXXV*. Leiden: Leiden University Libraries.

Langevin, Pierre. 1905. Magnétisme et théorie des électrons. *Annales de chimie et de physique*, 8(V), 70.

Lorentz, Hendrik A. 1914. Anwendung der Kinetischen Theorien auf Elektronenbewegung. In *Mathematische Vorlesungen an der Universität Göttingen: VI, Vorträge über die kinetische Theorie der Materie und der Elektrizität*, p. 189.

Lorentz, Hendrik A. 1921. Lessen over theoretische natuurkunde aan de Rijks-Universiteit te Leiden gegeven. In *Vol. 5, Kinetische problemen (1911–1912)*. Bruins E. D. and Reudler, J. (eds.). Leiden: Brill.

Lorentz, Hendrik A. 1927. *Vorlesungen über theoretische Physik an der Universität Leiden: Theorie der Strahlung, Vol I*. Fokker, A. D. (ed.); De Haas-Lorentz, G. L. (transl.). Leipzig: Akademische Verlagsgesellschaft.

Lorentz, Hendrik A. 1928. *Vorlesungen über theoretische Physik an der Universität Leiden: Kinetische Probleme, Vol II*. Bruins, E. D. und Reudler, J. (eds.); De Haas-Lorentz, G. L. (transl.). Leipzig: Akademische Verlagsgesellschaft.

Reudler, Johanna. 1912. *Over de Zwarte Straling in Ruimten van verschillende Vorm*. PhD Thesis. Amsterdam: N. V. Electrische Drukkerij 'Volharding.' https://www.lorentz.leidenuniv.nl/history/proefschriften/sources/Reudler_1912.pdf

Rowe, David E. 2018. An enchanted era remembered: interview with Dirk Jan Struik. In Rowe, D. E. (ed.), *A Richer Picture of Mathematics*. Cham: Springer. https://doi.org/10.1007/978-3-319-67819-1_32

Rubinowicz, Wojciech. 1963. Interview of Wojciech Rubinowicz by Theo Kahan and John L. Heilbron, May 18, 1963. In German. Niels Bohr Library & Archives. https://repository.aip.org/islandora/object/nbla%3A271348#page/1/mode/2up

Rubinowicz, Adalbert. 1917. Die Eigenschwingungen des Bohr–Debyeschen Wasserstoffmoleküls bei Berüksichtigung der Bewegung der Kern. *Physikalische Zeitschrift*, 18, 187–195.

Van der Heijden, Margriet. 2021. *Denken is verrukkelijk. Over de levens van Tatiana Afanassjewa en Paul Ehrenfest*. Amsterdam: Prometheus.

Van der Heijden, Margriet. 2024. More is known about him than about her: Tatiana Ehrenfest-Afanassjewa. *Physics Today*, 77(1), 40–47. https://doi.org/10.1063/PT.3.5381

Van Leeuwen, H. Johanna. 1916. Some remarks on the hydrogen molecule of Bohr–Debye. *Proceedings of the koninklijke Academie van wetenschappen te Amsterdam*, 18, 1071.

Van Leeuwen, H. Johanna. 1919. *Vraagstukken uit de electronentheorie van het magnetisme*. PhD Thesis. Leiden: Eduard IJdo. https://www.lorentz.leidenuniv.nl/history/proefschriften/sources/vanLeeuwen_1919.pdf

Van Leeuwen, H. Johanna. 1921. Problèmes de la théorie électronique du magnétisme. *Journal de Physique et le Radium*, 2 (12), 361–377. https://doi.org/10.1051/jphysrad:0192100201203610

Van Leeuwen, H. Johanna. 1925. Professor dr. H.A. Lorentz bij zijn gouden doctoraat. *Nieuwe Rotterdamsche Courant*, December 11, 9. https://resolver.kb.nl/resolve?urn=ddd:010028172:mpeg21:p009

Van Leeuwen, H. Johanna. 1947. De magnetisatiekromme. *Nederlands Tijdschrift voor Natuurkunde*, 13, 211. See also: Van Leeuwen, H. J. (1944). De vermindering der permeabiliteit bij toenemende frequentie. *Physica*, 11, 35. https://doi.org/10.1016/S0031-8914(44)80017-9

Van Steen, Agnes. 2003. Openbare Hogere Burgerscholen in Leiden: jongens en meisjes beginnen gescheiden. In *Leids Jaarboekje 2002*. Leiden: Vereniging Oud Leiden, pp. 83–118.

Van Vleck, John H. 1932. *The Theory of Electric and Magnetic Susceptibilities*. Oxford: Oxford University Press.

Van Vleck, John H. 1963. Interview of John H. Van Vleck by Thomas S. Kuhn, October 2, 1963. Niels Bohr Library & Archives. https://www.aip.org/history-programs/niels-bohr-library/oral-histories/4931-1

Van Vleck, John H. 1992. Quantum mechanics: The key to understanding magnetism. In Lindqvist, S. (ed.), *Nobel Lectures, Physics 1971–1980*. Singapore: World Scientific Publishing, 353–369.

Westerdijk, Jan B. 1952. *Natuurkunde als bijvak*. Delft: Delftsche Uitgevers Maatschappij. http://resolver.tudelft.nl/uuid:6923c0f8-34ea-431a-9080-3774868d9135

3

Hertha Sponer, Maven of Quantum Spectroscopy

ELISE CRULL

3.1 Introduction

Quantum theory owes its origins to spectroscopy. The role of experimental advances in the years spanning Niels Bohr's atomic model of 1913 and the complete quantum-mechanical formalisms of Werner Heisenberg (1925) and Erwin Schrödinger (1926) cannot be overstated. A veritable constellation of star experimental physicists worked diligently in this time period to improve spectroscopic techniques and to gather and interpret ever-better, wider-ranging results. These scientists, though gaining less attention in the narrative than theorists like Albert Einstein, Bohr, Heisenberg, and Schrödinger, still receive ample credit for their roles.

Within this shimmering constellation of early twentieth century experimentalists, the contributions of women have been almost entirely consigned to the interstitial spaces. Their names are omitted from the stories in which they played active roles. Their legacies, if mentioned at all, are trivialized due to the comparative fuss made over their male counterparts. These women were not merely stage-hands, quaint interlocutors, or assistants to Great Men, but were in fact often managing the show. This chapter concerns one such figure who certainly deserves more attention: Hertha Sponer.

Historians of physics are well aware that in the late 1910s James Franck frequently reported hot-off-the-press experimental findings to Bohr, as a consequence of which major quantum theoretical advances were made. These data – and indeed, the design and execution of the ground-breaking experimental work whence they came – were not entirely due to Franck himself (Lemmerich, 2011, pp. 111–113; Maushart, 2011, ch. 2). Much of the background work was done by Hertha Sponer, who oversaw teaching as well as the day-to-day operations in the spectroscopy labs at Göttingen in the 1920s. It was her experimental prowess that enabled a number of advances in theory as well as experimental technique, especially regarding research measuring dissociation phenomena in diatomic molecules. She received little to no credit for making or inspiring a number of important advances toward a fully formed quantum chemistry. She was one of the first physicists – if not the very first to explain a number of hitherto puzzling molecular phenomena through the framework of quantum mechanics, and to then use these findings to develop and implement improved experimental techniques – hence: the maven of quantum spectroscopy.

The aim of the present chapter is, by presenting initial findings with respect to three vignettes from the history of quantum physics in the years 1920–1930 wherein Sponer was influential, to bring her into the narrative of these developmental flashpoints. In addition, I hope in so doing to provide sufficient evidence to catalyze further, more detailed historical investigations into Sponer's scientific achievements. Section 3.2 offers a brief biographical introduction to Sponer. In this, I relied heavily on the only extant full-length biography of Sponer – Maushart (2011) – as well as transcripts of Sponer's interview alongside her then husband James Franck (whose name alone is listed on the archival tapes for their joint sessions!), with Thomas Kuhn and Maria Goeppert Mayer, for the Archive for the History of Quantum Physics (AHQP). Section 3.3 discusses Sponer's role in producing and interpreting various crucial experiments demonstrating the wave nature of matter. Sponer's trip to Berkeley to work with American physicists in 1924–1925 is the focus of Section 3.4. And last of the three vignettes, Section 3.5, demonstrates Sponer's early understanding of and application to molecular spectroscopy of the phenomenon now called quantum tunneling.

3.2 A Brief Biography of Sponer

Born in modern-day Nysa, Poland (formerly the German-speaking region of Silesia[1]) in late summer 1895, Sponer matriculated in 1917 at the University of Tübingen as one of very few female students. There she took the only physics classes available – experimental physics seminars and laboratories – all under Friedrich Paschen, a world-renowned spectrographer whose laboratories were shut down in 1918 due to trouble with staff and, more crucially, the on-going World War I (Maushart, 2011, pp. 7–9).[2] Women gained the right to qualify for *Habilitation* and to become a *Privatdozentin* (whereby they were qualified to teach and conduct examinations independently at the university level) only in 1920, by which time Sponer had completed her PhD studying spectroscopy under Debye – beginning her studies with him in Tübingen, then following him to Göttingen in

---

[1] Thanks to an anonymous reviewer for this note.

[2] In Duncan and Janssen (2023, pp. 153–156) we learn that one of Paschen's group members at Tübingen in 1917–1918 was Christian Füchtbauer, who together with Sponer's doctoral advisor, renowned physicist Peter Debye, and with Arnold Sommerfeld worked in these years to apply Bohr's quantum postulates to classical dispersion theory. Sponer's higher education was thus from the start directed by men who took Bohr's atomic theory seriously enough to attempt application to classical theories – this before many other physicists came on board.

1918. Because Sponer's timeline coincided with only the beginning of opportunities for women in academia, she encountered extreme opposition throughout her career. Sponer was blocked from various promotions and positions in Germany by the likes of Max Planck and Robert Pohl, and in the US by Robert Millikan and perhaps even by her famous collaborator, Raymond Birge (see Note 7).

While in Göttingen Sponer nevertheless occupied a number of increasingly prestigious roles. In her earlier years, there she served as an aide to David Hilbert; in her 1962 interview for the AHQP, Sponer recalled fondly her time working with Hilbert (Sponer and Franck, 1962b). She had a penchant for theoretical physics and mathematics, and happily her primary task as Hilbert's aide was to read the latest papers on his behalf, describe the main results to him, and then discuss them together. When asking Debye for advice at this early stage of her career, knowing of her talent for theoretical physics[3] and having instilled in her the fundamental importance of quantum theory,[4] he nevertheless discouraged her from pursuing theoretical physics as a career.[5]

Sponer left Göttingen for a short time in 1920–1921 to work in Berlin alongside renowned physicists Fritz Haber and James Franck – with whom she was a frequent collaborator and close friend (Figure 3.1). When Franck was offered a position in Göttingen in 1921, Sponer returned and became, in name only, his *Assistentin* (assistant). In name only, because Franck was head of the second Institute of Physics (experimental) at the university,[6] and this meant the duties of maintaining the laboratories, training laboratory workers, and conducting experiments was left to Sponer. Indeed, Franck himself rarely got his hands dirty, so to speak, leaving nearly all experimental matters in Sponer's extremely capable hands (Sponer and Franck, 1962a, b). In addition to running the laboratories, Sponer taught the basic physics seminars, and in this way had a lasting impact by introducing a generation of students (some of whom became physicists better known than she) to brand-new developments in quantum physics (Sponer and Franck, 1962b; Maushart, 2011, ch. 2). She also recalls having many rich discussions on both theory and experiment with folks like Heisenberg, Friedrich Hund, and Edward Teller, who joined the Göttingen community in 1931 (Sponer and Franck, 1962c). In the case of the latter,

---

[3] He even tested her on this by assigning her a very difficult paper by Karl Schwarzschild to read; she not only read and understood it, but explained it to Debye with perfect comprehension (Sponer and Franck, 1962c).

[4] As mentioned above, Debye was one of few physicists who had immediately and enthusiastically embraced Bohr's 1913 atomic model and the Bohr–Sommerfeld quantum theory (Segrè, 1980, p. 129); Sponer was thus introduced to Bohr's theory and thoroughly educated in this nascent physics as Debye's student during 1917–1920 (Sponer and Franck, 1962c; Maushart, 2011).

[5] During this point in the interview, in fact, Franck emphasizes how much better Sponer is at theoretical understanding of the physics than he, and provides a few anecdotes to this effect – one of which includes Franck's also reading, but not at all understanding, the same Schwarzschild paper Debye had used as a litmus test of Sponer's theoretical acuity (Sponer and Franck, 1962c).

[6] Max Born arrived at more or less the same time to take up his role as head of the first Institute of Physics (theoretical).

Figure 3.1 Sponer in like company, November 7, 1920. Seated, left to right: Hertha Sponer, Albert Einstein, Ingrid Franck, James Franck, Lise Meitner, Fritz Haber, Otto Hahn. Standing, left to right: Walther Grotrian, Wilhelm Westphal, Otto von Baeyer, Peter Pringsheim, Gustav Hertz. Image courtesy of the Archive of the Max Planck Society, Berlin.

she not only discussed theory and experiment, but also directed his research by posing problems for him to investigate. In the end Teller spent as much time with Sponer and Franck at Institute II as he did with Max Born at Institute I (Maushart, 2011, pp. 51–52).

During the Göttingen years she established herself as a leading expert in apply-ing quantum theory to physical chemistry. Her over 30 peer-reviewed publications and invited talks as well as a highly praised, two-volume textbook, *Molekülspektren und Ihre Anwendungen auf Chemische Probleme* (Molecular Spectra and their Application to Problems in Chemistry; Sponer, 1935, 1936) garnered international acclaim – a fact that would later aid her in finding a permanent academic home outside of Germany. In February 1936, she became Professor of Physics at Duke University – that university's first female physics faculty, a title she also claimed in Göttingen – where she remained until the mid-1960s.[7]

---

[7] I note that the sexism which nearly kept Sponer from receiving her academic title in Germany very nearly prevented her appointment at Duke University also. For instance, Margaret Rossiter's classic book, *Women*

Before Duke, her publications focused on the analysis of band spectra and measuring dissociative heats for diatomic molecules. While at Duke, she published over 50 papers and established herself also as a foremost expert on the benzene molecule, continued to work with absorption and emission spectra but now focusing in the near-ultraviolet range, and studied florescence and phosphorescence. At Duke she regularly offered courses titled "Theory of Atomic Spectra" and "Theory of Molecular Physics," as well as various general physics courses (advanced topics, modern physics, thermodynamics). In addition, she advised a dozen or so Masters theses and over 20 Doctoral theses.[8]

In sum, in the course of her career Sponer published over 80 articles in the leading physics and chemistry journals and collaborated frequently, though often unofficially, with prominent scientists worldwide. Nevertheless, her name nearly always arises – if it arises at all – in connection with two things. First, she co-authored an influential paper with American physicist Raymond Birge (more on which in Section 3.4), which is nowadays only known to historians of molecular physics or chemistry. Second, many years after Franck's wife Ingrid had died and he too had immigrated to the US (University of Chicago), he and Sponer married. Oftentimes what is said of her is outright incorrect. For instance, in the National Academy of Sciences *Biographical Memoir* on Franck, the authors write: "Among his [Franck's] students were Blackett, Hanle, Herzberg, von Hippel, Rabinowitch, and Sponer. His institute also attracted outstanding scientists to Göttingen as visiting professors, some of whom were K. T. Compton,[9] E. U. Condon, J. E. Mayer, and G. Scheibe" (Rice and Jortner, 2010, p. 7). Sponer was never Franck's student; she was his *colleague*, in fact rising to the rank of Professor of Physics at Göttingen in January 1932.[10] In addition, Edward Condon's choice to

---

Scientists in America, recounts how Sponer, "considered one of the most brilliant women physicists of the time, after Marie Curie and Lise Meitner," nevertheless faced explicit opposition by famous American physicist Robert Millikan in her appointment to Duke's faculty. Apparently he wrote a letter to the president of Duke University arguing that the future of physics in the US belonged to the "bright young man in the field" not to a woman, and that having female professors would diminish Duke's academic prestige (Rossiter, 1982, pp. 190–194). One also reads a similar sentiment in a letter from Birge asking how Sponer likes her new position at Duke. After telling Sponer that he had, upon being asked, written the strongest recommendation he could on her behalf, he concludes in a manner that suggests that whether or not Sponer likes her new situation, she ought not to complain. In a letter from Birge to Sponer on May 22, 1936 (Raymond Thayer Birge Papers, BANC MSS 73/79 c, Boxes 34 and 36. The Bancroft Library, University of California, Berkeley), he writes: "I also talked with Professor Franck about the position [at Duke] . We are having so much difficulty in getting positions for our PhD men that I presume one should be glad to have any position at all, even though it is not in all respects satisfactory."

[8] For details on her publications and talks, and their impact, see: chs. 2–3 and appendices 2–5 of Maushart (2011; translation of Maushart (1997) by Ralph A. Morris, with added material by Annette Vogt and edited by Brenda P. Winnewisser), Gutierrez (2006), the Hertha Sponer Papers at Duke University Libraries (https://find.library .duke.edu/catalog/DUKE005845532), and the Duke Physics Department website dedicated to Sponer (Duke University Department of Physics, n.d.).

[9] That is, Karl Taylor Compton, brother of Arthur Holly Compton

[10] In the preface to the English edition by translator Ralph Morris and editor Brenda P. Winnewisser, it is noted that this feat distinguishes Sponer as one of just three women (the others being Lise Meitner and Hedwig Kohn) ever to obtain that title in the history of pre-World War II Germany (Maushart, 2011, p. ix).

visit Göttingen was not only to see Franck and Born but also to visit Sponer, with whom he'd grown close while working on his dissertation at Berkeley, and whose influence upon him was significant. As we shall see in Section 3.4, Sponer herself was responsible for teaching new quantum applications to Condon but also persuading him to pursue the line of work which directly led to the "Condon" part of the Franck–Condon principle.

Or consider as another example the following excerpt from Brand's otherwise excellent history, *Lines of Light*. While introducing the Franck–Condon Principle, he writes (Brand, 1995, p. 205):

Sponer (1895–1968), a former student of Franck (later, his wife), spent the 1925–26 academic year as a Rockefeller Foundation fellow at the University of California at Berkeley, a sojourn that led to the Birge–Sponer studies of dissociation. In the same year Edward Condon (1902–1974), a graduate student with Birge, was preparing to write a thesis on spectral intensities. Birge, deeply involved in a program of measuring and analyzing intensities, was the link by which Condon learned from Sponer of Franck's ideas on photodissociation. Condon then spent a postdoctoral year, half with Born and Franck at Göttingen, half with Sommerfeld at Munich, in course of which he reworked Franck's ideas in the language of quantum mechanics.

Again, Sponer was not merely Franck's "student ... later wife," but one of his most esteemed colleagues and an international scientific figure in her own right. And while Brand's statements regarding Condon are closer to the truth than other accounts, he still manages to massively downplay Sponer's influence on Birge and Condon both (the extent to which, again, will become clear in Section 3.4).

As far as the author is aware, there exist just one book and one booklet dedicated to Sponer, one of which is a complete biography of her life only recently translated into English from the German (but which does provide excellent detail on some of her scientific contributions) – the book by Maushart (Maushart, 1997). The booklet, also a biography, is by Michael Gutierrez, but is only a few pages in length (Gutierrez, 2006). Most importantly for my purposes, the extant accounts which focus primarily on her US career make clear that Sponer was an important scientific figure, but in so doing neglect crucial aspects of Sponer's role in the development of quantum physics, happening as they did in Europe and in the 1920s to early 1930s.

Aside from these two books, the literature on Sponer is sparse. There are biographical sketches of her in a few collections on women in physics (e.g., Tobies (1996), Rechenberg (2006), Vogt (2007)), though mainly in German.[11]

---

[11] This adds grist to the mill of Götschel's excellent review and commentary on gender and physics scholarship in Götschel (2011). She concludes a section on human actors in physics with the following observations (p. 69): "In summary, female physicists are highly represented at the level of biographies. However, literature that is not written in English, is virtually unknown internationally ... In contrast, research on male physicists from a gender-aware perspective ... are rare." In the last analysis, however, "the quantitative and biographical approach ... is limited when it comes to understanding structures as well as processes on a larger scale."

And if Sponer is mentioned at all in historical texts – specifically those on early quantum theory, quantum chemistry, or molecular physics – she is at best a peripheral figure. This is true in the case of first-hand accounts like Hund (1967), who as Sponer's colleague and frequent interlocutor was certainly aware of her achievements – and also true of more recent scholarship along these lines.[12] I am unaware of any scholarly essays in English wholly dedicated to Sponer's scientific achievements; I am certain none exist regarding her part in the development of quantum theory.

3.3 First Vignette: Matter Is Wave-Like

It is an interesting feature of the history of quantum mechanics that the focus has tended to be on the theorists: Bohr, Einstein, Schrödinger, Heisenberg, Born, and Wolfgang Pauli to list a few. Perhaps this was inevitable, given the frequency of profound theoretical developments made in so brief a time by so small a number of revolutionaries (many of whom extraordinarily young) concentrated at so few, relatively proximate, universities and institutes. But as stated in the introduction, the advances in experimental techniques occurring in lock-step with quantum theory in particular cannot be overlooked. And though of course these histories mention crucial experiments and their results – for example, the Franck–Hertz experiment, Rutherford's atomic model, the Stern–Gerlach results, the Compton effect, and so on – the story of the experimentalists themselves is often overlooked. Telling the story of quantum physics without due attention to the close (often deeply personal) relationships and frequent interactions among the well-known theorists and the atomic and molecular physicists on whom they relied, has resulted in scholarship centered on *Knabenphysik* – "boys' physics."[13]

Because the women who contributed to quantum physics in its earliest days were for a variety of reasons (sexism first among them) discouraged from becoming theorists and instead relegated to laboratories and/or classrooms, the *Knabenphysik* canon in truth commits a double sin of omission. It ignores or downplays the names of the women like Sponer who did the real work of running laboratories, conducting experiments, and teaching basic physics (including quantum physics) seminars. As

---

[12] For instance, there is no reference to Sponer in Assmus (1992), written while an historian at Berkeley – the very place where Sponer taught Birge, Condon, and others how to use quantum theory. Regarding scholarship in the history of quantum chemistry (much of which is listed on Uno Anders' website (Anders, 2011)), Sponer is briefly mentioned – but only as a one-time collaborator of Birge – in Gavroglu and Simões (1994), but not referenced at all in their more recent work, Simões and Gavroglu (2001). Neither does she rate mention in the "detailed and unique historical narrative" of chemical physics by Califano (2012) – per the book's Springer Link website.

[13] As a reviewer pointed out, this term was of course primarily intended to convey the extreme precociousness of Heisenberg, Pauli, and others, and that their "maleness" was taken for granted.

feminist scholars have often pointed out,[14] the more hands-on roles as well as the roles of educators, because they were often the only sectors of science in which women were allowed, also tend to be the aspects of science considered less worthy of historical account.

The first vignette centers on the Ramsauer effect – an experimental result that is left out of many standard accounts of this history.[15] In a series of papers from 1921 to 1923 (Ramsauer, 1921a, b, c, 1923), Carl Ramsauer demonstrated that there was an unexpected relationship between the initial electron energy and the angle of electron scattering for certain gases. Shortly thereafter, Hertz would demonstrate the effect for argon while Sponer and Rudolf Minkowski – nephew of "space–time" Hermann Minkowski – would find the effect also for krypton (Jähnert, 2019, pp. 131–134): electrons with low energy were not scattered by atoms as expected, but instead passed through the gas undisturbed.

Some physicists considered the results spurious, but in Göttingen there were folks who took it as a serious explanatory challenge.[16] Sponer reports in her AHQP interview that while she and Minkowski did not immediately understand what to make of Ramsauer's work, they nevertheless understood it to be of great importance (Sponer and Franck, 1962b). Thus, they continued to tackle the issue experimentally, and not in isolation. Indeed, in the three or so years following Ramsauer – and close on his heels – a proliferation of slow-electron scattering experiments were run with equally confounding results (e.g., Townsend and Bailey, 1921). Next to Ramsauer the most reliable experimental data came out of America: Clinton Davisson and Charles Kunsman had found similar energy/ scattering angle dependencies in slow electrons colliding with nickel and published their findings in November 1921 (Davisson and Kunsman, 1921). Shortly thereafter, Sponer joined the trend from Göttingen, providing individual experimental results in Sponer (1923) and collaborating with Minkowski in Minkowski and Sponer (1923) as well as the article discussed in more detail below: Minkowski and Sponer (1924).

While the experimentalists attempted to find more and better data supporting the phenomena of negligible scattering of slow electrons by gas atoms, Franck and Hund were busy developing a series of theoretical explanations for the data, based largely on classical electron theory. Though eventually both Franck and Hund (in his 1922 dissertation and a published paper shortly thereafter; Hund, 1922, 1923)

---

[14] For a meta-perspective, see Götschel (2011) and the excellent references therein, as well as Rossiter (1982).

[15] Though see ch. 5 in Jähnert (2019, p.130) as an example of taking this history into consideration more thoroughly.

[16] Indeed, Jähnert (2019, p. 130) writes, "Bringing the problem to the classroom of his *Proseminar* in Göttingen, Franck and Max Born set students to work on the experimental and theoretical refutation of 'Ramsauer's crazy assertion'." There can be little doubt that Sponer was either in the proseminar at the time or heard about Franck and Born's challenge shortly thereafter.

would attempt to use the correspondence principle to relate Hans Kramers' work with X-ray scattering to the Ramsauer effect, this is about as far as quantum theory penetrates into their work. There were such significant problems arising even from their hybrid classical/quantum correspondence accounts that, by the end of 1922, both Franck and Hund had given up the project (Jähnert, 2019, ch. 5).

The hypothesis that matter might share light's duality, behaving both like a particle and a wave, is due to Louis de Broglie's work as a student in the period 1922–1924 (Medicus, 1974). He first hypothesized electron diffraction in a paper of 1923, and again in his 1924 dissertation (Bacciagaluppi and Valentini, 2009, pp. 51–52). Of course it is well known that the first public presentation on this hypothesis was not until de Broglie's talk at the Solvay Conference of 1927 (Bacciagaluppi and Valentini, 2009, p. 356),[17] but in the intervening years his papers were read – and seized upon – by Einstein (in particular, in Einstein (1925a) and Einstein (1925b)) – and through Einstein, Schrödinger (who was thus motivated to develop a wave mechanics for quantum theory) as well as Born's student, Elsasser (Bacciagaluppi and Valentini, 2009).

According to the common lore, Elsasser's paper was the first to connect the results of Ramsauer to those of Davisson–Kunsman, but decisive proof of his matter-wave thesis arrived only in 1927 and afterwards. An inexhaustive list of historians employing this narrative include Medicus (1974), Segrè (1980, p. 152), Beller (1999, pp. 43–44), Bacciagaluppi and Valentini (2009, ch. 2), and Jähnert (2019, ch. 5). Indeed, the analysis of Bacciagaluppi and Valentini echoes the lore endorsed in the primary source materials under consideration. Namely, in the Solvay reports by Born, Heisenberg, and de Broglie, as well as in the conference's general discussion, where it is clear that most if not all participants believed sufficient evidence for matter waves remained outstanding.

Thus it was seemingly only after Elsasser's paper of 1925 that Ramsauer's results were not only correctly interpreted but also linked with the results of the Davisson–Kunsman experiments, all of which were decisively proven in 1927 by Davisson with Kunsman's successor, Lester Germer (Davisson and Germer, 1927). These connections can be seen first-hand in a pair of letters Heisenberg sent to Pauli shortly after the publication of Elsasser's paper. I include segments of these two letters here, primarily because it is significant to note Heisenberg's immediate recollection that Minkowski and Sponer had anticipated Elsasser in appreciating the importance of the Ramsauer and Davisson–Kunsman experiments. On June 24, 1926, Heisenberg wrote to Pauli (Pauli, 1979, pp. 226–229; my translation): "You know the new Einstein paper on atoms moving in accordance with wave theory

---

[17] Interestingly, as the authors point out, de Broglie in the Solvay Report attributes the idea of electron diffraction to Walter Elsasser in the latter's 1925 paper (Elsasser, 1925) rather than, correctly, to himself in 1923!

[i.e., Einstein (1925a)]? Apply this theory to slow electrons and you get the Ramsauer noble gas curves." The editors of the Pauli correspondence here interject: "This [Einstein's 1925 paper] is the first attempt interpreting the Ramsauer effect as a phenomenon of diffraction of matter waves from atoms. Cf. on this Elsasser" – and then reference Elsasser (1925). This is in keeping with the tradition of naming Elsasser the first to connect the Ramsauer effect to the Davisson–Kunsman experiments. Medicus (1974) retells this story, but adds a few facts interesting for our purposes. Heinrich Medicus reports (p. 42) that Elsasser was in Göttingen studying with Franck when Einstein's paper came out, and it was through weeks of discussions with Franck and perusing de Broglie's thesis that Elsasser hit upon the connection between the wave theory of electrons (as presented by Einstein and de Broglie) and Ramsauer's "puzzling, and at that time quite novel, observations" as well as the Davisson–Kunsman experiments. Out of this was born Elsasser's famous 1925 *Naturwissenschaften* article.

Sponer and Franck remember this differently in their AHQP interviews, however, and there is evidence to support their version of events. They recall that not only had she alone and together with Minkowski produced experimental results along these lines in 1923 but, more importantly, had written a paper in 1924 in which they – not Elsasser – were the first to recognize that (i) the Ramsauer effect was indeed a similar phenomenon to what the Americans had found, and that (ii) both of these sets of results, along with their own from 1923 and Townsend–Bailey's and a few others' – all pointed to the same thing: matter was exhibiting behavior that could not be explained using classical electrodynamics. In short, they were the first to recognize – though without explicitly naming the behavior "wave-like" or using the term diffraction as Elsasser would a year later – that this was a decidedly quantum phenomenon.

Sponer was one of the main figures alongside Franck at Göttingen Physics Institute II, conducting many of the seminars and physics laboratories for students. Either she or Franck certainly discussed these issues with Elsasser, especially given that the vital insight – the interpretation of diffraction experiments like those of Ramsauer, Townsend–Bailey, and Davisson–Kunsman as evidence that matter behaved nonclassically – had already been made by Sponer and Minkowski, and to some extent had been evident in the failure of Franck's and Hund's individual, classical explanations. Indeed, going back to Heisenberg's letter to Pauli in the summer of 1926, Heisenberg appreciated the significance of Sponer's and Minkowski's 1924 paper. After connecting Einstein to Ramsauer, Heisenberg describes earlier experiments to the same effect, including this comment: " . . . or even better: you shoot slow electrons onto a crystal lattice and get a spectrum of 1st order, 2nd order, etc.; the experiments were done a long time ago and are in the article by Minkowski and Sponer" (Pauli, 1979, p. 226; my translation). A few days after this letter, Heisenberg sends a postcard to Pauli from Göttingen where he admits " . . . I think it is possible that both Ramsauer's experiments and Davisson and Kunsman's can be explained using Einstein's theory (which was Elsasser's sole claim), and ever since studying the Einstein paper, I'm actually pretty enthusiastic about it" (Pauli, 1979, p. 229).

Minkowski and Sponer (1924) not only describe Ramsauer's experimental results in connection with the findings of Davisson and Kunsman – recognizing not only that the same phenomena are examined in these and other studies, but furthermore explicitly stating the shared quantum properties (according to Bohr's second atomic theory) of those materials found to be reacting in an unexpected way. They provide the following interpretation of these considerations (Minkowski and Sponer, 1924, pp. 78–79; my translation):

If one compares the number of maxima in the various substances examined with the number of electron groups found in the relevant atoms according to Bohr's theory of atomic structure, one sees that the number of maxima is two fewer than the number of electron groups. This suggests that with a suitably idealized atomic model, these results can be derived under the assumption that the fast reflected electrons are able to penetrate into an atom and leave again without any noticeable loss of energy. If one assumes the validity of electrodynamics for describing the motion of an electron in an atom, then in the field of an atom, an electron experiences a deflection due to electrostatic attraction between the nucleus and electron and at the same time a loss of energy through radiation due to [that] acceleration.[18] This energy loss becomes greater the more the electron is deflected from its path. If the velocity of the electrons is sufficiently high, only a very small number of electrons whose paths come sufficiently close to the nucleus will experience a noticeable loss of energy, so that the practically negligible loss of energy through radiation can be ignored without fear of significantly influencing the outcome. At low speeds, on the other hand, the energy loss becomes increasingly considerable, ultimately . . . exceeding the kinetic energy of the electron. Then the electron can no longer leave the atom; it has to move on a path that gets closer and closer to the nucleus and finally falls into the nucleus. This conclusion is obviously in open contradiction to experience. Therefore, it is expected from the outset that a description of these processes at slow speeds cannot be achieved assuming the validity of electrodynamics, and that it will be necessary to impose restrictions on electrodynamics in order to bring it into agreement with experience.

Following these considerations the authors provide detailed and clear descriptions of the Ramsauer and Davisson–Kunsman experimental findings, and conclude that their initial interpretation indeed supports these data. Not only this, but other attempts to describe the behavior of electron diffraction at slow speeds do *not* support the new results. The authors consider in particular the then-current best explanation of this behavior – Hund (1923) – pointing out, for example, that Hund's theory makes it impossible for a large number of electrons to pass through an atom

---

[18] Here they are describing the phenomenon of Bremsstrahlung.

without noticeably influencing it, yet without noticeable influence one cannot explain the Ramsauer effect (Minkowski and Sponer, 1924, pp. 82–83). A few lines later they make the following significant statements:

The agreement between the results of e.g. Ramsauer and H. F. Mayer shows that there cannot be a constant accumulation around characteristic values [*ausgezeichnete Werte*] of deflection angles at a certain speed, but that a fraction of the electrons runs strictly straight ahead, while another fraction experiences deflections without a preferred direction. But this means that it is a discontinuous process that cannot be understood according to classical electrodynamics. . . . The curtailment of the validity of electrodynamics, which is unavoidable in any case, must occur in a more profound way than we initially assumed. . . . Given this, it is assumed that an electron entering the force field of an atom is not deflected from its rectilinear initial orbit into a rectilinear final orbit with the emission of a continuous spectrum, as is required by electrodynamics, but that the deflections can be described as a quantum transition from one rectilinear path to another rectilinear path. . . . The spectrum actually emitted differs from the classical spectrum in that it breaks off at a certain frequency.

When quantum assumptions are applied "which represent a consistent application of Bohr's correspondence principle" (Minkowski and Sponer, 1924, p. 84), Minkowski and Sponer's calculations do fit the data for the puzzling behavior of slow electrons impinging on atoms.

The paper ends as follows (Minkowski and Sponer, 1924, p. 84):

In summary, it can be said that, despite the validity of the energy and momentum [conservation] law, the collisions of electrons with atoms are completely nonmechanical and apparently can only be described according to the laws of quantum theory. This applies not only to the so-called inelastic collisions which stimulate quantum jumps in the struck atoms, but also to the so-called elastic collisions in which the incident electrons penetrate into the interior of the struck atom and dissipate there under the influence of the nuclear charge – effects which at high speeds indeed can be adequately described without the help of quantum theory, but which at very low speeds deviate completely from what is expected by mechanics.

Although Minkowski and Sponer do not appeal explicitly to wave-like behavior, this is certainly what they were on to. Thus arguably they were the first not only to see a connection between the results of Ramsauer and Davisson–Kunsman, but furthermore to examine the reasons for that connection – a line of thinking that led them to the unavoidable conclusion that classical electrodynamics was insufficient for describing the experimental data, which clearly revealed aspects of nonclassical behavior in matter. It is easy to connect the dots from here: because the classical (read: particle-like) theory of electrons fails in certain regimes, a (Bohrian) quantum description of electrons – one that can accommodate wave-like properties such as diffraction – must also be correct.

3.4 Second Vignette: Bringing Quantum Physics to America

Consider another excerpt from Segrè (1980, pp. 145–146), this time discussing the rich exchange of information among physicists in the pre-World War II era:

Professors from all the leading institutions met frequently and exchanged information not only on technical matters but also on promising new recruits. In the United States the Rockefeller Foundation financed many young scientists, choosing them with uncanny and admirable shrewdness. Among them were several of the founders of quantum mechanics, such as Heisenberg and Pauli.

And Sponer.

When Sponer traveled to UC Berkeley in 1925 on a year-long Rockefeller Foundation Fellowship to work with one of the most prominent American experimental physicists of the day – Raymond Birge – it was she, not Robert Oppenheimer, contrary to popular lore and previous historical accounts,[19] who taught Birge along with other American boys how to apply the new quantum theory to design, conduct, and interpret advanced experiments in molecular physics (Sponer and Franck, 1962d).[20]

Indeed, she ultimately convinced the much more senior Birge that quantum theory provided the correct grounds for explaining their excellent results (Sponer and Franck, 1962d; Kemble and Birge, n.d.). The joint paper they published (Birge and Sponer, 1926) discussing improved methods (and the better data attained thereby) for calculating the dissociation energies of diatomic molecules remains Sponer's best-known publication. Indeed, in a letter to Sponer from Birge in fall of 1941,[21] Birge has recently read a new paper on spectroscopy quoting an updated value for the heat of dissociation of carbon monoxide molecules of 11.05 eV (Kynch and Penney, 1941), vindicating the value of 11.18 eV they'd calculated already in their 1926 paper (wherein, note, they arrived at this value via experimental data obtained by Sponer alone, and using the extrapolation method whose theoretical underpinning Sponer provided and which Birge disliked; cf. Kemble and Birge (n.d., section 6-003) and Sponer and Franck (1962d)). This despite a number of textbooks and papers in the intervening years that reported a much smaller, safer value nearer to 9 eV.

Among those to whom she taught quantum theory at Berkeley was Edward Condon. Then Birge's doctoral student, it was nevertheless due to Sponer's expert

---

[19] Oppenheimer got his doctorate at Harvard only in 1925, at which point according to Segrè he was "among the first in America to grasp quantum mechanics" (Segrè, 1980, p. 215). A similar story about Oppenheimer is told in Rouzè (2024) and Bird and Sherwin (2005), especially focusing on his appointment to Berkeley's faculty in 1927. While he might have been the first to take quantum theory seriously on the East Coast, it seems clear from Sponer's story that she was the one who brought quantum mechanics to the West Coast.

[20] In the interview, Sponer herself makes this correction regarding the Oppenheimer lore. She also describes her (ultimately futile) attempts to convert Gilbert Lewis – one of the most influential chemists of the day, and whose seminar at Berkeley Sponer frequently attended – to the "quantum theory" side of life.

[21] Letter from Birge to Sponer, November 12, 1941. Raymond Thayer Birge Papers, BANC MSS 73/79 c, Boxes 34 and 36. The Bancroft Library, University of California, Berkeley.

technique and personal tutelage on quantum physics that Condon was inspired to change his doctoral thesis to a topic aligned with these new considerations. This work, eventually published after collaboration with Franck (and certainly also with Sponer!) while visiting Göttingen, would become his greatest claim to fame: the Franck–Condon Principle. This principle, crucial to the development of quantum chemistry, states that due to the extreme speed of electron transitions compared to atomic vibration frequencies in a diatomic molecule, one is justified in treating the molecule as being at rest. The principle bears the names of both physicists because, according to their own accounts, Franck provided a semi-classical explanation of the principle and later Condon made it a fully quantum one. Sponer had taught Condon quantum theory, and likely discussed Franck's principle with him. In this way the Franck–Condon principle came about due to Sponer's interactions with both men.

3.5 Third Vignette: Quantum Tunneling

Hund was by all accounts the first to publish on the phenomenon now called quantum tunneling, in his twin papers Hund (1927a) and Hund (1927b). Originally referred to by physicists as "barrier penetration" or "breakthrough effect," predicting such a phenomenon required one to take the wave picture of electrons seriously; given Hund's intense, earlier investigations of the Ramsauer effect, it is no surprise that he was the first to do so. Hund's reasoning in his 1927 papers was based on analogy with classical optics: light waves could, theoretically, be transmitted through a medium whose potential energy is nevertheless greater than the light's energy; this situation is described by assigning an imaginary index of refraction. Something similar must also be true for matter, then: when a matter wave encounters a potential well, or barrier, whose energy is greater than that of the electrons, there ought to be a nonzero probability of its being transmitted through the barrier (although with lessened intensity) (Merzbacher, 2002, p. 44). The quantum-mechanical explanation for the non-zero probability of tunneling for photons and electrons is due to the fact that superpositions of states are allowed states. A superposition of symmetric and antisymmetric energy states will result in a nonstationary state able to penetrate a potential barrier.

Interpreting experimental results in line with the possibility of quantum tunneling was first done, so the history goes, in 1928–1929 by three physicists, in quick succession: George Gamow, Ronald Gurney, and Condon each used quantum tunneling to describe alpha particle scattering (Merzbacher, 2002, p. 48). Describing certain atomic and molecular effects via quantum tunneling took until 1930, when Oscar Rice published a paper (Rice, 1930), in which he interpreted as a tunneling effect the dissociation by rotational energy of diatomic molecules in excited states (Merzbacher, 2002, p.47).

At the same time as Rice was working on his tunneling effect paper in Leipzig, Sponer was presenting on a similar theme at a conference organized by her *Doktorvater* Debye, also in Leipzig. Many luminaries of molecular physics as well as chemists like Walther Nernst were present and heard Sponer's new analysis, framed in the context of quantum mechanics, of her well-established topic area of expertise: the relationship between dissociation of diatomic molecules and band spectra (Maushart, 2011, pp. 50–51). There, Maushart describes Sponer's novel insights:

The fact that a molecule can sometimes take on more energy, in a quantized manner, than it needs for dissociation and then undergo a delayed dissociation was a mysterious quantum-mechanical phenomenon still not understood at the time. She [Sponer] had learned how to interpret the resulting broad, fuzzy lines in spectra, which usually had narrow, well-defined absorption features. . . . [H]er masterful and clear presentation, putting predissociation in the context of the wider context [sic] of the electronic energetics of molecules, earned her the nickname "How-do-I-explain-the-facts-of-predissociation-to-my-chemist?"[22] The chemists, especially Nernst, were not at all thrilled with the concept that she had refined and which was so hard for them to understand. She understood that the potential energy curves for molecules that show predissociation have a potential energy barrier to dissociation but nonetheless succeed in dissociating . . .

Maushart concludes her discussion of Sponer's 1930 talk/1931 paper by asserting that by this paper (along with her Franck and Birge collaborations) she became internationally renowned (Maushart, 2011, p. 51).

Looking at Sponer's paper, one can see Maushart's claims borne out. Though the majority of the paper concerns, as Maushart says, a review of experimental results – but framed now in terms of quantum mechanics – of particular interest for our purposes is her section on dissociation through rotational energy (Sponer, 1931, p. 126 ff.). The paper begins with Sponer's statement that there are only three ways that diatomic molecules can dissociate, one of them being through rotational energy. Now, after having discussed the other types of dissociation, in her final section she explains that when dissociation happens due to rotational energy, more energy can be absorbed by the molecule than is needed for dissociation in its given state. Rotational states below the level required for dissociation remain stable because they are "separated from the dissociated state by a potential maximum" (Sponer, 1931, p. 128; my translation). She continues on that same page with her analysis of dissociation through rotation (my translation):

The last rotation states before the limit curve become broadened. The explanation for this is based in quantum mechanics. According to the principles of quantum mechanics, there

---

[22] It sounds cooler in the original German, I'm sure.

always exists a finite probability for the transition between two states of equal energy which are represented by a potential mountain [*Potentialberg*], whereby according to Kronig, certain selection rules apply to the transition. But Franck's principle must also be satisfied, and an attempt has been made [referring to her 1928 paper with Franck] to attribute the unstable rotation of HgH and AlH to such a quantum-mechanical breakthrough effect [*Durchbrucheseffekt*] while maintaining Franck's principle.

She ends the paper by stating that various other band spectra results "have been given a quantum-mechanically precise theoretical representation by [Ralph] de Kronig [sic], Condon and [Felix] Villars, and Rice" – and then cites Rice (1930).

Though it may be subtle to see the novelty Maushart assigns to the 1931 paper, Sponer herself notes here that she and Franck had already in 1928 attempted to use the "quantum-mechanical breakthrough effect" to explain unstable rotations in mercury hydride and aluminum hydride molecules. This is in fact what one reads in their 1928 paper (Franck and Sponer, 1928), and this puts them ahead of the physicists usually credited with first using tunneling to explain hitherto confounding phenomena. One could in addition argue that Sponer's and Franck's application of quantum tunneling to molecular phenomena was a far more difficult and progressive step than application to alpha particle radiation, as the others had done.

3.6 Conclusion

Why was Sponer's role in these and other moments, if not ignored entirely, drastically minimized? Of course there's the question of gender. Maushart's biography describes in detail – and with clear evidence – the continual discrimination Sponer faced on her path towards Duke (where at last she received a stable position worthy of her achievements). For instance, the physics institutes in Göttingen were overseen by Pohl, who was an unabashed misogynist. While in position of leadership there, he stood in the way of a number of promotions and opportunities for Sponer and other women.

Though Sponer's gender is the primary reason for her neglect, there has also been a marked tendency in constructing the history of quantum physics to downplay the role of experimentalists. The several most famous *experiments* and their results are, of course, mentioned – but typically without detailed commentary on the experimentalists themselves, hastening instead to the story of how those results were then interpreted or utilized by theorists.

A third factor might be mentioned in this context, and that is Sponer's area of expertise – molecular physics – was the nexus for a new mingling of fields in physics: quantum mechanics, atomic and molecular physics, and chemistry all in the melting pot together. Much has been written from the perspectives of historiography as well as history and philosophy of science about the emergence of quantum chemistry as a field, but bringing this scholarship into closer contact with the standard theoretically focused accounts is crucial.

That's three strikes against Sponer: she was a woman, an experimentalist, and usually classified as a chemist rather than physicist (though she was, most truthfully, among the first *quantum* chemists). But as I hope to have shown above, her role within the history is a nontrivial one. I have focused on three moments in the development of quantum physics that clearly benefited from Sponer's participation, but there remains more to be uncovered regarding each of these stories, and more to Sponer than has been revealed within these stories.

References

Anders, Udo. 2011. *Early Ideas in the History of Quantum Chemistry.* quantum-chemistry-history.com, January 6. https://www.quantum-chemistry-history.com/index.html

Assmus, Alexi. 1992. The molecular tradition in early quantum theory. *Historical Studies in the Physical and Biological Sciences*, 22(2), 209–231. https://doi.org/10.2307/27757681

Bacciagaluppi, Guido and Valentini, Antony. 2009. *Quantum Theory at the Crossroads: Reconsidering the 1927 Solvay Conference.* Cambridge: Cambridge University Press.

Beller, Mara. 1999. *Quantum Dialogue: The Making of a Revolution.* Chicago, IL: University of Chicago Press.

Bird, Kai and Sherwin, Martin. 2005. *American Prometheus: The Triumph and Tragedy of J. Robert Oppenheimer.* New York: Knopf Doubleday Publishing Group.

Birge, Raymond T. and Sponer, Hertha. 1926. The heat of dissociation of non-polar molecules. *Physical Review*, 28, 259–283. https://doi.org/10.1103/PhysRev.28.259

Brand, John C. D. 1995. *Lines of Light: The Sources of Dispersive Spectroscopy, 1800–1930.* Amsterdam: Gordon & Breach.

Califano, Salvatore. 2012. *Pathways to Modern Chemical Physics.* Berlin, Heidelberg: Springer-Verlag.

Davisson, Clinton J. and Germer, Lester H. 1927. Diffraction of electrons by a crystal of nickel. *Physical Review*, 30(6), 705–740. https://doi.org/10.1103/PhysRev.30.705

Davisson, Clinton J. and Kunsman, Charles H. 1921. The scattering of electrons by nickel. *Science*, 54(1404), 522–524. https://doi.org/10.1126/science.54.1404.522

Duke University Department of Physics. n.d. *Hertha Sponer.* https://physics.duke.edu/about/history/historical-faculty/HerthaSponer

Duncan, Anthony and Janssen, Michel. 2023. *Constructing Quantum Mechanics, Volume 2: The Arch (1923–1927).* Oxford: Oxford University Press.

Einstein, Albert. 1925a. Quantentheorie des einatomigen idealen Gases, Zweite Abhandlung. *Sitzungsberichte der Preußischen Akademie der Wissenschaften, Physikalisch-mathematische Klasse*, 3–14.

Einstein, Albert. 1925b. Zur Quantentheorie des idealen Gases. *Sitzungsberichte der Preußischen Akademie der Wissenschaften, Physikalisch-mathematische Klasse*, 18–25.

Elsasser, Walter M. 1925. Bermerkungen zur Quantenmechanik freier Elektronen. *Die Naturwissenschaften*, 13(33), 711. https://doi.org/10.1007/BF01558853

Franck, James and Sponer, Hertha. 1928. Beitrag zur Bestimmung der Dissoziationsarbeit von Molekülen aus Bandenspektren. *Nachrichten von der Gesellschaft der Wissenschaften zu Göttingen, Mathematisch-physikalische Klasse*, 1929, 241–253.

Gavroglu, Kostas and Simões, Ana. 1994. The Americans, the Germans, and the beginnings of quantum chemistry: The confluence of diverging traditions. *Historical Studies in the Physical and Biological Sciences*, 25(1), 47–110. https://doi.org/10.2307/27757735

Götschel, Helene. 2011. The entanglement of gender and physics: Human actors, work place cultures, and knowledge production. *Science & Technology Studies*, 24(1), 66–80. https://doi.org/10.23987/sts.55270

Gutierrez, Michael. 2006. *Hertha Sponer*. New York: Great Neck Publishing.

Hund, Friedrich. 1922. *Versuch einer Deutung der grossen Durchlässigkeit einiger Edelgase für sehr langsame Elektronen*. PhD thesis. Göttingen: Georg-August-Universität Göttingen.

Hund, Friedrich. 1923. Theoretische Betrachtungen über die Ablenkung von freien langsamen Elektronen in Atomen. *Zeitschrift für Physik*, 13, 241–263. https://doi.org/10.1007/BF01328216

Hund, Friedrich. 1927a. Zur Deutung der Molekelspektren I. *Zeitschrift für Physik*, 40, 742–764. https://doi.org/10.1007/BF01400234

Hund, Friedrich. 1927b. Zur Deutung der Molekelspektren II. *Zeitschrift für Physik*, 42, 93–120. https://doi.org/10.1007/BF01397124

Hund, Friedrich. 1967. *Geschichte der Quantentheorie*. Mannheim: Bibliographisches Institut.

Jähnert, Martin. 2019. *Practicing the Correspondence Principle in the Old Quantum Theory. A Transformation through Implementation*. Archimedes: New Studies in the History and Philosophy of Science and Technology, Volume 56. Cham: Springer.

Kemble, Edwin C. and Birge. n.d. Raymond T. Correspondence, 1925–1927. Archive for the History of Quantum Physics, M/f 50 Sections 5–7.

Kynch, George J. K. and Penney, George. 1941. On the heat of sublimation of carbon and on some absorption bands of three mesomeric hydrocarbons. *Proceedings of the Royal Society of London A*, 179, 214–231. https://doi.org/10.1098/rspa.1941.0089

Lemmerich, Jost. 2011. *Science and Conscience: The Life of James Franck*. Stanford, California: Stanford University Press.

Maushart, Marie-Ann. 1997. *"Um mich nicht zu vergessen": Hertha Sponer – ein Frauenleben für die Physik im 20. Jahrhundert*. Bassum: Verlag für Geschichte der Naturwissenschaften und der Technik.

Maushart, Marie-Ann. 2011. *Hertha Sponer: A Woman's Life as a Physicist in the 20th Century, "So you won't forget me."* Durham, NC: Department of Physics, Duke University. (Translation of Maushart (1997) by Ralph A. Morris, with added material by Annette Vogt and edited by Brenda P. Winnewisser.)

Medicus, Heinrich A. 1974. Fifty years of matter waves. *Physics Today*, 27(2), 38–45. https://doi.org/10.1063/1.3128444

Merzbacher, Eugen. 2002. The early history of quantum tunneling. *Physics Today*, 55(8), 44–49. https://doi.org/10.1063/1.1510281

Minkowski, Rudolf and Sponer, Hertha. 1924. Über den Durchgang von Elektronen durch Atome. *Ergebnisse der Exakten Naturwissenschaften*, 67–85. https://doi.org/10.1007/978-3-642-94260-0_4

Minkowski, Rudolph and Sponer, Hertha. 1923. Über die freie Weglänge langsamer Elektronen in Gasen. *Zeitschrift für Physik*, 15, 399–408. https://doi.org/10.1007/BF01330489

Pauli, Wolfgang (ed.). 1979. *Wolfgang Pauli: Scientific Correspondence with Bohr, Einstein, Heisenberg a.o., Vol. 1, 1919–1929*, von Meyenn, Karl (ed.), Berlin: Springer Verlag.

Ramsauer, Carl. 1921a. Über den Wirkungsquerschnitt der Gasmoleküle gegenüber langsamen Elektronen. *Physikalische Zeitschrift*, 22, 613–615.

Ramsauer, Carl. 1921b. Über den Wirkungsquerschnitt der Gasmoleküle gegenüber langsamen Elektronen. *Annalen der Physik*, 64, 513–540.

Ramsauer, Carl. 1921c. Über den Wirkungsquerschnitt der Gasmoleküle gegenüber langsamen Elektronen I. Fortsetzung. *Annalen der Physik*, 66, 546–558.

Ramsauer, Carl. 1923. Über den Wirkungsquerschnitt der Gasmoleküle gegenüber langsamen Elektronen II. Fortsetzung und Schluss. *Annalen der Physik*, 72, 345–352.

Rechenberg, Helmut. 2006. Hertha Sponer. In Byers, Nina, and Williams, Gary (eds.), *Out of the Shadows: Contributions of Twentieth-Century Women to Physics*. Cambridge: Cambridge University Press, pp. 127–136.

Rice, Oscar K. 1930. A contribution to the quantum mechanical theory of radioactivity and the dissociation by rotation of diatomic molecules. *Physical Review*, 35, 1538–1550. https://link.aps.org/doi/10.1103/PhysRev.35.1538

Rice, Stuart A. and Jortner, Joshua. 2010. *James Franck 1882–1964: A Biographical Memoir*. National Academy of Sciences.

Rossiter, Margaret. 1982. *Women Scientists in America. Struggles and Strategies to 1940*. Baltimore, MD: Johns Hopkins University Press.

Rouzé, Michel. 2024. J. Robert Oppenheimer. *Encyclopedia Britannica*. https://www.britannica.com/biography/J-Robert-Oppenheimer

Segrè, Emilio. 1976. *Personaggi e Scoperte nella Fisica Contemporanea*. Milan: Mondadori.

Segrè, Emilio. 1980. *From X-rays to Quarks: Modern Physicists and Their Discoveries*. San Francisco, CA: W. H. Freeman and Company. (Translation of Segrè (1976).)

Simões, Ana and Gavroglu, Kostas. 2001. Issues in the history of theoretical and quantum chemistry, 1927–1960. In Reinhardt, C. (ed.), *Chemical Sciences in the 20th Century: Bridging Boundaries*. Wiley-VCH, ch. 3. https://doi.org/10.1002/9783527612734.ch03

Sponer, Hertha. 1923. Über freie Weglänge langsamer Elektronen in Edelgasen. *Zeitschrift für Physik*, 18, 249–257. https://doi.org/10.1007/BF01327705

Sponer, Hertha. 1931. Bandenspektren und Dissoziation. In Debye, Peter (ed.), *Molekülstruktur. Leipziger Vorträge 1931*. Leipzig: S. Hirzel, pp. 107–130.

Sponer, Hertha. 1935. *Molekülspektren und Ihre Anwendung auf Chemische Probleme. Vol. 1*. Berlin, Heidelberg: Springer.

Sponer, Hertha. 1936. *Molekülspektren und Ihre Anwendung auf Chemische Probleme. Vol. 2*. Berlin: Springer.

Sponer, Hertha and Franck, James. 1962a. Interview of James Franck and Hertha Sponer Franck by Thomas S. Kuhn and Maria Goeppert Mayer on 11 July 1962. Niels Bohr Library & Archives. https://www.aip.org/history-programs/niels-bohr-library/oral-histories/4609-3

Sponer, Hertha and Franck, James. 1962b. Interview of James Franck and Hertha Sponer Franck by Thomas S. Kuhn and Maria Goeppert Mayer on 12 July 1962. Niels Bohr Library & Archives. https://www.aip.org/history-programs/niels-bohr-library/oral-histories/4609-4

Sponer, Hertha and Franck, James. 1962c. Interview of James Franck and Hertha Sponer Franck by Thomas S. Kuhn and Maria Goeppert Mayer on 13 July 1962. Niels Bohr Library & Archives. https://www.aip.org/history-programs/niels-bohr-library/oral-histories/4609-5

Sponer, Hertha and Franck, James. 1962d. Interview of James Franck and Hertha Sponer Franck by Thomas S. Kuhn and Maria Goeppert Mayer on 14 July 1962. Niels Bohr Library & Archives. https://www.aip.org/history-programs/niels-bohr-library/oral-histories/4609-6

Tobies, Renate. 1996. Physikerinnen und spektroskopische Forschungen: Hertha Sponer (1895–1968). In Meinel, C. and Renneberg, M. (eds.), *Geschlechterverhältnisse in Medizin, Naturwissenschaft und Technik*. Bassum, Germany: Verlag für Geschichte der Naturwissenschaften und der Technik, pp. 89–97.

Townsend, John S. and Bailey, Victor A. 1921. The motion of electrons in gases. *The London, Edinburgh, and Dublin Philosophical Magazine and Journal of Science*, 42(252), 873–891. https://doi.org/10.1080/14786442108633831

Vogt, Annette. 2007. Die Lehrbeauftragte Margot Sponer. In Vogt, Annette (ed.), *Vom Hintereingang zum Hauptportal? Lise Meitner und ihre Kolleginnen an der Berliner Universität und in der Kaiser-Wilhelm-Gesellschaft*. Series Pallas & Athene, vol. 17. Stuttgart: Franz Steiner.

4

Angular and Career Momentum: What Lucy Mensing Contributed to Physics and Why She Left the Field

GERNOT MÜNSTER AND MICHEL JANSSEN

4.1 Introduction

In November 1927, Paul Ehrenfest wrote to Samuel Goudsmit, who had finished his PhD in Leiden under Ehrenfest's supervision earlier that year:

Please tell me – *in confidence*! – about the future plans of London and Mensing or of other young theorists who catch your attention and are *sympathetic human beings* (general idea: people one could invite to Leiden for a few months).[1]

Understandably, Ehrenfest, born in 1880, hoped that Goudsmit, born in 1902, could identify (agreeable) theorists of his generation to further develop the new quantum mechanics, jokingly called *Knabenphysik* ("boys' physics"), as many of its architects were born right after the turn of the century: Wolfgang Pauli in 1900, Werner Heisenberg in 1901, Paul Dirac and Pascual Jordan in 1902, and John von Neumann in 1903. While maybe not a household name, Fritz London, born in 1900, is well known to this day among both physicists and historians of science. There is even a full-scale biography of him (Gavroglu, 1995). The other young theorist mentioned by Ehrenfest, however, has largely been forgotten. Yet, as this letter suggests and this chapter will confirm, Lucy Mensing, born in 1901, was a highly respected member of the fraternity of young theorists in the early days of quantum mechanics. We used the word "fraternity" on purpose because of the gender bias it shares with

We owe a debt of gratitude to Dr. Dorothea Roloff, Lucy Schütz-Mensing's youngest daughter, for sharing manuscripts, letters, and photographs with us and for providing additional biographical information as well as detailed feedback on drafts of this paper. We also thank Dieter Hoffmann, Ted Jacobson, Arne Schirrmacher, an anonymous referee, and the editors of this volume, especially Daniela Monaldi, for helpful comments. The work of one of us (MJ) was supported by a Visiting Fellowship of the Lichtenberg Group for History and Philosophy of Physics led by Dennis Lehmkuhl at the University of Bonn.
[1] "Berichte mir bitte VERTRAULICH!!!!! über die nächsten Zukunftspläne von London und Mensing oder von anderen jungen Theoretikern, die Dir auffallen und die MENSCHLICH SYMPATISCH sind. (Allgemeine Idee: Leute, die man für einige Monate nach Leiden einladen könnte.)" Ehrenfest to Goudsmit, November 19, 1927 (we tried to preserve Ehrenfest's idiosyncratic orthography). Samuel A. Goudsmit papers. Niels Bohr Library & Archives, American Institute of Physics (hereafter: Goudsmit Papers), Box 2, Folder 11, Chronological correspondence, 1927.

the term *Knabenphysik*. While this term was used to refer to the age of this cohort, it was clearly taken for granted that all its members were male.

Mensing earned her PhD in Hamburg in early 1925 with Wilhelm Lenz, though she mostly worked with his assistant Pauli. Mensing then spent a year in Göttingen, where she interacted with Max Born and his young collaborators Heisenberg and Jordan during the gestation period of matrix mechanics. Mensing was the first to apply matrix mechanics to diatomic molecules, using the new rules for the quantization of angular momentum. As a byproduct, she found that, even though in general both integer and half-integer values are allowed for angular momentum, orbital angular momentum always takes on integer values (Mensing, 1926a). Impressed by her clear and masterful treatment of the problem, Pauli invited Mensing to collaborate with him on the susceptibility of diatomic gases, another problem involving the quantization of angular momentum (Mensing and Pauli, 1926).

After Göttingen, Mensing spent a year in Tübingen as assistant to Alfred Landé. Landé had made important contributions to the old quantum theory of Niels Bohr and Arnold Sommerfeld, his PhD adviser, but was then overshadowed by the young whippersnappers of the new quantum mechanics. In Tübingen, Mensing met the experimentalist Wilhelm Schütz, only about half a year older than she was, who had come from Frankfurt am Main, his home town, with Walther Gerlach, his PhD adviser. Gerlach and Landé soon had a falling-out but their assistants became a couple and married in 1928. The following year, Mensing, who saw little of Landé, as he worked mostly from home to avoid Gerlach, followed her husband and Gerlach to Munich. Mensing attended Sommerfeld's lectures on the new quantum theory and wrote what would turn out to be her last paper, published under the name Schütz-Mensing (1930).[2] In November 1930, the arrival of her first child (the couple would eventually have four children) marked the end of her independent research career, though she would occasionally still assist her husband in his research (see, e.g., Schütz, 1936).

So, did getting married and having children cause Mensing to give up her career in physics? Given the expectations of women in Germany and elsewhere in the 1920s, that would appear to be the obvious explanation. However, the scant documentary evidence we have suggests that in her case it was the other way around: the prospect of marriage and starting a family may have appealed to her because of mounting dissatisfaction with her own research and growing distaste of the ruthless attitude she saw in the physics circles she moved in. Even in fragmentary reminiscences written decades later, she still expressed frustration over her failed efforts in Tübingen to account for the Ramsauer effect, a research topic

---

[2] She later dropped her maiden name. We will refer to her as Lucy Schütz when describing her life (or identifying documents she wrote) after she got married.

suggested to her by Pauli. This frustration already comes through in the paper in which she reported the negative result of this project (Mensing, 1927). Moreover, one of her few surviving letters from the 1920s suggests that she did not have the stomach for the prevailing cutthroat competitive mentality of many of her male counterparts. Sometime in 1927, Landé demanded that his assistant work on the spectrum of bismuth even though he knew Goudsmit was working on the same topic. Mensing and Schütz had come to know Goudsmit when he spent several months in Tübingen in 1926–1927. While Mensing did not feel comfortable disobeying her boss, she was taken aback by this order and wrote to Goudsmit that she thought it was "outrageous that he demands of me to meddle in your handiwork with bismuth."[3]

In this chapter, we take a closer look at Mensing's short but remarkable career in the early years of quantum mechanics and briefly cover her life, now as Lucy Schütz, before, during, and after World War II, which would take her and her family from Königsberg and Jena to Gorodomlya, an island near the city of Ostashkov in the Soviet Union, and back again to Jena, which had meanwhile become part of the new German Democratic Republic. Lucy Schütz lived to ripe old age. She died in 1995 at age 94. In the late 1980s she jotted down some reminiscences about the 1920s. We will quote several passages from these fragmentary recollections, which are interesting both for what they do and do not mention. This chapter draws heavily on an earlier article by one of us (Münster, 2020a).[4] Lucy Schütz's youngest daughter, Dr. Dorothea Roloff, provided much of the documentary basis for that article and has since shared additional material in her possession shedding light on the life and career of her mother.[5]

4.2 Student in Hamburg

Lucy (officially: Lucy Luise Martha) Mensing (see Figure 4.1) was born on March 11, 1901 in Hamburg, the first of four daughters of the merchant Hermann (officially: Wilhelm Hermann) Mensing and his wife Martha Luise (née Beer).[6]

---

[3] See Section 4.4 and Note 58 for the full quotation.
[4] This article is the main source for the Wikipedia page on Lucy Mensing, which seems to have brought her some belated recognition and increased visibility, as illustrated by the following three examples. In 2021, the Helmholtz-Zentrum Dresden-Rossendorf announced a new development fund (*Förderfonds*) named after Mensing to support parents, especially women, seeking to return to the academic work force after giving birth and taking care of children. In 2022, Amazon Braket launched a new quantum processor named Lucy. In 2023, the Deutsches Museum presented an "augmented reality walk of fame" for important female scientists, which includes Mensing and features a sculpture by Anke Schiemann titled "Mother Lucy" (*Mutter Lucy*).
[5] We will refer to these materials as the "Mensing-Schütz Family Papers." They will eventually be deposited in the archives of the University of Jena.
[6] Lucy Schütz, *Lebenslauf*, undated but after August 1992. In the header of this document, she gives an address in Meiningen, where she moved into an assisted living facility in August 1992 to be close to her oldest daughter Cornelie.

Figure 4.1 Two (cropped) pictures of Lucy Mensing, demure in 1927 (left), grittier in 1928 (right). It would have better fitted our story had the latter been from 1925 to 1926. Mensing-Schütz Family Papers.

Martha had been sent to Hamburg from Costa Rica, where her father ran a coffee plantation, to attend the St. Johannis monastery school, especially known for its program to train women to become teachers (*Lehrerinnenseminar*). Lucy went to the same school as her mother, graduating in February 1920. Stimulated by her father and by an inspiring teacher, she decided to study physics, a subject she had already excelled at in school.[7] In April 1920, she matriculated at the University of Hamburg, which had been founded only the year before, registering for classes in physics, mathematics, and chemistry. Gifted in mathematics, she naturally gravitated toward theoretical physics during her studies.

Her most important teachers in Hamburg were Wilhelm Lenz and Wolfgang Pauli.[8] Lenz, a student and then an assistant of Sommerfeld in Munich, became the first professor for theoretical physics in Hamburg in 1921. He is known, among other things, for the Laplace–Runge–Lenz vector in the Kepler problem and the Lenz–Ising model of ferromagnetism. Lenz proposed this model the year before he came to Hamburg (Lenz, 1920) and had his student Ernst Ising work on it for his dissertation (Ising, 1924).

Pauli came to Hamburg as Lenz's assistant in the summer 1922. They had met in Munich, when Lenz was an assistant and Pauli a student of Sommerfeld. In 1924,

[7] Based on information provided by Dr. Dorothea Roloff in private correspondence. See also the short vita (*Lebenslauf*) in her dissertation (Mensing, 1925a).
[8] On Lenz and his collaborators in Hamburg, see Reich (2011).

Pauli did his habilitation (the final hurdle to be cleared in the German system, then and now, for a lectureship and eventually a professorship) and, in November 1926, received the title of professor.[9] While based in Hamburg, Pauli also spent considerable time in Copenhagen and Göttingen. In 1928, he left Hamburg to take up a proper professorship at Albert Einstein's alma mater, the Eidgenössische Technische Hochschule (ETH) in Zurich.

Mensing and Ising became friends during their time together in Hamburg. They lost touch in the 1930s but reestablished contact in 1959.[10] Because of his Jewish ancestry, Ising lost his job when the Nazis came to power and, in 1939, he and his wife relocated to Luxemburg. They were initially left alone but, in 1943, Ernst was forced into labor for the German army. In 1947, the Isings emigrated to the US, where Ernst ended up teaching physics at Bradley University in Peoria, Illinois. In 1972, a few months after the death of her husband, the Isings visited Mensing, now Schütz, in Jena.[11] She visited them in the US in 1981.

The following year, they met again in Kiel, where Lucy's sister Hildegard (the second of the four sisters) lived. In the Kunsthalle they admired paintings by Emil Nolde and others.[12] In a letter later that year, she told her friends about a visit to Museum Ludwig in Cologne, which had brought back memories of their student days in Hamburg:

The painting "Wild Boars" of Marc reminded me of the Hamburg "Mandrill". It used to be taunted and mocked in my youth. Then the professor for literature, in a class we, i.e., Ernst and I, attended once a week, discussed the Mandrill – the colorful animal in the lush jungle – and we saw it with different eyes. I assume that one was also by Franz Marc. The painting is said to have disappeared during the war. Had the Nazis already taken it earlier and sold it to the US?[13]

Franz Marc did indeed paint "The Mandrill" (see Figure 4.2) as well as "Wild Boars," both in 1913. In 1937, the Nazis confiscated "The Mandrill" from the Hamburger Kunsthalle, where Mensing saw it as a youngster, along with dozens of other paintings deemed "degenerate art" (*entartete Kunst*).

---

[9] Despite important differences between the German and the American university system, we will translate *Privatdozent* and *außerordentlicher Professor* as "lecturer" and "associate professor," respectively. The German professor is roughly the equivalent of an American full professor.

[10] Ernst Ising to Lucy Schütz, November 7, 1959. Mensing-Schütz Family Papers.

[11] Hanna and Ernst Ising to Lucy Schütz, August 10, 1972. Mensing-Schütz Family Papers.

[12] See Lucy Schütz to Wilhelm Hanle, November 3, 1982. Mensing-Schütz Family Papers. Emil Nolde (the artist name of Emil Hansen) was the model for the painter Max Ludwig Nansen in Siegfried Lenz's classic novel, *The German Lesson* (*Deutschstunde*), about obeying orders of the Nazi regime at all cost. Unlike the character in the novel, however, Nolde was a staunch supporter of the Nazi regime, even though his paintings were considered "degenerate art" (Fulda et al., 2019).

[13] "Bei einem Bild "Wildschweine" von Marc fiel mir der Hamburger "Mandrill" ein. Der wurde in meiner Jugend beschimpft und lächerlich gemacht. Dann hat der Lit.-Professor bei dem wir, d.h. Ernst und ich, einmal in der Woche ein Kolleg hörten, den Mandrill besprochen – das farbenprächtige Tier in dem üppigen Urwald – da sah man es mit anderen Augen an. Ich nehme jetzt an, daß der auch von Franz Marc war. Das Bild soll im Krieg verschwunden sein. Ob die Nazis es schon früher beseitigt und nach Amerika verkauft hatten?" Lucy Schütz to Ernst and Hanna Ising, August 1982. Mensing-Schütz Family Papers.

Figure 4.2 Franz Marc, "The Mandrill." Pinakothek der Moderne, Munich. Source: Wikimedia Commons.

After two semesters in Hamburg, Mensing spent two semesters at the University of Freiburg (see the vita mentioned in Note 7). She returned to Hamburg for the summer semester of 1922. In September 1922, she used a family visit to attend the *Naturforscherversammlung*, held in Leipzig that year.[14] During the winter semester 1923–1924, she wrote an article about diatomic molecules within the framework of the Bohr–Sommerfeld theory. She only submitted this article (Mensing, 1925b) to *Zeitschrift für Physik* in September 1925, along with an excerpt of her dissertation (Mensing, 1925c). By that time, the old quantum theory was already starting to give way to the new matrix mechanics.

The topic Mensing was assigned by Lenz for her dissertation (Mensing, 1925a) was the broadening of spectral lines due to the Stark effect of the atomic or molecular electric fields in a gas, again within the framework of the old quantum theory. There was one problem: Lenz's fragile health. In the summer semester of 1925 and the winter semester of 1925–1926, Lenz was on sick leave and Pauli and another assistant, Gregor Wentzel, had to take over his classes (Reich, 2011, pp. 129–130).

---

[14] Lucy Schütz to Walther and Ruth Gerlach, March 2, 1976 (see Note 87).

In his obituary for Lenz, Pascual Jordan[15] recalled the unconventional measures Lenz took to avoid any kind of draft, often to the secret amusement of students and colleagues (Jordan, 1957). With Lenz out of commission, it was Pauli who mostly supervised Mensing's work (Mehra and Rechenberg, 1982a, pp. 187–193). Even though they were close in age, Mensing thus always saw and talked about Pauli as a mentor rather than a peer.

On March 30, 1925, the doctoral degree was conferred upon Mensing. For her thesis and in all three of her oral exams, she obtained the highest possible grade. On Lenz's recommendation, she was awarded "1st prize" for her doctorate, which came with a stipend in the amount of 500 Mark. Earlier in 1925, she had presented the results of her dissertation at a meeting in Göttingen.[16] As she recorded in some handwritten recollections in 1989:

Since I had become acquainted with life in Göttingen at the meeting (a lot more contact between professors and students), I used the prize money for an additional year of study in Göttingen. I arrived at Easter 1925, right around the time that the new quantum theory started with Heisenberg. Interesting time! in physics![17]

Mensing "received some extra money from Prof. Born," she recalled in a letter she wrote in 1977, "and lived this one year at the outer frontier of progress."[18]

4.3 Postdoc in Göttingen

While Göttingen had long been a renowned center of science – maintaining its strong tradition in experimental physics with James Franck and Robert Pohl and an even stronger tradition in mathematics with David Hilbert and Richard Courant – its Institute for Theoretical Physics was younger still than Mensing's alma mater, the University of Hamburg. The institute had just opened its doors in 1922 with Max Born as its founding director, who came to Göttingen from Frankfurt (Greenspan, 2005). Within a few years, however, Göttingen rivaled Copenhagen

---

[15] Jordan replaced Pauli as Lenz's assistant before he himself became professor in Rostock in 1928. With Lenz's strong endorsement, he returned to Hamburg in 1947, where he spent the rest of his career (Ehlers and Schücking, 2002, p. 72). See Hoffmann and Walker (2007) for discussion of Jordan's involvement with the Nazis in between these two periods in Hamburg.

[16] Probably the meeting of a local chapter (Gauverein Niedersachsen) of the German Physical Society (Deutsche Physikalische Gesellschaft) in February 1925.

[17] "Da ich das Leben bei der Göttinger Tagung kennengelernt hatte (viel mehr Kontakte zwischen Professoren u. Studenten), benutzte ich das Geld (die Prämie) für ein weiteres Studienjahr in Göttingen, kam Ostern 1925 gerade zu der Zeit, als die neue Quantentheorie durch Heisenberg begonnen hatte. Interessante Zeit! in der Physik!" Lucy Schütz, handwritten manuscript with the heading "Recollection January 1989" (*Rückerinnerung Januar 89*), five pages, the last four numbered 2–5, followed by two more (both numbered 6) with a pair of afterthoughts (see Notes 70 and 71). Mensing-Schütz Family Papers (see Note 5).

[18] "bekam von Prof. Born noch etwas dazu und lebte das eine Jahr am vorderen Rand des Fortschritts." Carbon copy of part of Lucy Schütz to Hilda Mothes (the wife of Kurt Mothes, president of the Academy Leopoldina in the German Democratic Republic, 1954–1974), August 8, 1977. Mensing-Schütz Family Papers.

and Munich as a leading center for the development of quantum theory (Schirrmacher, 2019).

The grounds for the transition from the old quantum theory of Bohr and Sommerfeld to matrix mechanics, the first incarnation of the new quantum mechanics, were actually prepared in Göttingen by Born and his young collaborators Heisenberg and Jordan.[19] The work that would eventually lead to the breakthrough combined sophisticated techniques from celestial mechanics, a natural tool to use given that atoms were modeled as miniature solar systems in the old quantum theory, and an extended version of Bohr's correspondence principle. The basic strategy was to analyze some specific or generic mechanical system classically first and then translate the results into quantum formulas under the guidance of the correspondence principle. The most successful application of this strategy was a formula for optical dispersion, the phenomenon familiar from prisms and rainbows, first proposed by the Dutch physicist Hans Kramers (1924a, b), Bohr's right-hand man in Copenhagen, and then derived independently by Born (1924) in Göttingen and John Van Vleck (1924) in Minneapolis (Duncan and Janssen, 2007). Everyone involved, however, recognized the preliminary character of the theory as it stood.

In late 1924 and early 1925, on leave from his position as Born's assistant in Göttingen for an extended visit to Bohr's institute in Copenhagen, Heisenberg, who had earned his PhD with Sommerfeld in Munich, collaborated with Kramers on a lengthy exposition and expansion of the latter's dispersion theory (Kramers and Heisenberg, 1925). Back in Göttingen in April 1925, Heisenberg continued to develop the ideas suggested by his work in Copenhagen. In June 1925, during his now legendary trip to the barren island of Helgoland to seek relief from an attack of hay fever, he laid the basis for his *Umdeutung* (reinterpretation) paper (Heisenberg, 1925). Together with his Göttingen colleagues Born and Jordan, he then turned this into the theory superseding the old quantum theory that they had been groping toward.

The basic idea of the *seminal* (another gender-biased term!) *Umdeutung* paper was to retain the form of the laws of classical mechanics but to replace the numbers occurring in them, all referring to individual orbits, by arrays of numbers referring to *transitions between* orbits, labeled by the quantum numbers of the initial and final one. The unobservable orbits, which had become increasingly problematic in the early 1920s, were thus replaced by transition frequencies and transition probabilities, which could actually be observed as spectral lines and light intensities, respectively. Here Heisenberg was following Born and Jordan (1925a), who had

---

[19] For detailed analysis of the emergence and consolidation of matrix mechanics, see Duncan and Janssen (2023, chs. 10–12).

already insisted that any satisfactory theory should be formulated solely in terms of observable quantities.

A key feature of Heisenberg's new scheme was that his arrays of numbers satisfied an odd non-commutative multiplication rule (which means that $A \times B \neq B \times A$). Born quickly realized that these arrays of numbers were nothing but matrices and that this multiplication rule was nothing but the standard rule for matrix multiplication. Born and Jordan (1925b) thus reformulated Heisenberg's scheme in terms of matrices and replaced the quantization condition at the heart of it, based on the Kramers dispersion formula, by the now familiar commutation relations for position and momentum.[20] Later in 1925, the three men joined forces to systematize and extend the theory and produced what is known as the *Dreimännerarbeit* (Three-Man-Paper), published early the following year (Born et al., 1926). Whereas modern students of quantum mechanics are likely to be baffled by Heisenberg's *Umdeutung* paper, they will have no trouble recognizing many textbook results in the *Dreimännerarbeit*, which can be seen as the first authoritative exposition of matrix mechanics.

Mensing arrived in Göttingen in April 1925, right around the time that the new quantum theory was beginning to take shape. She also stayed in touch with Pauli, who closely followed and contributed to the proceedings. In fact, Heisenberg had stopped off in Hamburg on his way back from Helgoland to Göttingen to show his friend and fellow Sommerfeld student what he had wrought. In Göttingen, Mensing attended lectures on theoretical physics by Born and a talk by Heisenberg on the new matrix mechanics.[21] She thus learned about the latest developments as they were happening. Especially important for her own contributions was the treatment of angular momentum in the *Dreimännerarbeit* (Born et al., 1926, ch. 4). This treatment is based on the commutation relations for the components of angular momentum, which follow directly from the fundamental commutation relations for position and momentum.

Angular momentum played an important role in the interpretation of atomic and molecular spectra. At the suggestion of Jordan and probably Born, Mensing took it upon herself to apply the new rules for angular momentum in matrix mechanics to diatomic molecules such as oxygen (O_2), nitrogen (N_2), and carbon monoxide (CO). Given that she had already studied diatomic molecules in the context of the old quantum theory (Mensing, 1925b), it was only natural for her to revisit them in the context of the new quantum theory.

As Jordan told Thomas S. Kuhn in an interview in 1963:

---

[20] As Alfred Landé, who missed out on these developments, put it during a lunch break in an interview with Thomas Kuhn and John Heilbron in 1962 (see Note 22), with a touch of professional jealousy: "Heisenberg stammered something. Born made sense of it" (Duncan and Janssen, 2023, p. 276).

[21] Lucy Schütz, undated handwritten recollection, under the heading *Umschrift*. Mensing-Schütz Family Papers.

Back then Miss Mensing was a physicist who was also working in Göttingen and who, at our suggestion, made an application, a first application, of the new quantum mechanics, to, I believe, a diatomic molecule. Then Pauli applied it to hydrogen.[22]

The paper in which Pauli (1926) presented his derivation of the Balmer formula for the hydrogen spectrum, using purely algebraic methods, was published in January 1926. Mensing's (1926a) paper on diatomic molecules was submitted two months later. Yet, if Jordan's memory can be trusted, her application predated Pauli's. But even if Jordan got the order wrong, it is still true that the distinction of having made the *second* application of matrix mechanics to a physical system goes to Mensing.

The infrared spectrum of molecules had been known since the early twentieth century. It consists of a number of equally spaced spectral lines with the central line missing. Nernst (1911) and Bjerrum (1912) had already recognized in 1911 that these spectral lines originate from transitions, in which the molecule changes its vibrational and rotational state simultaneously. The energy gap between different vibrational states is much larger than the energy gap between different rotational states. Transitions in which both the vibrational and the rotational state change therefore result in what are called rotational bands.

Vibrations and rotations of a diatomic molecule can be treated by considering the molecule as a two-body system with a central potential $V(r)$, which has a minimum at an equilibrium distance a between the two atoms. For small oscillations the potential is approximated in leading order by a potential of the same form as that of a harmonic oscillator.

In the framework of the Bohr–Sommerfeld theory, the energy of a diatomic molecule is given, in leading order, by the sum of the vibrational (harmonic-oscillator) energy $E_{\text{vib}} = \hbar \omega n$ (where $\hbar \equiv h/2\pi$ is the reduced Planck constant, ω is 2π times the oscillation frequency and n is the quantum number characterizing the oscillator state) and the rotational energy $E_{\text{rot}} = L^2/2ma^2$ (where L is the angular momentum and m is the reduced mass). In the old quantum theory, the angular momentum is an integer multiple of $\hbar$:

$$L = \hbar l, \quad \text{with } l \in \{0, 1, 2, 3, \ldots\} \tag{4.1}$$

So the energy in leading order is

---

[22] "Frl. Mensing war damals eine Physikerin, die auch in Göttingen arbeitete, und die hat auf unsere Empfehlung hin eine Anwendung, eine erste Anwendung, gemacht von der neuen Quantenmechanik, [auf], ich glaube, ein zweiatomiges Molekül. Und Pauli hat dann die Anwendung auf Wasserstoff gemacht." Interview with Pascual Jordan by Thomas S. Kuhn, Hamburg, June 18, 1963, as part of the project to establish the Archive for the History of Quantum Physics (AHQP) (see Kuhn et al., 1967). The transcript has "*auch*" instead of "*auf*" but this is probably just an error, due perhaps to Jordan's well-known stutter.

$$E_{nl} = \hbar \omega n + \frac{\hbar^2 l^2}{2ma^2} \tag{4.2}$$

In transitions with electric dipole radiation, the quantum number n decreases by one unit, and the angular momentum quantum number l changes by one unit:

$$n \to n - 1, \, l \to l \pm 1 \tag{4.3}$$

The Bohr frequency condition states that the radiation emitted in a transition from one state to another is proportional to the energy difference between them. The transition in Eq. (4.3) should therefore result in two sets of spectral lines, on opposite sides of a (missing) central line, which would be produced by a (forbidden) transition in which only n changes. For transitions with $l \to l - 1$ ($l \geq 1$), the lines should be shifted from the center proportional to $l^2 - (l-1)^2 = 2l - 1$; for transitions with $l \to l + 1$ ($l \geq 0$), they should be shifted proportional to $l^2 - (l+1)^2 = -2l - 1$. This, however, did not match the experimental findings. Instead, l^2 in the expression for the rotational energy had to be replaced ad hoc by $(l + 1/2)^2$ (see, e.g., Reiche, 1920; Kratzer, 1922; Czerny, 1925). Thus, half-integer quantum numbers had to be used, which could not be explained.

It fell to Mensing in Göttingen in late 1925/early 1926 to sort out this problem in the framework of the new matrix mechanics.

The Bohr–Sommerfeld theory described the motion of particles in terms of classical trajectories subject to quantization conditions.[23] For bound orbits these imply the quantization of angular momentum, $L = \hbar l$, with $l \in \{1, 2, 3, \ldots\}$. However, in the period before the arrival of the new quantum theory, angular momenta of magnitude $\hbar/2$ had also been proposed in various contexts. Landé (1921) and Heisenberg (1922) had speculated about half-integer angular momenta to account for empirical observations in atomic spectra, in particular those concerning the so-called anomalous Zeeman effect (see below). They were both thinking in terms of orbital angular momenta and the considerations of both of them were in conflict with various accepted principles (Cassidy, 1979; Duncan and Janssen, 2019, ch. 7). And it was not just atomic spectra. As mentioned above, the rotation–vibration spectra of diatomic *molecules* likewise seemed to call for half-integer angular momenta.[24]

---

[23] See Duncan and Janssen (2019, Part 2 and Appendix A) for a detailed analysis.

[24] Half-integer quantum numbers were also proposed to account for the problem of the specific heat of molecular hydrogen (Gearhart, 2010).

With the advent of matrix mechanics, light was thrown on these puzzles. In their *Dreimännerarbeit*, Born et al. (1926) presented the commutation relations between the three components L_i of the angular momentum operator:

$$[L_1, L_2] \equiv L_1 L_2 - L_2 L_1 = i\hbar L_3, \text{ and cyclically permuted.} \tag{4.4}$$

From this algebra they were able to deduce the eigenvalues of the squared angular momentum $\vec{L}^2$:

$$\hbar^2 l(l+1)$$

where the quantum number l is allowed to take on integer and half-integer values:

$$l \in \left\{ 0, \frac{1}{2}, 1, \frac{3}{2}, 2, \frac{5}{2}, 3, \dots \right\} \tag{4.5}$$

At first blush, this might seem to open the door to half-integer orbital angular momenta and thereby solve the aforementioned problems.

However, Mensing, who learned about these results before they were published in Göttingen, showed that orbital angular momentum could only take on integer values. As Jordan (1975) recalled:

Lucy Mensing, who was working on first applications of the new quantum mechanical method at the Göttingen Institute at the time, was able to prove that an electron, which could be described as a mass point, could only have integer values.[25]

So, one has

$$l \in \{0, 1, 2, 3, \dots\} \text{ for orbital angular momentum.} \tag{4.6}$$

This result actually fits perfectly with the idea of electron spin, which gives the electron an *intrinsic* half-integer angular momentum of $\hbar/2$. Ralph Kronig first hit upon this idea in Tübingen in January 1925 but was talked out of it by Pauli, who happened to be visiting and thus robbed Kronig of the credit for a signature discovery. In November 1925, the same idea occurred to and was published by two students of Ehrenfest in Leiden, George Uhlenbeck and Samuel Goudsmit (1925).[26] Mensing's result and electron spin were the key ingredients needed to explain a number of hitherto puzzling properties of atomic and molecular spectra with and without external fields.

---

[25] In another article Jordan (1973, p. 297) similarly stated: "Already in the 'Dreimänner Arbeit,' Born–Heisenberg–Jordan, we saw that the eigenvalues of a component of the rotational impulse might be whole numbers or possibly half numbers $\pm\frac{1}{2}, \pm\frac{3}{2}, \dots$ It was from a study of Lucy Mensing that we learned, that from the model of the electron as a mass point only whole numbers $0, \pm 1, \pm 2$ and so on could arise. Indeed, Pauli came to this result also when he studied the hydrogen atom with the method of quantum mechanics."

[26] For discussion, see Duncan and Janssen (2023, sec. 9.2).

The restrictions on the allowed values of l result from algebraic relations that orbital angular momentum has to satisfy in addition to the algebra in Eq. (4.4). For instance, the angular momentum operator

$$\vec{L} \equiv \vec{Q} \times \vec{P} \tag{4.7}$$

where $\vec{Q}$ and $\vec{P}$ are the position and momentum operators, respectively, has to obey the algebraic relations $\vec{Q} \cdot \vec{L} = 0$ and $\vec{P} \cdot \vec{L} = 0$.

Mensing (1926a) obtained her results algebraically in the framework of matrix mechanics. In her article, the rule for the quantum number l of orbital angular momentum is not given as a theorem, but embedded in the matrix calculations and just stated in the text. For the case of the hydrogen atom, Pauli (1926) also found that only integer values of l occur, but his result is derived by means of the Laplace–Runge–Lenz vector and thus only holds for this specific system. Mensing's result is completely general.[27]

Equipped with her results for the necessary matrix elements and the properties of angular momentum, Mensing turned to the vibrational–rotational energy spectrum of diatomic molecules. In lowest order, the vibrational energy

$$E_{\text{vib}} = \hbar\omega(n + \tfrac{1}{2}) \tag{4.8}$$

just changes by a constant. In the formula for the rotational energy, however, the factor l^2 has to be replaced by $l(l+1)$, so that

$$E_{\text{rot}} = \frac{\hbar^2 l(l+1)}{2ma^2} \tag{4.9}$$

Put together, this yields

$$E_{nl} = \hbar\omega(n + \tfrac{1}{2}) + \frac{\hbar^2 l(l+1)}{2ma^2} + \text{const.} \quad \text{(leading order)} \tag{4.10}$$

When the selection rules in Eq. (4.3) are taken into account, the resulting spectrum fits the experimental results, and due to the relation $l(l+1) = (l+1/2)^2 - 1/4$ the explanation for the seemingly half-integer quantum numbers was found.

---

[27] It should be mentioned that in textbooks one often finds an argument for the integer nature of the orbital angular momentum that postulates the uniqueness of the wave function. This argument, however, is wrong, as explained, for example, by Pauli (1939). For a correct algebraic proof, see Born and Jordan (1930) and Münster (2020b, pp. 163–165).

Mensing did not just calculate the leading-order expression for the energies. She also calculated the intensities of the spectral lines, which are proportional to the squares of the corresponding matrix elements. These are important for the comparison with the experimental results. Furthermore, she went beyond the lowest order by means of quantum-mechanical perturbation theory. These calculations included corrections due to the coupling between oscillations and rotations and anharmonic-oscillator contributions. She also showed, again in agreement with the experimental data, that there would not be a linear Stark effect, the leading term in the splitting of spectral lines under the influence of an electric field. She submitted the article with all these results to *Zeitschrift für Physik* on March 9, 1926, where it was published the following November (Mensing, 1926a).

Apparently, using the new quantum mechanics to explain the spectrum of diatomic molecules was a natural thing to do. Several other physicists published on the topic in 1926. In his second paper on wave mechanics, Erwin Schrödinger (1926) chose it as one of the applications of his new theory. He calculated the energy eigenvalues, including some of the perturbative corrections without anharmonic terms. These were later added by his assistant, Erwin Fues (Fues, 1926). Igor Tamm (1926) obtained the rotational energies and line intensities to leading order by means of matrix mechanics. In his first scientific paper, the 18-year-old Lev Landau (1926) calculated energies and intensities in leading order with matrix mechanics as well and considered the Stark and Zeeman effects. Diatomic molecules were also the topic of J. Robert Oppenheimer's (1926) first paper. His starting point was Dirac's (1925) algebraic formulation of quantum mechanics, and he obtained the energies and intensities, including higher-order rotational terms but no anharmonic terms. His paper, however, contained unfounded assumptions about half-integer quantum numbers for angular momentum.

By contrast, Mensing's work not only came first, her investigation was also more systematic, complete, and thorough, taking into account contributions beyond leading order, calculating intensities, and demonstrating the absence of the linear Stark effect.[28]

Pauli – famous for his sharp and often sarcastic criticism of both his peers (Jordan, Kramers, . . .) and his elders (Born, Bohr, . . .) – was sufficiently impressed by Mensing's clear and expert handling of this subject that he invited her to collaborate with him on another problem in the quantum theory of molecules (Mehra and Rechenberg, 1982a, p. 191). This problem, the electric susceptibility of gases consisting of diatomic molecules, once again revolved around the quantization of angular momentum.

---

[28] For further discussion, see Mehra and Rechenberg (1982a, pp. 187–193, 1982b, pp. 236–244, 266–272, 1987, pp. 689–699, 842–850) and Cassidy (2007).

In classical physics, as Peter Debye (1912) showed, the contribution to the electric susceptibility of diatomic gases that depends on the temperature is proportional to $\mu^2 N/kT$ (with μ the molecular dipole moment, N the particle density, k Boltzmann's constant, and T the temperature). The proportionality constant C in the classical theory takes on the value 1/3. When Pauli (1921) calculated this susceptibility in the old quantum theory, however, he found a more complicated function, which gave a much higher value for the prefactor C at high temperatures, where the correspondence principle led one to expect quantum and classical theory to give the same result. In a 1963 interview with Kuhn (see Note 22), the mild-mannered Van Vleck derided Pauli's result as "wonderful nonsense" (Midwinter and Janssen, 2013, p. 140).

The matrix elements that Mensing (1926a) had obtained in her work on diatomic molecules could be employed to address the problem in the new quantum mechanics.[29] The calculations revealed some crucial cancelations, and Mensing and Pauli (1926) found an expression for the susceptibility, in which the prefactor C was restored to its classical value of 1/3 at high temperatures, so that the correspondence principle was saved.

Their joint publication represents another triumph for the new quantum mechanics. As Van Vleck wrote in the opening sentence of his classic book on susceptibilities: "The new quantum mechanics is perhaps most noted for its triumphs in the field of spectroscopy, but its less heralded successes in the theory of electric and magnetic susceptibilities must be regarded as one of its great achievements" (Van Vleck, 1932, p. vii).[30] Again, other physicists, in this case Van Vleck (1926) and Kronig (1926), independently found the same result but Mensing and Pauli were the first to publish it (Midwinter and Janssen, 2013, pp. 165–166). Moreover, both Van Vleck and Kronig cited Mensing's (1926a) earlier paper for the critical matrix elements.

Mensing had already left Göttingen when she and Pauli submitted their paper and they probably wrote most of it in Hamburg. Mensing was back in Hamburg by Easter, in early April 1926, and the paper was only submitted in late June. A month later, Mensing submitted yet another paper from Hamburg, the final one of this extremely productive period (Mensing, 1926b). It deals with the Paschen–Back effect, which had played an important role in the eventually unsuccessful attempts to explain the Zeeman effect in the old quantum theory (Duncan and Janssen, 2019, sec. 7.3).

---

[29] Mensing and Pauli (1926, p. 510) explicitly cited Mensing (1926a).
[30] Van Vleck felt so strongly about this result that it features prominently in his 1977 Nobel lecture (Midwinter and Janssen, 2013, p. 138).

What the experimentalist Pieter Zeeman had discovered in Leiden in 1896–1897 and what the theorist Hendrik Antoon Lorentz, his colleague and Ehrenfest's illustrious predecessor, had been able to explain on the basis of his classical electron theory was that a spectral line splits into three lines when the atom emitting the light is placed in a magnetic field (Kox, 1997). This won the two Dutch physicists the 1902 Nobel Prize. Not long thereafter, however, it was found that a magnetic field tends to split spectral lines into more than three components. This was called the *anomalous* Zeeman effect even though it was more common than the *normal* Zeeman effect with three components. In the context of the Bohr–Sommerfeld theory, experimentalists provided ever more accurate data on these line splittings in different atoms and theorists came up with ever more elaborate schemes to explain them. The Paschen–Back effect provided important guidance for these efforts. Friedrich Paschen and Ernst Back (1912, 1913) discovered that, in sufficiently strong magnetic fields, the complicated Zeeman splittings reduce to the simple triplets of Lorentz's original theory. As a result of all these efforts, the phenomenology of the Zeeman effect was well understood by the early 1920s, but a consistent mechanical model remained elusive. It would take the introduction of electron spin and the new quantum mechanics to successfully account for the Zeeman effect on the basis of spin–orbit coupling.

In a paper taking care of some of the failures of the old quantum theory and shoring up its successes, Heisenberg and Jordan (1926), using purely matrix-mechanical methods, were able to account for the anomalous Zeeman effect including the return of "normal" Lorentz triplets in strong magnetic fields found by Paschen and Back (Duncan and Janssen, 2023, sec. 15.3.1). Building on their paper, Mensing attacked the *partial* Paschen–Back effect, that is, the absence in some cases of lines in these normal triplets. Her experience with matrix mechanics enabled her to calculate the necessary matrix elements and provide a convincing explanation of the effect.[31]

All in all, it had been a wise decision by Mensing to use the prize money for her dissertation to spend a year in Göttingen. She had written one spectacular paper (Mensing, 1926a), co-authored a second one (Mensing and Pauli, 1926), and capped this extraordinarily productive period with another solid contribution (Mensing, 1926b).

---

[31] To mark the centenary of the births of both Paschen and Zeeman the year before, Wilhelm Schütz (1966), an expert on magneto-optics, published an article that focuses on the Zeeman effect. He did not mention the contributions of Heisenberg and Jordan, nor those of his wife.

4.4 Assistant in Tübingen

After her year in Göttingen, Mensing went back to Hamburg, where she planned to take the state exam needed to become a high-school teacher. She had considered this option even before she started working toward her doctorate (perhaps after her two semesters in Freiburg). In the 1977 letter from which we already quoted in Section 4.2, she explained why she had decided against it at that point. She had been led to believe "that it would take 36 years to get a permanent position at a school. That later turned out to be incorrect but I was missing the third subject for the state exam."[32] Instead, as we have seen, she earned her doctorate and spent a year in Göttingen "at the outer frontier of progress" (see Note 18).

Given that she had been an active participant in these developments, one may wonder why she did not consider habilitation, the next rung on the academic ladder, as many men with a comparable record probably would have. In fact, it made perfect sense for her to consider the state exam instead. As a woman, she stood little to no chance of ever being offered a professorship (not that this was easy for her male counterparts; see Kojevnikov, 2020). Taking the state exam and becoming a teacher would at least allow her to stay involved in physics.[33] So we can understand why Mensing insisted that "back home, I wanted to get serious about the state exam."[34] Then, however, an opportunity to continue working at the forefront of physics, at least for a while, presented itself.

Landé asked Mensing to become his assistant in Tübingen. The position was paid for by the *Notgemeinschaft der Deutschen Wissenschaft*. It was only for a year (see Note 36) and did not come with the prospect of habilitation. Yet, Mensing dropped her plan to take the state exam and accepted Landé's offer. Given her paper on the Paschen–Back effect (Mensing, 1926b), it was only fitting for her to go to Tübingen, where the effect had been discovered. Moreover, Landé himself had made his mark through work on the Zeeman effect, of which the Paschen–Back effect is a special case.

Under the directorship of Friedrich Paschen, Tübingen had become the most important center for spectroscopy in Germany. Had it not been for Pauli, it would also have been the birthplace of spin. In 1920, Paschen's student Ernst Back returned to Tübingen as his assistant. Two years later, on Paschen's recommendation, Landé came from Frankfurt to Tübingen as an associate (*ausserordentlicher*) professor (Brandmüller, 1989, p. 330). Landé had earned his doctorate in Munich

---

[32] "daß man erst in 36 Jahren im Schuldienst ankommen würde, Das hat nachher nicht gestimmt, mir aber fehlte fürs Staatsexamen das 3. Fach." Lucy Schütz to Hilda Mothes, August 8, 1977 (see Note 18).

[33] Another option would have been a job in industry. This was the route taken by Elisabeth Bormann, only six years older than Mensing. In the period 1919–1921, she had been an assistant of Born in Frankfurt. In 1921, Bormann got a job at the *Siemens-Schuckertwerke* in Berlin, where she spent the rest of her career.

[34] "Wieder zu Hause, wollte ich mich ernstlich ans Staatsexamen machen" (see Note 32). Mensing also mentioned this in two notes with reminiscences (see Notes 17 and 21).

with Sommerfeld right before the Great War and had, shortly after it ended, become an instructor (*Privatdozent*) in Frankfurt with Born, while teaching music in a nearby high school. In 1925, Paschen became president of the Physikalisch-Technische Reichsanstalt (Imperial Physical–Technical Institute) in Berlin and his former student Walther Gerlach, meanwhile associate professor in Frankfurt, succeeded him as full professor and director of the institute in Tübingen.

Pauli, Mensing's mentor, fully expected her to thrive in Tübingen. In June 1926, on a postcard to Landé, he wrote (rather condescendingly from today's perspective[35]): "I heard from Miss Mensing that you intend to get her as an assistant. I am very happy about that, it will be very good for her!" (Hermann et al., 1979, doc. 135). Landé also had high expectations of Mensing's appointment. This is clear from a letter he wrote to Goudsmit in July 1926. Unlike Pauli, who emphasized that Tübingen would be good for Mensing, Landé emphasized that her presence would enrich Goudsmit's planned research visit to Tübingen:

It is particularly fortunate that, in addition to you, Miss Dr. Mensing will spend a year in Tübingen, as she is really up to date on all new and old quantum methods; her computational skills will certainly fit well with your ideas.[36]

The first few months in Tübingen do indeed seem to have gone well for Mensing. As Landé envisioned, she and Goudsmit hit it off.[37] Home for the holidays in Hamburg in December 1926, she wrote to him to wish him happy new year and share some news of the profession:

Pauli already told me the other day that this spatial quantization business is already in Van Vleck. He gleefully pointed out that this sort of thing would happen to me more often, if I do not read foreign journals. He also told me a little about his own work. He has finished an article on electrons in metals. Heisenberg and Hund seem to have found a whole bunch of neat things about molecules. Heisenberg apparently also introduced "spinning" H nuclei, with which some fine structure and the specific heat of rotation[al degrees of freedom] can be taken care of. And it should also somehow enforce ⟨for H nuclei⟩ whether Fermi or Bose

---

[35] Despite his association with Chien-Shiung Wu, Pauli's attitude toward female physicists was hardly enlightened. According to Rudolf Peierls (Peierls, 1988, p. 73), Pauli "always suspected that they only studied physics to find a husband." He relates how at a conference Pauli "met a female theorist who had worked with him in the past" and, when he found out she was married, remarked: "So you succeeded after all?" (*Ist es Ihnen also doch noch gelungen?* (Peierls, 1988, p. 73). We strongly suspect Mensing was the physicist in question.

[36] "Es ist besonders günstig, dass ausser Ihnen noch Frl. Dr. Mensing für ein Jahr nach Tübingen kommt, die ja in allen neuen und alten Quantenmethoden sehr auf der Höhe ist, und deren Rechenkunst zusammen mit Ihren Ideen gewiss gut zusammenpassen wird." Goudsmit papers (see Note 1), Box 2, Folder 08, Chronological correspondence, 1926. In the same letter, Landé congratulated Goudsmit on the introduction of electron spin, "which I initially did not even consider worthy of discussion" (*das ich zuerst gar nicht für diskutierbar hielt*). As noted above, when Kronig made the same suggestion in Tübingen in January 1925, Pauli brusquely dismissed it.

[37] One of her daughters still owns the copy of a German edition of *One Thousand and One Nights* illustrated by Edmund Dulac that Goudsmit gave to Mensing with the handwritten dedication "Lucy Mensing Tübingen Dezember 1926 von Sam Goudsmit." Dorothea Roloff to Gernot Münster, February 3, 2020.

statistics apply. These last things I do not have from Pauli directly but by way of Mr. Estermann,[38] who does not understand any of it. So no guarantees! Pauli has urged me to keep doing calculations for the Ramsauer effect, but with a proper model, such as an idealized He atom. I hope it works; I have not tried it yet.[39]

The specifics do not matter for our purposes. We quote this letter – the only one we have of Mensing from this period – because it shows how well informed and how excited she was about these developments. In the coming year, 1927, she would lose the connection to the research frontier as well as her enthusiasm for her own research. An important factor in this was that the Ramsauer effect proved to be a much harder nut to crack than Pauli expected.

Looking back on her time in Tübingen decades later, Lucy Mensing, now Schütz, seems to have forgotten that her first few months there cannot have been all bad. In 1977 she wrote:

Then came the offer of the position in Tübingen paid for by the *Notgemeinschaft*. There I miserably muddled through alone for a year and a half. I still had the topic suggested to me by Pauli [the Ramsauer effect] and it was not going as planned. The idea to take a train to Zurich to pay him a visit never occurred to me and I did not have the money for it anyway. In 1928 I got married. As a woman one could then no longer have a job.[40]

In this section, we try to reconstruct, on the basis of what little source material we have, how the forward momentum in Mensing's career generated in Göttingen in 1925–1926 stalled in Tübingen in 1927.

Mensing arrived in Tübingen in August 1926. In her 1989 recollections, she sketched the constellation she encountered there (see Figure 4.3):

The situation in Tübingen was as follows: Gerlach ruled the institute; he had about 25–28 doctoral students. His only assistant was Schütz. Landé, the associate professor, had a room on the ground floor, in which a second desk was placed for me. As a common room there was the reading room next to the library, which was so small that sometimes two Japanese

---

[38] Immanuel Estermann, who was working with Otto Stern in Hamburg at the time.

[39] "Daß die Richtungsquantelungssache schon bei Van Vleck steht, sagte Pauli mir neulich schon! Er hat sehr schadenfroh erklärt, dergleichen würde mir öfter passieren, wenn ich keine ausländische Zeitschriften lese. – Ausserdem hat er mir von seinen Arbeiten einiges kurz angedeutet: Er selbst hat eine Arbeit über Elektronen in Metallen fertig. Heisenberg und Hund zusammen sollen allerhand hübsches über Moleküle gemacht haben. Dann soll Heisenberg "spinning" H-Kerne eingeführt haben, womit einige Feinstrukturen, dann aber auch die Rotationswärme von H_2 in Ordnung kommen sollen. Ferner soll sich damit auch irgendwie zwangsmässig ⟨für H-Kerne⟩ einmal die Fermi und einmal die Bose-Statistik ergeben. Die letzten Dinge weiss ich übrigens nicht von Pauli direkt, sondern auf dem Umweg über Herrn Estermann, der nichts davon versteht; also ohne Garantie! Pauli hat mir sehr geraten, am Ramsauereffekt weiter zu rechnen, aber mit einem richtigen Modell, etwa einen idealisierten He-Atom; hoffentlich geht es! Versucht habe ich es noch nicht." Lucy Mensing to Samuel Goudsmit, December 30, 1926. Goudsmit Papers (see Note 1), Box 2, Folder 09, Chronological correspondence, 1926.

[40] "Da kam das Angebot einer Notgemeinschaftsstelle in Tübingen. Dort habe ich 1 1/2 Jahr ziemlich kläglich allein gewurstelt. Ich hatte die Arbeit noch von Pauli, und sie lief nicht, wie erwartet. Auf die Idee, mal zu ihm nach Zürich zu fahren, bin ich nicht gekommen, hatte auch nicht das Geld dafür. 1928 habe ich geheiratet. Da durfte man als Frau schon nicht mehr berufstätig sein." Lucy Schütz to Hilda Mothes, August 8, 1977 (see Note 18).

Figure 4.3 Tübingen, c. 1928. Lucy Mensing (3rd from left), Wilhelm Schütz (2nd from right), Alfred Landé (right). American Institute of Physics, Emilio Segrè Visual Archives.

students studied their books in a squatting position, which I marveled at! Then there was some office space for Dr. Back.[41]

Wilhelm Schütz, born in 1900, the year before Mensing, was Gerlach's first PhD student in Frankfurt, Wilhelm's hometown. He earned his doctorate in 1923 and followed Gerlach to Tübingen two years later. Schütz was working on atomic spectroscopy. In the analysis of his spectroscopic measurements, Schütz (1926, p. 274) had referred to the PhD thesis of Mensing (1925c). Now that they were both in Tübingen, they presumably discussed spectroscopy with one another.[42] In any event, their contact soon blossomed into a romantic relationship.

For a while, Wilhelm Hanle, whom Mensing had met and befriended in Göttingen, where he had earned his doctorate with James Franck in 1924,[43] was

[41] "Die Situation in Tübingen war folgende: Gerlach beherrschte das Institut, hatte wohl etwa 25–28 Doktoranden. Sein einziger Assistent war Schütz. Landé als A[uß er]O[rdentlicher]-Professor hatte ein Zimmer im Erdgeschoß, in das für mich ein 2. Tisch gestellt wurde. Als gemeinsamen Raum gab es nur das Lesezimmer neben der Bibliothek, das so klein war, daß gelegentlich 2 japanische Studenten in Hockstellung Bücher studierten! was ich bestaunte! Dann hatte Dr. Back noch einen Arbeitsraum." Lucy Schütz, Recollection, January 1989 (see Note 17).
[42] Schütz (1927, p. 60) once again cited Mensing's PhD thesis. As is clear from a parenthetical remark ("as was pointed out to me by Miss Mensing"), the two of them discussed the paper's subject matter.
[43] In an interview with Brenda P. Winnewisser in 1979, Hanle recalled that Mensing, whom he described as a student of Born, often helped him with theoretical exercises (*Übungen*) (Hanle, 1979).

also in Tübingen. In a letter of condolence after Wilhelm Schütz's death in 1972, he wrote about this period:

I only remember too well working together with your husband in Tübingen. He was Gerlach's assistant back then, I was merely a kind of adjunct assistant. Given Gerlach's volatility, Schütz was an important partner and a kind of calm pole in the institute. Of course, we got separated after just half a year, after Gerlach threw me out of the institute.[44]

In a letter to Wilhelm Schütz in 1950, Hanle described Gerlach, by now the rector of the Ludwig Maximilian University in Munich, as "a veritable major bigwig" (*ein richtiger Grossbonze*), adding: "Since I don't care for *Grossbonzen*, I will never be on friendly terms with him again."[45] Schütz did not exactly rally to the defense of his *Doktorvater*: "Too bad about G.'s one-sided development. Your criticism, unfortunately, is all too justified."[46] Yet, despite these negative comments, Mrs. Schütz stayed in touch with the Gerlachs even after her husband's death (see Note 87).

Continuing her 1989 recollections, Lucy Schütz managed to paint, with just a few brush strokes, an endearing if not entirely accurate portrait of Back, with whom she had probably felt some solidarity at the time:

[He was] an overly modest person, definitely older than Gerlach, who would always apologize to the students whenever he had to give one of Gerlach's lectures. He had done an important experiment, in the field of optics I believe [interlineated: Paschen–Back effect]. The reason he was so inhibited was probably that he actually was a lawyer, who had developed an interest in physics and, even though he had not finished his physics education, was given a position by Paschen, which he then kept under Gerlach.[47]

On the next page, she wrote: "I now remember again: the Paschen–Back effect," which explains the interlineated amendment in the passage above. Given that, as we saw in Section 4.3, she had published on the Paschen–Back effect herself

---

[44] "Wie gut erinnere ich mich noch an die gemeinsame Tätigkeit mit deinem Gatten in Tübingen. Er war damals Assistent von Gerlach, ich nur ein Art Hilfsassistent. Bei Gerlachs Sprunghaftigkeit war mir Schütz ein wichtiger Helfer und eine Art ruhiger Pol im Institut. Freilich sind wir dann schon nach einem halben Jahr auseinander gekommen, nachdem mich Gerlach aus dem Institut geworfen hat." Wilhelm Hanle to Lucy Schütz, May 1, 1972. Mensing-Schütz Family Papers. The letter also contains an intriguing passage about Hanle's main claim to fame: "This Fall I want to go the USA again. I have many invitations for colloquia about the 'Hanle effect,' *about which you also once had some ideas that you discussed with me on the side (an dem Du ja auch einmal an Rande mit mir zusammen einige Überlegungen angestellt hast)*. But the time was not ripe for it back then. In fact, it has only become clear to me last Fall, on the occasion of my trip to the USA, how much attention it is getting now" (emphasis added).

[45] "Da ich Grossbonzen nicht mag, werde ich auch nie wieder mit ihm warm werden." Wilhelm Hanle to Wilhelm Schütz, May 3, 1950. Mensing-Schütz Family Papers.

[46] "Schade, daß sich G. so einseitig entwickelt hat. Ihre Kritik ist leider nur allzu berechtigt." Wilhelm Schütz to Wilhelm Hanle, August 31, 1950. Mensing-Schütz Family Papers.

[47] "ein ganz übertrieben bescheidener Mensch, sicher älter als Gerlach, der sich immer bei den Studenten entschuldigte, wenn er anstatt Gerlach die Vorlesung halten muß te. Er hatte ein wichtiges Experiment gemacht, ich glaube, auf dem Gebiet der Optik. Er war wohl so gehemmt weil er eigentlich Jurist war, sich dann für Physik interessiert hatte und wohl ohne abgeschlossenes Physikstudium diesen Posten von Paschen bekam und bei Gerlach behielt." Recollection, January 1989 (see Note 17).

(Mensing, 1926b), this is an odd memory lapse. And while it is true that Back started out as a lawyer, he gave up that profession to study physics and it was actually his work on the Paschen–Back effect that, contrary to what the passage above suggests, earned him a doctorate in 1913. He returned to Paschen and Tübingen in 1920 for his habilitation, which he obtained in 1923 (Brandmüller, 1989, p. 330).

Mensing seems to have been unaware of a conflict Landé, Back, and Sommerfeld had been embroiled in a few years earlier over Landé's use of Back's data on the Zeeman effect. This episode is worth recalling here, as it illustrates the competitive nature and the hierarchical command structure of the field that bothered Mensing. It is covered in much greater detail in a classic paper by Paul Forman (1970, secs. 5–7), which also corroborates Mensing's brief sketch of Back's character.[48]

Back had first met Landé in 1919, when he gave a colloquium in Frankfurt about the Zeeman effect. By 1921, the two men, independently of one another, had hit upon similar schemes to improve on Sommerfeld's efforts to bring order to the baffling array of spectroscopic data emanating from Tübingen on the Zeeman effect. Both had to worry about preserving their priority. As Forman (1970, pp. 157–158) emphasizes, their advancement in the academic physics community was at stake, habilitation for Back, the step from *Privatdozent* to a professorship for Landé.

The opening salvo of the resulting kerfuffle was a letter of March 3, 1921, in which Sommerfeld leaned hard on his former student Landé to abstain from publishing his new rules for the Zeeman effect until Back had finished his paper on the same topic. As can be inferred from this letter, Back (in a letter that is no longer extant) had asked Sommerfeld to help him secure his priority. Wanting to protect the flow of spectroscopic data from Tübingen to Munich, Sommerfeld obliged and fired off a letter to Landé. Within days of receipt of Sommerfeld's missive, Landé wrote to Back, who replied that he had no objection to Landé publishing first. Paschen told him the same thing, even signaling some impatience on his part with the pace of his protegé's progress. But Back's generosity was misleading. In another letter to Sommerfeld (also no longer extant), Back once again made it clear that he was distraught about Landé poaching on his preserves. Sommerfeld thereupon tried to press Born, until recently Landé's boss in Frankfurt, into service to help him bring Landé to heel. Writing to Born, Sommerfeld let on what can barely be read between the lines in his letter to Landé, namely that he felt bad about his own use of unpublished data from Tübingen in the past as Back might only reluctantly have given him permission to do so, just as he had in Landé's case.

---

[48] The appendix of Forman's paper helpfully collects all relevant correspondence. See Brandmüller (1989) for another concise version of this story.

While not prepared to do Sommerfeld's bidding, Born did inform Landé of his colleague's displeasure.

Landé meanwhile had found new and better rules for the Zeeman effect. He related these to Bohr and Sommerfeld in a pair of letters of March 17, 1921. He assured Sommerfeld that he had both Back's and Paschen's blessings to proceed. Unfortunately, when Sommerfeld found Landé's letter upon his return from a vacation, Landé's refusal to back down only enraged him more. In his response of March 31, he reiterated that Back, whatever he might have told Landé, was and should be concerned about his priority and added that he, Sommerfeld, would take it as a personal affront should Landé go ahead and publish anyway. To further put Landé in his place, Sommerfeld sneered that a first-semester student of his (Heisenberg, as it turned out) had found – but, as he added ominously, not published! – essentially the same results that Landé now reported. This intelligence, in fact, only made it more urgent for Landé to get his new rules in print, which he did in defiance of Sommerfeld (Landé, 1921).

By the time Landé took up his associate professorship in Tübingen in 1922, he and Back seem to have buried the hatchet that Sommerfeld had dug up for them (Brandmüller, 1989, pp. 330–331). In 1923, the latest results of their respective investigations of the Zeeman effect appeared in the same issue of *Zeitschrift für Physik* (Landé, 1923; Back, 1923). But while his conflict with Back had been resolved, the new associate professor soon found himself at loggerheads with the new director of the Tübingen institute. This must have been uncomfortable for Mensing and Schütz, their respective assistants, especially once they had become a couple.

For Mensing, a petty incident early on in the hostilities stood out:

The first collision between G[erlach] and L[andé] that I experienced happened after a meeting in Stuttgart. G. took a taxi on the way back and took Landé and me, among others,[49] with him. L. asked to get out at his place, which G. refused because of the many loose pieces of luggage still lying in the car. Some irritated back and forth ensued but G. did not relent. Whether there were any further, more substantial differences, earlier or later, I do not know. L. no longer set foot in the institute and worked from home.[50]

---

[49] Two more passengers are identified in another note: Wilhelm Schütz and Ruth Probst, who later became Gerlach's second wife (see Note 87). This note, dated January 27, 1989, and with the heading "Explanation to the letter from Landé" (*Erklärung zum Brief von Landé*), is about a letter from Landé of May 5, 1969, addressed to Mr. and Mrs. Schütz (see Note 72). Lucy Schütz, Mensing-Schütz Family Papers.

[50] "Der erste Zusammenstoß zwischen G. u. L., den ich miterlebte, geschah nach einer Tagung in Stuttg. G. hatte ein Taxi für die Rückfahrt genommen, nahm u.a. Landé u. mich mit. L. bat, an seiner Wohnung aussteigen zu dürfen, was G. wegen des vielen Gepäcks, das noch lose im Auto lag, ablehnte. Es gab ärgerliches hin u. her. G. gab nicht nach. Ob noch weitere, wesentlichere Differenzen vorher oder nachher vorlagen, weiß ich nicht. L. betrat das Inst. nicht mehr, arbeitete zu Hause." Recollection, January 1989 (see Note 17). In another note (see Note 49), she recalled that Landé "came only through a special entrance to give his lectures" (*kam nur durch den Spezial-Eingang zu seiner Vorlesung*).

One can still sense her indignation in the comment (in quotation marks) added to that last sentence: "Never mind that he could not talk to his assistant this way!"[51]

Landé, she recalled, "rarely demanded anything of me. Occasionally he would ask me to read some pages of his book. We would then discuss it. But it was minimal."[52] In her 1989 recollections, she wondered: "Did the book appear? Around 27/28/29?" (see Note 17). Landé (1926) actually did publish a book on recent developments in quantum theory during Mensing's time in Tübingen, even if it did not make a big impression on his assistant (or on other quantum physicists for that matter). She remembered mostly working alone (she had Landé's office to herself) on the Ramsauer effect, which was causing her considerable grief. But first we turn to the one other task we know Landé assigned to Mensing, much to her chagrin.

During his research visit to Tübingen, which lasted from October 1926 to April 1927,[53] Goudsmit had worked with Back on the hyperfine structure of the spectrum of bismuth, interpreting it as the result of the interaction of the total electronic angular momentum with an angular momentum that they attributed to the nucleus (Goudsmit and Back, 1927). Landé was enthusiastic about this work. As he wrote to Goudsmit, now in Copenhagen, in May 1927:

I heartily congratulate you on your beautiful joint papers with Back. It is really wonderful that the Zeeman effect now also provides information about the angular momentum of the nucleus. When you get to America, I am sure the spectroscopists there will immediately start working in this field and supply you with new material.[54]

Back and Goudsmit (1928) would submit another paper on their joint project in December 1927, which would explain the plural "papers". By that time, Goudsmit had already moved to Ann Arbor, Michigan, where he spent the rest of his career.

After Goudsmit had left Tübingen, Landé asked Mensing to look into an idea that had occurred to him in connection with the work of Back and Goudsmit. Mensing was reluctant to do so. As she wrote to Goudsmit:

With regard to your and Back's work on the fine structure of bismuth, Prof. Landé has suggested that, even without the Zeeman effect in a weak field, which Back now wants to measure, one can draw conclusions about the size of the magnetic moment of the nucleus

---

[51] "kein Zustand, dass er seine Assistentin nicht sprechen könnte!" Source as for Note 50.

[52] "Prof. Landé verlangte selten etwas von mir. Gelegentlich bat er mich, Seiten seines Buches zu lesen. Darüber sprachen wir dann. – Aber das war bescheiden!" Lucy Schütz, untitled two-page document with recollections of 1925–1927. Mensing-Schütz Family Papers.

[53] See the biographical sketch in the preamble to Goudsmit's (1927) dissertation.

[54] "Zu den schönen gemeinsamen Arbeiten von Ihnen und Back beglückwünsche ich Sie sehr, es ist wirklich wunderschön, dass der Zeemaneffekt jetzt auch noch über die Kernmomente Auskunft gibt; wenn Sie nach Amerika kommen, werden gewiss die dortigen Spektroskopiker sich auch sofort auf dieses Gebiet stürzen und Ihnen neues Material liefern." Alfred Landé to Samuel Goudsmit, May 20, 1927. Goudsmit papers (see Note 1), Box 2, Folder 14, Chronological correspondence, 1927.

from the absolute size of the fine-structure splitting. Following Landé's wishes, I have started to calculate this but, before I spend more time on it and before something short might be published about it, I wanted to ask you whether you have not long calculated the same thing yourself. That actually seems likely to me since you already used most of the necessary formulas for the intensity calculations you planned to do in Copenhagen, as you wrote to Back.[55]

In the next paragraph, Mensing recorded some serious doubts about Landé's idea "because the effect of spin is not included and the model is thus almost certainly wrong."[56] In conclusion, she wrote: "I have discussed with Landé that I would write to you about the matter and, following his wishes, have done so immediately."[57]

Mensing shared this letter with Landé but not a second one she mailed to Goudsmit the very same day. In this second letter, she gave free reign to her resentment over Landé's orders:

For obvious reasons I am writing two letters to you today. The physics one I wrote in the presence of my boss, making sure he saw it. For the same reason, please be so kind as to answer these letters on separate postcards or sheets. About the physics I just want to add that I find it outrageous that L. demands of me to meddle in your handiwork with bismuth, Sam. Had it been up to me, I would never have done that. I almost had a fight with L. because I didn't want to do it and because I explained that I did not believe that with the complicated bismuth anything sensible would come out of it. But enough about physics.[58]

One wonders whether Landé would have been equally insensitive had his assistant been a man. Gender may well have been a factor in preventing him from seeing that he was abusing his position of power to lord it over Mensing, not unlike the way Sommerfeld had abused his to lord it over him back in 1921. (Their actual orders, of course, went in opposite directions: Sommerfeld told Landé to stay off Back's turf, at least for the time being; Landé told Mensing to get onto Goudsmit's.)

---

[55] "Zu deiner und Backs Wismut-feinstruktur Arbeit hat Prof. Landé sich überlegt, dass man auch ohne Zeemaneffekt bei schwachem Feld, was Back jetzt messen will, aus der absoluten Grösse der Feinstruktur Aufspaltung Schlüsse auf die Grösse des magnetischen Momentes des Kerns ziehen kann. Ich habe auf Landé's Wunsch die Sache zu rechnen angefangen, möchte aber, ehe ich viel mehr Zeit darauf verwende, und ehe etwa kurz drüber publiziert würde, Dich fragen, ob Du nicht schon längst dasselbe gerechnet hast. Mir scheint das eigentlich wahrscheinlich, da Du die hierfür nötigen Formeln doch grösstenteils auch für die Intensitätsberechnung gebraucht hast, die Du in Kopenh. machen wolltest, wie Du an Back schriebst." Lucy Mensing to Samuel Goudsmit, July 2, 1927 (first letter). Goudsmit papers (see Note 1), Box 2, Folder 14, Chronological correspondence, 1927.

[56] "da aber noch die Spinwirkung fehlt und das Modell doch sicher falsch ist." Source as for Note 55.

[57] "Ich habe mit Landé besprochen, dass ich Dir über die Sache schreiben würde und habe auf seinen Wunsch hin gleich geschrieben." Source as for Note 55.

[58] "Aus naheliegenden Gründen schreibe ich Euch heute lieber gleich 2 Briefe, den physikal. habe ich in Gegenwart meines Chefs geschrieben und auf alle Fälle so, dass er ihn sehen konnte. Ihr seid aus demselben Grunde vielleicht auch so nett, die Briefe auf getrennten Postkarten bzw. Bogen zu beantworten. Zu der Physik nur noch: Ich finde es empörend von L., dass er von mir verlangt Dir, Sam, bei dem Wism. ins Handwerk zu pfuschen. Ich hätte das aus freien Stücken nie getan. Ich hab schon fast Krach mit L. gehabt, weil ich es nicht wollte und weil ich es erklärte, bei dem komplizierten Bi glaube ich nicht, dass was vernünftiges raus kommen könne. Nun aber Schluss mit der Physik." Lucy Mensing to Samuel and Jaantje (neé Logher) Goudsmit, July 2, 1927 (second letter). Goudsmit Papers (see Note 55). In a later article on the topic, Goudsmit and Bacher (1930, p. 14) actually cited Mensing's (1926b) paper on the Paschen–Back effect.

The bismuth affair was not the only thing bothering Mensing about physics at the time. In the next paragraph of this second letter, she painted a rather disturbing picture of her current state of mind:

Except for physics, I am doing very well here. For a while I just lived from one Sunday to the next and only 'sat out' the time in between in the institute.[59]

After this blunt confession, she told the Goudsmits about a few outings that had lifted her spirits. One of these she described in strikingly euphoric terms:

In the meantime there also was a very nice physics excursion. I think Schütz may have sent you pictures of it. I was in high spirits and enjoyed it so much that I didn't know what to do with myself; good thing it lasted only three days, as those hardly could have been topped.[60]

We can infer from Landé's letter to Goudsmit of May 20 that this excursion had taken place about two months earlier, at the beginning of May:

Two weeks ago the whole physics institute had a three-day excursion to the Black Forest, which was very beautiful and a lot of fun but also pretty strenuous. The 100 kilometers and more would not have been up your alley and Back also did not join us in anticipation of the agonies; Miss Mensing, however, valiantly kept up with the 27 men.[61]

Figure 4.4 shows a beaming Mensing carrying a hefty backpack.[62] This picture was probably among the ones (she assumed) Schütz sent to the Goudsmits.

After describing this outing, Mensing concluded her letter to the Goudsmits a little more upbeat:

Right now, I would also enjoy some physics again, but I can't think of anything sensible to do. I just confided in Pauli about my physics woes and am waiting for his response.[63]

Note how sharply this letter contrasts with the exuberant letter she wrote to Goudsmit half a year earlier, when she was visiting Hamburg for the holidays (see Note 39). As we saw, Pauli had urged her at that point to keep working on the

---

[59] "Mir geht es bis auf die Physik hier recht gut. Zeitweilig habe ich nur von einem Sonntag zum anderen gelebt und die Zeit dazwischen im Institut nur 'abgesessen'." Source as for Note 58.

[60] "Inzwischen war auch einmal eine sehr nette physik. Exkursion. Schütz hat euch, glaube ich, Bilder davon geschickt. Ich wusste kaum mehr wohin vor Übermut und Lebenslust, und es war nur gut, dass es nur 3 Tage waren, da es kaum noch eine Steigerung gegeben hätte." Source as for Note 58.

[61] "Vor 2 Wochen hat das ganze phys. Institut einen 3 tägigen Ausflug in den Schwarzwald gemacht, der sehr schön und lustig, aber anstrengend war; für Sie wären die über 100 km nichts gewesen, und auch Back war nicht dabei in Vorahnung der Strapazen, dagegen hat Frl. Mensing mit den 27 Männern tapfer Schritt gehalten." Landé to Goudsmit, May 20, 1927 (see Note 54). Another picture of this trip allows us to determine its exact destination. This picture, which shows Mensing reclining in the grass in the foreground, has a shed in the background with a sign on its doors on which we can make out "Freudenstadt" and "Igelsberg" (now part of Freudenstadt).

[62] In the interview mentioned in Note 43, Hanle also recalled biking and hiking with Mensing during their time together in Göttingen: "I was also good friends with an assistant of Born, Lucy Mensing [the transcript has 'Menzing']. In the weekends we took long bike rides in the beautiful area around Göttingen and in the summer vacation we hiked in the Ortler region and the Rosengarten in South Tirol."

[63] "Augenblicklich hätte ich auch schon zu Physik etwas Lust, doch weiss ich eben nichts Vernünftiges zu machen. Ich habe gerade Pauli meine physikalische Not geklagt und warte auf Antwort von ihm." Source as for Note 58.

Figure 4.4 Excursion of Tübingen physicists to the Black Forest, early May 1927. In the foreground: Lucy Mensing and Wilhelm Schütz (with backpacks and their hands in their pockets on opposite sides of an unidentified individual). Second from the left: Alfred Landé (with glasses). Mensing-Schütz Family Papers.

Ramsauer effect. In 1989, she recalled that Pauli had suggested this topic to her even before she went to Tübingen in August 1926 (Recollection, January 1989, see Note 17). She must still have been at it when she wrote to the Goudsmits in July 1927. She would not submit her paper on the Ramsauer effect for another two months (Mensing, 1927).

The effect discovered by and named after Carl Ramsauer (1921) pertains to the extremely high permeability of gases for low-energy electrons and has its origin in a strong reduction of the relevant scattering cross section of slow electrons. The effect could not be explained classically and also failed to yield to attempts by Friedrich Hund and James Franck in Göttingen in 1922–1923 to account for it on the basis of the old quantum theory (Jähnert, 2019, ch. 5). Mensing had become familiar with Schrödinger's wave mechanics in the meantime, so in Tübingen she investigated the scattering of slow electrons off atoms by means of wave-mechanical scattering theory. Unfortunately, the problem turned out to be much more difficult than expected. As she vividly recalled in the two-page recollections quoted above (see Note 52):

I worked on the Ramsauer effect. That led to an essentially unknown mathematical function. I would start bright and fresh in the morning, calculating a piece of this function; in the evening, the result would be $x = \infty/\infty$. Fall 1927: I wrote up this negative attempt and sent it to *Zeitschrift für Physik* (Pauli was in Switzerland, I could not visit him there. But I could have written to him!). My note appeared in the same issue in which the successful paper of a Swedish team appeared on the same topic! (With the help of a calculating machine).[64]

She recorded the same story, including her self-admonishment for not contacting Pauli, in the seven-page document from which we already quoted repeatedly above:

It led to complicated mathematical functions (not tabulated). It was a tedious work calculating this and it did not lead to any usable result. Pauli was in Switzerland at the time, so it was impossible to visit him and get his advice. Landé did not understand any of it. Why did I not write to Pauli???[65]

Recall, however, that she told the Goudsmits in July 1927 that she *had* just written to Pauli (see Note 63)! Then again, his response, if there was one, may have been of little help, which would explain why, decades later, she no longer remembered having asked him for advice in the first place.

With Landé clueless and mostly absent, Mensing went home to Hamburg to write up her results and submit the paper. In her later recollections, she misremembered the year. She understandably thought she went home when "early 1928, Landé had an invitation to the USA" (see Note 17) and that her paper only appeared in the summer of 1928 rather than in the fall of 1927.

Mensing's paper admittedly did not provide a satisfactory explanation of the Ramsauer effect. Nevertheless, she did fully develop the methodology for tackling the problem within the framework of partial wave analysis and formulated an approximation for the cross section using hypergeometric functions. However, the numerical evaluation was only rudimentary and did not lead to useful results. The Swedish team she mentioned (Faxén and Holtsmark, 1927) got further than she did by using calculators, but also did not achieve a quantitative reproduction of the Ramsauer effect. Part of the failure was that they used the same unrealistic effective potential as Mensing. As she warned in her abstract: "the shell model proves to be

---

[64] "Ich x-te an dem Ramsauereffekt. Das führte auf eine nicht näher bekannte mathematische Funktion. Ich begann morgens mit frischem Mut, ein Stück davon auszurechnen, abends war das Ergebnis $x = \infty/\infty$! Herbst 27 Schrieb ich diesen negativen Versuch zusammen und schickte ihn an die Zsch. f. Phys. (Pauli war in der Schweiz, dort konnte ich ihn nicht aufsuchen. Aber ich hätte ihm doch schreiben können!) Meine Notiz erschien, im gleichen Heft die erfolgreiche Arbeit eines schwedischen Teams über das gleiche Thema! (Arbeit mit Rechenmaschine)." Recollections of 1925–1927 (see Note 52).

[65] "Es führte auf komplizierte mathem. Funktionen (nicht tabuliert). Es war langwierige Rechenarbeit und führte zu keinem brauchbaren Ergebnis. Pauli war damals in der Schweiz, so war es unmöglich, ihn zu besuchen u. Rat zu holen. – Landé verstand nichts von der Sache. – Warum schrieb ich nicht an Pauli???" Recollection, January 1989 (see Note 17).

much too crude an idealization for this problem" (Mensing, 1927, p. 603). So it is not as if she narrowly missed out on another major success. Yet, as the anguished recollections quoted above show, Mensing was quite crestfallen about this episode.

Coming right on the heels of the bismuth affair, the Ramsauer debacle must have made it even harder for Mensing to recover a sense of joy and satisfaction in physics, which she had clearly lost when she wrote to the Goudsmits in July 1927. It may thus not be a coincidence that early the following year, on April 8, 1928, she and Wilhelm Schütz got engaged. The prospect of getting married and starting a family may have struck her as an appealing alternative to a career in physics at this point. Economic circumstances and the general uncertainty in Weimar Germany at the time may also have played a role in this decision, even though they did not leave any traces in her later reminiscences. But disenchantment with physics seems to have been the more important factor. Independently of us, her youngest daughter came to a similar conclusion:

It is becoming increasingly clear to me that she did not leave science primarily out of a desire to start a family, but rather, conversely, that she started a family because she could not cope with whatever kind of fighting was going on in the scientific world – or did not take up the gauntlet to begin with.[66]

Goudsmit figured that Mensing would continue to do physics even after getting married and, as Goudsmit must have realized, without being employed in the field. When he heard about the engagement, he wrote to Gerlach: "I believe I should also congratulate you, since in this wonderful way you secured the services of such a capable theoretician."[67] As we will see, the capable theoretician (*die tüchtige Theoretikerin*) did manage to keep working on physics at first, but this would prove unsustainable.

The couple got married on September 1, 1928. The next day, she recalled,

I read in the paper that Wilhelm Wien had died and said: "now Gerlach will become his successor." And that indeed happened. At the end of 1929, Gerlach and the two of us moved to Munich. Whether or when Landé returned to Tübingen in 1929 I don't remember. I did hear, however, that he got a call to the USA.[68]

---

[66] Dorothea Roloff to Gernot Münster, February 16, 2024 (email message, our translation). In this message, Dr. Roloff also related two memories of her older siblings. Her sister Cornelie remembered that their mother "left Tübingen because she wanted no part in a scientific indecency (*Unsauberkeit*) that she rejected but felt forced to accept." This probably refers to the bismuth affair. Her oldest brother, Jürgen, now deceased, remembered that their mother had an offer to come to the USA in 1928 but decided to get married instead. We have been unable to find any traces of such an offer. Maybe she was not talking about herself but about Landé.

[67] "Ich glaube dass ich auch Ihnen damit gratulieren muss, denn jetzt haben Sie sich in so schöner Weise die Mitarbeit einer so tüchtigen Theoretikerin versichert." From Samuel Goudsmit to Walther Gerlach, June 7, 1928, copied in Ruth Gerlach to Lucy Schütz, March 30, 1980. Mensing-Schütz Family Papers.

[68] "Am nächsten Tag las ich in der Zeitung den Tod von W. Wien und sagte: `da wird Gerlach sein Nachfolger.' Das war dann wirklich so, Ende 29 zogen Gerl. u. wir nach München. Ob u. wann Landé 29 nach Tübingen zurück gekommen ist, erinnere ich nicht. Ich hörte aber, daß er eine Berufung nach den USA hatte." Lucy Schütz, recollection, January 1989 (see Note 17).

So the newlyweds left Tübingen without seeing Landé again. Yet, Lucy Schütz does not seem to have harbored hard feelings toward her former supervisor. And she realized that, in hindsight, it had been a stroke of good fortune for Landé that he had emigrated when he did – because of Gerlach rather than anti-Semitism.[69] In a note added to the passage above, she recorded a thought she said had occurred to her before (probably in the mid-1930s): "How lucky for the Landés. They are already accustomed to country and language! Other Jews still have to find a position etc."[70] In a second note she expressed regret that she had not made more out of her position as Landé's assistant:

Why did I not take theoretical physics again with Landé? I had taken it in Hamburg, with Lenz, freshly arrived from Sommerfeld, [interlineated: and with Born] and also with Pauli, the kinetic theory of gases. It could have led to a good contact between Landé and me, and, for sure, inspiration.[71]

These regrets late in life may well have been colored by the trials and tribulations visited upon her family during and after World War II. Maybe Mensing would have fared better in Tübingen had she and Landé seen more of each other, she by attending his lectures, he by working in their shared office. Then again, maybe not.

In 1969, the Schützes reached out to Landé, who was still at Ohio State University in Columbus, which he had first visited in 1929–1930 and made his permanent home after a second visit the following year. Landé was delighted to hear from them. In his response, he still mentioned the "Gerlach affair" but recognized its silver lining: "I am, however, very grateful to him, because he strongly contributed to me accepting the call to the USA."[72] The most telling passage in Landé's letter, however, is the following: "So the young couple from Tübingen is already emeritus; hard to believe when one looks at the old photos of the young physicist and the charming Lucy."[73]

Given that Landé was still buying into such gender stereotypes in 1969, one may well wonder how this affected his attitude toward his female assistant (*die charmante Lucy*) in the 1920s. A passage we left out when we quoted Landé's letter to Goudsmit of May 1927 above suggests that the attitude toward female physicists of Mensing's supervisor was similar to that of her teacher and mentor Pauli (see Note 35):

---

[69] See Häffner (1995) for a discussion of anti-Semitism at the University of Tübingen.

[70] "Ich hatte schon vorher gedacht: 'Was haben die Landés für ein Glück gehabt!['] Sind jetzt in Land u. Sprache eingelebt! Die anderen Juden müssen erst eine Arbeitsstelle suchen u.s.w.!" Source as for Note 68.

[71] "Ich denke jetzt: 'Warum habe ich nicht die theoret. Physik bei Landé nochmal gehört?' Ich hatte sie in Hamb. bei Lenz gehört, der frisch von Sommerfeld kam, [interlineated: u. bei Born] und daneben bei Pauli 'kinetische Gas-Theorie['] . Es hätte einen guten Kontakt geben können zwischen Landé u. mir, und sicher Anregung!" Source as for Note 68.

[72] "Aber ich bin ihm sehr dankbar, denn er hat stark dazu beigetragen, dass ich den Ruf nach USA annahm." Alfred Landé to Wilhelm and Lucy Schütz, May 5, 1969. Mensing-Schütz Family Papers.

[73] "Also das junge Paar aus Tübingen ist nun auch schon emeritiert, kaum zu glauben, wenn man die alten Photos mit dem jungen Physiker und der charmanten Lucy ansieht." Source as for Note 72.

I greatly envy you for the close collaboration with Bohr and Heisenberg [in Copenhagen]. Especially in the new quantum theory, one runs into unsolved issues every day, which one would love to discuss with a physicist likewise interested in these questions. I'll probably be traveling to Zurich for that purpose at Whitsuntide, to have a sensible conversation again with Mr. London.[74]

Whereas Ehrenfest,[75] as we saw at the beginning of this chapter, mentioned London and Mensing in the same breath, Landé wanted to travel to Zurich to discuss physics with the former while the latter was sitting at a desk in his own office in Tübingen! In July 1926, Landé had assured Goudsmit that Mensing was "really up to date on all new and old quantum methods" (see Note 36). The only thing about Mensing that impressed Landé enough to mention her in his letter to Goudsmit of May 1927 was that she had been able to keep up with the men during a strenuous hike in the Black Forest.

Lucy Schütz stoically accepted the prevailing views on the respective roles of men and women. As she wrote matter of factly in 1977 in the passage we quoted at the beginning of this section (see Note 18): "In 1928 I got married. As a woman one could then no longer have a job."

4.5 Married with Children in Munich, Königsberg, and Jena

In late 1929, the Schützes followed Gerlach to Munich. Wilhelm had done his habilitation in Tübingen; Lucy had not. Even if she had, she would probably not have been offered a position. Initially, as Goudsmit expected (see Note 67), she was able to keep her hand in as a physicist. As she wrote decades later:

I went to Sommerfeld's lectures and to his seminar, and in the morning attended Gerlach's classes and worked out the lecture notes (twice actually). My husband dictated and I wrote down parts of his handbook or I dictated the barely legible text to an assistant. When I was writing myself, I was dreaming under the red setting sun about Freiburg, where my husband was on the list [of candidates for a professorship], and tried to conjure the call with the full force of wishful thinking.[76]

---

[74] "Um die enge Zusammenarbeit mit Bohr und Heisenberg und den andern beneide ich Sie sehr; gerade in der neuen Quantentheorie trifft man täglich auf ungelöste Fragen, über die man sich gern mit einem dafür ebenfalls interessierten Physiker aussprechen möchte, und ich reise zu diesem Zweck Pfingsten wahrscheinlich nach Zürich, um mit Herrn London wieder mal ein vernünftiges Wort sprechen zu können." Landé to Goudsmit, May 20, 1927 (see Note 54).

[75] In sharp contrast to his close friend Einstein and the latter's ETH classmate and first wife Mileva Marić, Ehrenfest and his physicist–mathematician wife Tatiana Afanassjewa had a productive scientific partnership (see the double biography of the couple by Margriet van der Heijden (2021)). On Einstein and Marić, see Stachel (1996) and, more recently, Esterson et al. (2019). Stachel's paper appears in *Creative Couples in the Sciences* (Pycior et al., 1996). We do not consider Lucy Mensing and Wilhelm Schütz another example of a creative couple. As we saw, Mensing's most important papers were written before she met Schütz.

[76] "Ich ging in die Vorlesung von Sommerfeld und in sein Seminar, hörte morgens die Vorlesung von Gerlach und arbeitete sie aus (wohl 2 mal) schrieb nach dem Diktat meines Mannes Teile seines Handbuchs oder diktierte den kaum lesbaren Text einer Schreibkraft. Wenn ich selbst schrieb träumte ich bei der rot untergehenden Sonne

Her own project at the time had to do with the broadening of spectral lines. She thus returned to the topic of her dissertation and one of her first papers (Mensing, 1925c), which she cites a few times in the short new paper she published, using both her married and maiden names (Schütz-Mensing, 1930). This paper would be her last. After the birth of her first son, Jürgen, in November 1930, taking care of the family became her primary responsibility (see Figure 4.5). In 1934, a second son,

Figure 4.5 Wilhelm and Lucy Schütz with their first son Jürgen (born 1930), Niendorf/Timmendorfer Strand, Schleswig-Holstein, 1933. Picture by Esther Goudsmit, the daughter of Samuel Goudsmit. American Institute of Physics, Emilio Segrè Visual Archives.

von Freiburg, wo mein Mann auf der Liste stand, suchte den Ruf herbeizuziehen mit allen Wunsch-Kräften." Lucy Schütz to Hilda Mothes, August 8, 1977 (see Note 18).

Ulrich, was born. The young mother nevertheless found time and energy to do some physics on the side. She wrote the section on the quantum theory of the Faraday effect for her husband's handbook article on magneto-optics, which appeared as a separate volume and is clearly the handbook she was referring to in the passage quoted above (Schütz, 1936).

The call to Freiburg, where Mensing had spent two semesters as a student, never materialized, but in 1936 Wilhelm Schütz got a call to Königsberg as the successor of Walter Kaufmann, an experimental physicist best known for his measurements of the velocity-dependence of electron mass in the early 1900s (Duncan and Janssen, 2019, pp. 19 and 34). So in late 1936, the family moved from Munich to Königsberg, where two more children were born, daughters Cornelie (1937) and Dorothea (1939).

Königsberg in East Prussia (now Kaliningrad in Russia) lies at the mouth of the river Pregel. Many years later, Lucy Schütz still remembered

a sleigh ride across the solidly frozen Pregel with the Mothes family [see Note 18]. Did we have two sleighs? Did we take the children? I only remember a niece in a wonderful fox fur coat and a live fox crossing the Pregel. We saw it for a long time. (As a city child I had never seen one except at the zoo.)[77]

She probably also never saw Franz Marc's Cubist painting "The Foxes," from the same year (1913) as "The Mandrill" and "Wild Boars," which she did see, in Hamburg and Cologne, respectively, more than 60 years apart (see Section 4.2 and Figure 4.2).[78]

The Nazis' rise to power had shaken up the physics faculty at the University of Königsberg, where in the late 1880s its native son Sommerfeld had still studied with a young Hilbert and other luminaries. In the early 1930s, the professors for experimental and theoretical physics were both Jewish. The experimentalist Kaufmann was forced into early retirement at age 64; the theorist, Richard Gans, who had done his habilitation with Paschen in Tübingen in 1903, was dismissed at age 55. They were replaced by Schütz and Fritz Sauter, respectively (Gebhardt, 2016, pp. 7–8).

Schütz joined the Nazi party in May 1937, which put him in the orbit of Arthur Stuart. A molecular physicist of some renown, Stuart was also an assiduous party member and a virulent anti-Semite, which clearly aided his ascent up the academic

---

[77] "Eine Schlittenfahrt über den fest zugefroren Pregel mit Familie Mothes. Waren es 2 Schlitten? waren die Kinder mit? Ich erinnere nur eine Nichte in einem wundervollen Fuchsmantel, und einen lebendigen Fuchs, der den Pregel überquerte. Wir sahen ihn lange (ich als Stadtkind hatte außer im Zoo noch nie einen gesehen)." Lucy Schütz, recto of a sheet (with page number 3) with fragmentary memories of Königsberg (numbered 7, 9–13), undated but probably from the 1980s. Mensing-Schütz Family Papers.

[78] A Jewish banker bought "The Foxes" in 1928 and was able to hang on to it even though the Nazis forced him to sell much of his art collection. It found its way to the Kunstpalast in Düsseldorf in the 1960s but was returned to the heirs of its original owner in 2022 and sold at auction for £42.6 million (see the Wikipedia page for the painting).

ladder.[79] He had started in Königsberg as Gans's assistant and was made associate professor there in 1935, only to leave for Berlin the following year to temporarily fill the vacancy left by Schrödinger's departure in 1933. By 1939, Stuart was full professor in Dresden (Rammer, 2006, pp. 410–411). A letter from Schütz to Stuart in April that year strongly suggests that the two men were of like mind.[80]

Though he could ill afford to renounce his membership, Schütz, as opposed to Stuart, would sour on the Nazi party over the next few years. One clear indication comes from another of his wife's later recollections of their time in Königsberg. This one is about the celebration of Königsberg's two most famous citizens on May 24, 1943, the 400th anniversary of the death of one of them. That Bernhard Rust, Minister of Science, Education and National Culture (*Reichserziehungsminister*), gave a speech on the occasion adds poignancy to Lucy Schütz's story:

The Kant–Copernicus fest. My husband was later always proud that he had nominated two non-party-members as prize winners and prevailed: Heisenberg and [Albrecht] Unsöld. We gave a rare reception with dinner. In addition to the guests of honor we had the Mothes and Sauter couples over. The former came at 12 pm instead of 1 pm when we were still rearranging the furniture [interlineated: I only had the Polish maid]. And then the wild boar roast (which we had finally gotten that morning and for three times the usual number of vouchers after it had been promised for days) was hard as a rock. I was embarrassed when Heisenberg left a few vouchers. Unfortunately, I don't remember any of the conversations.[81]

This passage stands out, not only for what it says about her husband's attitude toward the Nazi party in 1943, but also for how seriously she took her duties as a hostess at the time. Four decades after the fact, she still remembered those duties but not the conversations. Because of the former, she may, in fact, have had little opportunity to participate in the latter.

During the war, Königsberg came under attack from both the east and the west. The Red Army bombed it repeatedly between September 1941 and July 1943. The

---

[79] Stuart is the first of five physicists mentioned by name in an extraordinary article published not long after the war by a female physics student with Jewish ancestry, Ursula Maria Martius (1947), in which she took the German Physical Society to task for turning a blind eye to the Nazi actions and pronouncements of its members (Rammer, 2006, pp. 389–409). Disillusioned about the reluctance of the German Physical Society to clean house, Martius emigrated to Canada in 1949, where she had a distinguished career, based at the University of Toronto, under the name Ursula Franklin.

[80] This letter is quoted and discussed in Hoffmann (2006, pp. 173 and 176). Here is a translation of the incriminating passage: "Without a doubt, the definitive integration of the German Physical Society into the Third Reich is urgently needed, but it is a delicate matter. I was not too happy when [Peter] Debye became chairman of the society [in 1937] . . . The handling of the Jewish question by the society showed, as was to be expected, that he lacked the necessary understanding of political issues. Back then I tried in vain to bring about a definite position by the chairman and thereby a definitive solution to the problem."

[81] "Die Kant–Kopernikus Feier. Mein Mann war später immer stolz darauf, daß er als Preisträger zwei Nicht-Partei-Genossen vorgeschlagen und durchgesetzt hatte: Heisenberg u. Unsöld. Wir gaben eine der ganz seltenen Mittagseinladungen: außer den beiden Gästen die Ehepaare Mothes u. Sauter. Erstere kamen um 12h statt um 1h und wir waren noch beim Möbel-Umräumen [interlineated: ich hatte ja nur die poln. Hilfe]. Und dann war der Wildschweinbraten (auf 3-fache Marken erst am Morgen bekommen aber von Tag zu Tag vertröstet) steinhart. Ich schämte mich, als Heisenberg noch Marken hinlegte! Leider erinnere ich nichts von den Gesprächen." Lucy Schütz, memories of Königsberg (see Note 77). Mensing-Schütz Family Papers.

Royal Air Force bombed it twice in the span of four days in August 1944. The damage of the first raid was limited but the second destroyed much of the city's historic center, including university buildings and the Schützes' apartment. After the second British bombardment, the Schützes left Königsberg. Lucy Schütz managed to salvage her sewing machine along with the volumes of their *Brockhaus* encyclopedia and *Propyläen* world history (so named after their respective publishers).[82] They first went to Wilhelm's family in Frankfurt, then to Jena, where Wilhelm had established a branch of his institute of which he provisionally became director. These were hard times. Both in Frankfurt and in Jena, they experienced further bombings. Their daughter, however, remembers that her mother used to say, even when those around her did not want to hear this, that the Germans only got what they deserved.[83] To feed and sustain the family in Jena, Lucy Schütz collected crops from the fields and worked as a cleaning lady, another step down from the lone Polish maid they could still afford in Königsberg. In March 1946, she got a job as trainee assistant at the mathematical institute, which brought some relief.

Before Wilhelm Schütz could officially be appointed in Jena after the war, he needed a so-called *Persilschein* (after the name of a well-known laundry detergent) to "whitewash" his involvement with the Nazi party (see Schmidt, 2015, p. 106). Like several other colleagues, Hans Hermann Weber, former professor of physiology in Königsberg, vouched for him. In a document dated April 9, 1946, Weber stated:

Testimonial. Even a bitter enemy and active opponent of National Socialism such as myself could at best know, but not detect, that he was a party member. He was my neighbor in the Cäcilienallee in Königsberg, Prussia, and he has – at least in private conversations – never made a secret out of his criticism and rejection of National Socialism. The same applies, perhaps in an even stronger and broader sense, to his wife. It goes without saying that Prof. Schütz never reported anybody for being an opponent of the regime or for passing on materials, the distribution of which was forbidden in Germany (e.g., materials from foreign radio stations), although everyone in Germany and especially party members had a duty to denunciate.[84]

It is interesting that Weber, presumably only called upon to exonerate the husband, not the wife, made a point of comparing her favorably to him.

[82] Dorothea Roloff to Gernot Münster, 2020 (see Note 100).   [83] Dorothea Roloff, private communication.
[84] "*Zeugnis.* Auch ein erbitterter und aktiver Gegner des Nationalsozialismus wie ich konnte von Herrn Prof. Dr. Wilhelm Schütz höchstens wissen, aber nicht bemerken, dass er Pg [Parteigenosse] war. Er war mein Nachbar in der Cäcilienallee in Königsberg Pr[eußen] und hat – wenigstens im Privatgespräch – aus seiner Kritik und Ablehnung [added in the margin: des Nationalsocialismus] kein Hehl gemacht. Dies gilt in vielleicht noch gesteigertem Umfang von seiner Frau. Selbstverständlich hat Herr Prof. Schütz keinem Menschen um oppositioneller Einstellung willen oder auf Grund von Tatsachenmaterial, dessen Verbreitung in Deutschland verboten war (z B. Material aus fremden Sendern) angezeigt, obwohl in Deutschland jeder und zwar erst recht jeder Pg. zum Denunzieren verpflichtet war." This and other "Persilscheine" are part of the Mensing-Schütz Family Papers.

Lucy Schütz never seems to have had any sympathy for the Nazis. We already saw that she detested the notion of "degenerate art." The denouement of the story about the confiscation of Franz Marc's paintings (see Section 4.2 and Note 12), however, may serve as a reminder that one can be a Nazi and still appreciate German expressionist art. Marc's giant painting "Tower of Blue Horses" – created, like those of the mandrill, the wild boars, and the foxes, in 1913 – was put on display in the notorious exhibit of *entartete Kunst* in Munich in 1937. It was removed shortly after the exhibit opened, when a group of World War I veterans protested that Marc had been a decorated officer, killed in action in 1916. None other than Hermann Göring then took possession of the painting. And while the Nazi leader sold several other works of "degenerate art" at a handsome profit, he seems to have kept the "Tower of Blue Horses" for his own private collection.

We already saw that Lucy Schütz remained friends or at least stayed in touch with several Jewish physicists, such as Ising, Goudsmit, and Landé. Of course, the same can be said about many former party members and Nazi sympathizers. Goudsmit, however, is a special case. As noted above, shortly after his time in Tübingen with Mensing and Schütz, Goudsmit had emigrated to the US. Toward the end of the war, he was appointed the scientific head of the Alsos mission charged with finding out how far the Germans had gotten in developing an atomic bomb (Goudsmit, 1947; Calmthout, 2016). In that capacity he turned the tables on Gerlach, his former host in Tübingen, who was among those rounded up in May–June 1945 and then "detained at Her Majesty's pleasure" for half a year at Farm Hall, a country house outside of Cambridge (Bernstein, 2001; Hoffmann, 2023).

As head of the Alsos mission, Goudsmit came into possession of a wealth of documents revealing the Nazi sympathies of many a German physicist. The compromising letter from Wilhelm Schütz to Stuart mentioned above (see Note 80) actually ended up among those (Goudsmit papers, Box 28, Folder 53, see Note 1). So did a similar letter from Stuart to the Austrian physicist Georg Stetter of March 1939, in which Stuart mentions Schütz as someone in their camp. In 1951, when Stuart was being considered for a position in Munich, the Göttingen physicist Hans Kopfermann protested, given Stuart's Nazi past. When Stuart demanded proof, Kopfermann turned to Goudsmit (as many American physicists did before inviting German colleagues after the war), who supplied him with a copy of the letter to Stetter (Rammer, 2006, pp. 412–413).

So Goudsmit must have known that Wilhelm Schütz had been a party member. Yet this did not stop him from remaining on friendly terms with the Schützes. They met at least once after the war, in October 1965 in Frankfurt am Main, at the annual meeting of the German Physical Society. The Schützes had obtained permission to travel to West Germany with a small East German delegation, because, as Goudsmit noted in a short report on the meeting, Wilhelm Schütz had reached retirement age and the couple had relatives in West Germany (recall that Frankfurt was Schütz's hometown). Goudsmit was pleased to see them, noting in his report that "both were friends of mine in the mid-twenties" and explaining that Mrs. Schütz was "the former Lucy Mensing, one time assistant of Pauli and of Land[é]."[85]

An incident Lucy Schütz related in a letter in 1976 suggests that, unlike her husband, she had acquired immunity against Nazi ideology early on:

As a very young student I was at the physics meeting in Leipzig in 1922 or 1923[86] in connection with a family visit. There pamphlets against relativity theory were distributed, signed by about 10–20 names, most of whom were unknown to me. But one of the 'big ones' said: 'How could Lenard sign something like this?' Somebody other than Einstein, probably von Laue, then gave a lecture about it. So the evil already started that early. Could we – and should we – have prevented it at that point?[87]

This sentiment is echoed in a 1988 letter from Ising, the friend from their student days in Hamburg, to her son Jürgen. Ising apparently shared a memoir he was working on, thanked Jürgen Schütz for his comments, and promised to make some corrections. "In particular," he wrote, "I would like to insert a page about the fate of *your* family and about the thoughts that occupied your mother so strongly: what could or should we have done to prevent the horrors?"[88]

4.6 Deportation to Gorodomlya and Return to Jena

In July 1945, the US withdrew their forces from Thuringia, taking many scientists and engineers with them in Operation Overcast, and the Soviet Union took over the occupation regime. The following year – in the early hours of October 22, 1946, to be precise – the Schütz family in Jena was in for a shock. Russian soldiers showed up on their doorstep and told them they only had a few hours to pack their belongings for a long journey.[89]

---

[85] Goudsmit papers (see Note 1), Box 40, Folder 01, Foreign travel: Germany, Holland, Israel, 1965.

[86] The *Naturforscherversammlung*, the traditionally annual but by 1922 bi-annual meeting of the *Gesellschaft Deutscher Naturforscher und Ärzte*, was held in Leipzig in 1922. Schütz remembered correctly that von Laue gave a talk about relativity theory.

[87] "als ganz junge Studentin war ich 1922 oder 23, zur Physikertagung in Leipzig in Verbindung mit Verwandtenbesuchen. Dort wurden Zettel gegen die Relativitätstheorie verteilt, unterzeichnet von etwa 10–20 Namen, die mir weitgehend unbekannt waren. Aber einer der 'Groß en' sagte: 'Wie konnte nur Lenard sowas unterschreiben?' Es trug dann statt Einstein jemand anders, wohl v. Laue, drüber vor. So früh fing das Übel schon an! Hätte man damals schon vorbauen können? und müssen?" Copy of Lucy Schütz to Walter and Ruth (née Probst) Gerlach, March 2, 1976. Mensing-Schütz Family Papers.

[88] "Vor allem möchte ich auch eine Seite über das Schicksal *Ihrer* Familie einfügen und über die Gedanken, die Ihre Mutter so sehr beschäftigten: Was hätten wir tun können oder sollen, um die Schrecknisse zu verhindern?" Ernst Ising to Jürgen Schütz, December 1988. Mensing-Schütz Family Papers.

[89] The Mensing-Schütz Family Papers contain a short note by Jürgen Schütz, dated October 22, 1946, informing an aunt in Frankfurt (Anni Himmelreich) that they had been taken to a railway station in a Russian military truck, as well as four typed and numbered pages, with the handwritten note "not sent, copied later" (*nicht abgeschickt, später abgeschrieben*) at the top of p. 1, in which Lucy Schütz describes, diary-style, the ensuing journey, constantly guarded by Russian soldiers and plagued by bugs. They departed a station at the Russian border on October 28 and, after several long stops, arrived at their final destination (the island of Gorodomlya)

In Operation Osoaviakhim, mirroring Operation Overcast of the Americans, the Soviets deported about 2,000 scientists and engineers and their families from the Soviet occupied zone to the Soviet Union to contribute to the reconstruction of its industry and science.[90] Wilhelm and Lucy Schütz and their four children, ranging in age from 7 to 16 years in October 1946, were among them. They were brought to the island of Gorodomlya in the Seliger Lake, near the city of Ostashkov, some 300 km northwest of Moscow. The island was about 3.5 square kilometers and surrounded by a barbed-wire fence guarded at some stretches by dogs kept on long leashes tied to the fence.[91] Together with other scientists, Wilhelm Schütz worked on the Soviet rocket project, based on the German A4. He was leading a section responsible for measurement instruments to be placed in the rockets.[92] Meanwhile, his wife volunteered as a teacher of German and history at a school for the children of the German families, making good use of the encyclopedias she salvaged in Königsberg. It would be six years before they were allowed to return to Jena.

Lucy Schütz detailed some of the privations they suffered during their first year on the island in two long letters to Wilhelm Hanle, her old friend from Göttingen (see Notes 43 and 62), meanwhile professor in Gießen.[93] The first of these was written at the height of summer, the second in the dead of winter. "Now that it is summer," she wrote in the first one, "some things are easier, others harder."[94] While her children, despite the barbed-wire fence, could play on the beach in the nice weather, she was dependent on a ferry, which only made a few trips a week and could only carry a limited number of passengers, to reach the mainland and go shopping or see a doctor in Ostashkov, always guarded by soldiers. Once the lake froze in the winter, transportation became easier, though she also reported a case of someone who had tried to cross the lake after a few days of thaw and drowned.

For the same reason, mail delivery was easier in the winter than in the summer. The first of two letters from Hanle to which Lucy Schütz was responding, sent February 25, 1947, from Gießen, still reached Gorodomlya in mid-April before the thaw set in. It was the first letter the Schützes received after they arrived on the island half a year earlier. Hanle's second letter, sent March 22, arrived shortly before she wrote back to him in July.

---

in the wee hours of November 7. For another account of these events and the subsequent detention on Gorodomlya, see Albring (1991), summarized in the dissertation by Nadin Schmidt (2015, pp. 140–150).

[90] See the Wikipedia entry "Operation Osoaviakhim," (Schmidt, 2015) and references therein. On page 14, Schmidt explicitly warns that the term "deportation," used throughout her dissertation, should not be taken in the same sense as when used to describe the Nazi deportation of Jews.

[91] Dr. Dorothea Roloff, private communication.

[92] CIA Information reports 25X1 of June 30, 1953, and 25X1A of August 26, 1953, approved for release to the public.

[93] Lucy Schütz to Wilhelm Hanle, July 7, 1947 and December 12, 1947. Copies of typed letters, both two pages, single-spaced. Mensing-Schütz Family Papers.

[94] "Jetzt im Sommer ist manches leichter, manches schwieriger" (see Note 93).

In the letter to Hanle six months later, in response to three further letters sent between August and December, she told Hanle that they finally received letters from two Jena physicists, Hund[95] and Buchwald.[96] Especially Buchwald's letter struck a nerve with Lucy Schütz:

The second letter is so life-affirming and full of wonderful activity that one feels quite bitter here in banishment. It is like peeking into a room all lit up for Christmas, but not being allowed to enter. But then I tell myself that this Christmas room is nothing but a chimera and that in reality everything is difficult and dreary over there as well.[97]

She then closed the letter with the consoling thought that "here at least it is not our own ones who are doing this to us."[98]

In June 1952, four and a half years after this Christmas letter, the family could finally return to Jena,[99] where they rebuilt their lives after the dual nightmare of war and banishment. Wilhelm Schütz got a chair for experimental physics, which he held until his retirement in 1965 (see Schmidt, 2015, pp. 108–109). Although no longer employed as a scientist, Lucy Schütz occasionally assisted her husband in his work, for example, by helping him prepare lecture notes. As their daughter Dorothea remembers, the conversation during family meals typically revolved around physics. Her brothers Jürgen and Ulrich (who passed away in 2018 and 2022, respectively) both went into (medical) physics (Cornelie and Dorothea, now both in their late eighties, went into pharmacy and medicine, respectively). As we have seen, their mother stayed in touch with several physicists she had met during her own brief but remarkable career in the 1920s.

In the fragments with recollections of that period, which she wrote in the late 1980s and which form the basis for much of our reconstruction of a career cut short, she described her negative experiences in Tübingen in 1927 in considerably more detail than her positive experiences in Göttingen in 1925–1926. This also fits with what her youngest daughter remembers: "In terms of what we as children heard, Tübingen was more important than Göttingen. Objectively, it's probably the other way around."[100] While the paper on the Ramsauer effect (Mensing, 1927) is mentioned prominently in Lucy Schütz's reminiscences, the papers she wrote

---

[95] Friedrich Hund to Wilhelm and Lucy Schütz, November 8, 1947. Mensing-Schütz Family Papers.

[96] Eberhard Buchwald to Wilhelm and Lucy Schütz, May 5, 1947. Mensing-Schütz Family Papers. Among these papers is also a short note by Buchwald dated October 24, 1946, a couple of days after the Schützes were deported. In this note, Buchwald bids farewell to the Schützes in case he will not see them before their train departs the station in Göschwitz. A post-it note attached to Buchwald's note by Dorothea Roloff says that Buchwald actually did still see them at this station.

[97] "Der letztere Brief ist so lebensbejahend un[d] voll schöner Betriebsamkeit, dass man recht bitter wird hier in der Verbannung[.] Wie ein Blick in ein erleuchtetes Weihnachtszimmer ist das, in das man selbst nicht eintreten darf. Aber dann sage ich mir, dass das ganze Weihnachtszimmer nur eine Fata Morgana ist, dass in Wirklichkeit auch dort alles schwer und trostlos ist!" Source as for Note 93.

[98] "Hier sind es wenigstens nicht die eigenen die uns das antun!" Source as for Note 93.

[99] The Mensing-Schütz Family Papers contain several more letters from Hanle from this period.

[100] Dorothea Roloff to Gernot Münster after reading a draft of Münster (2020a).

Figure 4.6 Joint meeting of the Physical Societies of East and West Germany in Berlin, 1952. Werner Heisenberg (left) and Lucy Schütz (center). Mensing-Schütz Family Papers.

during her year in Göttingen are not, not even the one on the elaboration and application of the new rules for the quantization of angular momentum (Mensing, 1926a) for which she deserves a place among the pioneers of quantum mechanics.

Some recollections of her time in Göttingen can be found in her letter to the Gerlachs of March 1976, from which we already quoted in Section 4.5. This was one month after the death of Heisenberg whom she had probably last seen in person in 1952 (see Figure 4.6). She wrote:

I saw an interview with him on television that probably took place a few years ago. His voice sounded as if it came from the time of his youth. I read again in his book "The Part and the Whole" [Heisenberg, 1969] and rejoiced about the two semesters we spent together in Göttingen, experiencing the new quantum mechanics from up close and contributing a little to it, and also about skiing together and occasional bobsledding [added in the margin: and, finally, being allowed to listen during Heisenberg's chamber music evenings (e.g., Schubert's Trio in B-flat)]. I once started to write a letter to Heisenberg about his book but did not finish it because of the illness of my husband.[101]

---

[101] "Im Fernsehen sah ich ein Intervieux [sic] mit ihm, das wohl vor einigen Jahren stattgefunden hat. Die Stimme klang wie aus der Jugendzeit herüber. Ich las wieder in seinem Buch "Der Teil und das Ganze" und freute mich über die zwei gemeinsamen Semester in Göttingen 1925/26, das Mit-erlebt-Haben der neuen Quantenmechanik aus nächster Nähe und ein klein wenig Mitarbeit, und auch über gemeinsame Schifahrten und gelegentliches Bobrodeln [Handwritten in the margin of this typewritten letter:] und schließlich über das Zuhören-Dürfen bei Heisenbergs Kammermusik-Abenden (u.a., das Schubert B-Dur-Trio). Zu dem Buch

Wilhelm Schütz died in 1972,[102] so she started this letter (which she does not seem to have kept) not long after Heisenberg's famous memoir came out (Heisenberg, 1969). Four years later, she appears to remember the skiing, the bobsledding, and the musical evenings (*Schubert's B-Dur trio*)[103] more vividly than the physics, which she describes in rather generic and reverent terms, conveying gratitude and modesty that she had been in Göttingen to witness the birth of quantum mechanics and make a small contribution (*ein klein wenig Mitarbeit*) herself.

Lucy Schütz might have appreciated the importance of her own contribution to quantum mechanics better, had she been interviewed by Kuhn and his collaborators in the 1960s as part of the project to establish the AHQP.[104] In Section 4.3, we saw that her name came up in Kuhn's interview with Jordan, which took place in June 1963. This may be why Kuhn's curiosity was piqued in December 1963, when Goudsmit also mentioned Mensing in his AHQP interview. Talking about his time in Tübingen, Goudsmit told Kuhn that "Fräulein Mensing was the one I discussed physics with; she was the theorist. She worked for Landé, but she was more communicative."[105] "Are you in touch with Fräulein Mensing, now Mrs. something or other?," Kuhn wanted to know. "Mrs. Schütz," Goudsmit could help him but, no, he was not in touch with her: "she is in the Soviet zone now." "Could you write?," Kuhn asked. "Ja, ja," said Goudsmit, but nothing came of it. Goudsmit does not seem to have broached the topic when he met Lucy and Wilhelm Schütz in Frankfurt two years later (see Section 4.5). One can't help but wonder whether Mensing's contributions to quantum mechanics would have been better known had she been interviewed for the AHQP project, which could have been arranged even though she was living behind the Iron Curtain.

Lucy Schütz survived her husband by more than two decades. On April 28, 1995, she died in Meiningen, where she had moved to be close to her oldest daughter. She thus still had almost half of her long life ahead of her when she and her family were finally allowed to leave the Soviet Union in 1952. The return to Jena is nonetheless a good point to end our account of her life. We leave the reader with a motto she shared with her children: "Everything is interesting; one just has to start."[106]

---

hatte ich einen Brief an Heisenberg angefangen, der über der Krankheit meines Mannes nicht fertig wurde." Lucy Schütz to Walter and Ruth Gerlach, March 2, 1976 (see Note 87).

[102] He left behind an unfinished manuscript for a biography of Galileo. Lucy Schütz approached Ernst Schmutzer, one of his younger colleagues in Jena, who finished the book with her help, though only Wilhelm was listed as co-author (Schmutzer and Schütz, 1975). Lucy Schütz contributed a foreword (Schütz, 1975), in which she expressed her regret that the book failed to realize her husband's original plan to focus on Galileo's experiments, which he felt had not received adequate attention.

[103] This piece is mentioned repeatedly in *Der Teil und das Ganze* (Heisenberg, 1969, pp. 33, 38, 41).

[104] Ironically, Esterman, whom Mensing told Goudsmit in 1926 "did not understand any of it" (see Note 39), *was* interviewed for the project!

[105] Interview with Samuel Goudsmit by Thomas S. Kuhn, Rockefeller Institute, New York, December 7, 1963 (see Note 22).

[106] "Es ist alles interessant; man muss nur anfangen." As recalled by Cornelie Schütz in an interview for an article on the occasion of her mother's 94th birthday published in the Meininger Tageblatt of March 11, 1995.

References

Albring, Werner. 1991. *Gorodomlia. Deutsche Raketenforscher in Russland.* Hamburg: Luchterhand.

Back, Ernst. 1923. Der Zeemaneffekt des Bogen- und Funkenspektrums von Mangan. *Zeitschrift für Physik*, 15, 206–243. https://doi.org/10.1007/BF01330474

Back, Ernst and Goudsmit, Samuel. 1928. Kernmoment und Zeemaneffekt von Wismut. *Zeitschrift für Physik*, 47, 174–183. https://doi.org/10.1007/BF02055794

Bernstein, Jeremy (ed.). 2001. *Hitler's Uranium Club. The Secret Recordings at Farm Hall*, 2nd ed. New York: Springer.

Bjerrum, Niels. 1912. Über die ultraroten Absorptionsspektren der Gase. Festschrift W. *Nernst zu seinem fünfundzwanzigjährigen Doktorjubiläum gewidmet von seinen Schülern.* Halle: Verlag von Knapp, pp. 90–98 (English translation, "On the infrared absorption spectra of gases," in Niels Bjerrum, *Selected Papers* (Copenhagen, E. Munksgaard, 1949).)

Born, Max. 1924. Über Quantenmechanik. *Zeitschrift für Physik*, 26, 379–395. (English translation in van der Waerden (1968, pp. 181–198).) https://doi.org/10.1007/BF01327341

Born, Max and Jordan, Pascual. 1925a. Zur Quantentheorie aperiodischer Vorgänge. *Zeitschrift für Physik*, 33, 479–505. https://doi.org/10.1007/BF01328329

Born, Max and Jordan, Pascual. 1925b. Zur Quantenmechanik. *Zeitschrift für Physik*, 34, 858–888. (English translation of chs. 1–3 in van der Waerden (1968, pp. 277–306).) https://doi.org/10.1007/BF01328531

Born, Max and Jordan, Pascual. 1930. *Elementare Quantenmechanik.* Berlin: Springer.

Born, Max, Heisenberg, Werner, and Jordan, Pascual. 1926. Zur Quantenmechanik II. *Zeitschrift für Physik*, 35, 557–615. (English translation in van der Waerden (1968, pp. 321–385).) https://doi.org/10.1007/BF01379806

Brandmüller, Josef. 1989. Tübingen: Ein "Schauplatz der Quantentheorie." *Physikalische Blätter*, 45(8), 327–332. (Slightly different English version in Brandmüller (1990).) https://doi.org/10.1002/phbl.19890450805

Brandmüller, Josef. 1990. Tübingen: another scene of quantum theory. *European Journal of Physics*, 11, 313–322. (Slightly different English version of Brandmüller (1989).) https://doi.org/10.1088/0143-0807/11/6/001

Calmthout, Martijn van. 2016. *Sam Goudsmit. Zijn jacht op de atoombom van Hitler.* Amsterdam: Meulenhoff.

Cassidy, David C. 1979. Heisenberg's first core model of the atom: the formation of a professional style. *Historical Studies in the Physical Sciences*, 10, 187–224. https://doi.org/10.2307/27757390

Cassidy, David C. 2007. Oppenheimer's first paper: Molecular band spectra and a professional style. *Historical Studies in the Physical and Biological Sciences*, 37, 247–270. https://doi.org/10.1525/hsps.2007.37.2.247

Czerny, Marianus. 1925. Messungen im Rotationsspektrum des HCl im langwelligen Ultrarot. *Zeitschrift für Physik*, 34, 227–244. https://doi.org/10.1007/BF01328469

Debye, Peter. 1912. Einige Resultate einer kinetischen Theorie der Isolatoren. *Physikalische Zeitschrift*, 13, 97–100.

Dirac, Paul A. M. 1925. The fundamental equations of quantum mechanics. *Proceedings of the Royal Society of London A*, 109, 642–653. (English translation in van der Waerden (1968, pp. 307–320).) https://doi.org/10.1098/rspa.1925.0150

Duncan, Anthony and Janssen, Michel. 2007. On the verge of *Umdeutung* in Minnesota: Van Vleck and the correspondence principle. Two parts. *Archive for History of Exact*

Sciences, 61, 553–624, 625–671. https://doi.org/10.1007/s00407-007-0010-x (Pt. 1), https://doi.org/10.1007/s00407-007-0009-3 (Pt. 2).

Duncan, Anthony and Janssen, Michel. 2019. *Constructing Quantum Mechanics. Volume 1, The Scaffold: 1900–1923*. Oxford: Oxford University Press.

Duncan, Anthony and Janssen, Michel. 2023. *Constructing Quantum Mechanics. Volume 2, The Arch: 1923–1927*. Oxford: Oxford University Press.

Ehlers, Jürgen and Schücking, Engelbert. 2002. "Aber Jordan war der Erste": Zur Erinnerung an Pascual Jordan (1902–1980). *Physik Journal*, 1(11), 71–74.

Esterson, Allen, Cassidy, David C., and Lewin Sime, Ruth. 2019. *Einstein's Wife. The Real Story of Mileva Einstein-Marić*. Cambridge, MA: The MIT Press.

Faxén, Hilding and Holtsmark, Johan Peter. 1927. Beitrag zur Theorie des Durchganges langsamer Elektronen durch Gase. *Zeitschrift für Physik*, 45, 307–324. https://doi.org/10.1007/BF01343053

Forman, Paul. 1970. Alfred Landé and the Anomalous Zeeman Effect, 1919–1921. *Historical Studies in the Physical Sciences*, 2, 153–261. https://doi.org/10.2307/27757307

Fues, Erwin. 1926. Das Eigenschwingungsspektrum zweiatomiger Moleküle in der Undulationsmechanik. *Annalen der Physik*, 80, 367–396. https://doi.org/10.1002/andp.19263851204

Fulda, Bernhard, Ring, Christian, and Soika, Aya (eds.). 2019. *Emil Nolde. The Artist During the Third Reich*. New York: Prestel.

Gavroglu, Kostas. 1995. *Fritz London: A Scientific Biography*. Cambridge: Cambridge University Press.

Gearhart, Clayton A. 2010. "Astonishing successes" and "bitter disappointment": the specific heat of hydrogen in quantum theory. *Archive for History of Exact Sciences*, 64, 113–202. https://doi.org/10.1007/s00407-009-0053-2

Gebhardt, Wolfgang. 2016. *Erich Kretschmann. The Life of a Theoretical Physicist in Difficult Times*. Max Planck Institute for the History of Science. Preprint 482. https://www.mpiwg-berlin.mpg.de/sites/default/files/Preprints/P482.pdf

Goudsmit, Samuel. 1927. *Atoommodel en structuur der spectra*. PhD Thesis. Leiden: Leiden University.

Goudsmit, Samuel. 1947. *Alsos*. New York: Henry Schuman.

Goudsmit, Samuel and Bacher, Robert F. 1930. Der Paschen–Back-Effekt der Hyperfeinstruktur. *Zeitschrift für Physik*, 66, 13–30. https://doi.org/10.1007/BF01397522

Goudsmit, Samuel and Back, Ernst. 1927. Feinstrukturen und Termordnung des Wismutspektrums. *Zeitschrift für Physik*, 43, 321–334. https://doi.org/10.1007/BF01397446

Greenspan, Nancy T. 2005. *The End of the Certain World. The Life and Science of Max Born. The Nobel Physicist Who Ignited the Quantum Revolution*. New York: Basic Books.

Häffner, Michaela. 1995. Schlägereien und Berufsverbote. Antisemitismus an der Universität. In Geschichtswerkstatt Tübingen e.V. (ed.), *Zerstörte Hoffnungen. Wege der Tübinger Juden*, Stuttgart: Theiss, pp. 173–190.

Hanle, Wilhelm. 1979. Interview of Wilhelm Hanle by Brenda P. Winnewisser on May 23 to June 2, 1979. Niels Bohr Library & Archives. https://repository.aip.org/islandora/object/nbla:266550

Heisenberg, Werner. 1922. Zur Quantentheorie der Linienstruktur und der anomalen Zeemaneffekte. *Zeitschrift für Physik*, 8, 273–297. https://doi.org/10.1007/BF01329602

Heisenberg, Werner. 1925. Über quantentheoretische Umdeutung kinematischer und mechanischer Beziehungen. *Zeitschrift für Phyik*, 33, 879–893. (English translation in van der Waerden (1968, pp. 261–276).) https://doi.org/10.1007/BF01328377

Heisenberg, Werner. 1969. *Der Teil und das Ganze*. Munich: R. Piper & Co. Verlag. (English translation: *Physics and Beyond: Encounters and Conversations* (New York, Harper & Row, 1971).

Heisenberg, Werner and Jordan, Pascual. 1926. Anwendung der Quantenmechanik auf das Problem der anomalen Zeemaneffekte. *Zeitschrift für Physik*, 37, 263–277. https://doi.org/10.1007/BF01397100

Hermann, Armin, Meyenn, Karl von, and Weisskopf, Victor F. (eds.). 1979. *Wolfgang Pauli: Wissenschaftlicher Briefwechsel mit Bohr, Einstein, Heisenberg u.a.* Band I: *1919–1929/Scientific Correspondence with Bohr, Einstein, Heisenberg, a.o. Vol. I: 1919–1929*. New York, Heidelberg, Berlin: Springer. https://doi.org/10.1007/978-3-540-78798-3

Hoffmann, Dieter. 2006. Die Ramsauer-Ära und die Selbstmobilisierung der Deutschen Physikalischen Gesellschaft. In Hoffmann and Walker (2006, pp. 173–215).

Hoffmann, Dieter (ed). 2023. *Operation Epsilon: Die Farm-Hall-Protokolle erstmals vollständig, ergänzt um zeitgenössische Briefe und weitere Dokumente der 1945 in England internierten deutschen Atomforscher*. Diepholz: GNT-Verlag GmbH.

Hoffmann, Dieter and Walker, Mark (eds.). 2006. *Physiker zwischen Autonomie und Anpassung: die Deutsche Physikalische Gesellschaft im Dritten Reich*. Weinheim: Wiley.

Hoffmann, Dieter and Walker, Mark. 2007. Der gute Nazi: Pascual Jordan und das Dritte Reich. In Hoffmann, Dieter, Ehlers, Jürgen, and Renn, Jürgen (eds.), *Pascual Jordan (1902–1980). Mainzer Symposium zum 100. Geburtstag*. Max Planck Institute for History of Science, Preprint 329, pp. 83–112. https://www.mpiwg-berlin.mpg.de/preprint/pascual-jordan-1902-1980-mainzer-symposium-zum-100-geburtstag

Ising, Ernst. 1924. *Beitrag zur Theorie des Ferro- und Paramagnetismus*. PhD Thesis. Hamburg: Hamburg University.

Jähnert, Martin. 2019. *Practicing the Correspondence Principle in the Old Quantum Theory*. Cham: Springer.

Jordan, Pascual. 1957. Wilhelm Lenz †. *Physikalische Blätter*, 13, 269–270. https://doi.org/10.1002/phbl.19570130604

Jordan, Pascual. 1973. Early years of quantum mechanics: some reminiscences. In Mehra, Jagdish (ed.), *The Physicist's Conception of Nature*. Dordrecht: Reidel, pp. 294–299.

Jordan, Pascual. 1975. Die Anfangsjahre der Quantenmechanik: Erinnerungen. *Physikalische Blätter*, 31, 97–103. https://doi.org/10.1002/phbl.19750310301

Kojevnikov, Alexei. 2020. *The Copenhagen Network. The Birth of Quantum Mechanics from a Postdoctoral Perspective*. Cham: Springer.

Kox, Anne J. 1997. The discovery of the electron: II. The Zeeman effect. *European Journal of Physics*, 18, 139–144. https://doi.org/10.1088/0143-0807/18/3/003

Kramers, Hendrik A. 1924a. The law of dispersion and Bohr's theory of spectra. *Nature*, 113, 673–676. (Reprinted in van der Waerden (1968, pp. 177–180).) https://doi.org/10.1038/113673a0

Kramers, Hendrik A. 1924b. The quantum theory of dispersion. *Nature*, 114, 310–311. (Reprinted in van der Waerden (1968, pp. 199–201).) https://doi.org/10.1038/114310b0

Kramers, Hendrik A., and Heisenberg, Werner. 1925. Über die Streuung von Strahlung durch Atome. *Zeitschrift für Physik*, 31, 681–707. (English translation in van der Waerden (1968, pp. 223–252).) https://doi.org/10.1007/BF02980624

Kratzer, Adolf. 1922. Störungen und Kombinationsprinzip im System der violetten Cyanbanden. *Sitzungsberichte der mathematisch-physikalischen Klasse der Bayerischen Akademie der Wissenschaften*, 107–118.

Kronig, Ralph. 1926. The dielectric constant of diatomic dipole-gases on the new quantum mechanics. *Proceedings of the National Academy of Sciences*, 12, 488–493. https://doi.org/10.1073/pnas.12.8.488

Kuhn, Thomas S., Heilbron, John L., Forman, Paul, and Allen, Lini. 1967. *Sources for History of Quantum Physics. An Inventory and Report.* Philadelphia, PN: The American Philosophical Society.

Landau, Lew D. 1926. Zur Theorie der Spektren der zweiatomigen Moleküle. *Zeitschrift für Physik*, 40, 621–627. https://doi.org/10.1007/BF01390460

Landé, Alfred. 1921. Über den anomalen Zeemaneffekt. I. *Zeitschrift für Physik*, 5, 231–241. https://doi.org/10.1007/BF01335014

Landé, Alfred. 1923. Termstruktur und Zeemaneffekt der Multipletts. II. *Zeitschrift für Physik*, 19, 112–123. https://doi.org/10.1007/BF01327550

Landé, Alfred. 1926. *Die neuere Entwicklung der Quantentheorie.* Dresden, Leipzig: T. Steinkopff.

Lenz, Wilhelm. 1920. Beiträge zum Verständnis der magnetischen Eigenschaften in festen Körpern. *Physikalische Zeitschrift*, 21, 6136–15.

Martius, Ursula M. 1947. Videant consules... *Deutsche Rundschau*, 70(11), 99–102. Reprinted in Hoffmann and Walker (2006, pp. 636–640).

Mehra, Jagdish and Rechenberg, Helmut. 1982a. *The Historical Development of Quantum Theory. Vol. 3. The Formulation of Matrix Mechanics and Its Modifications 1925–1926.* New York: Springer.

Mehra, Jagdish and Rechenberg, Helmut. 1982b. *The Historical Development of Quantum Theory. Vol. 4. Part 1. The Fundamental Equations of Quantum Mechanics. 1925–1926. Part 2. The Reception of the New Quantum Mechanics. 1925–1926.* New York: Springer.

Mehra, Jagdish and Rechenberg, Helmut. 1987. *The Historical Development of Quantum Theory. Vol. 5. Erwin Schrödinger and the Rise of Wave Mechanics. Part 1. Schrödinger in Vienna and Zurich 1887–1925. Part 2. The Creation of Wave Mechanics. Early Response and Applications 1925–1926.* New York: Springer.

Mensing, Lucy. 1925a. *Beitrag zur Theorie der Verbreiterung von Spektrallinien.* PhD Thesis. Hamburg: Hamburg University.

Mensing, Lucy. 1925b. Zur Störungsmechanik der Molekülmodelle. *Zeitschrift für Physik*, 34, 602–610. https://doi.org/10.1007/BF01328505

Mensing, Lucy. 1925c. Beitrag zur Theorie der Verbreiterung von Spektrallinien. *Zeitschrift für Physik*, 34, 611–621. https://doi.org/10.1007/BF01328506

Mensing, Lucy. 1926a. Die Rotations-Schwingungsbanden nach der Quantenmechanik. *Zeitschrift für Physik*, 36, 814–823. https://doi.org/10.1007/BF01400216

Mensing, Lucy. 1926b. Die Intensitäten der Zeemankomponenten beim partiellen Paschen-Back-Effekt. *Zeitschrift für Physik*, 39, 24–28. https://doi.org/10.1007/BF01321897

Mensing, Lucy. 1927. Zur Theorie des Zusammenstoßes von Atomen mit langsamen Elektronen. *Zeitschrift für Physik*, 45, 603–609. https://doi.org/10.1007/BF01331923

Mensing, Lucy and Pauli, Wolfgang. 1926. Über die Dielektrizitätskonstante von Dipolgasen nach der Quantenmechanik. *Physikalische Zeitschrift*, 27, 509–512.

Midwinter, Charles and Janssen, Michel. 2013. Kuhn losses regained: Van Vleck from spectra to susceptibilities. In Badino, Massimiliano and Navarro, Jaume (eds.), *Research and Pedagogy: A History of Early Quantum Physics through its Textbooks*. Berlin: Max Planck Institute for the History of Science, ch.7. https://www.mprl-series.mpg.de/studies/2/8/index.html

Münster, Gernot. 2020a. (K)eine klassische Karriere? *Physik Journal*, 19(6), 30–34.

Münster, Gernot. 2020b. *Quantentheorie*, 3rd ed. Berlin: De Gruyter.

Nernst, Walther. 1911. Zur Theorie der spezifischen Wärme und über die Anwendung der Lehre von den Energiequanten auf physikalisch-chemische Fragen überhaupt. *Zeitschrift für Elektrochemie und angewandte physikalische Chemie*, 17, 265–275.

Oppenheimer, J. Robert. 1926. On the quantum theory of vibration–rotation bands. *Mathematical Proceedings of the Cambridge Philosophical Society*, 23, 327–335. https://doi.org/10.1017/S0305004100009221

Paschen, Friedrich and Back, Ernst. 1912. Normale und anomale Zeemaneffekte. *Annalen der Physik*, 39, 897–932. https://doi.org/10.1002/andp.19123441502

Paschen, Friedrich and Back, Ernst. 1913. Normale und anomale Zeemaneffekte. Nachtrag. *Annalen der Physik*, 40, 960–970. https://doi.org/10.1002/andp.19133450507

Pauli, Wolfgang. 1921. Zur Theorie der Dielektrizitätskonstante zweiatomiger Dipolgase. *Zeitschrift für Physik*, 6, 319–327. https://doi.org/10.1007/BF01327993

Pauli, Wolfgang. 1926. Über das Wasserstoffspektrum vom Standpunkt der neuen Quantenmechanik. *Zeitschrift für Physik*, 36, 336–363. (English translation in van der Waerden (1968, pp. 387–415).) https://doi.org/10.1007/BF01450175

Pauli, Wolfgang. 1939. Über ein Kriterium für Ein- oder Zweiwertigkeit der Eigenfunktionen in der Wellenmechanik. *Helvetica Physica Acta*, 12, 147–168. https://doi.org/10.5169/seals-110936

Peierls, Rudolf. 1988. Was ich von Pauli lernte. In Enz, Charles P. and von Meyenn, Karl (eds.), *Wolfgang Pauli: Das Gewissen der Physik*. Braunschweig, Wiesbaden: Vieweg, pp. 68–74.

Pycior, Helena M., Slack, Nancy G., and Abir-Am, Pnina G. (eds.). 1996. *Creative Couples in the Sciences*. New Brunswick: Rutgers University Press.

Rammer, Gerhard. 2006. Sauberkeit im Kreise der Kollegen. Die Vergangenheitspolitik der Deutschen Physikalischen Gesellschaft. In Hoffmann and Walker (2006, pp. 359–420).

Ramsauer, Carl. 1921. Über den Wirkungsquerschnitt der Gasmoleküle gegenüber langsamen Elektronen. *Annalen der Physik*, 64, 513–540. https://doi.org/10.1002/andp.19213690603

Reich, Karin. 2011. Der erste Professor für Theoretische Physik an der Universität Hamburg: Wilhelm Lenz. In Schlote, Karl-Heinz and Schneider, Martina (eds.), *Mathematics Meets Physics*. Frankfurt: Verlag Harri Deutsch, pp. 89–143.

Reiche, Fritz. 1920. Zur Theorie der Rotationsspektren. *Zeitschrift für Physik*, 1, 283–293. https://doi.org/10.1007/BF01326909

Schirrmacher, Arne. 2019. *Establishing Quantum Physics in Göttingen: David Hilbert, Max Born, and Peter Debye in Context, 1900–1926*. Cham: Springer.

Schmidt, Nadin. 2015. *Die Deportation der wissenschaftlichen Intelligenz an den Universitäten der SBZ nach 1945 und deren Re-Integration an den Universitäten der Bundesrepublik Deutschland und der Deutschen Demokratischen Republik*. PhD Thesis. Leipzig: University of Leipzig.

Schmutzer, Ernst and Schütz, Wilhelm. 1975. *Galileo Galilei*. Leipzig: Teubner.

Schrödinger, Erwin. 1926. Quantisierung als Eigenwertproblem. (Zweite Mitteilung). *Annalen der Physik*, 79, 489–527. (English translation in Schrödinger (1982, pp. 13–40).)

Schrödinger, Erwin. 1982. *Collected Papers on Wave Mechanics*. Providence, RI: American Mathematical Society Chelsea Publishing.

Schütz, Lucy. 1975. Hinweis. In Schmutzer, Ernst and Schütz, Wilhelm, *Galileo Galilei*. Leipzig: Teubner, pp. 6–7.

Schütz, Wilhelm. 1926. Ein experimenteller Beitrag zur Kenntnis des Wirkungsquerschnitts angeregter Atome. *Zeitschrift für Physik*, 35, 260–275. https://doi.org/10.1007/BF01380296

Schütz, Wilhelm. 1927. Über natürliche Breite und Verbreiterung der D-Linien des absorbierenden Natriumdampfes durch Dampfdichte und Druck fremder Gase. *Zeitschrift für Physik*, 45, 30–66. https://doi.org/10.1007/BF01341824

Schütz, Wilhelm. 1936. Magnetooptik (ohne Zeeman-Effekt). In Wien, Wilhelm and Harms, Friedrich (eds.), *Handbuch der Experimentalphysik*, Band 16, Teil 1. Leipzig: Akademische Verlagsgesellschaft.

Schütz, Wilhelm. 1966. Friedrich Paschen und Pieter Zeeman zum 100. Geburtstag. *Jenaer Rundschau*, 11, 219–224.

Schütz-Mensing, Lucy. 1930. Zur Theorie der Kopplungsverbreiterung von Spektrallinien. *Zeitschrift für Physik*, 61, 655–659. https://doi.org/10.1007/BF01341175

Stachel, John. 1996. Albert Einstein and Mileva Marić: a colloboration that failed to develop. In Pycior, Helena M., Slack, Nancy G., and Abir-Am, Pnina G. (eds.), *Creative Couples in the Sciences*. New Brunswick: Rutgers University Press, pp. 207–219.

Tamm, Igor. 1926. Zur Quantenmechanik des Rotators. *Zeitschrift für Physik*, 37, 685–698. https://doi.org/10.1007/BF01403242

Uhlenbeck, George E. and Goudsmit, Samuel. 1925. Ersetzung der Hypothese vom unmechanischen Zwang durch eine Forderung bezüglich des inneren Verhaltens jedes einzelnen Elektrons. *Die Naturwissenschaften*, 13, 953–954.

Van der Heijden, Margriet. 2021. *Denken is verrukkelijk: het leven van Tatiana Afanassjewa en Paul Ehrenfest*. Amsterdam: Prometheus.

Van der Waerden, Bartel L. (ed.). 1968. *Sources of Quantum Mechanics*. New York: Dover.

Van Vleck, John H. 1924. The absorption of radiation by multiply periodic orbits, and its relation to the correspondence principle and the Rayleigh–Jeans Law. Part I. Some Extensions of the Correspondence Principle. Part II. Calculation of Absorption by Multiply Periodic Orbits. *Physical Review*, 24, 330–346 (Part I). https://doi.org/10.1103/PhysRev.24.330; 347–365 (Part II). https://doi.org/10.1103/PhysRev.24.347 (Part I reprinted in van der Waerden (1968, pp. 203–222).)

Van Vleck, John H. 1926. Magnetic susceptibilities and dielectric constants in the new quantum mechanics. *Nature*, 118, 226–227. https://doi.org/10.1038/118226c0

Van Vleck, John H. 1932. *The Theory of Electric and Magnetic Susceptibilities*. Oxford: Oxford University Press.

5

Discouraging Jane: Dewey Among the Lucky Generation of US Physicists

ADRIANA MINOR

5.1 Introduction

Jane Mary Dewey (1900–1976) was the only US woman involved in quantum physics research in the 1920s. In that sense, she was a trailblazer in science at a time of deep intellectual and institutional transformation. Within the US during this decade, physics departments rapidly improved their research programs as a result of multiple investments by universities, private foundations, and government agencies. Such improvements included inviting leading European physicists, organizing specialized seminars, endowing high-level universities with grants, giving fellowships to promising students to train in the main European research centers, and hiring them when they returned to the US (Coben, 1971; Kevles, 1979; Schweber, 1986). Dewey was the first US woman to be awarded this type of support for postdoctoral research in quantum physics. She received one fellowship from Barnard College and another from the National Research Council (NRC). Her research on the Stark effect in helium (the splitting of its spectral lines when placed in an external electric field), performed during her stay in the Niels Bohr Institute for Theoretical Physics in Copenhagen, produced such promising results that it should have sufficed to guarantee her a position in a US physics department. Thereafter, Dewey had a few opportunities to continue her work in academia and physics research, but the difficulties she encountered outweighed those opportunities, thus distancing her from quantum physics. She therefore failed in her efforts to consolidate a scientific career in her chosen field.

In light of this, why is her case meaningful in a book about women in the history of quantum physics? This chapter reclaims a place for Jane Dewey in the history of quantum physics, considering her involvement in Bohr's research program, which

I thank the reviewers and editors of this volume for so many helpful suggestions. I also thank Christian Joas (Niels Bohr Archive) for providing me with digital copies of Jane Dewey's and John Stuart Foster's files. Last, but not least, my appreciation to the whole WiHQP group for sharing this journey.

focused on the so-called *new quantum theory*, as well as her efforts to continue in this field of research in the US. It also sheds light on *her*stories of women physicists who found themselves unable to continue in the physics profession because of patterns of gender discrimination in science, whether explicit or subtle, but almost always structural and systemic. Their remarkable abilities for research, promising results, and outstanding expertise were not sufficient to guarantee a position that would allow them to continue a life in science. Moreover, they found themselves excluded not only from science, but from history of science as well.

This chapter is based on a mix of primary and secondary sources, including archival collections located at the Niels Bohr Archive and the Rockefeller Archive Center that are directly related to Dewey's initial research in quantum mechanics. Besides focusing on her work in this field, this chapter also considers Dewey's prior education and her incursion into the professional world outside of quantum research and academia. For these time periods, I have used digitized collections of the women's colleges at which she studied (Barnard College) and worked (Bryn Mawr College).

Among secondary sources, I have found great inspiration in the work of three pioneering historians of physics: Margaret W. Rossiter, Katherine Russell Sopka, and Shannon Melinda Davies. In particular, Sopka and Davies focused on the first generation of US physicists who specialized in quantum mechanics, the former giving a general view of this generation and the latter studying specifically those who went to Bohr's institute. All three authors mention Dewey, providing some valuable – albeit, limited – details about her background and scientific work. Only Sopka approached Dewey directly as part of her historical research (Sopka, 1988, p. 166). Dewey was not interviewed by other historians, nor selected for inclusion in the Archive for the History of Quantum Physics (Davies, 1985, p. 147). More recently, Carlos Stroud, emeritus physicist at the University of Rochester, dedicated a brief chapter to Dewey in a volume devoted to commemorating Rochester's Institute of Optics. There, he emphasized Dewey's pioneering work in quantum optics (Stroud, 2004). With the exception of these works, the history of physics has paid scant attention to Dewey's scientific career.

This chapter reflects, first, on the common characteristics of the generation of US physicists to which Dewey arguably belonged, and the crucial differences that affected her scientific career (Section 5.2). Second, it provides some background details regarding her family and education (Section 5.3) and discusses her interest in studying at the University of Copenhagen, her research expectations, involvement in Bohr's research program, and efforts to continue funded research work in Europe (Section 5.4). Third, it analyzes the significance of Dewey's results concerning the Stark effect in helium and a priority competition in which she was unconsciously involved, that ended up overshadowing her contributions to quantum mechanics (Section 5.5). Finally, I refer to the opportunities and difficulties she faced to secure a position in the US, and thereby how she became excluded from quantum mechanics research (Section 5.6). There are many fascinating aspects of Dewey's life that this chapter explores only briefly and in relation to its main focus. These include her involvement in progressive educational endeavors fostered by the Dewey family, her sources of inspiration and influence (her parents and, in particular, the philosophical and scientific thought of her father, John Dewey),[1] as well as her research for the US Army in the field of ballistics. Each of these topics calls for further research.

5.2 Foundations of the Lucky Generation of US Physicists

Inspired by an article describing *the lucky generation* of US writers (those associated with literary modernism), physicist John Slater wrote in his autobiography that there was an analogously *lucky generation* of US physicists.[2] This generation was mainly characterized by the fact that their professional development took place at an exciting time of profound change in physics, and accordingly, by their major contributions to the development of quantum mechanics. Furthermore, the lucky generation had the opportunity to travel to Europe in the 1920s to train at important physics research centers with the most remarkable physicists of the time, which of course included Bohr.[3]

Dewey shared many characteristics with Slater and the generation he described. For instance, both were born in Illinois at the turn of the century: Dewey on July 11, 1900, in Chicago, and Slater on December 22 of the same year in Oak Park, a suburb of the very same city. Both physicists completed their undergraduate studies in New York State, Dewey at Barnard College (a women's college associated with Columbia University, where her father was Professor of Philosophy at Teachers College), and Slater at the University of Rochester (where his father was head of the English Department). Slater and Dewey both continued their graduate studies in Cambridge, Massachusetts, he in physics at Harvard University; she in physical chemistry at the Massachusetts Institute of Technology (MIT). Both then went to Bohr's institute, supported by fellowships from their respective universities – he from Harvard, she from Barnard – and they both were granted an honorary appointment (without stipend) from the American–Scandinavian Foundation, an opportunity which introduced them to Bohr. Last but not least,

---

[1] In this chapter, "Dewey" refers to Jane Dewey. Her father is referred to as John Dewey.
[2] John Clarke Slater, "A physicist of the lucky generation," pp. 1–2. MIT Libraries Institute Archives and Special Collections (from now on, MIT Archives), John C. Slater Papers, Box 1.
[3] Information about US scholars who studied in European centers of quantum physics (1926–1929) in Sopka (1988, pp. 166–167). An exhaustive list of visitors at Bohr's institute up to 1927 in Kojevnikov (2020, pp. 107–109).

both physicists conducted research in Copenhagen in the field of quantum mechanics at the precise time that has been identified as an historiographically relevant turning point between the old and the new quantum theory.

However, there are also important differences regarding their scientific careers. Slater, and virtually all US scientists who worked at Bohr's institute in the 1920s, found a permanent position in a physics department when returning to the US. According to Davies, "Until Dewey, all the physicists who had been to Copenhagen to study with Bohr had returned to academic positions in America that ensured rapid promotions and the opportunity to participate in the formation of an American core of teaching and research in quantum theory" (Davies, 1985, p. 121). Slater returned to Harvard and in 1930 was then invited to become the head of the Department of Physics at MIT by Karl T. Compton. At the time, Compton (the brother of Arthur H. Compton, the discoverer of the Compton effect) had recently been appointed as MIT President, and he wanted to make its physics department a leader in training and research, just as he had previously done at Princeton University. At Princeton, Dewey arrived as an NRC fellow after her stay in Copenhagen. Slater remained at MIT until his mandatory retirement at 65, and then continued his career at the University of Florida (Morse, 1982).[4] Dewey, however, never got a permanent faculty position.

5.3 The Early Twentieth-Century World of Women in Education and Science

Margaret Rossiter has conclusively identified the structural difficulties faced by US women in higher education and science in the nineteenth century and beyond. Despite this discouraging reality, Rossiter also points out that in the early twentieth century there were people – mostly women – concerned with providing high-quality education for women beyond traditional housekeeping-related subjects. Some were able to have scientific careers of their own, becoming models for younger generations (Rossiter, 1984, pp. 1–28). Progressive ideas about female education became an especially significant influence in Dewey's early life.

Jane Mary Dewey was the youngest daughter of the famous pragmatist philosopher and educator, John Dewey, and the educator and feminist, Alice Chipman, a couple profoundly committed to changing traditional educational models. Although her father gained prominent public notoriety in these matters, Dewey later remarked that her mother "was undoubtedly responsible for the early widening of [John Dewey's] philosophic interests from the commentative and classical to the field of contemporary life ... things which had previously been matters of theory

---

[4] Slater, "A physicist of the lucky generation." See Note 2.

acquired through his contact with her a vital and direct human significance" (Dewey, 1939, p. 21). The couple's common ideas on education reform crystalized into the Laboratory School – an elementary school that they created at the University of Chicago in 1896, four years before their daughter Jane was born. Dewey's father directed this school until 1904, while her mother and the Chicago educator Ella Flagg Young served as its official administrators (Durst, 2010, p. 50). For John Dewey, both female educators were "the greatest influence in educational matters of those days" (Dewey, 1939, p. 29). The new and experimental school was revolutionary in its organization and educative methods, featuring a democratic practice in teaching and learning, collaborative work among teachers and students, and a curriculum strongly connected with applications in society. The Dewey family became part of a community of school reformers in Chicago, with whom they shared ideals about philosophy, education, and democracy as a way of life. Especially influential for the family were Jane Addams, a member of the settlement movement of social reform, and the Hull House, an organization that Addams co-founded and managed, offering educational and other opportunities to working class people on a similar philosophical basis as the Dewey School. The Deweys expressed their admiration by naming their youngest daughter after Addams – "one of the most remarkable women of her day" – and her life partner Mary Rozet Smith. Jane Mary grew up surrounded by important female figures who implemented progressive initiatives relating to education, social inclusion, and the role of women in society. She would later note that John Dewey attributed "much of the enthusiasm of his support of every cause that enlarged freedom of activity of women to his knowledge of the character and intelligence of his wife, of Ella Flagg Young, and of Jane Addams" (Dewey, 1939, p. 30).

Dewey's childhood was also marked by family trips to Europe, including one when her father resigned from the University of Chicago in 1904. It meant the end of the Laboratory School and the closing of the Dewey family residence in Chicago (Dewey, 1939, pp. 34–35). In the opinion of her mother, travels would benefit the growth, culture, and language learning of their children, while they attended schools and experienced educational models existing in Europe (Hall, 2005, pp. 44–45, p. 112). The Deweys wanted their children and their pupils to be educated with methods that promoted independence and self-reliance (Hall, 2005, p. 37). When the family returned to the US and settled in New York City in 1906, Jane and her sister Lucy registered at the Ethical Culture School, which they attended until 1909. The girls' new school promoted ideals of secular humanism, communitarianism, and social justice similarly to the Deweys' Laboratory School. Its founder and rector, Felix Alder, was among John Dewey's contacts, and they met regularly at the New York Philosophical Club (Dewey, 1939, pp. 37–38;

Martin, 2002, p. 215 and p. 219).[5] Between 1914 and 1917, Jane attended the Spence School, a high school for girls founded and headed by the innovative educator Clara B. Spence. Provided with encouragement and options to develop her talents, Jane stood out among her siblings for her mathematical abilities (Martin, 2002, pp. 219–220).[6] Distancing herself from the world of progressive education enterprises that attracted many in the Dewey family, Jane gradually leaned into a scientific path (to a certain extent also cultivated by her father).

In 1919, after a year of undergraduate courses at the University of California, Berkeley, Dewey joined her sisters, Evelyn and Lucy, as an undergraduate student at Barnard College, part of the educational model for women associated with Columbia University. Her father had joined the faculty at Columbia after leaving Chicago (Martin, 2002, p. 221). At the time Dewey studied at Barnard, physicist Margaret Eliza Maltby was departmental chair. Maltby was the first woman to receive a BSc at MIT and to graduate with a PhD in physics from the University of Göttingen in Germany (Behrman, 2020). She might have had Dewey in her class, and might have been a source of inspiration, given Maltby's conscious role in encouraging students interested in science (Behrman, 2020, pp. 49–50). Further research in the archives of Barnard College might provide specific information about Dewey's performance there, and the inspirations and supports that oriented her towards science, but that avenue is left for future consideration.[7]

After graduating from Barnard in June 1922 (see Figure 5.1), Dewey continued her graduate studies at MIT in the field of physical chemistry. Three years later, she was awarded a PhD for a thesis supervised by Duncan MacInnes, an expert in chemical analysis with X-rays. MacInnes was associate professor of the MIT Research Laboratory of Physical Chemistry, the laboratory created by the famous physical chemist Arthur Noyes. Under Noyes, the laboratory had achieved enough prestige to attract a very important group of chemists, MacInnes among them (Servos, 1990, p. 84). Dewey was appointed as research assistant in the laboratory for the last year of her PhD (Massachusetts Institute of Technology, 1925, xii).

At that time, a very small group of MIT physicists, engineers, and mathematicians – Manuel Sandoval Vallarta and Norbert Wiener, among others – began to work on theoretical physics, writing the first theses and articles on the old quantum theory (Sopka, 1988, p. 94; Minor García, 2019, pp. 70–71). Also notably, in 1924 there were

---

[5] J. Robert Oppenheimer is possibly the most famous student from the Ethical Culture School in New York City, which he attended from 1911 to 1921 (Schweber, 2000, p. 42). Although this school had several students coming from Jewish families, the Deweys were not Jewish themselves. Presumably, this school attracted the Deweys because of its core values.

[6] The Rockefeller Archive Center (from now on, RAC), International Education Board, series 1: Appropriation, subseries 3: Fellowships in science, box 47, folder 696, "Jane Dewey 1925–1928, Personal History Record Submitted in Connection with Application for a Fellowship, Jane Mary Dewey (Clark), April 24, 1926."

[7] The yearbook, available in digitized format, includes two mentions of Dewey. She was described as "cold" and "most bored" (see Figure 5.1) (The Class of 1922 at Barnard College (1921, p. 150 and p. 171)).

JANE DEWEY
New York City

Figure 5.1 Jane Dewey, c. 1921 in The Class of 1922 at Barnard College (1921, p. 150).

two women instructors in the MIT Department of Physics (Massachusetts Institute of Technology, 1923, p. 33). According to one of them, Dorothy Weeks, female instructors were hired only because physics professors were scarce after World War I (Weeks, 1978).[8]

In 1925, when Dewey earned her PhD, her first research paper, titled "The transference numbers of sodium and potassium chlorides and of their mixtures," appeared in the *Journal of the American Chemical Society* (Dewey, 1925). At the time, she was also preparing a paper on the experimental determination of gauge constants of a cylindrical piston under high pressures, co-authored with Frederick G. Keyes, director of the Research Laboratory of Physical Chemistry and head of the MIT Department of Chemistry, with whom she worked as a research assistant

---

[8] In this interview, Weeks commented on the circumstances of her employment at MIT: "This is one thing that I've always resented, that my opportunities have come through wars, which is no way for women to get their opportunities."

(Keyes and Dewey, 1927). Interestingly, Dewey signed these publications and the rest of her articles as *Jane M. Dewey*, even after getting married in 1925 to MIT physics instructor John Alston Clark. In some institutional letters, she added her husband's surname in parentheses. Certain scientists with whom she worked insistently referred to her as *Mrs. Clark* (see Section 5.4); yet she clearly preferred to be known as a Dewey in the scientific world.

5.4 A US Woman Scientist in Copenhagen

In April 1925, Dewey sent a letter to Bohr expressing interest in studying at the University of Copenhagen the following winter. She proposed a

study of the absorption spectrum of a gas, such as carbon dioxide or ammonia, which is supposed to be associated with the object of determining the presence or absence of double molecules at various temperatures. If this particular problem does not work in with [sic] the research problems under way in one of the laboratories at Copenhagen I should like to take up some problem more convenient to the laboratory in the general field of radiation chemistry. I have not worked in this field but have done research in connection with the theory of solutions and with equations of state in the Research Laboratory of Physical Chemistry here during the past three years.[9]

Having received a positive response from Bohr, Dewey would travel to Copenhagen on a fellowship from Barnard College, courtesy of a newly created fund that aimed to promote advanced study abroad among Barnard's graduates. As the first awardee, she expressed her appreciation, writing: "I am deeply grateful for the opportunity which it affords me to study with men prominent in science abroad and to come in contact with the European method of approach to scientific problems" (The Barnard Bulletin, 1925a, p. 2).

In addition, Dewey received an honorary appointment from the American–Scandinavian Foundation, which also awarded a fellowship to her husband. It is not clear where and what kind of research Dewey's husband conducted in Copenhagen (Davies, 1985, p. 119). According to Charles Ladd Norton, a contemporary physicist at the MIT Department of Physics, Clark was not a good experimentalist (Davies, 1985, p. 118). The only letter preserved in the Niels Bohr Archive about him is one in which the American–Scandinavian Foundation introduces him as one of its fellows.[10] By contrast, Dewey's association with Bohr's institute is well documented.

When Bohr agreed to host Dewey, he demurred as to the type of research that could be done in his institute:

---

[9] Jane M. Dewey to Niels Bohr, April 18, 1925. Niels Bohr Archive, Niels Bohr General Correspondence, series 1: People and Institutions, box 8: Da–Dz, folder 23: Dewey, Jane M. (Clark), 1925–1928.
[10] The American–Scandinavian Foundation to Niels Bohr, July 3, 1925. See Note 9.

I understand that you wish primarily to study the absorption spectra in associated gasses. In this institute, however, we are not well equipped for such work as we are especially prepared for spectral studies from the purely physical point of view, but I think it will be the most practical to postpone a discussion of the subject of your work until your arrival in Copenhagen.[11]

Bohr's institute had been created only four years earlier, in 1921. At the time Dewey intended to visit, the premises were being remodeled and expanded as a result of a sizable grant from the International Education Board (IEB) after Bohr's visit to the US in 1923 (Cassidy, 2009, p. 135). Created as a program of the Rockefeller Foundation, the IEB supported US and European scholars who wished to study abroad and provided substantial funds to European scientific institutions. The grant to Bohr's institute stood out as one of the first and most important. Despite its youth, Bohr's institute soon became one of the leading quantum physics research centers of its time, along with Arnold Sommerfeld's in Münich and Max Born's in Göttingen (Kragh, 1999, p. 159).

Dewey adapted her research as soon as she arrived at Copenhagen, as she reported to the Dean of Barnard College (The Barnard Bulletin, 1925b, pp. 1–2):

I have talked to Dr. Bohr twice and understand that there will be apparatus and space for me later, although there is no space now. . . . He had one or two ideas for problems he would like me to work on and it may be better to take up one of them than the one I had planned on myself in order to get the necessary apparatus and help. In any case, they are all along the same line so that I can decide on that while I am doing some reading on the subject and the laboratory is being [prepared]. There are some advanced courses given at the institute, one by Bohr and one by his assistant [Hans Kramers]. I think I shall take Bohr's course in atomic theory and a course in differential geometry and attend the physics colloquium, and for the rest confine myself to experimental work.

Some months later, in April 1926, she detailed the type of work she was pursuing and the research expectations she had for the following months. She was measuring the intensities of the various components of spectral lines split by the Stark effect:[12]

I have been measuring the intensities of lines in the spectrum emitted by helium in an electric field, a research which permits an experimental check on certain of the recently developed ideas for the quantum theoretical treatment of atomic problems. I should like to continue these measurements next fall and extend them to the hydrogen spectrum. I have also been studying the quantum theory in general and applied to this research. Another year I should like to devote more time to theoretical work, particularly in the quantum mechanics. My general purpose in this study is to obtain a sufficient command of the theory to be able to work on applications of it to chemical theory, particularly to energy emission and absorption in chemical reactions.[13]

---

[11] Bohr to Dewey, May 10, 1925. See Note 9.
[12] For discussion of the Stark effect in both the old and new quantum theory, see Duncan and Janssen (2014, 2015).
[13] "Personal History Record Submitted in Connection with Application for a Fellowship, Jane Mary Dewey (Clark), April 24, 1926." RAC, International Education Board, series 1: Appropriation, subseries 3: Fellowships in science, box 47, folder 696 Jane Dewey 1925–1928.

This description, which was included in Dewey's application for a fellowship to the IEB, contrasts with the vague goal of working "in some European University in Physical Chemistry next year," that she had drafted when she first approached the IEB in November 1925.[14] Since the time of her initial request of information about the IEB fellowship program, her plans had crystalized into continuing her research in Copenhagen and spending additional time at the University of Göttingen in Germany "in order to study with men who are giving particular attention to applications of quantum theory to chemistry," specifically with Max Born and James Franck.[15]

Dewey's IEB application was sponsored by MacInnes and seconded by Bohr. In their respective support letters, both stated that *Mrs. Clark* had remarkable abilities for research. Bohr recommended her emphatically, writing:

I do not hesitate in supporting this proposal with my heartiest recommendation. Mrs. Clark has worked in this institute during the last half year and has during her stay here shown unusual ability for research work. She has been engaged in an investigation on the intensities of Stark effect components, a problem which through the recent development of the quantum theory has become of great importance for the theory of atomic structure. In her experimental work she has shown great perseverance and skill and also as regards the theoretical discussion of the results she has shown ability of penetration into quite complicated lines of thought and possession of independent judgment.[16]

Three weeks later, he sent another letter:

I allow me [sic] to write again about the proposal of a fellowship for Mrs. Clark. The reason is that when writing my letter of April 9, I was not informed about Mrs. Clark's plans for the coming year. ... Recently Mrs. Clark has obtained some interesting results as regards the polarisation and intensity of anomal [sic] lines in the helium spectrum, excited by electric field, of which she hopes soon to publish an account. However, she has spent much time in setting up and mastering the experimental arrangement, and she ought now with this to be able within a short time to investigate several problems of considerable theoretical interest.[17]

These words from Bohr lend themselves to conflicting interpretations. Bohr might have pointed out Dewey's difficulties in setting up her experimental arrangement as a strategy to convince the IEB to provide funds to allow Dewey to complete her promising research. However, IEB officials read them in less favorable terms, interpreting Bohr's letter as diminishing Dewey's accomplishments. In an exchange between two members of the selecting committee, Augustus Trowbridge to Wickliffe Rose, one section reads:

I must say I am not tremendously impressed with the information about the candidate so far as I have it. [Bohr says] that now she is over this difficulty and may use her time to

---

[14] Dewey to the International Education Board, November 7, 1925. See Note 13.
[15] "Personal History Record Submitted in Connection with Application for a Fellowship, Jane Mary Dewey (Clark), April 24, 1926." See Note 13.
[16] Bohr to Augustus Trowbridge, April 6, 1926. See Note 13. For MacInnes' letter to Trowbridge, see Note 20.
[17] Bohr to Trowbridge, April 26, 1926. See Note 13.

advantage, if appointed. Altogether this is a rather weak letter from Bohr, whose style I have come very thoroughly to know. . . . In an earlier letter he is somewhat more enthusiastic . . . Bohr's two letters are written in somewhat different tone and MacInnes is so cold and formal that he throws no light on the subject at all.[18]

MacInnes had endorsed her application expressing his conviction that she "will make a definite contribution to science. She has unusual originality, energy, and perseverance."[19] However, he attributed some of these abilities to inheritance from her father.

Such doubts about Dewey's merits may help explain the IEB decision to deny a fellowship to *Mrs. Clark*. The decision-makers noted, "our inclination is to act upon the principle of not providing a fellowship in any case where there is doubt. Unless the case should be convincing to you we should be disposed to let the matter rest. I assume that if the case were thoroughly meritorious the letters of the proposers would have been sufficiently convincing."[20]

After Dewey's unsuccessful application to IEB, Bohr obtained a supplementary fellowship for her from the Danish Rask–Ørsted Foundation, thus allowing her to continue working with him for a total of 18 months (Davies, 1985, p. 120). Notwithstanding Bohr's support in this capacity, Davies suggests he might have intentionally delayed the publication of Dewey's second article resulting from her research, instead favoring another physicist, John Stuart Foster, who was also studying the Stark effect and had obtained similar results to hers (Davies, 1985, p. 148; see Section 5.5).

Davies' interpretation is perfectly plausible within the framework of Bohr's general policy of requiring his personal approval for all publications by fellows of his institute. For instance, in 1922 Wolfgang Pauli had to wait to publish his calculations of the helium spectrum until Bohr's personal assistant Hendrik Kramers finished his own (Kojevnikov, 2020, pp. 54–55). In this sense, it is difficult to interpret Bohr's caution regarding Dewey's research and his preference for a competing male physicist as having been explicitly motivated by sexism; indeed, it might have been attributable to Bohr's intellectual ambitions about quantum theory. To be sure, the ambivalent support he gave Dewey contrasts with his unequivocal promotion of Foster. Certainly, Bohr's judgment impacted the two physicists' respective futures in the field.

5.5 Struggles for Priority and Recognition

When Bohr first expressed an opinion on Dewey's IEB application, he also signaled that he would soon send the application of John Stuart Foster. Foster had worked on

---

[18] Trowbridge to Rose, May 18, 1926. See Note 13.   [19] MacInnes to Rose, May 13, 1926. See Note 13.
[20] Rose to Trowbridge, June 2, 1926. See Note 13.

the Stark effect since his doctoral studies at Yale University and had maintained an active correspondence with Bohr about his research since 1924. After graduating that same year, Foster had been hired at McGill University in Montreal (Bell, 1966; see also Chapter 6). In early 1926, he expressed to Bohr his hope to work in Copenhagen. Bohr reacted enthusiastically and advised Foster to apply for an IEB fellowship to fund his stay.[21] Foster's application, sponsored by Bohr, was successful, and the grant afforded him several months in Copenhagen during the second half of 1926, which was ample time to discuss his previous results on the Stark effect in helium, as well as to perform calculations using matrix mechanics and to design a theory–experiment comparison.[22] There, he overlapped with Dewey.

After Foster's return to Canada, he exchanged letters with Bohr about an article that he was writing and that Bohr planned to present to the Royal Society. From one of his letters, it can be inferred that he had asked Bohr to delay the publication of an article that Dewey was preparing: "I cannot ask you to further delay the publication of Mrs. Clark's paper and I regret my inability to get my paper in final form."[23] One month later, in June 1927, he wrote:

Mrs. Clark kindly sent a copy of her paper to me. As my results are of a qualitative nature only, I think there is no disagreement. *I was surprised at her statement about displacements being greater when the transition was to the 2s level.* My measurements on 2S–5Q are in that direction, but they are <u>rough</u>, and I thought it accidental (emphasis added).[24]

Foster wrote this before Dewey published her second paper reporting the last results she obtained at Copenhagen. As such, it might seem natural to assume that Foster is referring here to her first paper, sent to the *Physical Review* in August 1926 and published that December, when both were at Bohr's institute (Dewey, 1926). However, his mention of Dewey's results regarding the 2s level rather points to her second paper, in which she asserted (Dewey, 1927, pp. 778–779):

The deviation of the measured intensities of the orthohelium lines from the calculated intensities is considerably larger than for the parhelium lines. Particularly in the groups with the final state 2s the deviations are very large. The polarization of the weak lines does not agree with that given by the theory and the deviations cannot be regarded as depending only on the initial states. The deviations are in the same direction for lines with the same initial and different final state but are much larger for those with the final state 2s.

In this regard, she concluded that (Dewey, 1927, pp. 779–780):

Calculations from the quantum theory of perturbations on the assumptions used here and to this approximation give only approximate values for the intensities of the new lines

---

[21] Foster to Bohr, May 31, 1926. Niels Bohr Archive, Niels Bohr Scientific Correspondence, 1903–1962, folder 90, "Foster, John Stuart."
[22] Foster to Bohr, June 24, 1926. See Note 21.   [23] Foster to Bohr, May 5, 1927. See Note 21.
[24] Foster to Bohr, June 27, 1927. See Note 21.

appearing in the Stark effect. The deviations are such as to suggest that they are partly due to differences in the number of atoms excited to different states of almost the same energy. Apparently, factors come into play in the case of orthohelium other than those which influence the parhelium lines. No explanation of the deviations suggests itself. They may be due to the influence of the atoms on each other, which is here entirely neglected, and to the fact that we have neglected higher approximations of the theory in making the calculations.

We can thus assume that Dewey shared with Foster a draft of her second paper and that, after reading it, he was more confident about his own results. What also emerges is that Foster was explicitly competing with Dewey for priority and recognition. Coincidently or not, Dewey sent her second paper to *Physical Review* the following month, in July, and Bohr communicated Foster's article to the Royal Society in early August. Both articles were published in December 1927.

Dewey prepared two articles regarding her research on the Stark effect in helium while in Copenhagen. As mentioned above, she tackled this problem at Bohr's suggestion at a crucial moment in the transition from the old quantum theory of Bohr and Sommerfeld to the new quantum theory of Werner Heisenberg, Erwin Schrödinger, and others. Her first paper is on the intensity in the Stark effect in helium. She used the Lo Surdo method for obtaining emissions in an electric field.[25] Theoretically, she referred to very recently published articles (Pauli, 1925; Kramers and Heisenberg, 1925; Kronig, 1925; Born et al., 1926; Heisenberg, 1926a, 1926b; Schrödinger, 1926a). Regarding previous research on the Stark effect in helium, she mentioned Foster (1924) and asserted that her research was a continuation of Takamine and Werner (1926). Foster, in his article published in the *Proceedings of the Royal Society* (1927), recognized that "The details of the application [of quantum mechanics] to helium, as well as the quantitative measurement of intensities with which the theoretical estimates are compared, were initiated by Takamine and Werner and greatly extended by Dewey" (Foster, 1927, p. 139). This is the only reference to Dewey's work in Foster's article.

By the time Dewey published her second paper, she had sufficiently mastered the new quantum theory to calculate the intensities of the various lines in the Stark effect in helium. She could therefore conclude that there was "a qualitative but not a quantitative agreement between theory and experiment" (Dewey, 1927, p. 776). Her conclusions were cautious in contrast to Foster's. In her paper, Dewey further referred to Foster's paper as "in press," and she acknowledged that he "was kind enough to show her his results on calculations of the wave-lengths in advance of publication" (Dewey, 1927, p. 780). Thus, Foster and Dewey seemed to have shared their research results. Foster acknowledged this in his private letters to Bohr, but Dewey did so in her publication.

---

[25] On this method, see Leone et al. (2004). See also Chapter 6.

Foster's 1927 article was decisive for his career (Bell, 1966, p. 148). It was a long paper in which he summarized his research on the Stark effect in helium and offered an interpretation of his results in terms of the new quantum theory, drawing on essentially the same articles as Dewey. He concluded (Foster, 1927, pp. 162–163):

On the whole, *the theoretical intensities in helium are in very good agreement with the present qualitative observations*. The agreement is possibly even better than in the case of hydrogen as presented in a recent paper by Schrödinger [1926b].[26] Lo Surdo photographs of the Balmer lines published by the writer show slight qualitative variations from Prof. Stark's experimental results, and agree better with the new theory. It seems most probable to the writer that the principal cause of variations in the experimental values is a small change in the field strength during the very long exposures. Many workers on the Stark-effect have noticed that such fluctuations are not uncommon, and unfortunately they are rather abrupt when the field and light intensity are unusually high (emphasis added).

The Stark effect in hydrogen had been a success for the old quantum theory. Problems with that theoretical description were recognized only in hindsight, when Stark's intensity measurements were found to be wrong (of utmost relevance for the work of Dewey, and of Foster and his PhD student, Laura Chalk[27]) and brought out a puzzling coordinate dependence of the shape of the electron orbits (Duncan and Janssen, 2014; Duncan and Janssen, 2023, pp. 523–529). Discussions about these problems and the new theoretical approaches around the crucial year of 1925 would have animated Bohr's institute, where a group of clever and ambitious physicists gathered, among them Heisenberg. He spent the summer and fall of that year in Copenhagen and starting in mid-1926 was appointed as Bohr's assistant, replacing Kramers. Heisenberg also overlapped with Dewey and with Foster (arriving that crucial year as well). Foster and Dewey both acknowledged the role of Heisenberg in their research, but in different terms: the former "for many helpful and friendly discussions," and the latter "for his assistance and suggestions on the theoretical part" (Foster, 1927, p. 163; Dewey, 1927, p. 780). Foster and Heisenberg, in fact, developed a friendship during their time together in Copenhagen.

Heisenberg retrospectively remarked that Foster's article was essential insofar as it demonstrated the power of the new quantum theory by explaining the Stark effect in helium. In so doing, Heisenberg reinforced credit to Foster and overlooked other results, notably Dewey's, writing (Heisenberg, 1967, p. 102):

Foster's beautiful measurements of the Stark-effect in the helium spectrum played an important role . . . Foster had come to Copenhagen for a short stay from Canada to compare his results with those of the new theory. . . . The agreement was perfect, and we were happy to see how many of the most complicated and apparently unconnected details resulted more or less automatically from the formulae of quantum mechanics. Bohr too was glad to note

---

[26] For an analysis of this paper, see Duncan and Janssen (2014, pp. 74–76).   [27] See Chapter 6.

how the Stark-effect once again, just as ten years earlier with the hydrogen atom, proved one of the most beautiful confirmations that one was on the right road to an understanding of the atom.

As mentioned above, Foster had done research on this topic since the time of his PhD at Yale University, and he continued in the following years, assembling a group of students and collaborators at McGill University. According to his letters to Bohr, he pursued this experimental research in an environment which had been hostile towards quantum physics at McGill.[28] Possibly, his urgency to secure priority in reporting his results about the Stark effect in helium came as a consequence of the pressure to demonstrate the capabilities of his group, as Monaldi suggests (Chapter 6). In counterpoint, Dewey was a newcomer when she studied the Stark effect, following Bohr's suggestion, as she acknowledged in her own articles. Yet, she did it skillfully. Her research, considered promising by Bohr, contributed to deciphering the Stark effect in helium, as even Foster himself recognized. Dewey's work showed a qualitative agreement between experiment and quantum mechanics, as she – perhaps too cautiously – stated. Afterwards, she returned to this research only intermittently. We cannot say that she was less interested in it, but it is clear that she did not find the right conditions to continue in this field.

5.6 A Quantum Physics Woman in the Professional World

When Dewey returned to the US in 1927 (close to her mother's passing), she did so with a fellowship from the NRC to work at Princeton University. The fellowship was renewed for a second year starting in October 1928.[29] Dewey was the first woman to pursue a postdoctoral position in physics at Princeton (Rossiter, 1984, p. 365). William Francis Magie (one of the founders of the American Physical Society and the first professor of physics at Princeton) strongly supported her admission. Because of that, she was nicknamed *Magie's folly* by other members of the physics faculty, who endorsed the opinion that "no woman could do physics" (Kevles, 1979, p. 207).[30] During this period, Dewey prepared five articles: the first two co-authored with physicist Howard P. Robertson (another NRC fellow) on the theory of the Stark effect and spectral lines; the next two dealing with an experimental problem suggested by Karl Compton, then head of Princeton's Department of Physics; and the last one as a sole author again on the Stark effect (Dewey and Robertson, 1928; Robertson and Dewey, 1928; Dewey, 1928, 1929, 1930). It seems

---

[28] Foster to Bohr, May 5, 1927. See Note 21.
[29] National Research Council, Minutes, April 26, 1928. See Note 13.
[30] Princeton's Department of Physics was the home institution of Trowbridge, the director of the IEB Division of Natural Sciences who had declined Dewey's application. See Section 5.4.

that she had gained momentum in her research activities, having developed a strong record of publication during the years immediately following her return to the US. In recognition of her achievements, Compton tried to help her secure a permanent job. However, his efforts proved unsuccessful, not only due to the ongoing economic crisis, but also because of gender prejudices. Compton "wrote all over the country," but the only expression of interest came from someone at Berkeley, "who regretted to report that his colleagues simply refused to have a woman on the staff" (Kevles, 1979, p. 207).

In the fall of 1929, Dewey thus moved to the University of Rochester as a research fellow, while her husband took a position at Cornell University as an instructor in mathematics (Stroud, 2004). In Rochester, she worked at the Institute of Applied Optics directed by Thomas Russell Wilkins. Stroud explains the difficulties she faced in securing appropriate space for setting up her laboratory and continuing her experiments. In addition, Dewey wrote in one of her letters to her father, quoted by Stroud, that every move she made required the approval of Wilkins; she noted that he was not especially dominant, but just couldn't seem to conceive of a different way of doing things (Stroud, 2004, p. 26).

In 1931, Dewey moved to Bryn Mawr College, a women's college in Pennsylvania, and she took a position as an associate lecturer in physics (Bryn Mawr College, 1931, pp. 84–86). There, she taught quantum mechanics. In 1932, she started to give courses on spectroscopy and atomic theory, using Arnold Sommerfeld's *Atomic Structure and Spectral Lines* (Sommerfeld, 1923) as one of the textbooks.[31] In addition, she served on the Laboratory Committee and was in charge of advanced courses in physics (including an introductory course of theoretical physics, a seminar in theoretical physics that included quantum mechanics, and a physics journal club for discussing recent articles), at times teaching jointly with Walter Michels, also a former Princeton NRC fellow (Figure 5.2) (Bryn Mawr College, 1932a, p. 24 and p. 80; Bryn Mawr College, 1932b, p. 73). After two years of teaching, Dewey was promoted to associate professor (Bryn Mawr College, 1933, p. 16). For the 1935 academic year, she added to her teaching schedule a free elective course on the structure of matter, which addressed "the evidence for the existence of atoms and of the elementary particles, the structure of atoms, and the physical basis of the periodic system" (Bryn Mawr College, 1935, p. 88).

At that time, however, even women's colleges were reluctant to hire married women. It was expected that when a woman on the teaching staff married, she would resign, with only a few exceptions. As Rossiter has pointed out, men on the boards of trustees of these colleges believed married women could not dedicate themselves fully to their academic work. They also judged that a married woman

---

[31] Presumably, Dewey also used the "wave-mechanical supplement" to this book (Sommerfeld, 1930).

Figure 5.2 Walter Michels (left) and Jane Dewey (right) sitting outside an unknown location in Michigan, c. 1932. Both were then associate professors of physics at Bryn Mawr College. Photograph by Samuel Goudsmit, courtesy AIP Emilio Segrè Visual Archives, Goudsmit Collection.

who was not willing to dedicate herself fully to her home duties could not be trusted (Rossiter, 1984, p. 16). Dewey, however, faced a different type of difficulty at Bryn Mawr, arising from her divorce around 1935.[32] She then requested a leave of absence due to health problems, apparently precipitated by the separation (Martin, 2002, p. 222). During this period, she was supported by her father, who had become a widower in 1927, and with whom she traveled to South Africa. After her recovery, Dewey returned to Bryn Mawr only to find that she had been replaced by the physicist Arthur Lindo Patterson (Bryn Mawr College, 1936). In a 1936 letter to Robertson (her Princeton collaborator), she wrote of a rumor at Bryn Mawr that the authorities had a preference for "appointing men for all vacancies left by women" (quoted in Rossiter, 1984, p. 176).

Dewey was never hired into a physics department again, except for a few evening classes she taught from 1940 to 1942 at Hunter College, at that time a women's college in New York City. Apparently, she suffered from a fragile health, both physically and mentally (Martin, 2002, pp. 222–223). In later years, she was employed in industrial

[32] Divorce in the US was abided until the 1970s by a fault-based legal regime. The more common legal grounds to decree a divorce in 1930 were: cruelty, desertion or abandonment, neglect or nonsupport, adultery, drunkenness, conviction of crime, separation or absence, bigamy and fraud, and incompatibility (US Department of Health, Education, and Welfare, 1973).

research laboratories. She published two articles in the second half of the 1940s, one on the theory of filler reinforcement and another about the elastic constants of materials (her most cited article);[33] both articles reflected her affiliation with the United States Rubber Company of New Jersey and were published in the *Journal of Applied Physics* (Dewey, 1945, 1947). In the spring of 1947, she finally found permanent employment and continued performing research as Chief Physicist at the Terminal Ballistic Laboratory, one of the branches of the Ballistics Research Laboratory at the Aberdeen Proving Grounds, in Maryland.[34]

Eventually, Dewey found a quiet and lovely place to live, close to her place of work, in a farm in Havre de Grace. Her close friend, the poet Elizabeth Bishop, once described it saying: "I am down here visiting an old friend for a few days – she goes to work for the Army at 7:30 every morning and I am left in imaginary possession of a huge estate, thirty-some steers, a housekeeper, a flower garden, etc. – and the entire spring, which is perfectly beautiful" (Bishop, 1995, p. 219). Dewey's home inspired one of Bishop's most famous poems, "A Cold Spring," which she dedicated to her friend Jane Dewey (Bishop, 1952). In a 1950 letter, Bishop tried to describe Dewey's job at the time, writing (Bishop, 1995, p. 205):

At present she is in charge of "Terminal Ballistics" at the Aberdeen proving ground, and when I stay at her farm on weekdays the rural scene shakes slightly once in a while as Jane practices her art about 15 miles away, and then there is a faint "boom". It seems there are three kinds of ballistics: Internal, External & Terminal.

After roughly two decades of ballistics research, Dewey retired in 1970. Then, she moved to her family home in Key West, Florida (the same place her father had chosen for retirement). There, she passed away in September 1976. That same year, John Slater died in Sanibel Island, also in Florida. Both members of *the lucky generation* of US physicists had been initiated to science under similar conditions, but they experienced very different outcomes. Their differences in this regard can be traced to both implicit and explicit gender expectations and patterns of discrimination, a structural feature of science that manifested in varying degrees and was reinforced directly by some physicists themselves.

5.7 Final Remarks

The year she retired, Dewey sent a letter to the editor of the *Barnard Alumnae* magazine, in which she argued against the coeducational model. She wrote (Dewey, 1970, p. 21):

---

[33] According to Google Scholar, it has 246 citations as of 2024.

[34] During World War II, the Ballistic Research Laboratory sponsored the construction of the Electronic Numerical Integrator and Computer (ENIAC), the first programmable and general-purpose electronic digital computer. In 1947, the ENIAC was transferred to the Aberdeen Proving Grounds facility in Maryland.

After nearly 50 years in graduate work and working for a living, I feel [that the opposition to] coeducation expresses my opinions as well as I could after hours of work. I suggest that [those in favor are] so naive as to think that they [women] enter a man's university on an equal basis ... Do Barnard girls who go on into graduate work want to be instructors with a possibility of a final promotion to associate professor just before retirement? If they are active in extracurricular clubs in college, do they want to be vice-president or secretary or whatever office has the most work and the least prestige? Girls and young women should be encouraged to think they are of some importance as long as this is possible. The working world will teach them [otherwise] soon enough. Seventeen is too impressionable to start the lesson.

The way Dewey saw it, she had proven herself as a research scientist, beginning in graduate school at MIT, only to be denied access to a university position where she could have continued her research. Later, she worked for a living at the Terminal Ballistic Laboratory. For Dewey, it was crystal clear that women did not receive the same opportunities as men did, neither during their university studies nor in academic jobs, and even when they were accepted, they would be relegated to less prestigious positions than their male peers. She seems to have supported women's colleges as secure places to instill a sense of self appreciation in female students. In Dewey's case, her sense of self was inculcated not only through the formal education she received, but also by her family and by the intellectual circles of which they were part. Later, the working world would impose its gender hierarchies, as she herself experienced.

This chapter brings out the circumstances faced by a woman of Slater's *lucky generation* of US physicists. Dewey had the good fortune of coming from a socially and economically privileged upbringing, growing up in a remarkable family known for its promotion of progressive ideals on social justice and education. Dewey was also surrounded by major inspirational female figures who encouraged the education of women at different levels and worked for expanding opportunities for women in education and society. Having the freedom to develop her talents, she demonstrated her value as a scientist from the very beginning of her scientific career. Dewey skillfully conducted experimental research on the Stark effect in helium that was a promising contribution to the new developments of quantum mechanics. However, as this chapter shows, she was discouraged from continuing in the field of quantum research and academia by the accumulation of subtle, yet persistent and structural, gender discriminatory practices. A counterfactual exercise may illustrate how systematic bias manifested itself and impacted her career. Had she not been a woman, the IEB might have read the letters of support with less suspicion; her intelligence might have been attributed to her and not to inheritance from her father; and her colleagues might have recognized her skills and contributions to scientific research more prominently. Had she been a man, her applications for faculty positions might have been successful given the support of prominent scientists, thus opening up research opportunities well beyond what the women's college where she ended up could offer. At the very least, had she been a man, she would have been able to retain her Bryn Mawr position. A gender approach reveals all these subtle and not so subtle hurdles for the only woman who belonged to *the lucky generation* of US physicists.

A certain degree of conservatism in science – and in the history of science as well – has been sustained by a discourse of excellence, exceptionality, and success. The history of physics, in its constant search for scientific genius and great milestones, has contributed to the invisibility of less successful, even marginal figures. The case of Dewey illustrates that it is still necessary to broaden the field through more inclusive *her*stories in science. As physicists and historians of physics, we have an excellent opportunity to expand our historical narratives of quantum mechanics in the context of the centenary celebration. Of course, this book shows our intention to do so.

References

Behrman, Joanna. 2020. The personal is professional: Margaret Maltby's life in physics. In Forstner, C. and Walker, M. (eds.), *Biographies in the History of Physics*. Cham: Springer, pp. 37–57. https://doi.org/10.1007/978-3-030-48509-2_3

Bell, Robert E. 1966. John Stuart Foster, 1890–1964. *Biographical Memoirs of the Fellows of the Royal Society*, 12, 147–161. https://doi.org/10.1098/rsbm.1966.0006

Bishop, Elizabeth. 1952. A cold spring: for a friend in Maryland. *New Yorker*, May 31. https://archives.newyorker.com/?i=1952-05-31#folio=030

Bishop, Elizabeth. 1995. *One Art: Letters*. Giroux, Robert (ed.). New York: Farrar, Straus and Giroux.

Born, Max, Heisenberg, Werner and Jordan, Pascual. 1926. Zur Quantenmechanik II. *Zeitschrift für Physik*, 35, 557–615. https://doi.org/10.1007/BF01379806

Bryn Mawr College. 1931. *Bryn Mawr College Calendar Undergraduate Courses 1931*. Vol. XXIV. 1. Bryn Mawr, PA: Bryn Mawr College. https://repository.brynmawr.edu/bmc_calendars/44/

Bryn Mawr College. 1932a. *Bryn Mawr College Calendar Undergraduate Courses 1932*. Vol. XXV. 1. Bryn Mawr, PA: Bryn Mawr College. https://repository.brynmawr.edu/bmc_calendars/43/

Bryn Mawr College. 1932b. *Bryn Mawr Calendar Graduate Courses 1932*. Vol. XXV. 2. Bryn Mawr, PA: Bryn Mawr College. https://repository.brynmawr.edu/bmc_calendars/43/

Bryn Mawr College. 1933. *Bryn Mawr College Calendar Undergraduate Courses 1933*. Vol. XXVI. 1. Bryn Mawr, PA: Bryn Mawr College. https://repository.brynmawr.edu/bmc_calendars/43/

Bryn Mawr College. 1935. *Bryn Mawr College Calendar Undergraduate Courses 1935*. Vol. XXVIII. 1. Bryn Mawr, PA: Bryn Mawr College. https://repository.brynmawr.edu/bmc_calendars/42/

Bryn Mawr College. 1936. *Bryn Mawr College Calendar Graduate Courses 1936*. Vol. XXIX. 2. Bryn Mawr, PA: Bryn Mawr College. https://repository.brynmawr.edu/bmc_calendars/42/

Cassidy, David C. 2009. *Beyond Uncertainty: Heisenberg, Quantum Physics, and the Bomb*. New York: Bellevue Literary Press.

Coben, Stanley. 1971. The scientific establishment and the transmission of quantum mechanics to the United States, 1919–32. *The American Historical Review*, 76(2), 442–466. https://doi.org/10.2307/1858707

Davies, Shannon M. 1985. *American Physicists Abroad: Copenhagen, 1920–1940*. PhD Thesis. Austin, TX: University of Texas.

Dewey, Jane M. 1925. The transference numbers of sodium and potassium chlorides and of their mixtures. *Journal of the American Chemical Society*, 47(7), 1927–1932. https://doi.org/10.1021/ja01684a021

Dewey, Jane M. 1926. Intensities in the Stark effect of helium. *Physical Review*, 28, 1108–1124. https://doi.org/10.1103/PhysRev.28.1108

Dewey, Jane M. 1927. Intensities in the Stark effect of helium: II. *Physical Review*, 30, 770–780. https://doi.org/10.1103/PhysRev.30.770

Dewey, Jane M. 1928. Spectral excitation by recombination in the electric arc. *Physical Review*, 32, 918–921. https://doi.org/10.1103/PhysRev.32.918

Dewey, Jane M. 1929. Temperatures of positive ions in a uniformly ionised gas. *Nature*, 123(3105), 681. https://doi.org/10.1038/123681a0

Dewey, Jane M. 1930. The intensity maxima in the continuous helium spectrum. *Physical Review*, 35, 155–157. https://doi.org/10.1103/PhysRev.35.155

Dewey, Jane M. (ed.) 1939. Biography of John Dewey. In *The Philosophy of John Dewey*. New York: Tudor Publishing Co., pp. 3–45.

Dewey, Jane M. 1945. Theory of filler reinforcement. *Journal of Applied Physics*, 16(1), 55–55. https://doi.org/10.1063/1.1707501

Dewey, Jane M. 1947. The elastic constants of materials loaded with non-rigid fillers. *Journal of Applied Physics*, 18(6), 578–581. https://doi.org/10.1063/1.1697691

Dewey, Jane M. 1970. Coeducation: Con. *Barnard Alumnae*, Fall 1970. https://digitalcollec tions.barnard.edu/do/8044ec7c-1927-4086-82db-d916b7e4a1c9

Dewey, Jane M. and Robertson, Howard M. 1928. Stark effect and series limits. *Nature*, 121(3053), 709–710. https://doi.org/10.1103/PhysRev.31.973

Duncan, Anthony and Janssen, Michel. 2014. The trouble with orbits: the Stark effect in the old and the new quantum theory. *Studies in History and Philosophy of Science Part B*, 48, 68–83. https://doi.org/10.1016/j.shpsb.2014.07.008

Duncan, Anthony and Janssen, Michel. 2015. The Stark effect in the Bohr–Sommerfeld theory and in Schrödinger's wave mechanics. In Aaserud, Finn and Kragh, Helge (eds.), *One Hundred Years of the Bohr Atom: Proceeding from a Conference*. Copenhagen: Royal Danish Academy of Sciences and Letters, pp. 217–271.

Duncan, Anthony and Janssen, Michel. 2023. *Constructing Quantum Mechanics. Volume 2: The Arch: 1923–1927*. Oxford: Oxford University Press.

Durst, Anne. 2010. *Women Educators in the Progressive Era: The Women behind Dewey's Laboratory School*. New York: Palgrave Macmillan. https://doi.org/10.1057/9780230109957

Foster, John S. 1924. Observation of the Stark effect in hydrogen and helium. *Physical Review*, 23, 667–684. https://doi.org/10.1103/PhysRev.23.667

Foster, John S. 1927. Application of quantum mechanics to the Stark effect in helium. *Proceedings of the Royal Society A*, 117, 137–163. https://doi.org/10.1098/rspa.1927.0171

Hall, Irene. 2005. *The Unsung Partner: The Educational Work and Philosophy of Alice Chipman Dewey*. PhD thesis. Cambridge, MA: Harvard University.

Heisenberg, Werner. 1926a. Mehrkörperproblem und Resonanz in der Quantenmechanik. *Zeitschrift für Physik*, 38, 411–426. https://doi.org/10.1007/BF01397160

Heisenberg, Werner. 1926b. Über die Spektra von Atomsystemen mit zwei Elektronen. *Zeitschrift für Physik*, 39, 499–518. https://doi.org/10.1007/BF01322090

Heisenberg, Werner. 1967. Quantum theory and its interpretation. In Rozental, S. (ed.), *Niels Bohr: His Life and Work as Seen by His Friends and Colleagues*. Amsterdam: North–Holland Publishing Company, pp. 94–108.

Kevles, Daniel J. 1979. A new center of physics. In *The Physicists: The History of a Scientific Community in Modern America*. New York: Vintage Books, pp. 200–221.

Keyes, Frederick G. and Dewey, Jane. 1927. An experimental study of the piston pressure gage to six hundred atmospheres. *Journal of the Optical Society of America*, 14(6), 491–504. https://doi.org/10.1364/JOSA.14.000491

Kojevnikov, Alexei. 2020. *The Copenhagen Network: The Birth of Quantum Mechanics from a Postdoctoral Perspective*. Cham: Springer. https://doi.org/10.1007/978-3-030-59188-5

Kragh, Helge. 1999. *Quantum Generations: A History of Physics in the Twentieth Century*. Princeton, NJ: Princeton University Press.

Kramers, Hendrik A. and Heisenberg, Werner. 1925. Über die Streuung von Strahlung durch Atome. *Zeitschrift für Physik*, 31, 681–708. https://doi.org/10.1007/BF02980624

Kronig, Ralph de L. 1925. Über die Intensität der Mehrfachlinien und ihrer Zeemankomponenten. *Zeitschrift für Physik*, 31, 885–897. https://doi.org/10.1007/BF02980643

Leone, Matteo, Paoletti, Alessandro, and Robotti, Nadia. 2004. A simultaneous discovery: the case of Johannes Stark and Antonio Lo Surdo. *Physics in Perspective*, 6, 271–294. https://doi.org/10.1007/s00016-003-0170-2

Martin, Jay. 2002. *The Education of John Dewey: A Biography*. New York: Columbia University Press.

Massachusetts Institute of Technology. 1925. *Catalogue Academic Year 1925–26*, Cambridge: The Technology Press, vol. 60. 6.

Massachusetts Institute of Technology. 1923. *Technique 1924: The Yearbook of the MIT Published by the Junior Class*. Boston: Perry & Elliott Co.

Minor García, Adriana. 2019. *Cruzar fronteras: Movilizaciones científicas y relaciones interamericanas en la trayectoria de Manuel Sandoval Vallarta (1917–1942)*. Mexico City: UNAM-CISAN & El Colegio de Michoacán.

Morse, Philip M. 1982. A biographical memoir of John Clarke Slater. In *Biographical Memoir*. Washington, DC: National Academy of Sciences, pp. 297–321.

Pauli, Wolfgang. 1925. Ueber die Intensitäten der im elektrischen Feld erscheinenden Kombinationslinien. *Det Kongelige Danske Videnskabernes Selskab. Skrifter. Naturvidenskabelig og Matematisk Afdeling*, 7(3), 3–20.

Robertson, Howard M. and Dewey Jane M. 1928. Stark effect and series limits. *Physical Review*, 31, 973–82. https://doi.org/10.1103/PhysRev.31.973

Rossiter, Margaret W. 1984. *Women Scientists in America. Volume 1: Struggles and Strategies to 1940*. Baltimore: Johns Hopkins University Press.

Schrödinger, Erwin. 1926a. Quantisierung als Eigenwertproblem. Erste Mitteilung. *Annalen der Physik,* 79, 361–376. https://doi.org/10.1002/andp.19263851302

Schrödinger, Erwin. 1926b. Quantisierung als Eigenwertproblem. Dritte Mitteilung: Störungstheorie, mit Anwendung auf den Starkeffekt der Balmerlinien. *Annalen der Physik*, 80, 437–490. http://dx.doi.org/10.1002/andp.19263851302

Schweber, Silvan S. 1986. The empiricist temper regnant: theoretical physics in the United States 1920–1950. *Historical Studies in the Physical and Biological Sciences* 17, 55–98. https://doi.org/10.2307/27757575

Schweber, Silvan S. 2000. *In the Shadow of the Bomb: Oppenheimer, Bethe and the Moral Responsibility of the Scientist*. Princeton, NJ: Princeton University Press.

Servos, John W. 1990. *Physical Chemistry from Oswald to Pauling: The Making of a Science in America*. Princeton, NJ: Princeton University Press.

Sommerfeld, Arnold. 1923. *Atomic Structure and Spectral Lines*. London: Methuen.

Sommerfeld, Arnold. 1930. *Wave-Mechanics*. London: Methuen.

Sopka, Katherine R. 1988. *Quantum Physics in America 1920–1935*. New York: AIP Publishing.

Stroud, Carlos. 2004. Jane Dewey: pioneer in quantum optics. In Stroud, C. (ed.), *A Jewel in the Crown: 75th Anniversary Essays, The Institute of Optics, University of Rochester.* Rochester: University of Rochester Press, pp. 25–26.

Takamine, Toshio and Werner, Sven. 1926. Intensitätsmessungen im Stark-Effekt. *Naturwissenschaften,* 14, 47–48. https://doi.org/10.1007/BF01506833

The Barnard Bulletin. 1925a. Letter from Jane M. Dewey to Miss Virginia C. Gildersleeve, April 4. *The Barnard Bulletin,* April 24, 2. https://digitalcollections.barnard.edu/do/b0e8f17f-3fbe-433c-906b-91899f0617a2

The Barnard Bulletin. 1925b. Winner of student fellowship writes. *The Barnard Bulletin,* November 6, 1–2. https://digitalcollections.barnard.edu/do/4ab52af9-8260-4ac7-a547-7ac144d9821f

The Class of 1922 at Barnard College. 1921. *The Mortarboard 1922. Volume 28.* New York: Charles L. Willard. https://digitalcollections.barnard.edu/do/75df7928-23d2-4596-b33b-18f9b31ca06e

US Department of Health, Education, and Welfare. 1973. *Marriage and Divorce Statistics United States, 1867–1947.* Vital and Health Statistics Publication Series 21. US Department of Health, Education, and Welfare.

Weeks, Dorothy. 1978. Interview of Dorothy Weeks by Katherine Sopka on July 19, 1978. Niels Bohr Library & Archives. https://www.aip.org/history-programs/niels-bohr-library/oral-histories/4943

6

Laura Chalk and the Stark Effect

DANIELA MONALDI

6.1 Introduction

In 1926, Laura Mary Chalk (later, Laura Rowles, 1904–1996) produced the first data to validate Erwin Schrödinger's wave mechanics. Today, she is remembered at McGill University together with her husband William Rowles for their joint contributions to campus life, and in Canada for her endowment of the Chalk–Rowles Graduate Fellowships in Physics.[1] She is also occasionally celebrated for being the first woman to complete a PhD in physics at McGill (Chalk, 1928).[2] How did her contribution to quantum physics come to be nearly forgotten?

Schrödinger wrote in 1952 that his calculations of frequencies and intensities of the Stark spectral lines in hydrogen in 1926 were the "earliest quantitative achievement of wave mechanics," and that they "agreed very satisfactorily with observations," but he did not name the researchers who produced these observations (Schrödinger, 1952, p. 235). Noting this "early verification of Schrödinger's theory," Yves Gingras attributed it to Chalk's PhD supervisor and co-author, John

I am deeply grateful to Patrick Charbonneau, Michelle Frank, and Margriet van der Heijden for their invaluable collaboration and feedback. Special thanks to Michel Janssen for sparking this research project. The WiHQP group has been a source of inspiration, support, and advice. I am thankful to my colleagues in the STS Department at York University for their feedback on an early draft of this chapter. Thanks also to the director and staff of the Niels Bohr Archive and the staff of the McGill University Archive for their assistance, and to Laura Chalk Rowles's relatives for their kind response and permissions.

[1] In 1995, a building named The Rowles House was dedicated to her and her husband, William Rowles (1897–1987), who directed the Department of Physics of the Faculty of Agriculture from 1930 and was then appointed Professor Emeritus (Snell, 1963, p. 127). She also created the William and Laura Rowles Endowment Fund for the Dean of the Faculty of Agricultural and Environmental Sciences and an annual visiting lectureship for the McGill Centre for Studies of Aging (The Montreal Gazette, 1996; Marshall, 2000). In this chapter, I am naming my subject Laura Chalk, or simply Chalk, when referring to events before her marriage (as for example her scientific publications, which she signed as "M. Laura Chalk") and Chalk Rowles when referring to her life more in general. Her autobiographical memoir is cited as Rowles (1995) because she signed it "Laura Rowles."

[2] McGill University awarded a PhD to the astrophysicist Alice "Allie" Vibert Douglas in 1926 for research she carried out as a volunteer research assistant at the Yerkes Observatory in Wisconsin (Crossfield 1997, pp. 73–74; Gosztonyi Ainley, 2012, pp. 85–87). Laura Chalk was the first woman to earn a PhD in physics from McGill, in 1928, for research conducted within the McGill Physics Department under the supervision of McGill faculty.

Stuart Foster (1890–1964) (Gingras, 1981, p. 24).[3] Likewise, Stephen G. Brush initially folded Chalk's work into Foster's research program when discussing the experimental test of Schrödinger's Stark effect calculations as part of an extended study of the dynamics of theory change (Brush, 1994, p. 137).

Canadian historians Margaret Gillett and Ann Beer made a pioneering effort to document Chalk Rowles' life as part of the movement to bring to light women's histories. In fact, the main documentary source we have about Chalk Rowles is a memoir that she wrote for Gillett and Beer's collection, *Our Own Agendas. Autobiographical Essays by Women Associated with McGill University* (Gillett and Beer, 1995; Rowles, 1995). A section based on the memoir is also included in Marianne Gosztonyi Ainley's *Creating Complicated Lives. Women and Gender at English Canadian Universities* (Gosztonyi Ainley, 2012). The neglect of Chalk's work in quantum mechanics has begun to be addressed thanks to these efforts. In his more recent book on theory change, Brush cited Chalk Rowles' memoir and reframed his account to highlight her role in the 1926 measurements, while contrasting the significance of her work to its obscurity (Brush, 2015, p. 495):

The Foster–Chalk experiment was certainly one of the first tests of a prediction of quantum mechanics (if not *the* first …). Has anyone heard of it? Chalk seems to be completely unknown to most historians of physics and to physicists interested in publicizing the achievements of women.

Brush had his reasons to stress this contrast, but the blackout has started to lift. Anthony Duncan and Michel Janssen, in particular, have more deliberately highlighted Chalk's contribution in their in-depth analysis of the transition from the old quantum theory to quantum mechanics (Duncan and Janssen, 2014, 2015, 2019).[4]

This chapter seeks to recover more fully Laura Chalk's scientific achievement, and to investigate the reasons for her absence from physics historiography until now. It begins with an overview of Chalk Rowles' life up to her graduation (Section 6.2), then moves to an examination of the place of the Stark effect in the development of quantum physics (Section 6.3) and of Foster's research program (Section 6.4). After describing Chalk's graduate work and results (Section 6.5), it discusses her postdoctoral experiences and choices, in particular, her decision to leave research (Section 6.6). It further considers other factors that, compounded with her choices, prevented the Foster–Chalk experiment from gaining lasting recognition (Section 6.7). Finally, it returns to a more detailed discussion of her experiences, inquiring what role gender (Section 6.8) and the quantum revolution (Section 6.9) played in the self-determining choices of a young woman who had

---

[3] "Cette précoce vérification de la théorie de Schrödinger[.]" My translation.
[4] An entry on Laura Chalk Rowles has appeared on Wikipedia since the start of this project.

wanted to be "one of the boys" in a colonial physics laboratory of the late 1920s (Rowles, 1995, p. 42).

6.2 Laura Chalk

Laura Chalk was the model daughter of a family of teachers from a privileged social group. Her father, Walter Chalk had moved to Canada from England as a private-school teacher, and then he had become a principal and a school superintendent in an affluent anglophone district of Montreal. Her uncle was the principal of the high school that she attended, and her mother, Nina Howe, had been a high-school teacher before marriage. Upon completing high school as a top-standing and medal-winning student, Chalk earned a Macdonald Entrance Scholarship to Montreal's McGill University. She was already versed in mathematics, and she developed an interest in physics when her father took her along to the Men's Society of his church for a lecture on radioactivity given by the director of the McGill Physics Department, Arthur S. Eve (1862–1948). Since she was taking a first-year course with Eve at the time, Eve asked her after the lecture how well she was doing and encouraged her to move to the second-year course. When she then wanted to study yet more physics, Eve helped her to transfer to the honours course in mathematics and physics, despite contrary advice from her family. Eve's own family relation-ships may have contributed to his favourable outlook on women in physics. He was the brother-in-law of Harriet Brooks (1876–1933), the first woman to earn a master's in physics in Canada, a protégée of Ernest Rutherford, and one of the first women scientists in the emerging field of radioactivity (Rayner-Canham and Rayner-Canham, 1992; Gosztonyi Ainley, 2012). Furthermore, in 1921–1923, Eve had supervised the master's theses of at least two women, Alice "Allie" Vibert Douglas and Anna Isobel McPherson (Crossfield, 1997, pp. 73–74; Gosztonyi Ainley, 2012, pp. 76–77 and pp. 85–87).[5]

There were comparatively higher numbers of women in physics, in Canada and elsewhere, in the 1920s–1930s than in the following decades. When Chalk started her honours course, she found herself in a class composed of four women and one man (Rowles, 1995, p. 37). Even though this "rather surprising ratio" was a fluke – her three female classmates were advised to switch to a general course after one year – group photographs in the MacPherson Collection of the Macdonald Physics Laboratory show that in graduate school she was not the only female student (Figure 6.1).[6]

---

[5] McPherson went on to earn her PhD at the University of Chicago, and returned to McGill to become, after some years of struggle, a faculty member of the physics department from 1940 until her death in 1979 (Gosztonyi Ainley, 2012, pp. 76–77). For Douglas, see Note 2.

[6] Unfortunately, except for Douglas, only the surnames of the other women in the photograph are known at the moment.

MACDONALD PHYSICS BUILDING STAFF 1927-28

J·FOSTER D·A·KEYS A·N·SHAW H·T·BARNES A·S·EVE L·V·KING H·E·REILLEY E·S·BIELER

H·G·WATSON H·LANE G·TWEEDALE A·GILSON L·CHALK H·PYE N·CAM A·DOUGLAS C·YOUNG E·TAYLOR

F·DAVIES M·CROWE I·LAWSON E·JAHN L·SMITH S·AMESSE J·YOUNG C·LANE

M·McLEOD R·CURRIE J·HENDERSON P·McDONALD D·McCRAE T·WHITE A·PATTERSON H·HARKNESS

Figure 6.1 The Macdonald Physics Laboratory 1927–1928. Laura Chalk is the fourth from the left in the first standing row. McPherson Collection, McGill Physics Department. Reproduced with permission.

After graduating as a prize-winning student in 1925, Chalk decided "to stay on at home and do research at McGill" (Rowles, 1995, p. 37).[7] She became involved with the Stark effect because she chose as her supervisor a rising star in the McGill physics department, J. Stuart Foster (1890–1964), who had just arrived from Yale and "presented what seemed the most interesting programs," she later recalled (Rowles, 1995, p. 37).

---

[7] Chalk won the Anne Molson Gold Medal, a recognition awarded each year to the best student in physics, mathematics, and physical science (see also Note 35). She also obtained a National Research Council bursary of about $700 for the year (Rowles, 1995, p. 37). The National Research Council was instituted in 1917 as part of the construction of a national system of scientific research in Canada, and offered master's and doctoral scholarships. McGill started a PhD programme in 1907 and instituted a Faculty of Graduate Studies in 1922 (Gingras, 1991, p. 38).

6.3 The Stark Effect, the Lo Surdo Technique, and Quantum Theory

The Stark effect is the splitting and shifting of the spectral lines of atoms under the effect of an external electric field.[8] It is named after the German physicist Johannes Stark, who first detected it in hydrogen and helium at the Rheinisch-Westfälische Technische Hochschule in Aachen in 1913. Independently and nearly concurrently, the Italian physicist Antonino Lo Surdo also detected it, accidentally, while investigating the Doppler effect of atomic spectra in discharge tubes at the Institute of Experimental Physics of the University of Florence in Arcetri. As a result, it is sometimes called the Stark–Lo Surdo effect, even though Stark always insisted on claiming sole credit for the discovery. Stark did conduct a more intentional and extensive program of research, but Lo Surdo devised an experimental configuration that was more effective and versatile, and his method soon turned out to be more useful to spectroscopists, including Stark himself (Rowles, 1928, p. 1; Leone et al., 2004).[9] Thanks, in large measure, to the Lo Surdo method, observing the Stark effect in increasing detail and in a widening set of atomic spectra became a staple of spectroscopy and a testing workbench of quantum physics, on which experimental and theoretical apparatus were forged and evolved symbiotically with one another.

Theoretical study of the Stark effect started in 1914 with Niels Bohr, who investigated the phenomenon in the framework of his quantum model of the atom. Bohr obtained an agreement with the available data that he judged to be qualitatively good, even though he could not explain all the details (Jammer, 1966, p. 107). Two years later, Paul Sophus Epstein, a former student of Arnold Sommerfeld, and Karl Schwarzschild, the director of the Astrophysical Observatory at Potsdam, independently applied the Hamilton–Jacobi method to Sommerfeld's generalization of Bohr's model. Using perturbation theory, they calculated the Stark components to the first order of approximation in the electric field intensity, termed the linear Stark effect. The results were regarded as being in excellent quantitative agreement with Stark's observations, and they established the Hamilton–Jacobi method as the preferred tool of the Bohr–Sommerfeld theory. This calculation method, however, only provided the frequencies of the radiation emitted by atomic systems in quantum transitions, not its intensity and polarization. In 1919 – the year Stark was awarded the Nobel Prize in Physics for his discovery – Hendrik A. Kramers, a doctoral student of Bohr, applied the correspondence

---

[8] The Stark effect is now understood to result from the interaction of an atomic or molecular electric dipole moment (permanent or induced) with an external electric field. Thanks to Bretislav Friedrich for this definition.

[9] While Stark applied an external electric field to a canal-ray tube, Lo Surdo modified the tube in such a way that an intense electric field was produced internally. This change improved the resolving power of the instrument, and permitted the precise comparison of observations at different field strengths. Stark applied Lo Surdo's technique to observe the splitting of the spectral lines of helium in 1918, and a finer separation of hydrogen lines that the quantum theory predicted as a second-order effect in high electric fields in 1927 (Leone et al., 2004, p. 284 and p. 289).

principle, which Bohr had introduced a year earlier, to estimate the relative intensities of the fine-structure spectral lines and of the Stark effect of the first four Balmer lines in the hydrogen spectrum (Jammer, 1966, p. 115). Once again, the theoretical calculations were judged to be in excellent agreement with available data, and that agreement, in turn, validated the correspondence principle as an effective, if hazily defined, means to further develop quantum theory.

From 1914 to 1925, the theoretical treatment of the Stark effect was widely regarded as a resounding success of the Bohr–Sommerfeld theory (Jammer, 1966, p. 109; Kragh, 2012, pp. 168–170; Duncan and Janssen, 2014; 2015, pp. 219–220; 2019, pp. 296–298). Duncan and Janssen reckon it to be a part of the "winning streak" of the old quantum theory, together with the relativistic fine structure and X-ray spectra (Duncan and Janssen, 2019, p. 31). Nevertheless, they argue that these alleged successes were incomplete and largely accidental, and that a truly satisfactory theoretical treatment, free of ambiguities and additional hypotheses, was only achieved by quantum mechanics (Duncan and Janssen, 2014, 2015, 2019).

From the point of view of an experimental spectroscopist in the mid-1920s, however, *both* theory and experiments were tentative and uncertain. The interaction of theoretical calculations and experimental observations was one of reciprocal stimulus and support, rather than one of straightforward empirical verification of theoretical hypotheses. Chalk summarized the interplay of theory and experiment concerning the Stark effect in the period 1913–1926 in her PhD thesis as follows (Chalk, 1928, p. 3):

Thus, the theoretical interpretations and experimental facts have developed simultaneously and with very great benefit. While the theory has supplied what has grown to be an entirely adequate formal interpretation of all essential details of the experiments in hydrogen, it is also true that many theoretical points were either suggested or given practical test through contact with observations on the Stark effect. For some years, in fact, this effect has served as a leading exercise in the development of quantum methods.

The creators of quantum mechanics also road-tested their new theory on the spectrum of the hydrogen atom and its alterations by external fields. Wolfgang Pauli discussed the effects of electric and magnetic fields in a paper in which he calculated the Balmer terms of the hydrogen atom in matrix mechanics, but he left out the calculation of the intensities (Pauli, 1926). In the third installment of his *Quantisierung als Eigenwertproblem* series (completed in late spring 1926 and published July 18, 1926), Schrödinger managed to calculate the intensities and polarizations of the Stark components of the first four Balmer lines, and compared them with the "known measurements and estimates of intensity that Stark made [in 1915] with a field strength of about 100,000 volts per centimetre" (Schrödinger, 1926, p. 467). He concluded that "the agreement is fairly good for almost all strong components, and all considered it is somewhat better than the estimates made from correspondence considerations" (Schrödinger, 1926, pp. 470–471). In other words, he considered the theory–observation agreement encouraging, though still incomplete and affected by uncertainties. Notably, however, this was the first time the new quantum theory diverged from the old one; thus, the intensities calculations could be used as a crucial test of the new theory (Brush, 1994, 2015).

A few months later (in October 1926), Epstein recalculated the displacements and calculated the intensities of Stark components applying Schrödinger's formalism. In contrast to Schrödinger, he assumed that the observed intensities were proportional to the amplitudes of the emitted waves rather than to the squares of the amplitudes. His figures differed from Schrödinger's, but he, too, found the agreement with Stark's measurements to be "fair, and decidedly better than that obtained from Bohr's correspondence principle in Kramers' work" (Epstein, 1926, p. 710). Schrödinger and Epstein challenged the hitherto accepted agreement between the Bohr–Sommerfeld theory and experiments, and claimed the superiority of the new theory; at the same time, they made diverging claims regarding how best to fit the new theory to the data. Their claims therefore provided experimental spectroscopists with a fresh problem of realigning theory and experiment. This research opportunity arose right as Laura Chalk was beginning her PhD.

6.4 J. Stuart Foster and the Stark Effect

When Chalk started her graduate studies, her supervisor, Foster, was building a scientific reputation as a master of "the black art of making Lo Surdo tubes work properly," as his Royal Society biographer, Edward R. Bell later noted (Bell, 1966, p. 148). Foster, having acquired this special skill for his doctoral research at Yale University, brought it to McGill when he became assistant professor in the physics department in 1924, and carried it on for about 15 years, training many students along the way (Bell, 1966, p. 153; Thomas, 1984, p. 361; Gingras, 1991, p. 74). Despite lacking the then canonical Cambridge credentials, Foster went on to have a very distinguished career.[10] He became a Fellow of the Royal Society of Canada in 1929 and of the Royal Society of London in 1935. But his career also epitomized the shifting of Canadian physics from the British to the American sphere of influence that

---

[10] For details on Foster's life and career, see Bell (1966), Thomas (1984), Gingras (1991), and Ramos (2019). Today, he is chiefly remembered for leading the country into the nascent world of nuclear physics with the construction of Canada's 100 MeV cyclotron in the 1930s–1940s, the establishment of a major program of nuclear physics during the war, and the concomitant training of the first generation of Canada's nuclear physicists. Foster's sons, John Stuart Jr. and L. Curtis also became nuclear physicists, but they went to work in the US. John Stuart Jr. completed his PhD at the University of California, Berkeley, and became the director of the Lawrence Livermore National Laboratory and an advisor to the US Secretary of Defense during the Cold War. L. Curtis Foster obtained his PhD from McGill and became Vice-President and Director of Research at Zenith Radio Research in Menlo Park, California (Bell, 1966, p. 158; Ramos, 2019).

started in earnest in the 1930s, and the changes that came with it (Thomas, 1984). An able scientific entrepreneur, Foster was successful in procuring funds from the university and from the National Research Council in Ottawa (Thomas 1984, p. 358).

Foster's most valuable asset in the growing phase of Canadian physics was his commitment to the building of up-to-date apparatus. When he arrived at McGill, "the physics department was not only hiring staff [...] but also buying equipment" (Thomas, 1984, pp. 360–361). His first accomplishment was to secure a donation to purchase a state-of-the-art spectrograph to carry on with his work on the Stark effect (Bell, 1966, pp. 147–148). His research was chiefly experimental but unfolded in close interaction with theoretical developments. Gingras highlights Foster's work as part of an analysis of the hesitant reception of quantum mechanics at McGill (Gingras, 1981). The physics department had been on the forefront of microphysical research in the very first years of the twentieth century, when Rutherford and Frederick Soddy, then faculty members, conducted their ground-breaking work on radioactivity. But they left without creating a research school. Between Rutherford's departure in 1907 and Foster's arrival in 1924, work in the department was chiefly dedicated to experimental and applied classical physics in areas such as electromagnetism, acoustics, and geophysics. Foster's new Stark-effect laboratory was the institution's first foray into "modern" physics since Rutherford's time, and it represented a bid to regain status in the world of cutting-edge research.

In fact, Foster was determined to stay active on the forefront of modern physics. He had met Bohr during the latter's 1923 visit to the US and had since corresponded with him about the Stark effect in helium. He developed a veneration for Bohr and would continue to depend on his advice for many years.[11] When Foster started working on quantum mechanics in 1926, he met resistance from a senior faculty member, Louis Vessot King (1886–1956), who was developing a rival non-quantum theory of his own.[12] Despite that initial hurdle, Foster's efforts started the penetration "of quantum mechanics as well as the philosophical discourse that accompanied it" at McGill (Gingras, 1981, p. 26). Foster also markedly increased the number of doctoral theses produced, and all the theses he directed until 1939 were on the Stark effect. Laura Chalk and William Rowles were Foster's first two doctoral students. She, in particular, fondly remembered him as "an inspiration to us all" (Rowles, 1995, p. 37).

6.5 Laura Chalk's Graduate Work

The plan for Chalk's graduate work was to attempt a new and improved determination of the relative intensities of the Stark components of the Balmer series of

---

[11] Bohr–Foster letters, Niels Bohr Scientific Correspondence, Niels Bohr Archive, folder 90 (Bell, 1966, p. 148).
[12] Foster to Bohr, May 5, 1927. See Note 11. See also Gingras (1981, pp. 17–20).

hydrogen (Chalk, 1926, 1928). This continuation of Foster's line of research was an experimentally challenging "mopping up" operation (Kuhn, 1962, p. 24). Even before learning of the new calculations by Schrödinger, Foster knew that potentially interesting discrepancies lurked behind the apparent agreement between theory and data. Stark's 1915 photographic plates, for example, showed several components of the H_β line of hydrogen that were forbidden by Kramers' theory; furthermore, the relative intensities of a pair of adjacent components appeared "inverted" with respect to the theoretical estimates. New observations that Foster had carried out using a Lo Surdo tube of his own design appeared to confirm the theory regarding the forbidden components but not regarding the intensities. Kramers estimated the ratio of intensities of two of the H_β components to be nearly 1:2, while the observations yielded a rough estimate of nearly 2:1. Foster did not judge this discrepancy to be threatening to Bohr's theory; after all, H_β represented a transition between states of very small quantum numbers, while Kramers' estimates were expected to increase in accuracy with the quantum numbers. "The present photographs," he wrote, "show that no serious experimental departure from the expectations based on the correspondence principle occurs in the Stark effect for H_β" (Foster, 1925, p. 236). Nevertheless, he planned for Chalk to repeat the experiment using an improved tube and a more accurate method for measuring relative intensities.

This type of measurement was challenging because experimentalists had to estimate the intensity of the emitted light from the blackening of the photographic plates, which depended on many variables, from the source type (intermittent or steady) and the exposure time to the characteristics of the photographic plates and the process of development (Chalk, 1928, p. 28). Instead of relying on the blackening of the plates, Foster and Chalk decided to adopt the "neutral wedge" method, which determined the relative intensities from the lines' lengths on a photograph of the spectrum after passing through a neutrally tinted glass wedge.[13] Chalk presented this work as having "begun in order to obtain more accurate results than those of Stark with which to compare the estimates given by Kramers, and to use the results to determine the corrections in the theory required to make it agree with experiment" (Chalk, 1928, p. 40).

---

[13] The neutral wedge method was devised in 1916 by John William Nicholson and Thomas Ralph Merton, who explained it as follows: "An accurate wedge of neutral-tinted glass, cemented to a similar wedge of clear glass so as to form a plane parallel plate, was mounted immediately in front of the slit of a spectroscope, in the manner commonly used for determining the sensibility curves of photographic plates. Under these conditions the spectrum of a discontinuous source thrown on to the slit through the neutral wedge is seen to consist of lines which are bright at one end, corresponding to the thin end of the wedge, and gradually fade off towards the region corresponding to the thick end of the wedge. The apparent length of a line depends on its intensity, and the relative intensity of two adjacent lines can be determined by measuring the lengths at which they can just be seen" (Nicholson and Merton, 1916, p. 463).

Pride in her experimental skills colored Chalk Rowles's memories of graduate work (Rowles, 1995, pp. 37–38):

So to the Stark laboratory in the sub-basement of the old physics building. It was dark . . . It was also so radioactive that we had to be careful always to use fresh photographic plates . . . We graduate students were expected to build our own vacuum systems, McLeod gauges, and charcoal traps to make basic changes to our Stark effect tubes [. . .] which were all designed by Dr. Foster. . . . In due course, the necessary vacuum equipment got made and, using the six-prism glass spectrograph, photographs were achieved but not before dozens of attempts.

Oddly, she was not as effusive about her actual research work, which she described rather summarily. But other sources allow us to retrace it more in detail: her master's and PhD theses, her publications, and Foster's correspondence. She began her graduate studies in 1925 and, by the summer of 1926, had managed to make the necessary equipment and obtain high-quality photographs. Her master's thesis, titled *Potential Distributions in the Crookes Dark Space*, was intended to be an exercise in the construction and calibration of the instruments to be used for her PhD work (Chalk, 1926, 1928). But as she tested the setup, she found that it worked well enough to produce interesting results. "At this point in the work," she wrote, "it was thought advisable to drop the investigation of potential distributions in other types of tubes and to make a direct application of the results obtained" (Chalk, 1926, p. 14). In this first application, she was then able to confirm the intensity anomaly noted earlier by Foster and measure it more precisely. The ratio of intensities of two pairs of components of H_β was "clearly in total disagreement with Kramer's [sic] theory which gives the ratio 1:2, the experimental results giving it 1.6:1" (Chalk, 1926, p. 34).

In July 1926, as Chalk was completing her master's thesis and starting her PhD work, Foster left Montreal for a six-month stay as a fellow of the International Education Board at the Niels Bohr Institute for Theoretical Physics in Copenhagen (Bell, 1966, p. 148; Kojevnikov, 2020, p. 108 and p. 115). During his leave, Foster corresponded regularly with his departmental director, Eve, to report on his activities abroad as well as to provide guidance to his students.[14] Anxious to prove that he was building up a world-class laboratory, from the ship that took him across the Atlantic, Foster assured Eve that his students' new apparatus worked even better than expected. "I have not had time to study the plates with any care. The general features are, however, very clear and beautiful."[15] The quality of the images was so good that it was not necessary to ask Bohr for help with the interpretation of even the most complicated spectra, as apparently the hesitant Eve had suggested. Foster added, "I built a tube for Miss Chalk's work, of new design, but the hot spell simply

[14] Arthur S. Eve Fonds, MG1035, Accession 454, Item 293, Date 1926, Box 16, McGill University Archives.
[15] Foster to Eve, July 24, 1926. See Note 14.

knocked me out and I could not push this work through. She has sent me an account of her work to date, which I plan to read during the time on ship."[16] Thus, Foster learned of Chalk's sharp measurement of the H_β discrepancy from reading her account while on the ship. By the end of the voyage, he had decided that the result deserved publication.

His decision was confirmed the moment he stepped into Bohr's institute, for Foster wasted no time inspecting a similar experiment that was being conducted there by "Mrs. Clark from Boston," that is, Jane Dewey, who was a postdoctoral fellow with Bohr and was then married to J. Alston Clark (see Chapter 5). Foster was doubly interested in Dewey's experiment, as she was also working on helium, which was his primary interest. In fact, in parallel to supervising his students, Foster was busy completing a paper of his own on Stark-effect patterns in helium, which Eve would present for him to the Royal Society of London in November (Foster, 1927a). Having found that Chalk's setup was "a little better" and her plates "much better" than Dewey's, he informed Eve that he and Chalk "observed a slight general correction to Kramers' estimate," which Chalk could confirm by further observations. He advised Chalk to "make a careful effort to extend her observations to include H_γ and H_δ," and implement a "slightly improved technique developed last month." In any case, they had to send the paper to the publisher by October 1, whether new plates showed marked improvements or not.[17]

Until his arrival in Copenhagen, Foster's plan for Chalk's doctorate had consisted of having her compare her data with Stark's earlier measurements and with Kramers' estimates. Once there, however, he was drawn into the excitement about the birth of quantum mechanics. He interacted closely not only with Bohr but also with Werner Heisenberg, who had recently replaced Kramers as Bohr's assistant.[18] Heisenberg had just published a groundbreaking paper on the quantum mechanics of many-body systems and on the helium spectrum, which was then the cutting edge of research in quantum mechanics (Heisenberg, 1926). The following October, Schrödinger arrived as a visitor and engaged in intense debates with Bohr and Heisenberg on the physical interpretation of wave mechanics. Thus, Foster soon learned of Schrödinger's new results, including his claim that his wave-mechanical calculations of Stark-effect intensities agreed more closely with experiment than Kramers'. This claim gave a whole new relevance to the Foster–Chalk experiment, for Schrödinger's numbers agreed unequivocally with Chalk's observations of H_β.

Upon learning about the Foster–Chalk experiment, Bohr urged Foster to publish it promptly.[19] On September 6, Foster mailed Eve the handwritten draft of a letter,

---

[16] Foster to Eve, July 24, 1926. See Note 14.   [17] Foster to Eve, August 25, 1926. See Note 14.
[18] Bohr to Foster, March 31, 1926. See Note 11.
[19] Bohr's micromanaging of all publications originating from his institute is well known. See, for example, Kojevnikov (2020, p. 72).

signed jointly by himself and Chalk, that he intended to send to *Nature*, along with a new experimental plan tellingly titled, "Suggestions re Schrödinger Test."[20] Foster framed this new plan conventionally as a test of the new theory. The novel theory–experiment agreement, however, also reassured him as to the validity of the experiment and, by extension, to the prospects for pursuing cutting-edge research at his university. He wrote to Eve, "There is no doubt about the nature of the result. If now the work can be carried out successfully with a high standard it should encourage research people at McGill."[21] He followed up with a telegram two days later announcing that the "joint Chalk letter Bohr suggestion strong beta component [sic]" had been sent.[22] The handwritten draft shows a few wording corrections, and bears on the margin the annotation, "Original copy with changes arranged by Prof. Bohr and myself."

Now that Bohr had brought home to him the significance of Chalk's work for the nascent quantum mechanics, Foster grew confident. He anxiously asked Eve, "Shall we say that Miss Chalk's programme is settled?"[23] It was evidently necessary to persuade the senior McGill faculty of the value of his students' work, for Foster pressed the point again a couple of weeks later: "Certainly Miss Chalk should stay with the intensity measurements until they are finished. There are comparatively few jobs quite so definite & I suggest that the Staff considers the possibility of accepting her completed programme in that line as a thesis for a PhD."[24]

The joint Foster–Chalk letter appeared in *Nature* in October. It reported both a discrepancy of their data from Kramers' estimates and an agreement with "the new theoretical calculations by Schrödinger" for the strong components of the spectral line H_β. Noting "the increased interest of this research," it announced that work was in progress to extend the measurements "to the weaker components and to H_γ and H_δ with such modification in method as to permit the detection of any secondary spectrum lines" (Foster and Chalk, 1926, p. 592).

Foster befriended Heisenberg, and for the rest of his leave he became deeply immersed in quantum-mechanical calculations of the helium pattern using matrix mechanics. This work resulted in what he and others would regard as the most important publication of his career (Foster, 1927b; Bell, 1966, p. 152). He nevertheless did not neglect his students during that period. He sent more laboratory instructions for Chalk, encouraging her to extend the observations to H_ε as well. When Schrödinger arrived, Foster informed Eve, "Schrödinger is a little

---

[20] Foster to Eve, September 6, 1926; Foster to Eve, "Suggestions re Schrödinger Test," n.d. See Note 14.
[21] Foster to Eve, "Suggestions re Schrödinger Test," n.d. See Note 14.
[22] Foster to Eve, September 8, 1926. See Note 14.   [23] Foster to Eve, September 6, 1925. See Note 14.
[24] Foster to Eve, September 23, 1926. See Note 14. Foster had to work even harder to convince Eve that his plans for William Rowles, who was measuring the Stark effect in complex spectra using neon, were sufficient for a PhD (Rowles, 1928).

disappointed with the failure of his physical interpretations, but greatly pleased with the intensity measurements. He did not realize that the calculations could be so good as they are."[25] Foster also wrote to Eve that he was trying to arrange a visit of Schrödinger to McGill. Schrödinger would be travelling to the University of Wisconsin during that winter and was, according to Foster, keen to give a lecture in Montreal, as he wanted "to see something of a Canadian winter" and was "greatly interested in Miss Chalk's work."[26]

Epstein's wave-mechanical calculations of displacements and intensities of Stark components appeared in print in October 1926, in time for Chalk to include them in the analysis of her data along with the "Schrödinger Test." She completed her doctoral research carrying out observations of the relative intensities of the Stark spectra of the four main lines of the Balmer series, and systematically compared her results with Stark's original observations and with all the available theoretical estimates. Chalk obtained remarkably sharp measurements of the first four terms of the Balmer series, H_α, H_β, H_γ, and H_δ, and was therefore able to compare her results to Stark's previous measurements as well as to Kramers', Schrödinger's, and Epstein's theoretical estimates (Chalk, 1928).[27] The comparison revealed the clear superiority of Schrödinger's figures, which was especially evident on the intensity ratios for the outer components of H_α and H_β, as shown graphically in Figure 6.2.

Moreover, Chalk noticed that the differences between Schrödinger's and Epstein's numbers could not be accounted for by the difference in their theoretical approaches alone. She commented, "the results have not been checked by the writer, but it is thought that an error has been introduced into the expression obtained for the amplitudes" (Chalk, 1928, p. 43).[28] Her conjecture was proven correct a year later by Walter Gordon and Rudolph Minkowski at the University of Hamburg (Duncan and Janssen, 2015, p. 249 and note 17 on p. 222). She concluded, "Finally, it must be stated that for all the components observed by the writer, the intensities correspond to those calculated by Schrödinger" (Chalk, 1928, p. 43).

Chalk and Foster published a second note in *Nature* in the spring of 1928 to confirm that the "numerous plates obtained by the junior author as an extension to the earlier experiments" showed that "on this most outstanding point Schrödinger is correct" (Foster and Chalk, 1928, pp. 830–831). They followed up with a complete report of the experiments and results, which was communicated to the Royal

---

[25] Foster to Eve, October 13, 1926. See Note 14.   [26] Foster to Eve, November 28, 1926. See Note 14.

[27] Chalk obtained her data using several configurations of Foster's modification of Lo Surdo tubes, which she characterized as the "most successful" to obtain "a sharp analysis" (Chalk, 1928, p. 15). The discharges were produced by hydrogen mixed in small amounts with helium or neon at low pressures, as such mixtures were found to enhance the Balmer series. Photographs of the spectra were obtained with the neutral wedge method and exposures from 30 min to four hours (Chalk, 1928, p. 17).

[28] See also Foster and Chalk (1929, p. 109).

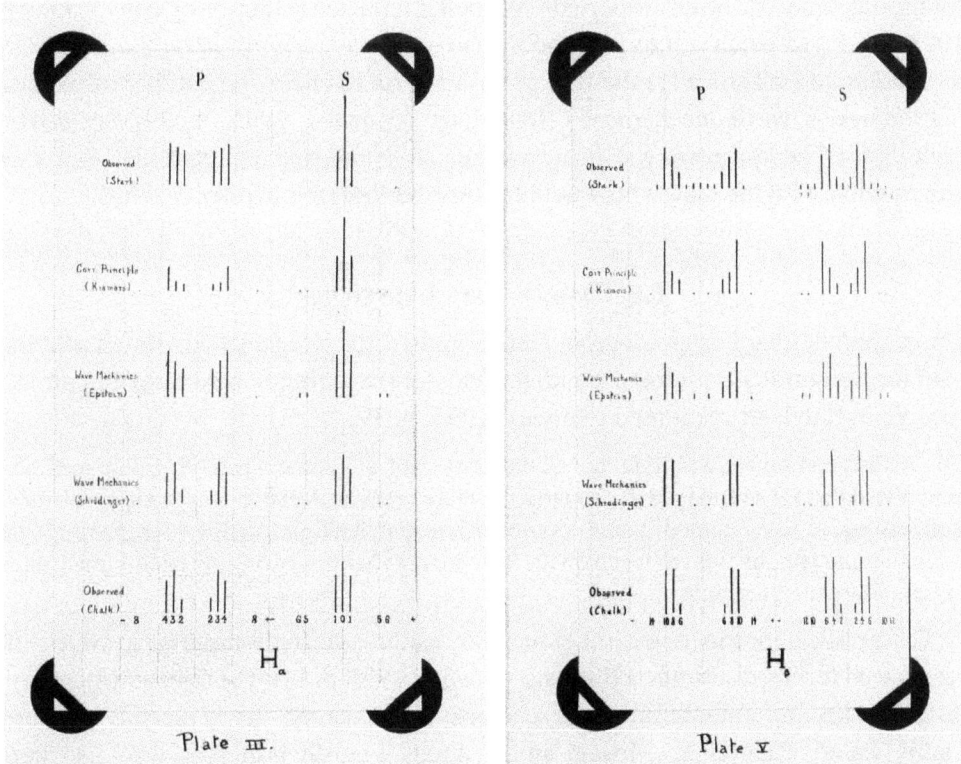

Figure 6.2 Diagrams representing the observed and calculated intensities of the Stark components of the first two lines of the Balmer series. From Laura Chalk's PhD thesis (Chalk, 1928). Reproduced with permission from executor of Laura Chalk Rowles' estate.

Society of London by Eve the following November (Foster and Chalk, 1929). Before the paper appeared in print, Foster shared with Bohr that "Miss Chalk got good support for Schrödinger as far as H_δ."[29]

The excitement transpiring from Foster's Copenhagen letters contrasts with the dryness of Chalk Rowles' recollection of her graduate work nearly seven decades later. In her 1995 memoir, she wrote that in the first year of her PhD, while Foster was on leave, she and William Rowles were busy "with passing French and German language requirements and preparing for the dreaded general PhD exams in physics and mathematics." Upon his return, Foster engaged both his graduate students to do "bits of the calculations [of the Stark effect of helium] with hand-operated calculating machines." According to her memoir, only in their third year did the two students apply their efforts "to getting data for our PhD theses" (Rowles, 1995, pp. 38–39). As

---

[29] Foster to Bohr, January 3, 1929. See Note 11.

for the outcome, she briefly reported, "We both graduated at the spring convocation in 1928 and our research results were published in the *Proceedings of the Royal Society of London*. My results were the first published data to check the validity of the new Schrödinger's wave mechanical calculations" (Rowles, 1995, p. 39). Instead of highlighting her professional achievement, she preferred to express a sense of communion with the man who would become her lifelong partner.

6.6 Postgraduate Experience

Upon approaching her graduation, Chalk applied for a recently instituted scholarship for postgraduate studies abroad, the Moyse Travelling Scholarship, but did not receive it. She later recounted (Rowles, 1995, p. 39):

The awards committee was in favour of me up to the last minute when one of the members announced that I was engaged to be married, whereupon the decision was made in favour of a psychologist. As a matter of fact, I was not engaged at that time, nor did I become engaged until after my appointment to teach on the Macdonald Campus two years later. I must admit disappointment at the time.

This episode appears twice in her memoir, and it is the only instance in which she recognized feeling discriminated against (Rowles, 1995, p. 42). The rejection must have tasted particularly bitter because the scholarship was named after a recently deceased family friend, Charles E. Moyse, an illustrious English professor who had been responsible for bringing her father to Canada, and at whose galleried house on a posh Montreal street she remembered playing as a girl (Rowles, 1995, p. 34 and p. 36).

Partly consoled by Foster's offer of a one-year research assistantship, she worked on the helium spectrum, then reapplied for a Moyse Scholarship and this time received it. She thus spent the year 1929–1930 working with the recent Nobel laureate Owen W. Richardson at King's College London, UK. Richardson gave her the task of continuing his measurements of the band spectrum of molecular hydrogen. This "proved to be an uninspiring project," but she diligently complied (Rowles, 1995, p. 39). She discovered a few more bands and wrote her only research paper as a single author (Chalk, 1930). Richardson presented her work to a meeting of the Royal Society of London, to which she was "fortunately allowed to attend, along with Rutherford, Sir Oliver Lodge, Aston, Dirac, and others" (Rowles, 1995, p. 39).

Evidently, however, the postdoctoral experience did not inspire Chalk to pursue a career in science. Upon her return to Montreal, she took a teaching position at the Macdonald College of Agriculture of McGill University, where meanwhile William Rowles had become the head of the physics department.[30] It "seemed

---

[30] The McGill Faculty of Agriculture (now Agricultural and Environmental Sciences) had its own physics department and was located in the satellite Macdonald College campus, while the main physics department, called the Macdonald Physics Laboratory, was in the downtown campus. Both college and department are

ordained that we were to be together again," she wrote (Rowles, 1995, p. 39). She converted her full-time position to half-time after a year because in June 1931 she got married to Rowles.[31] Gender norms required women of Chalk's social class to leave their paid jobs upon marriage. As Gosztonyi Ainley reports, "archival sources certainly indicate that most women scientists who were employed at Canadian universities during the 1920s and 1930s resigned on marriage, and unfortunately we do not know how they felt about it" (Gosztonyi Ainley, 2012, p. 96). But instead of leaving the workplace altogether, Chalk Rowles chose to scale down her professional role, and Gillett and Beer have given us the opportunity to know how she felt about it, at least in hindsight (Rowles, 1995).

Chalk Rowles' memoir conveys retrospective contentment. Although she recognized that "odds" were against women when their aspirations "involved displacing men in fields that had 'belonged' to them," she emphatically denied having experienced gender discrimination in academia except for the loss of a scholarship due the false engagement rumor (Rowles, 1995, p. 45). "I always felt myself to be just one of the boys," she asserted (Rowles, 1995, p. 42). Her choice to keep teaching part-time was presumably not financially motivated, as she implied that she did not have to worry about earning a living (Rowles, 1995, p. 42). This suggests that she loved her job enough not to want to be a full-time housewife (Figure 6.3). Since the Rowles never had children, we do not know whether she would have quit teaching altogether if she had become a mother.

Even so, gender discrimination awaited her once more. In 1936 she was laid off because of so-called anti-nepotism rules: "After I had spent about five years in this position, McGill decreed that, because of the state of the economy, no wives could hold positions in their husband's departments. So I lost my job and thereafter was employed only intermittently 'as needed'" (Rowles, 1995, p. 41). In Canada, there were formal rules against nepotism as well as restrictions on the employment of married women, but these rules were limited to the civil service. In universities, such rules were informal and applied discretionally, frequently "as a way to keep out women" (Gosztonyi Ainley, 2012, p. 107). For two decades, women in Canada and elsewhere had increasingly pursued higher education and academic employment. However, the economic crisis of the 1930s reversed this positive trend, even though the overall number of students and the number of academic positions continued to grow. Universities responded to financial strictures by reducing professorial salaries and by leveraging gender norms and "anti-nepotism rules" to

---

named after William Christopher Macdonald (1831–1917), a businessman and philanthropist who funded the construction of a large part of the university in the 1870s–1900s (Snell, 1963).

[31] A historian of Macdonald College recorded, "Dr. M. Laura Chalk gave the courses in 1930–1931 and again, as Mrs. W. Rowles, from the beginning of the session of 1931–1932 to the end of that of 1935–1936, except that for the first year of that period" (Snell, 1963, p. 141).

Figure 6.3 Portrait of Laura M. Chalk Rowles c. 1931, captioned "Laura Rowles as her students knew her." From her 1995 memoir, "Long experience and a happy existence," p. 42.

push women out (Gingras, 1991, p. 79). Chalk Rowles nevertheless insisted that this did not amount to gender discrimination. In her view, she did not lose her job because she was a woman, but because she was "drawing a second salary" in her husband's department (Rowles, 1995, p. 45).

Being employed "as needed" meant working occasionally as a replacement instructor and laboratory demonstrator. It also meant being called to teach "a crash course in electricity theory to five hundred Canadian Air Force men" who were being trained as radar officers during World War II, and then teaching "the big Department of Veterans' Affairs grant classes" of engineering students at an improvised "overflow campus at what had been an Air Force training centre in St. Jean"[32] (Rowles, 1995, p. 41). A historian of Macdonald College recorded that "Dr. and Mrs. Rowles [sic] and Dr. Oliver [another physics faculty member] took a very active part in the courses given in the McGill Physics Department for the training of enlisted men as radiotechnicians" (Snell, 1963, p. 153). That she was glad for these opportunities even in circumstances that she described as *"very*

---

[32] Now the Royal Military College Saint-Jean, outside Montreal.

hectic" and "strenuous" is a testimony to her dedication to physics and to teaching (Rowles, 1995, p. 41, emphasis in the original).

6.7 The *Experimentum Crucis* That Wasn't

Despite its potential to be an *experimentum crucis* for the new quantum mechanics, the Foster–Chalk experiment has nearly disappeared from the historiography of physics. Analyzing it as a case study of the complex dynamics between theory and experiments, Brush argued against the idea that scientific revolutions depend on novel predictions. As he emphasized, quantum mechanics gained acceptance thanks to its logical coherence and explanatory power rather than empirical tests (Brush, 1994; 2015, p. 241 and pp. 494–495). An additional, more pragmatic reason why the Foster–Chalk experiment did not attain symbolic significance in the collective narrative of the quantum revolution is the neglect of the protagonists, starting with Foster himself. In Foster's view, his work on helium, which he carried out in the framework of matrix mechanics with Heisenberg's collaboration, completely overshadowed Chalk's "Schrödinger Test" as an experimental confirmation of the new theory (Foster, 1927b, pp. 139–140).[33] Schrödinger, on his part, even though he was greatly pleased with Chalk's work in 1926 (see Section 6.5), later failed to acknowledge the experimenters who had provided empirical verification of "the earliest quantitative achievement of wave mechanics" (Schrödinger, 1952, p. 235). Heisenberg also contributed to eclipse the Foster–Chalk experiment in physics' collective memory by fondly reminiscing, in 1967, about collaborating with Foster on the latter's work on the helium spectrum in the fall of 1926 in Copenhagen (Heisenberg, 1967, p. 102).[34] He stressed that Foster's data on helium showed perfect agreement with his quantum-mechanical theory. He further stated that Bohr as well was glad to see that the Stark effect in helium provided "one of the most beautiful confirmations" of the new theory just as the hydrogen spectrum had done 10 years earlier for the old quantum theory (Heisenberg, 1967, p. 102). Of Foster and Chalk's earlier confirmation of Schrödinger's intensities calculations in

---

[33] Foster mentioned his and Chalks' 1926 letter to *Nature* only to characterize Schrödinger's treatment of the Stark intensities as a mere "quantitative extension of the qualitative estimate given earlier by Kramers" (Foster, 1927b, p. 139). While carefully positioning himself in his department as a prudent champion of cutting-edge European theoretical advances, at the level of spectroscopic practice he did not regard the matrix mechanics of the helium spectrum as revolutionary either. He described it as computationally simpler because it permitted a perturbative treatment of Stark displacements and intensities without "further information of the inner field" of the atom, but otherwise "quite analogous to that carried out by Kramers on the basis of the earlier formulation of the quantum theory" (Foster, 1927b, p. 139). From then on, on the rare occasions he referred to his experiment with Chalk it was more to deflate than to extol it, framing it as a lucky coincidence of experimental conditions and theoretical idealizations within a context of bewildering complexity (Foster, 1930, pp. 591–592; Foster and Snell, 1937, p. 568; Foster, 1938, p. 239).

[34] See also Heisenberg (1963). In the 1963 interview with Kuhn, Heisenberg misremembered having co-authored Foster's paper on the helium spectrum.

hydrogen he said nothing at all. As a result, also Foster's Royal Society biographer spotlighted the work on helium as Foster's most important paper and overlooked the Foster–Chalk experiment entirely (Bell, 1966, p. 153). Moreover, Foster's focus on establishing a research center at McGill led him to prioritize training, updating apparatus, and sustaining open-ended research programs over banking on any single instance of theory confirmation. It is also possible that the unfavorable attitude toward quantum mechanics of a senior member of McGill's department dissuaded him from showcasing his and Chalk's experimental support to the new theory (see Section 6.4). In any event, Chalk's decision to leave research after her postdoctoral year also doubtlessly contributed to obscuring her contribution to quantum mechanics.

6.8 Gender and Physics: Laura Chalk's Choices

Laura Chalk Rowles chose to quietly exit the stage of physics research at the point when young scientists normally start their journey toward research independence. Her case is an early example of the "leaky pipeline," the high rate of women leaving promising scientific career paths at early junctures. Thanks to Gillett and Beer, we have the rare opportunity to gain insight into the role that gender played in the choices made by one of these women in the formative phase of modern physics, and also in her retrospective self-narrative. When Gillett and Beer asked Chalk Rowles – then 91 years old – to look back and record her memories, she chose to write in a light and unassuming register, highlighting certain aspects of her personal journey that were evidently most meaningful to her at that point in her life, while skimming over others. Not only did she blur her own role in an experiment that Bohr and Schrödinger had judged to be of great interest at a pivotal moment of the development of physics, but she did not even mention that she had been the first woman to complete a PhD in the McGill Physics Department. The contrast between the laconic account of her work and the vivid descriptions of her upbringing and education, the physical environment of the laboratory, and her married life on campus suggests that she made a deliberate choice of self-representation. For her years as a graduate student, Chalk Rowles fashioned a narrative of her and her future husband's joint experiences. She chose to downplay her abdication from a potentially brilliant career and instead foregrounded the success of her marital relationship. In her story, the significance of her data for the development of quantum mechanics was barely notable. "Clearly I was never a famous scientist," she demurred, "but I kept busy in my field a lot of the time and was glad to be needed" (Rowles, 1995, p. 42).

Chalk Rowles made clear that her decision to return home and concentrate on teaching had been a free choice, and that she did not regret it. Knowing that her essay would be part of a feminist anthology, she gave it a title that was an unequivocal statement, "Long experience and a happy existence." Moreover, as we have seen, she dismissed the question of gender discrimination, admitting one exception only and pinning her disappointment on the false rumor, not on women's unfair treatment: "I never felt discriminated against in any way except, perhaps, when turned down for a scholarship because I was reported to be engaged to be married; but I might well have voted with the committee, had I been a member. I still am sympathetic towards their decision" (Rowles, 1995, p. 42). She did not see systemic discrimination and overtly approved of the committee's decision.

She refused to present herself as a victim of sexism even though it had hit her directly at least twice. Looking back at her life across the twentieth century, she remained at arm's length from feminism. She had "watched with interest the agitation of women as they have gradually been relieved of many of the duties of their forebears by modern inventions." As such inventions and higher education became available, women had "broadened their interests and horizons" (Rowles, 1995, p. 45). Even though she recognized that the academic playfield was uneven, she opined that "so-called 'equity' is fundamentally incompatible with logic when it comes to choosing people for specific positions." And even though she agreed with "the feminists in our ranks who have achieved some useful ends for women at large," she dismissed as "picayune fussing" the concerns about androcentric language such as "chairman" and "similar activities" (Rowles, 1995, p. 45). Systemic sexism, in language as in institutional norms and practices, was invisible to her.

Chalk Rowles conformed to the gender norms of her milieu without any apparent strain. Teaching was the employment of choice for educated women of her time, and leaving their paid jobs upon getting married was the norm for women of her social class. She accepted this traditional arrangement unquestioningly, to the point that she professed it to be right to deny a woman a scholarship because she was engaged. At the same time, Chalk Rowles embodied a set of contradictions. She considered it natural that women wanted to enjoy the advantages of modernity and broaden their interests and horizons. Her milieu included women who took leading roles in the movement for women's education.[35] Her sociocultural situation admitted some space for women's non-traditional choices, especially in the field of education. Her decision to study physics, a subject that she perceived as both male-dominated and masculine, was emblematic of this ambivalence. Just as she embraced traditional femininity, she had enjoyed feeling "one of the boys" in the laboratory (Rowles, 1995, p. 42).

---

[35] The best man at her parents' wedding was Percival Molson, a famous member of the prominent Molson family (owners of Canada's largest brewing company). The prize that Chalk won upon graduation (see Note 7) had been instituted by Percival Molson's grandaunt, Anne Molson (1824–1899), the first president of the Montreal Ladies' Educational Association, an organization that had pioneered women's education in Canada and opened the way for the admission of women to McGill University in 1884 (Bradbury, 1990).

Women physicists of Chalk Rowles' generation lived in a liminal space. Although she knew that physics "had 'belonged' to men," she did not need to intentionally break any gender barrier or feel like an oddity for pursuing it (Rowles, 1995, p. 45). Women who earned scientific doctorates were still infrequent but no longer exceptional in her day. Being the first woman to complete her graduate studies and earn a PhD entirely in the McGill Physics Department was so insignificant to her as to not deserve a mention in her memoir.[36] She did not express any sense of connection with the other women in her field, nor did she react to the injustices that they routinely experienced. Like many women of her generation and upbringing, she had internalized gender subordination to the point that she experienced it as natural and unremarkable.[37]

Notably, Chalk Rowles assertion, "I always felt myself to be one of the boys" is both metaphorical and literal (Rowles, 1995, p. 42). It communicates that she felt as an equal in the physics department, but also that to be an equal she had to perform the masculine gender. It is no original observation that physics, the "culture of extreme objectivity: a culture of no culture," is a heavily masculine field that projects a disembodied and gender-neutral image (Traweek, 1988, p. 162; Ottemo et al., 2021). In such a field, gender subordination takes up a special character. Participants of both genders are trained to pretend, to themselves and to others, that gender does not matter to success, while, in fact, they live in a social world where only individuals of the dominant group are presumed capable of transcending their gender and becoming unburdened by it. This dissociation sustains the mirage of meritocracy in the discipline.[38] It enabled Chalk Rowles to recognize that women were not given a fair chance and, at the same time, to flatly deny discrimination and pointedly reject equity policies.

Chalk Rowles' self-narration touches also on a related paradox. She claimed that most women avoided physics because they "were more concerned with social problems in the world than with mathematics and abstract science" (Rowles, 1995, p. 45). Characteristically, she attributed to physics the gender-transcending

---

[36] The first woman to earn a PhD in physics in Canada was Mattie Levi Rotenberg at the University of Toronto in 1926 (Prentice, 2006, p. 18); the first woman to earn a PhD at McGill was Annie Louise Macleod, in chemistry, in 1910 ("Blazing trails: McGill's women," https://www.mcgill.ca/about/history/features/mcgill-women).

[37] I am deliberately using the term "subordination" instead of "oppression." While the term "oppression" emphasizes the victimization of the oppressed, subordination encompasses more of the historical experiences of women, as it includes downright oppressive cases as well as the willing acceptance of subordinate status in exchange for protection and class privilege (Lerner, 1986, pp. 233–235). Talking of internalized subordination thus helps to express the effects of the gender hierarchy while also asserting women's agency.

[38] For example, Hertha Mark Ayrton famously said to a journalist, "I do not agree with sex being brought into science at all. The idea of 'woman and science' is completely irrelevant. Either a woman is a good scientist, or she is not" (quoted in Fara, 2006, p. 458). The defensive strategy of denying the relevance of gender to success in professional fields is not limited to physics. Gillett and Beer report, for instance, that one "highly successful woman" declined to participate in their autobiographical anthology "because she believed that a collection of essays by women only is unnecessary – that to single out members of her sex is inappropriate, or amounts to reverse sexism" (Gillett and Beer, 1995, p. xxiv).

character that she denied to women. Caring for society was for her a feminine trait, and it must have mattered a great deal to someone like her – who, as a girl, had wanted to become a Christian missionary and, as an adult, had chaired the volunteer Red Cross hospital visiting committee (Rowles, 1995, pp. 44–45). At the same time, she had been drawn to a scientific field that she regarded as mathematical and abstract, hence simultaneously masculine and genderless. Training hundreds of servicemen to become radar officers for war did not appear to register in her mind as an intervention in the social world, and it did not cloud the crystalline abstractness of the discipline. In the strangely dissociated world of physics, it was evidently possible to insulate the disciplinary image of unworldly detachment from the very earthly lived experience of the laboratory and the classroom. This spectrum of contradictions – the splitting and shifting of a woman's identity in a masculine field that is culturally construed as disembodied, transcendent, value-neutral, and gender-free – was not unique to Chalk Rowles. It has long been detected and studied by scholars of gender and physics (Keller, 1985; Harding, 1986; Lloyd, 1993; Schiebinger, 1999; Götschel, 2014; Gonsalves and Danielsson, 2020). She only displayed it with extraordinary clarity, just as her hydrogen atoms did with the Stark spectrum.

6.9 Final Remarks: Laura Chalk Rowles and the Quantum Revolution

The Foster–Chalk experiment occupies an awkward position in the history of quantum physics. It was philosophically crucial but historically irrelevant, theoretically revolutionary but experimentally incremental, and valued by Bohr and Schrödinger but underappreciated by Foster. The position of the junior author who carried it out, like that of many scientifically inclined women of the twentieth century, was also awkward in the social world of the time, caught between the pull of social progress and the push of traditional norms. Her gender clearly played a role in the historical eclipse of the experiment both because of systemic discrimination and of her self-effacing choices. Did the quantum revolution play any role in Chalk's choices? About this question, she left us only one fleeting remark (Rowles, 1995, p. 46):

As the years rolled by, I became more and more exposed to science and found that the more I learned, the less I seemed to know; and what I "believed" became more and more cloudy – e.g. quantum theory and wave theory of light were each true in describing certain phenomena, but never both applicable for any specific case.

It is impossible to draw an answer from this brief passage, but two points call out for attention. First, Chalk Rowles did not feel that her choice to dedicate herself to teaching had been a retreat from science. Second, the intellectual impact of quantum mechanics was for her destabilizing and disempowering. Self-doubt and feelings of inadequacy are well-known effects of internalized subordination, and a female postdoctoral researcher in a masculine field is doubly subordinate, because of her gender and because of the academic hierarchy. It is easy to conjecture that the intersection of Chalk's juniority and gender amplified her insecurities, and that the unsettling consequences of the theory she saw confirmed in her spectroscope contributed to dampen her desire to be a researcher.

The local environment at McGill University probably did not give her a boost. During Chalk's doctorate, Foster was the only faculty member working on the newfangled theory, and even he felt insecure about openly supporting it in his department (Gingras, 1981, p. 26). He complained to Bohr that the situation at McGill was unsympathetic to his work on helium due to King's "rigid opposition" (see Section 6.4).[39] It is tempting to ask whether Chalk's choices could have been different, had she received more validation from senior faculty and scholarship committees. When she went to England, she certainly found more support for quantum mechanics; however, she was evidently not enthused by her experience as a postdoc there either.

Perhaps, even a stronger mentorship would not have been sufficient to keep her in research, as the example of Harriet Brooks suggests. There are interesting parallels and contrasts between the cases of Brooks and Chalk. They belonged to the same social world. Brooks was one generation older, but she was more of a free spirit and a trailblazer than Chalk. Moreover, Brooks could count on the mighty mentorship of Ernest Rutherford. When Brooks lost her teaching position because of her engagement, she protested firmly, albeit in vain. Yet even she eventually became fatigued with the struggle of carving a professional path for herself, and she abandoned her scientific vocation to become a wife and mother (Rayner-Canham and Rayner-Canham, 1992; Gosztonyi Ainley, 2012, pp. 92–94). The "leaky pipeline" effect, at least for most of the twentieth century, has been not only due to the obstacles in women's career paths, but also to the readiness of an alternative for them, which was mostly unavailable to men. Women could afford to be "less persistent in their careers than men because they [could] rely on the socially sanctioned safety net of marriage" (Schiebinger, 1999, p. 59). Still, I cannot help but wonder whether Chalk might have chosen to persist as a researcher if she had been mentored by Brooks, or if she had seen at least one woman sitting among Rutherford, Lodge, Aston, and Dirac at the Royal Society meeting that she was allowed to attend.

---

[39] Foster to Bohr, March 15, 1927; Foster to Bohr, May 5, 1927. See Note 11.

References

Bell, Robert E. 1966. John Stuart Foster. Biographical memoir. *Biographical Memoirs of Fellows of the Royal Society*, 12, 147–161. https://doi.org/10.1098/rsbm.1966.0006

Bradbury, Bettina. 1990. Molson, Anne. In *Dictionary of Canadian Biography, Volume XII 1891–1900*. https://www.biographi.ca/en/bio/molson_anne_12E.html

Brush, Stephen G. 1994. Dynamics of theory change. *Proceedings of the Biennial Meeting of the Philosophy of Science Association*, 2, 133–145. https://www.jstor.org/stable/192924

Brush, Stephen G. 2015. *Making 20th Century Science: How Theories Became Knowledge*. New York: Oxford University Press.

Chalk, M. Laura. 1926. *Potential Distributions in the Crookes Dark Space and Relative Intensities of Stark Effect Components of H_β and $He_{\lambda\ 4922}$*. MSc Thesis. Montreal: McGill University Library.

Chalk, M. Laura. 1928. *Observed Relative Intensities of Stark Components in Hydrogen*. PhD Thesis. Montreal: McGill University Library.

Chalk, M. Laura. 1930. The spectrum of H_2: the bands ending on $2p^1\Pi$. *Proceedings of the Royal Society of London A*, 128(808), 579–587. https://doi.org/10.1098/rspa.1930.0132

Crossfield, E. Tina. 1997. Allie Vibert Douglas (1894–1988): astrophysicist, astronomer. In Shearer, Benjamin F. and Shearer, Barbara S. (eds.), *Notable Women in the Physical Sciences. A Biographical Dictionary*. Westport, CT: Greenwood Press, pp. 69–76.

Duncan, Anthony and Janssen, Michel. 2014. The trouble with orbits: the Stark effect in the old and the new quantum theory. *Studies in History and Philosophy of Modern Physics*, 48, 68–83.

Duncan, Anthony and Janssen, Michel. 2015. The Stark effect in the Bohr–Sommerfeld theory and in Schrödinger's wave mechanics. In Aaserud, Finn and Kragh, Helge (eds.), *One Hundred Years of the Bohr Atom. Proceedings From a Conference. Scientia Danica. Series M, Volume 1*. Copenhagen: Royal Danish Academy of Sciences and Letters, pp. 217–271. Also available at arXiv:1404.5341v1. https://doi.org/10.48550/arXiv.1404.5341

Duncan, Anthony and Janssen, Michel. 2019. *Constructing Quantum Mechanics. Volume 1. The Scaffold, 1900–1923*. Oxford: Oxford University Press.

Epstein, Paul S. 1926. The Stark effect from the point of view of Schrödinger's quantum theory. *Physical Reviews*, 28, 695–710. https://doi.org/10.1103/PhysRev.28.695

Fara, Patricia. 2006. Women or just good scientists? *Nature*, 44(30), 548. https://doi.org/10.1038/444548a

Foster, J. Stuart. 1925. The Stark effect for H_β and He $_{\lambda4686}$. *Astrophysical Journal*, 63, 229–237.

Foster, J. Stuart. 1927a. Stark patterns observed in helium. *Proceedings of the Royal Society A*, 114, 47–66. https://doi.org/10.1098/rspa.1927.0024

Foster, J. Stuart. 1927b. Application of quantum mechanics to the Stark effect in helium. *Proceedings of the Royal Society A*, 117, 137–163. https://doi.org/10.1098/rspa.1927.0171

Foster, J. Stuart. 1930. Some leading features of the Stark effect. *Journal of the Franklin Institute*, 209, 585–624. https://doi.org/10.1016/S0016-0032(30)90851-4

Foster, J. Stuart. 1938. Spectroscopy III: the Stark effect and some related phenomena. *Reports on Progress in Physics*, 5, 233–241. https://doi.org/10.1088/0034-4885/5/1/320

Foster, J. Stuart and Chalk, M. Laura. 1926. Observed relative intensities of Stark components in hydrogen, *Nature*, 118(2973), 592. https://doi.org/10.1038/118592b0

Foster, J. Stuart and Chalk, M. Laura. 1928. Observed relative intensities of Stark components of H_α. *Nature*, 121(3056), 830–831. https://doi.org/10.1038/121830b0

Foster, J. Stuart and Chalk, M. Laura. 1929. Relative intensities of Stark components in hydrogen. *Proceedings of the Royal Society of London A*, 123(791), 108–118. https://www.jstor.org/stable/95096

Foster, J. Stuart and Snell, Hawley. 1937. The Stark effect in hydrogen and deuterium. *Proceedings of the Royal Society of London A*, 162, 349–356. https://doi.org/10.1098/rspa.1937.0186

Gillett, Margaret and Beer, Ann (eds.). 1995. *Our Own Agendas. Autobiographical Essays by Women Associated with McGill University.* Montreal: McGill-Queen's University Press.

Gingras, Yves. 1981. La physique à McGill entre 1920 et 1940: la réception de la mécanique quantique par une communauté scientifique périphérique. *History of Science and Technology in Canada Bulletin*, 5, 1, 15–39. https://doi.org/10.7202/800094ar

Gingras, Yves. 1991. *Physics and the Rise of Scientific Research in Canada.* Translated from the French by Peter Keating. Montreal: McGill-Queen's University Press.

Gonsalves, Allison J. and Danielsson, Anna T. 2020. Introduction: why do we need identity in physics education research? In Gonsalves, Allison J. and Danielsson, Anna T. (eds.), *Physics Education and Gender: Identity as an Analytic Lens for Research.* New York: Springer, pp. 1–8. https://doi.org/10.1007/978-3-030-41933-2_1

Gosztonyi Ainley, Marianne. 2012. *Creating Complicated Lives. Women and Science at English-Canadian Universities, 1880–1980.* Montreal: McGill-Queen's University Press.

Götschel, Helene. 2014. No space for girliness in physics: understanding and overcoming the masculinity of physics. *Cultural Studies of Science Education*, 9, 531–537. https://doi.org/10.1007/s11422-012-9479-y

Harding, Sandra, *The Science Question in Feminism*. Ithaca, NY: Cornell University Press, 1986.

Heisenberg, Werner. 1926. Mehrkörperproblem und Resonanz in der Quantenmechanik. *Zeitschrift für Physik*, 38, 411–426. https://doi.org/10.1007/BF01397160

Heisenberg, Werner. 1963. Interview of Werner Heisenberg by Thomas S. Kuhn, Session VI, February 19, 1963 and Session VIII, February 25, 1936. Niels Bohr Library & Archives. https://www.aip.org/history-programs/niels-bohr-library/oral-histories/4661-6 and https://www.aip.org/history-programs/niels-bohr-library/oral-histories/4661-8

Heisenberg, Werner. 1967. Quantum theory and its interpretation. In Rozental, Stefan (ed.), *Niels Bohr. His Life and Work as Seen by His Friends and Colleagues.* New York: Interscience Publishers.

Jammer, Max. 1966. *The Conceptual Development of Quantum Mechanics.* New York: McGraw-Hill.

Keller, Helen F. 1985. *Reflections on Gender and Science.* New Haven, CT: Yale University Press.

Kojevnikov, Alexei. 2020. *The Copenhagen Network. The Birth of Quantum Mechanics from a Postdoctoral Perspective.* New York: Springer.

Kragh, Helge. 2012. *Niels Bohr and the Quantum Atom: The Bohr Model of Atomic Structure, 1913–1925.* Oxford: Oxford University Press.

Kuhn, Thomas S. 1962. *The Structure of Scientific Revolutions.* Chicago, IL: University of Chicago Press.

The Montreal Gazette. 1996. Laura Rowles. Obituary. *The Montreal Gazette*, February 17, E:14.

Leone, Matteo, Paoletti, Alessandro, and Robotti, Nadia. 2004. A simultaneous discovery: the case of Johannes Stark and Antonino Lo Surdo. *Physics in Perspective*, 6, 271–294. https://doi.org/10.1007/s00016-003-0170-2

Lerner, Gerda. 1986. *The Creation of Patriarchy*. Oxford: Oxford University Press.

Lloyd, Genevieve. 1993. *The Man of Reason: "Male" and "Female" in Western Philosophy*, 2nd ed. London: Methuen.

Marshall, Joan. 2000. News from the Macdonald Campus (MAC). *The MSE Newsletter*, 1, 4–5.

Nicholson, John W. and Merton, Thomas R. 1916. On the distribution of intensity in broadened spectrum lines. *Philosophical Transactions A*, 216, 459–488. https://doi.org/10.1098/rsta.1916.0010

Ottemo, Andreas, Gonsalves, Alison J., and Danielsson, Anna T. 2021. (Dis)embodied masculinity and the meaning of (non)style in physics and computer engineering education. *Gender and Education*, 33, 8, 1017–1032. https://doi.org/10.1080/09540253.2021.1884197

Pauli, Wolfgang. 1926. Über das Wasserstoffspektrum vom Standpunkt der neuen Quantenmechanik. *Zeitschrift für Physik*, 36, 336–363. https://doi.org/10.1007/BF01450175

Prentice, Alison. 2006. A blackboard in her kitchen: women and physics at University of Toronto. *Scientia Canadiensis*, 29(2), 17–44.

Ramos, Thomas F. 2019. *Call Me Johnny*. Lawrence Livermore National Laboratory LLNL-BOOK-783447, July 26. https://www.osti.gov/servlets/purl/1576166

Rayner-Canham, Marelene and Rayner-Cahnam, Geoff. 1992. *Harriet Brooks, Pioneer Nuclear Scientist*. Montreal: McGill-Queen's University Press.

Rowles, William. 1928. *The Stark Effect in Complex Spectra*. PhD Thesis. Montreal: McGill University Library.

Rowles, Laura. 1995. Long experience and a happy existence. In Gillett, Margaret and Beer, Ann (eds.), *Our Own Agendas. Autobiographical Essays by Women Associated with McGill University*. Montreal: McGill-Queen's University Press, pp. 33–46. https://doi.org/10.1515/9780773565593-006

Schiebinger, Londa. 1999. *Has Feminism Changed Science?* Cambridge, MA: Harvard University Press.

Schrödinger, Erwin. 1926. Quantisierung als Eigenwertproblem. Dritte Mitteilung: Störungstheorie, mit Anwendung auf den Starkeffekt der Balmerlinien. *Annalen der Physik*, 80(13), 437–490.

Schrödinger, Erwin. 1952. Are there quantum jumps? Part II. *British Journal for the Philosophy of Science*, 3(11), 233–242. https://www.jstor.org/stable/685266

Snell, John F. 1963. *Macdonald College of McGill University. A History from 1904–1955*. Montreal: McGill University Press.

Thomas, Jerry. 1984. John Stuart Foster, McGill University, and the renascence of nuclear physics in Montreal, 1935–1950. *Historical Studies in the Physical Sciences*, 14(2), 357–377. https://doi.org/10.2307/27757537

Traweek, Sharon. 1988. *Beamtimes and Lifetimes: The World of High Energy Physicists*. Cambridge, MA: Harvard University Press.

7

Elizabeth Monroe Boggs: From Quantum Chemistry to the Manhattan Project

PATRICK CHARBONNEAU

7.1 Introduction

In 2018, *The New York Times* started publishing belated obituaries of remarkable people whose deaths had been overlooked at the time of their passing. Through a new section titled "Overlooked No More," the famous daily wants to surpass its white-male bias, recognizing that obituaries express "how society valued various achievements and achievers" (Padnani and Bennett, 2018). Perhaps unsurprisingly, many remarkable people to have been so recognized are women and people of color. Some were also scientists. They notably include nuclear chemist Elizabeth Rona (Greenwood, 2019), radio astronomer Ruby Payne-Scott (Schwartz, 2020), and computer programmer Ada Lovelace (Miller, 2018). Obituaries, however, can also contribute to unbalances by highlighting certain achievements and ignoring others. One such example is Elizabeth Monroe, married Boggs, a remarkable scientist who took part in bringing quantum physics to chemistry and whose death was not contemporaneously overlooked by *The New York Times* (Saxon, 1996). Unlike many of the other women featured in this book, Monroe Boggs *did* receive broad public recognition during her lifetime. That visibility, however, came from her relentless advocacy on behalf of the rights of people with disabilities, not from her scientific contributions. In fact, she has thus far been left out of most of the reviews of women scientists of her epoch (Rayner-Canham and Rayner-Canham, 1998; Howes and Herzenberg, 1999; Byers and Williams, 2006; Green and LaDuke, 2009; Abbate, 2012; Rayner-Canham and Rayner-Canham, 2008).

I thank the late Simon Bauer, John Lennard-Jones, and Pat Kenschaft as well as Anne Cooper, Ruth Doherty, Pam Murphy, Deborah Spitalnik, and Tom Underwood for sharing various materials and recollections. I also thank Robyn Carroll for help with The Boggs Center archives, Daphne Klotsa for early assistance, Lillian Hoddeson for help accessing classified material, and Jessica Bright, Zach Brodt, Liz Milewicz, and especially Jodi Psoter for archival assistance. I further thank Benoit Charbonneau for continuously supporting this effort, Ashley Elrod and Seymour Mauskopf for encouraging its restart, and Katherine Brading, Evan Hepler-Smith, and WiHQP for motivating its finish. *Bibliographical note:* Many of the primary sources currently in personal collections as well as The Boggs Center archives on Monroe Boggs should be transferred to the Rutgers University Library Special Collection in 2025. See also Elizabeth Monroe Boggs Collection, Duke Research Data Repository. https://doi.org/10.7924/r4v128x9p

Figure 7.1 Elizabeth Monroe in 1928, before leaving for a European summer vacation with her McNairy grandparents. P. M. Murphy Personal Collection. Reproduced with permission.

Through primary and oral history sources, many of which previously inaccessible, this chapter retraces Monroe Boggs' journey from early quantum chemistry enthusiast to key public figure of the disability rights movement in the second half of the twentieth century. By doing so, this chapter also restores the value given to her various achievements.

7.2 Family Background

Elizabeth Monroe was born on April 5, 1913, in Cleveland, Ohio, the older of two surviving children from an affluent household (Figure 7.1). Her father, Francis Adair Monroe Jr. (1878–1969), came from a prominent Louisianan Southern Democrat family. The son of a famed state judge and Tulane faculty, he trained as a chemical engineer and steadily climbed managerial ranks in the sugar industry. Beet sugar had brought him north to Cleveland. Her mother, Elizabeth McNairy (1888–1974), descended from George Willis Pack, whose fortune from Michigan timber made him and his descendants leading citizens of Cleveland after settling there in 1870 (Wykle, 2006).[1]

Following Adair Monroe's advancement, the family relocated to a wealthy suburb of New York City a few years later. There, Elizabeth attended the progressive (and private) Rye Country Day School (1919–1929), and she completed her schooling in Massachusetts, as a boarder at the all-girl Concord Academy (1929–1931) (Bryn Mawr College, 1935). Meanwhile, she developed a curiosity for science, which her family wholly supported,[2] as she explained in a 1979 interview (Boggs, 1979, p. 2):

As I was growing up, he [my father] accepted my intellectual interests in exactly the same frame of reference that he would have accepted them had I been a man. In retrospect, I realize that this had a great deal to do with the ease with which I accepted my own desire to move into fields which not only involved higher education, but fields of endeavor which were not at that time widely sought by women.

Through these years of familial and financial comfort, she nevertheless struggled socially. "I felt like an odd ball for many reasons; I was taller than all the boys and I enjoyed doing school work, definitely not the 'in' thing."[3]

7.3 Bryn Mawr College and Emmy Noether (1931–1935)

It was only after she arrived at Bryn Mawr College, Monroe later recalled, that "for the first time in my life, I did not feel very atypical. . . . I was surrounded by contemporaries, girls of my own age who were equally interested in whatever it was they were interested in" (Boggs, 1979, p. 4). Monroe's main interest was chemistry, and she was particularly intrigued by the new quantum theory. As early as her first-semester philosophy class, she was considering applying quantum physics to the science of molecular attraction.

Modern theories of combination – the molecule [and] the question of valence – can only be held in the light of the electron theory and with knowledge of electrical charges and the forces connected to them. . . . The whole science of wave mechanics and the ether offers problems which were not dreamed of . . . in Democritus' time.[4]

It is unclear how much Monroe knew of wave mechanics at the time – her mention of the ether suggests that she had yet to solidly grasp modern physics – but Bryn Mawr certainly offered opportunities to learn more. For instance, Jane Dewey, who had been a postdoctoral researcher with Niels Bohr and Werner Heisenberg, taught a class on quantum mechanics there, starting in 1932 (see Chapter 5).

---

[1] For instance, Elizabeth Monroe's mother gave her own mother's collection of historical embroideries to the Cleveland Museum of Art (Underhill, 1943).

[2] The Monroe family's sensibility to women's education was deep-seated. Adair Monroe's five sisters attended Newcomb College, then the women's college associated with Tulane University (Murphy, 2023).

[3] "Elizabeth Monroe Boggs," 1991 Concord Academy 60th Anniversary Reunion Book. The Boggs Center archives.

[4] Elizabeth Monroe, "Ancient and Modern Atomic Theories," (December 1931). The Boggs Center archives.

In parallel to finishing the chemistry coursework, Monroe completed a mathematics major. William W. (Bill) Flexner, a young faculty member who taught the introductory calculus class and who would remain a life-long acquaintance, nudged her in that direction "partly because he thought that I was interested in mathematics, and partly because he knew that it was desirable if I wanted to go on for graduate work in science" (Boggs, 1979, p. 5), she later recalled. Bryn Mawr was also particularly well-suited to study the subject. Its long tradition of mathematical excellence notably brought in Emmy Noether (Parshall, 2015), who as a Jewish woman had struggled to secure an academic position after being forced out of Göttingen (Rowe, 2021). Monroe, having completed all upper-level mathematics courses by the end of her junior year (spring 1934), got to be the first – and last – Bryn Mawr undergraduate to study with the recently landed mathematician.[5] During her senior year, she made good progress through Bartel van der Waerden's *Moderne Algebra* (van der Waerden, 1930),[6] the same German-language textbook Noether assigned to her graduate students. Following her mentor's untimely passing in April 1935, however, one of Noether's postdocs, Olga Taussky, had to oversee Monroe's actual thesis writing (Boggs, 1979; Bryn Mawr College, 1935).[7] The experience of formulating a new algebraic proof for an existing theorem certainly helped build Monroe's mathematical confidence. It is unclear, however, whether either Noether or Taussky ever advised her on subsequent career steps.

7.4 Cambridge and John E. Lennard-Jones (1935–1939)

As recipient of the 1935 Bryn Mawr European Fellowship,[8] Monroe headed to the University of Cambridge for graduate studies. She later reasoned that (Boggs, 1979, p. 9):

[a]t that time, the area of study I was focusing on, which was sometimes referred to as theoretical chemistry and sometimes as mathematical chemistry, was developed [in the US] in only three or four universities and most of them did not accept women; thus, my decision to go to Cambridge was confluent with my further career objectives.

This retrospective assessment over the status of women in US universities seems overly bleak. In the 1930s, the leading Americans pursuing quantum chemistry – as the field is now known – were Linus Pauling at Caltech, Robert Mulliken at the University of Chicago, and John Slater at MIT (Gavroglu and Simões, 2011). While

---

[5] P. C. Kenschaft, "Emmy Noether's Only U.S. Undergraduate Dies," unpublished (1996), Patrick Charbonneau Personal Collection.

[6] The book was based in part on Emmy Noether's lectures (van der Waerden, 1997).

[7] Monroe Boggs was notably interviewed for the Noether chapter in MacGrayne (1993).

[8] European fellowships were established in the late nineteenth century to enhance the Bryn Mawr graduate experience (Parshall, 2015, p. 76). In 1938, Monroe also received a Margaret E. Maltby fellowship from American Association of University Women (New York Times, 1938).

Caltech was clearly not welcoming to women – they only joined as undergraduates in 1970 (Bix, 2014) – Chicago awarded the most science, technology, engineering, and mathematics (STEM) doctorates to women of all US schools in the 1930s (including four in physics and 17 in chemistry in 1938), and MIT granted a graduate degree in chemistry to a woman as early as 1873 (Rossiter, 1982, p. 68 and p. 150). At the same time, Cambridge was no progressive bastion either. Women obtained the right to attend lectures only a dozen years before Monroe joined the student body and even then were solely given the title of their degrees, not the degree itself (Tullberg, 1998). Her reasoning might thus have had less to do with particular institutions of higher learning as with a general assessment of gender relations in the US, coupled with a certain European curiosity (see Figure 7.1).

In any event, who was she to work with in Cambridge? In the 1930s, the sole faculty member pursuing quantum chemistry at the British university was John E. Lennard-Jones (Figure 7.2).[9] His applied mathematics approach to the problem

Figure 7.2 Social tennis at Lennard-Jones' home in Cambridge, UK, c. 1937 (Lennard-Jones, 2011): (left to right) Corner, Monroe, Lennard-Jones, and Devonshire. The Papers of Sir John Edward Lennard-Jones, GBR/0014/LEJO/13, Churchill Archives Centre, Churchill College, Cambridge. Reproduced with permission.

---

[9] Upon marrying Kathleen Mary Lennard in 1925, John Edward Jones hyphenated his wife's surname to his. The change was a gesture of respect toward the Lennard family who had lost their two sons to World War I (Lennard-Jones, 2011).

of determining the electronic structure of molecules using wave mechanics was a solid match for Monroe's mathematical training and interests. Lennard-Jones himself had first studied mathematics at Manchester University and later completed a DSc at Cambridge in 1924 under the aegis of Ralph Fowler, a mathematical physicist working at the interface of statistical mechanics and quantum theory (Mott, 1955; Gavroglu and Simões, 2011). There, Lennard-Jones grew more specifically interested in intermolecular interactions, thus eventually leading him to formulate the potential form that famously bears his name. He subsequently joined the University of Bristol as a reader in mathematical physics in 1925 and was quickly granted the newly created chair of theoretical physics.

Lennard-Jones' exposure to the new quantum theory came largely thanks to a Rockefeller Foundation fellowship that brought him to Max Born's Institute of Theoretical Physics at the University of Göttingen for part of 1929. Although Born was out sick for most of that time, Lennard-Jones could nevertheless interact with chemical bond theorist Walter Heitler and with one of Born's postdoctoral researchers, the molecular spectroscopist Gerhard Herzberg (Herzberg, 1989; Stoicheff, 2002). Lennard-Jones became so excited by the application of quantum mechanics to chemistry that he recruited Herzberg to spend the following academic year in Bristol. His ensuing 1929 seminal work, "The electronic structure of some diatomic molecules" (Lennard-Jones, 1929), laid the basis for a quantitative molecular orbital theory – based on quantum mechanics – that offered a more general description of chemical bonding than Heitler's earlier work with Fritz London (Hall, 1991). Notes from his 1931 lecture "Quantum Mechanics of Atoms and Molecules" further reflect his quantum-physics-based excitement about "the pairing of electrons [being] brought into close connection with the valency rules of the chemists" (Lennard-Jones, 1931).

Lennard-Jones' enthusiasm and pioneering work in the emerging field of quantum chemistry led to his recruitment, the following year, to Cambridge University as the first Plummer Professor of Theoretical Chemistry. His first two graduate students there carried the program along. Charles Coulson's thesis on the electronic structure of molecules reported "fairly exact knowledge of the simplest molecules, so that it might be possible to infer properties of other molecules which are less tractable mathematically," while Albert Devonshire's thesis considered a simple quantum-mechanical model hoping to apply it "to the determination of molecular orbitals" (Cambridge University, 1937; Altmann and Bowen, 1974; Nethercott, 1983). The focus of both projects on simple systems implicitly highlighted the need for new numerical approaches and resources to advance quantum chemistry toward greater chemical relevance.

Progress along that direction came via a graduate school friend, Douglas Hartree, then Professor of Applied Mathematics at the University of Manchester (Fischer, 2003). In 1934, Hartree – himself inspired by a visit to Vannevar Bush's MIT laboratory (Zachary and Bush, 2022, p. 19) – built a differential analyzer using Meccano construction toys. Lennard-Jones followed the work closely, aware that a machine capable of solving differential equations by integration could help to develop molecular physics starting from quantum mechanics. As Monroe would later note in her own 1939 PhD thesis (Monroe, 1939, p. 19):

the labour involved in carrying out numerical integration or in evaluating series has heretofore been prohibitive in several interesting cases. Fortunately, though perhaps not entirely fortuitously, just at a time when interest in them has been so stimulated, several mechanical aids . . . have become available.

In fact, one of these aids would become available in Cambridge as early as 1935, when Lennard-Jones, with guidance from Hartree, engaged a mechanically skilled undergraduate student, John Bernard Bratt, to build and run an (improved) analyzer (Wilkes, 1985; Croarken, 1992; Robinson, 2005). That project fostered broader campus interest. In particular, one physics graduate student, Maurice V. Wilkes, learned from Bratt how to use the analyzer and agreed to take over its maintenance and support after the latter joined the National Physical Laboratory (Nature, 1955; Wilkes, 1985). Wilkes was motivated by the prospect of using it for his own research on electromagnetic waves; see, for example, Wilkes (1939, 1940).

It is in this context that Monroe joined the Lennard-Jones research group in early 1936. Her thesis project naturally entailed using the analyzer to solve the Schrödinger equation for the asymmetric two-center problem, that is, a heteronuclear diatomic ion. That particular problem could be solved analytically, and hence its numerical study offered a proof of concept for the eventual consideration of more complex molecules (Monroe, 1938; Cambridge University, 1941). Wilkes, who closely assisted her with the machine, later fondly recalled Monroe's enthusiasm with the problem, even though he himself found "little personal interest" in it (Wilkes, 1985, p. 26). Admittedly, "nothing startlingly new was to be expected from this investigation" (Monroe, 1939, p. iv), as Monroe herself recognized in her thesis. Novelty largely rested on tool and method development, notably the construction of an additional integrator (Figure 7.3), which, as Monroe explained, "obviated the necessity for an input and thus rendered the operation automatic" (Monroe, 1939, pp. 315–316). In short, the project was a clear technical success and an important instrumental step forward, but it had – as designed – not produced any new chemical insight.

Why did Lennard-Jones assign a study with limited immediate visibility to an ambitious and promising graduate student? A hint can be found in a 1936 report to the faculty for the planning of a new type of laboratory dedicated to automated computation, which led to the founding of the Cambridge Mathematical Laboratory the following year:

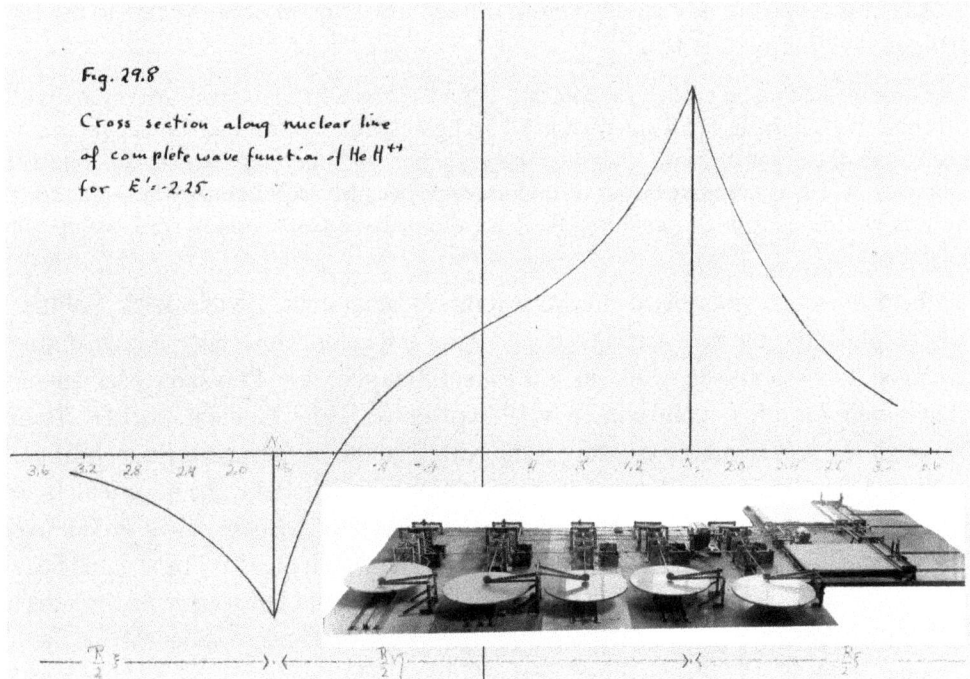

Figure 7.3 Wavefunction of the HeH^{2+} ion excited electronic state along its axial direction. (Inset) Differential analyzer with the five units in place (Monroe, 1939, Figs. 29.8 and 36.1). The Boggs Center archives. Reproduced with permission.

It might be thought that a staff of trained computers . . . could solve the same problems at lower costs, but . . . [t]he machine is more reliable in its results and quicker in time. In many problems it is essential that reliable results should be obtained speedily if further developments . . . are not to be held up.[10]

At the time, *computers* were typically *women* (Grier, 2013). One might further fathom that Monroe was assigned a subservient task (to science), just like the calculators' tasks were subservient, and that subservience was attributed to women rather than to men. It may therefore have seemed natural to Lennard-Jones to assign a computational problem to the only woman doctoral student he ever supervised![11] Whether that reasoning was deliberate or not is unclear. Monroe, for one, never brought up the matter.

---

[10] "Report of the Faculty Board of Mathematics on the Need for a Computing Laboratory," [1936], Churchill Archives Center, Cambridge, The Papers of Sir John Edward Lennard-Jones, GBR/0014/LEJO/6. See also (Cambridge Reporter, 1937).

[11] Beryl May Dent was Lennard-Jones' research assistant (and part-time departmental librarian) in Bristol from 1925 to 1930, during which time she co-authored six papers with him. She obtained a MSc in physics in 1927 and left (without a doctorate) to work as a technical librarian at Metropolitan-Vickers, where she eventually led the computation section (Plant, 2005).

As her thesis work advanced, a more immediate concern of hers came to the fore (Boggs, 1979, pp. 12–13):

I became aware that by going to school in England as I had, I had to some extent cut myself off from the American old boy network. . . . [A] good deal depends on the inter-referral by peers, the faculty members, and their peers either on other faculties or in industrial laboratories. I was completely out of that network because J. E. Lennard-Jones, although he was well-known in this country [the US], didn't have that working relationship with other people.

To help overcome this disadvantage, Monroe's great-aunt, Beulah Pack Rollins,[12] whose philanthropy had acquainted her with Princeton University administrators, arranged for Monroe to work in the Fine Library of the Princeton Mathematics Department for a few months in early 1939 (Monroe, 1939; Graham, 2018).[13] There, although she was regularly crossing paths with Institute of Advanced Study affiliates, such as Albert Einstein and quantum physics pioneer Niels Bohr, she nevertheless felt like "a very small and inconspicuous person with no status" (Boggs, 1979, p. 11). Even other junior researchers marginalized her. When she won a popular board game heavy on strategy against a French mathematician, bystanders did not ascribe her success to skill but to "several moves which were so naive that nobody should have made them" (Boggs, 1979, p. 13). Professional networking clearly required more than mere US presence. In the end, through the intervention of Bill Flexner – Monroe's former Bryn Mawr professor and by then a Cornell faculty member – she secured a research assistantship at Cornell with an up-and-coming theoretical chemist, John G. Kirkwood, starting the following September (Boggs, 1979, p. 13).

First though, Monroe returned to Cambridge for a last bout of thesis writing. While there, World War II formally broke out on September 1st, but the brewing conflict had begun to be felt weeks prior. In early August 1939, the Cambridge Mathematical Laboratory's premises and equipment were leased out to the Directorate of Scientific Research of the Ministry of Supply, which connected military requirements with various scientists and engineers. Lennard-Jones, who also joined the directorate, was then tasked with assembling a research team attached to the External Ballistics Department of the Ordnance Board (Croarken, 1990). He recruited affiliates, including John Corner and Devonshire (Croarken, 1992; Challens, 1996), and similarly offered Coulson, his former student, to join him for computation-based work.[14] Given that Wilkes was called to the Royal Air Force, Monroe remained as sole expert

[12] Beulah Brewster Pack, daughter of George Willis Pack, married Philip Ashton Rollins in 1895 (Bentley, 1948; New York Times, 1950, 1957).
[13] Lennard-Jones also visited Princeton during her stay (Daily Princetonian, 1939a, b, c). See also, Letter from Lennard-Jones to his wife, March 14, 1939, Churchill Archives Center, Cambridge, The Papers of Sir John Edward Lennard-Jones, GBR/0014/LEJO/88.
[14] Coulson, a staunch Methodist and pacifist, declined. Letter from Lennard-Jones to Coulson, August 10, 1939; letter from Coulson to Lennard-Jones, August 26, 1939; letter from Lennard-Jones to Coulson, September 1,

on the differential analyzer. She was naturally invited to join, and she gladly accepted, promptly notifying Kirkwood that she had to cancel her Cornell plans. That wartime arrangement, however, turned out to be short-lived. Lennard-Jones could not work out an exception from the blanket exclusion of foreigners from secret work, and had to let Monroe go in December 1939 (Boggs, 1979).[15] Fortunately, the Cornell position was still available to her.

7.5 Cornell and John G. Kirkwood (1940–1942)

Once Monroe finally made it to Cornell, in February 1940, she was assigned "to work out problems he [Kirkwood] was working on, ... like other research assistantships," she later recalled (Boggs, 1979, p. 14). Monroe might have felt sorry to leave quantum chemistry, but she nevertheless found her assigned problem reasonably interesting. She stayed at Cornell her first summer – even though she was not contractually required to do so – and moved her new project along.

Kirkwood came to theoretical chemistry more traditionally, and also more precociously than Lennard-Jones did (Rice and Stillinger, 1999). After completing a BSc in physics at the University of Chicago in December 1926, at age 19, Kirkwood obtained a PhD in experimental physical chemistry from MIT a mere 30 months later. He then worked on the theory of intermolecular forces with John Slater at MIT (1930–1931) and on that of ionic solutions with Peter Debye in Leipzig (1931–1932). After a couple of more years as a researcher at MIT, he joined the Cornell faculty in 1934. There, his standing rapidly grew. Within only four years, he had received a couple of national awards and been granted a named professorship.

Although the young theoretical chemist did not pay much attention to quantum chemistry, he shared with Lennard-Jones an interest in liquids.[16] Berni Alder, a later graduate student of Kirkwood, described his mentor's key contributions to the topic from this period as going beyond earlier theories in which (Alder, 1968):

[a] representative particle was thought of as moving in some average potential due to all the other particles in the system. ... By considering the behavior of pairs of particles in the potential due to all the other particles Kirkwood took the next logical step in the development of a theory of the many-body problem.

In other words, Kirkwood – starting from his seminal 1935 paper (Kirkwood, 1935) – shifted the focus from single particles to pair correlations, described as a function of intermolecular distance r by the pair (or radial) distribution function $g(r)$.

1939, Bodleian Library, Oxford University, The Papers of Charles Alfred Coulson, F.R.S. (1910–1974), "CAC's appearance before Conscientious Objectors Tribunal CSAC 60.4.78/A.3.6 1939–40."

[15] Letter from Wilkes to Monroe, January 14, 1940. P. M. Murphy Personal Collection.
[16] During Monroe's time in Cambridge, Lennard-Jones and Devonshire published a cell theory of liquids to explore the physics of the critical point (Lennard-Jones and Devonshire, 1937). Corner then applied this model to certain other properties of liquids (Corner, 1939).

The following year, crystallographers J. D. Bernal and Jan Prins separately proposed that $g(r)$ for liquids was "a blurred copy of that of the corresponding crystalline solid," in Bernal's words, thus suggesting a structural continuum between fluids and crystals (Bernal, 1937; Prins and Petersen, 1936). Kirkwood, by contrast, took an approach that did not explicitly refer to the crystal state. In an early 1939 paper, he instead estimated the average work $W(r)$ needed to bring two particles together in a fluid, and then obtained the pair distribution function $g(r)$ using the reversible work theorem, $g(r) = \exp[-\beta W(r)]$, where β is the inverse of temperature (Kirkwood, 1939). He further considered attractive forces between molecules as mere perturbations to the structure of simple liquids. This assumption naturally led him to consider a model fluid of *hard spheres* (Alder, 1968, p. xii), which are impenetrable, frictionless, and perfectly elastic (hence, idealized) balls. His ensuing analysis puzzlingly gave rise to a limiting density "above which a liquid type of distribution and a liquid structure cannot exist. Above this density, only structures with crystalline long-range order would then be possible," he reported (Kirkwood, 1939, p. 924) In other words, these purely repulsive particles transformed from a disordered fluid to an ordered crystal upon simple compression.[17] The continuous or discontinuous nature of that thermodynamic transformation, however, remained unclear, and pushing the inquiry further required a more dedicated research effort than Kirkwood alone likely could muster. Despite Kirkwood's own theoretical prowess, he had thus far only supervised students doing experimental work, hence finding the right research affiliate for this problem might not have been obvious.

In this context, hiring Monroe, a postdoctoral mathematical chemist with some exposure to the theory of liquids must have seemed like a golden opportunity. Kirkwood and Monroe indeed quickly managed to submit a joint note, "On the theory of fusion," in September 1940, a mere eight months after her arrival. That analysis, which considered how $g(r)$ for the crystal and the liquid phases controls their relative stability, could more concretely propose the existence of "a first-order [discontinuous] transition from a completely disordered density distribution to one of crystalline order" (Kirkwood and Monroe, 1940, p. 846), unlike the perfect continuity assumed by the crystallographers.[18] Two longer papers, comparing predictions for the melting and the structure of the hard sphere model with experimental results for liquids, as well as a clarifying note were submitted over the following 18 months (Kirkwood and Monroe, 1941; Kirkwood and Boggs, 1942a, b).

---

[17] Lev Landau had already predicted in 1937, based on symmetry arguments, that "unlike the behavior of liquid and gas above the critical point, no continuous change of state between fluid and crystal is possible" (Hoddeson et al., 1992a, p. 523). Kirkwood, who was focused on a specific microscopic model, however, seemed unaware of that more general result, which was published in German and in Russian.

[18] Although qualitatively correct, the Kirkwood–Monroe prediction was later found to be quantitatively off. This early treatment indeed neglected the density dependence of $g(r)$, whose significance was only understood decades later (Baus, 1984).

The remarkable prediction that a phase transition could proceed without inter-molecular attraction, which now lies at the heart of soft matter physics, became a key concern of the burgeoning field of computational physics after the war (Battimelli et al., 2020; Royall et al., 2024). Wartime interest, however, was limited. Neither Kirkwood nor Monroe gave talks at national meetings about their joint work. Monroe's only conference invitation was to discuss mechanical calculators, in December 1941.[19] War concerns were also seeping into university life. By fall 1941 Kirkwood was already thinking about the relationship between shock waves and explosives,[20] and by spring 1942 he was deeply involved with war-related research. Under contract from the Office of Scientific Research and Development (OSRD), he started to collaborate with his Cornell colleague, Hans Bethe, and with another postdoctoral scholar, Elliott Montroll, on the study of underwater explosives (Kirkwood and Bethe, 1942; Kirkwood and Montroll, 1942). He even organized a set of lectures on fluid dynamics for the broader Cornell community.[21]

More immediately impactful on Monroe than her hard sphere work was her sharing an office with one of Kirkwood's graduate students, Fitzhugh W. Boggs. Boggs was born in Vermont, but schooled in France, where his father spent nearly a decade writing an expansive religious history (Boggs, 1931; Gunn, 2009). Upon moving back to the US in 1934, he enrolled as a mature student in chemistry at Columbia and carried on with graduate studies at Cornell in 1939. Monroe and Boggs had a lot to share, having both followed atypical paths and being "strangely suited – tall, lanky, serious and dedicated to do theoretical work in physics/chemistry,"[22] as a Cornell colleague later commented. Affinity led to engagement in June 1941 (Figure 7.4) and marriage the following September (New York Times, 1941a, b). The change in marital status, however, had little immediate impact on their scientific day-to-day (except for now signing scientific articles as E. M. Boggs). As Monroe Boggs wrote to Wilkes: "My life is changed less than most people's [is] by this step, for I go on living in the same apartment, do the same job and have lunch with the same man as I did last year."[23] With Boggs completing his thesis work and looking for an industrial position for the fall of 1942 (Boggs,

---

[19] Neither Kirkwood nor Monroe presented their hard sphere findings at pre-war American Chemical Society (ACS) meetings. She gave no ACS talk; he presented his work on dielectric loss in polar polymers in August 1941 (ACS, 1941). Her computational expertise with another Cambridge mechanical calculator, the Mallock machine, captured more contemporary interest. She wrote to Wilkes that "Mallock has written asking me to talk at the meeting [of the American Statistical Association] about the machine. As you know I am scarcely an authority, but at least I know what it looks like, which is more than most people over here." She hence gave a talk about it (JASA, 1942). Letter from Monroe Boggs to Wilkes, October 2, 1941, St. John's College Library, Cambridge, Papers of Maurice Wilkes, Wilkes/I/G/10.

[20] Letter from John von Neumann to George B. Kistiakowsky, September 1941, cited in Macrae (1992, pp. 203–204).

[21] Simon H. Bauer, email to author (October 2, 2011).   [22] Simon H. Bauer. Source as for Note 21.

[23] Letter from Monroe Boggs to Wilkes, October 2, 1941, St. John's College Library, Cambridge, Papers of Maurice Wilkes, Wilkes/I/G/10.

Figure 7.4 Engagement of Elizabeth Monroe and Fitzhugh Boggs at Bunrannoch, the Monroe–McNairy house in Manchester, Vermont, June 21, 1941. P. M. Murphy Personal Collection. Reproduced with permission.

1944), things eventually did change. Through Kirkwood, Boggs found a job as a research engineer at the Westinghouse Research Laboratory, in East Pittsburgh, then headed by Edward Condon, who was very much "part of the old boy network," Monroe Boggs later noted (Boggs, 1979, p. 18).[24] It's unclear whether Kirkwood offered to help Monroe Boggs as well, but Condon, eager to build a stable research team for war work on the resnatron – a microwave-based radar countermeasure device for initial wartime use (Lassman, 2018; Boggs, 1947) – did assist her finding a position nearby her husband's. For the academic year 1942–1943, she hence joined the University of Pittsburgh as lecturer. Teaching introductory physics to pre-medical and pre-dental students, however, left her unsatisfied (Boggs, 1979, p. 19). It also did not accommodate research. Condon, aware of various wartime opportunities in the Pittsburgh area, soon devised another plan for Monroe Boggs: a research position at the Explosives Research Laboratory in nearby Bruceton.

---

[24] Condon and Kirkwood knew each other since at least early 1931. At that time, Slater – Kirkwood's supervisor at MIT – tried to recruit Condon to the faculty, and Condon duly visited. Letter from Slater to Condon, March 17, 1931, American Philosophical Society, John Clarke Slater Papers (Mss.B.SL2p), Condon, Edward Uhler – Folder 1.

7.6 Explosives Research Laboratory
and the Manhattan Project (1943–1945)

The laboratory that Monroe Boggs joined in July 1943 was a fruit of the war. Early June 1940, before even the fall of France to the Nazi forces, efforts were undertaken to mobilize US scientific talent. The ensuing National Defense Research Committee (later OSRD) included a Division B: Bombs, Fuels, Gases, Chemical Problems. Within that organization the need for a Division for Explosives quickly emerged (Connor and Kistiakowsky, 1948). The Department of Defense was deemed by some to be "way behind the times" (MacDougall, 1986, p. 4), especially concerning *high explosives*, which detonate rather than deflagrate, and therefore propagate faster than the speed of sound. A subsequent reorganization led to the creation of the Division 8: Explosives and to the establishment in 1941 of the Explosives Research Laboratory (ERL) on a site of the Bureau of Mines, in Bruceton, outside Pittsburgh. George B. Kistiakowsky, a Harvard University physical chemist and champion of the topic, was initially put in charge of ERL.[25]

Technical expertise, however, was initially lacking. Explosives research had not been a topic of much academic interest in the US up to that point. Understanding how to make efficient use of high explosives lay at a thinly covered interface between chemistry, physics, and mathematics. One of the early ERL researchers who later became its Deputy Director, Duncan MacDougall, recalled in a 1986 interview spending "a lot of time in the Widener library at Harvard, reading up about explosives" during fall 1940, "because, while Kistiakowsky and I both had chemistry PhDs,[26] neither of us knew particularly anything about explosives" (MacDougall, 1986, p. 2). Despite the steep learning curve, MacDougall, who had complex familial relationships with Kistiakowsky that may have affected Monroe Boggs' later work,[27] was initially put in charge of the fundamental research group at ERL.

Physical chemist George Messerly[28] was another early ERL group leader. His team carried out extensive studies of shock waves in air and other gases by optical methods. Monroe Boggs recounted in a 1979 interview (Boggs, 1979, p. 20) that

---

[25] In the context of this book, it is noteworthy that the quantum chemistry taught by Kistiakowsky at Harvard University in 1932–1933 was outdated. Frank H. Westheimer, a graduate student at the time, later remembered that this course only covered Bohr's old quantum theory, even though quantum mechanics had already been making headways in chemistry for a few years (Westheimer, 1979).

[26] MacDougall trained as a low-temperature physical chemist at the University of California, Berkeley, under the supervision of William Giauque (MacDougall, 1933).

[27] While being affiliated with Harvard, MacDougall fathered a child with Kistiakowsky's then-wife, Hildegard (née Moebius). Although the precise nature of their relationship at that time is unclear, the two later got married, following her divorce from Kristiakowsky. John C. Warner, who was chemistry faculty at the Carnegie Institute of Technology in Pittsburgh during the war, did note household tensions in those years (Warner, 1984).

[28] Messerly trained as an experimental physical chemist at Pennsylvania State College, under the supervision of John G. Aston (Messerly, 1938).

Messerly had designed a rather interesting type of camera, the purpose of which was to study explosions, detonation rates and so forth, by photographing a moving trace. ... This camera had a rotating drum and a slip which enabled you to track that moving bang stripe ... to track a point on it and create a record on moving film, which if you knew the rate at which the film was moving past the slot, you could calculate the rate at which the detonation was moving in the explosive. ... It also permitted for the study of certain other phenomena like the shape of shock waves.

As a theoretical and mathematical chemist, Monroe Boggs was equally if not more prepared for an ERL position than MacDougall or Messerly were when she joined during the summer of 1943.[29] She initially "did some of the theoretical work that attempted to interpret what were the experimental results" in Messerly's group, she later recalled (Boggs, 1979, p. 22), and quickly got to author a report (Martin et al., 1943). Years later, an ERL colleague, detonation expert Sigmund J. Jacobs (Doherty et al., 2006), highlighted that "a number of important ... experimental observations on the transition from shock to detonation" were reported by Monroe Boggs during the war (Jacobs et al., 1963, p. 517).[30] MacDougall further commented that despite being a theorist she was essentially "in charge of Messerly's group" for both theory and experiments, adding that "she was very, very capable" (MacDougall, 1986, p. 20). Putting the only woman researcher at ERL – at least, based on the enumeration by Howes and Herzenberg (1999) – formally in charge of the group might have raised eyebrows. It might have also been unnecessary from an organizational standpoint. ERL work tended to be collaborative, and hence the hierarchy was likely fairly flat. As MacDougall later described while discussing authorship of some other ERL-related work:

All of the activities at Bruceton were definitely group effort. Many of the fundamental ideas [for this particular project] were my own but they were certainly influenced by discussion with other members of the staff .... For the experimental part, the size of the group participating was even larger.[31]

As war progressed, work at ERL evolved. Although many details remain classified to this day,[32] some of that evolution can be reconstructed from indirect sources. In early 1944, the Messerly group significantly reorganized. Kistiakowsky moved full-time to Los Alamos and eventually became to head the X (explosives) Division in charge of the implosion program for the nuclear bomb. Although he took along a few ERL staff members (Kistiakowsky, 1980; Linshitz, 1988), he still found his

---

[29] Letter from MacDougall to Monroe Boggs, September 1, 1945, "Elizabeth M. Boggs Research Associate." The Boggs Center archives.

[30] Interestingly, another group working contemporaneously on these problems was that of Herzberg, Lennard-Jones' former affiliate, then faculty at the University of Saskatchewan (Jacobs et al., 1963).

[31] Letter from MacDougall to Emerson M. Pugh, May 24, 1946, Carnegie Mellon University Archives, Emerson M. Pugh Papers, 1983–0001/1/5, "MacDougall, D. P., 1946–1948."

[32] As Kistiakowsky noted shortly after the war that "there exist ... very direct military applications of these fundamental studies [on detonation done at ERL]. They are of great importance, and it is disappointing that military security prohibits their discussion" (Connor and Kistiakowsky, 1948, p. 65). Access to National Archives and Records Administration series of ERL-related materials remains restricted.

new team lacking in some respects. In a 1982 interview, he indeed recalled assigning certain tasks to ERL researchers (Kistiakowsky, 1982). Per MacDougall's recollection: "At that time, [the ERL in Bruceton] had a good deal more capability in what you might call explosive or detonation physics, detonation phenomena than [Los Alamos]. So, we agreed to put … [about] 20% of the Bruceton effort on explosives … to things of interest [to Los Alamos]" (MacDougall, 1986, p. 20). One of the goals of Project Q at ERL, which MacDougall headed, was later described by Monroe Boggs (Boggs, 1979, pp. 21–22):

[T]he nature of the fissionable material is such that it will only explode when a critical mass is brought together. A certain minimum amount of it has to be compacted in one place, and at that point it will explode. … The problem was to bring the critical mass together in a controlled way. That is to say, not inside the airplane that was carrying the bomb but somewhere outside it at a time that was appropriate to the target. We had developed in the course of the work in our particular group, techniques for shaping detonations. … They didn't really want to tell us what they were going to do with the product they wanted us to design, so they told us what they wanted us to do. … What they wanted to do was to cast as a hollow sphere of explosive and coat the interior with the fissionable material and then cause what they called an implosion, that is to say an explosion from the outside going inward toward the center, which would have the effect of collapsing that shell of fissionable material that coated the interior of the hollow sphere of explosive.

In response to this challenge, Monroe Boggs designed the logarithmic lens needed to control the implosion. MacDougall, who went back and forth between ERL and Los Alamos, would later claim that "the first log spiral lens was designed by Elizabeth Boggs [at Bruceton] and fired at Los Alamos" (MacDougall, 1986, p. 20). Jacobs also reported that in early 1944 Monroe Boggs made the "first efforts to study wave shaping in this country" (Jacobs, 1956, p. 46). She eventually recognized the outcome from her work (Boggs, 1979, p. 25):

Although we had not been told very much about what it was we were doing, and why we were doing it, we had enough knowledge of it so that when that bomb was dropped, I knew that that had been the target against which we were working.

Despite Monroe Boggs' involvement, the above is not how the sequence of events that led to atomic gadgets is generally understood (Rhodes, 1986; Brown and Borovina, 2021). In histories of the Manhattan Project, the logarithmic lens is usually traced back to a suggestion by a British researcher, James Tuck (Szasz, 1998), and its ensuing development to the genius of John von Neumann (Macrae, 1992). Hoddeson et al., who authored an authoritative description of wartime science in Los Alamos and were able to access some of its classified material, report that MacDougall had indeed communicated Monroe Boggs' earlier (and comparable) design to Los Alamos (Hoddeson et al., 1992b, p. 168).[33] Why did her

---

[33] See also Jacobs (1956).

work not influence the work of Division X despite MacDougall's regular back and forth? Should it be blamed on the research equivalent of the fog of war, on a particularly tense moment in the MacDougall–Kistiakowsky relationship (see Note 26), or on the limited attention paid to the work of a woman researcher? Without access to primary sources, a definitive answer is challenging to tease out.

What is clear from Monroe Bogg's time at ERL is that she was an exceptional research scientist, who could jump from one hard problem to another, while also acting as an intellectual leader. By a strange coincidence, her last day there, August 6, 1945, was the day a fission bomb was dropped over Hiroshima.[34] Her departure from the group, however, was unrelated to that otherwise consequential event. She was expecting, and her son, David (New York Times, 2000), was born only a couple of weeks later (Figure 7.5). Maternity ended up cutting her research career short.

7.7 Conclusion

Monroe Boggs was a liminal scientist: a mathematician pursuing the burgeoning field of quantum chemistry, a mechanical computer researcher, a Manhattan Project researcher in Pennsylvania, and a woman in science to boot. That liminality resulted in her missing out on various opportunities. Had she been allowed to remain in the UK, she might have worked on the British war effort, as did Corner and Wilkes. The former designed explosives for the Armament Research Department, which led him to work on the British hydrogen bomb and for which he was made Commander of the Order of the British Empire in 1958 (Challens, 1996; Gowing and Arnold, 1974). After working on radar during the war, the latter took over the leadership of the Cambridge Mathematical Laboratory, where he innovated in the burgeoning field of electronic computing, leading him to receive the prestigious Turing Award in 1967. Alternatively, Monroe Boggs might have become one of the early researchers to develop and use electronic computers. Like one of Lennard-Jones' postwar PhD students, John A. Pople, she might have further developed computational quantum chemistry and received a Nobel Prize for it; like one of Kirkwood's postwar PhD students, Alder, she might have developed molecular simulation to solve the hard sphere phase behavior and eventually have a computational physics prize named after her (CECAM, n.d.). MacDougall would later presume Monroe had pursued a science career of some sort in that "her life story, I would guess, is in *American Men of Science*" (MacDougall, 1986, pp. 20–21). She, herself, assumed she would get "a research position at Esso Corporation" or somewhere similar (Boggs, 1979, p. 26). Job opportunities for

---

[34] Letter from MacDougall to Monroe Boggs, September 1, 1945. Source as for Note 29.

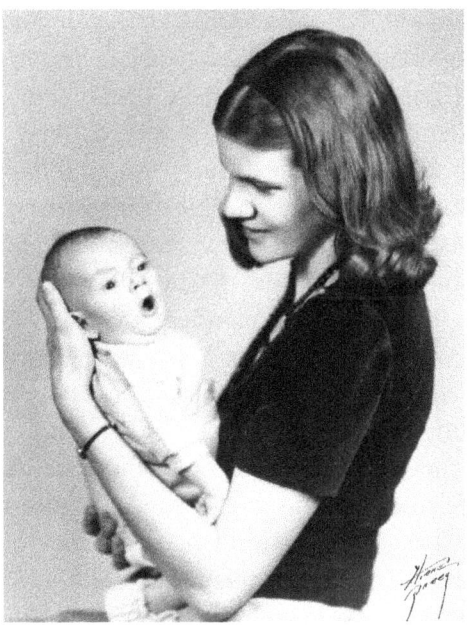

Figure 7.5 Monroe Boggs and her son, Jonathan David (1945–2001), shortly after his birth on August 25, 1945, taken at the Horne-Parry Studio of Horne's department store in Pittsburgh, Pennsylvania. The Boggs Center archives. Reproduced with permission.

US research scientists in the immediate aftermath of World War II were plentiful, graduate education having essentially stopped during the winter of 1940–1941 (Martin, 2018, p. 84).

Yet none of that was to be. Monroe Boggs was a woman living at a time when most women understood their responsibilities to lie primarily within their household. As she later explained to fellow Bryn Mawr alumnae, she was to find work "wherever else her husband's career would have taken them. His greater [financial] responsibility carried with it the privilege of first choice of job; my choice would be circumscribed less by financial consideration than his and more by geographical limits associated with his choice" (Boggs, 1971, p. 36). In light of her son David's growing developmental challenges, she elected to take care of him for as long as she could. Although she still attended scientific meetings for a few years after the war (Tukey, 1947; Martin and Hollcroft, 1948), when David was institutionalized, she definitively turned the page on research. She became invested in local and then national activities of the nascent Association for Retarded Children (now the Arc). That involvement led her to play a key advocacy role for people with disabilities, notably co-chairing a taskforce of President Kennedy's Panel on

Mental Retardation (Hills, 1982; Mercer, 1993). Her singular capability to interact with researchers in that field, no doubt grounded in her own research experience, was repeatedly noted (Spitalnik and Carroll, 2022). Over the years, Monroe Boggs' engagement was recognized with numerous awards, and The Boggs Center on Developmental Disabilities from the Robert Wood Johnson Medical School is named in her honor. Hence, although her pre-war research achievements and her contributions to the Manhattan Project were remarkable, it is her postwar science-grounded policy accomplishments that warranted an obituary in *The New York Times* only a few days after her passing, on January 27, 1996. In this form, she was overlooked as a scientist. Being affected by another kind of expression of over-looking women in science, this chapter brings balance to the obituary of an undoubtedly outstanding woman.

Ultimately, it was the tenuous position of women researchers within the pre-World War II scientific enterprise, no matter how exceptional or financially secure, that diverted Monroe Boggs away from mathematical (or quantum) chemistry. Despite her having trained with leaders in her field – Noether, Lennard-Jones, and Kirkwood – her mentors were either unable or unwilling to help advance her career. Over the decade during which she pursued scientific research, her gender only compounded the complex mixture of geopolitical and personal difficulties she lived through. The resulting professional barriers were too large even for her to surmount.

References

Abbate, Janet. 2012. *Recoding Gender: Women's Changing Participation in Computing.* Cambridge, MA: MIT Press.

ACS. 1941. Program of the Atlantic City Meeting of the American Chemical Society. *Chemical & Engineering News*, 19(August 25), 895–931. https://doi.org/10.1021/cen-v019n016.p895

Alder, Berni J. 1968. Preface. In Alder, Berni J. (ed.), *Collected Works of John G. Kirkwood: Theory of Liquids*. New York: Gordon and Breach, pp. xi–xxiv.

Altmann, Simon L. and Bowen, Edmund J. 1974. Charles Alfred Coulson, 1910–1974. *Biographical Memoirs of Fellows of the Royal Society*, 20, 74–134. https://doi.org/10.1098/rsbm.1974.0004

Battimelli, Giovanni, Ciccotti, Giovanni, and Greco, Pietro. 2020. *Computer Meets Theoretical Physics: The New Frontier of Molecular Simulation*. Cham: Springer.

Baus, Marc. 1984. A comment on the Kirkwood–Monroe theory of freezing. *Molecular Physics*, 53(1), 183–186. https://doi.org/10.1080/00268978400102201

Bentley, Esther F. 1948. A conversation with Mr. Rollins. *The Princeton University Library Chronicle*, 9(4), 178–190. http://www.jstor.org/stable/26400287

Bernal, John D. 1937. An attempt at a molecular theory of liquid structure. *Transactions of the Faraday Society*, 33, 27–40. https://doi.org/10.1039/TF9373300027

Bix, Amy S. 2014. *Girls Coming to Tech!: A History of American Engineering Education for Women*. Cambridge, MA: MIT Press.

Boggs, Elizabeth M. 1971. Elizabeth Monroe Boggs '35. *Bryn Mawr Alumnae Bulletin*, 51, 36.

Boggs, Elizabeth M. 1979. Transcript of an oral history interview of Elizabeth M. Boggs by Tom A. Underwood in Trenton, NJ, March 3. Patrick Charbonneau Personal Collection.

Boggs, Fitzhugh W. 1944. *I. A theory of the ionization constants of the benzene-dicarboxylic acids; II. A method for measuring dielectric loss of solids and liquids at low and moderately high frequencies*. PhD Thesis. Ithaca, NY: Cornell University.

Boggs, Fitzhugh W. 1947. The resnatron: a generator of microwaves. *Westinghouse Engineer*, 7, 57–60.

Boggs, Norman T. 1931. *The Christian Saga*. New York: The Macmillan Company.

Brown, Eric N. and Borovina, Dan L. 2021. The Trinity high-explosive implosion system: the foundation for precision explosive applications. *Nuclear Technology*, 207(sup1), S204–S221. https://doi.org/10.1080/00295450.2021.1913954

Bryn Mawr College. 1935. Elizabeth Monroe is awarded European Fellowship. *The College News*, 21(25), 1&3.

Byers, Nina and Williams, Gary. 2006. *Out of the Shadows: Contributions of Twentieth-century Women to Physics*. Cambridge: Cambridge University Press.

Cambridge Reporter. 1937. Report of the General Board on the establishment of a Computing Laboratory. *Cambridge University Reporter*, 57(February 2), 625–629.

Cambridge University. 1937. *Abstracts of Dissertations Approved for the Ph.D., M.Sc. and M.Litt. Degrees in the University of Cambridge During the Academic Year 1936–37*. Cambridge: Cambridge University Press.

Cambridge University. 1941. *Abstracts of Dissertations Approved for the Ph.D., M.Sc. and M.Litt. Degrees in the University of Cambridge During the Academic Year 1940–41*. Cambridge: Cambridge University Press.

CECAM. n.d. Berni J. Alder CECAM Prize. Centre Européen de Calcul Atomique et Moléculaire (CECAM). https://www.cecam.org/awards

Challens, John. 1996. Obituary: John Corner. *The Independent*, July 26. https://www .independent.co.uk/news/people/obituary-john-corner-1330514.html

Connor, Ralph A. and Kistiakowsky, George B. 1948. History of Division 8. In Noyes, William A. Jr. (ed.), *Chemistry: A History of the Chemistry Components of the National Defense Research Committee, 1940–1946*. Boston: Little, Brown and Company.

Corner, John. 1939. The Joule–Thomson inversion curves of recent equations of state. *Transactions of the Faraday Society*, 35, 784–791. https://doi.org/10.1039/TF9393500784

Croarken, Mary. 1990. *Early Scientific Computing in Britain*. Oxford: Clarendon Press.

Croarken, Mary G. 1992. The emergence of computing science research and teaching at Cambridge, 1936–49. *IEEE Annals of the History of Computing*, 14(4), 10–15. https://doi.org/10.1109/85.194050

Daily Princetonian. 1939a. Theoretical chemist will present speech in 138 Frick tonight. *The Daily Princetonian*, 64(March 6), 1.

Daily Princetonian. 1939b. Chemists to hold forums on liquids. *The Daily Princetonian*, 64 (March 8), 1.

Daily Princetonian. 1939c. Symposium on liquids to be concluded today. *The Daily Princetonian*, 64(March 9), 1.

Doherty, Ruth, Forbes, Jerry W., and Short, James. 2006. *Mr. Detonation Science for DOD: Sigmund J. Jacobs*. White Oak Laboratory Alumni Association. https://wayback.archive-

it.org/1284/20090304111225/http://www.wolaa.org/files/Oral_History_Jacobs_Tribute
    .pdf

Fischer, Charlotte F. 2003. *Douglas Rayner Hartree: His Life in Science and Computing.*
    Singapore: World Scientific.

Gavroglu, Kostas and Simões, Ana. 2011. *Neither Physics nor Chemistry: A History of
    Quantum Chemistry.* Cambridge, MA: MIT Press.

Gowing, Margaret and Arnold, Lorna. 1974. *Independence and Deterrence: Volume 2:
    Policy Execution.* London: Palgrave Macmillan.

Graham, Elyse. 2018. Adventures in Fine Hall. *Princeton Alumni Weekly*, January 10.
    https://paw.princeton.edu/article/adventures-fine-hall

Green, Judy and LaDuke, Jeanne. 2009. *Pioneering Women in American Mathematics: The
    Pre-1940 PhD's.* Providence, RI: American Mathematical Society.

Greenwood, Veronique. 2019. Overlooked no more: Elizabeth Rona, pioneering scientist
    amid dangers of war. *The New York Times*, September 2, A18. https://www.nytimes
    .com/2019/08/28/obituaries/elizabeth-rona-overlooked.html

Grier, David A. 2013. *When Computers Were Human.* Princeton, NJ: Princeton University
    Press.

Gunn, Virginia B. 2009. *Bébé Ginette.* lulu.com

Hall, George G. 1991. The Lennard-Jones paper of 1929 and the foundations of molecular
    orbital theory. *Advances in Quantum Chemistry*, 22. 1–6. https://doi.org/10.1016/
    S0065-3276(08)60361-5

Herzberg, Gerhard. 1989. Interview of Gerhard Herzberg by Brenda P. Winnewisser at the
    National Research Council, Ottawa, Canada, February 28. Niels Bohr Library &
    Archives. https://www.aip.org/history-programs/niels-bohr-library/oral-histories/5029-1

Hills, Jo A. 1982. Guide to the Elizabeth Boggs Personal Papers (#32). John F. Kennedy
    Presidential Library. https://www.jfklibrary.org/asset-viewer/archives/EBPP

Hoddeson, Lillian, Schubert, Helmut, Heims, Steve J., and Baym, Gordon. 1992a.
    Collective phenomena. Hoddeson, Lillian, Braun, Ernest, Teichmann, Jurgen, and
    Weart, Spencer (eds.), *Out of the Crystal Maze: Chapters from the History of Solid
    State Physics.* Oxford: Oxford University Press, pp. 489–615. https://doi.org/10.1093/
    oso/9780195053296.003.0008

Hoddeson, Lillian, Henriksen, Paul W., Meade, Roger A., and Westfall, Catherine L. 1992b.
    *Critical Assembly: A Technical History of Los Alamos during the Oppenheimer Years
    1943–1945.* Cambridge: Cambridge University Press.

Howes, Ruth H. and Herzenberg, Caroline L. 1999. *Their Day in the Sun: Women of the
    Manhattan Project.* Philadelphia, PA: Temple University Press.

Jacobs, Sigmund J. 1956. Principles of wave shaping. *Proceedings of Detonation Wave
    Shaping Conference Held at Jet Propulsion Laboratory California Institute of
    Technology Pasadena, California 5–7 June 1956.* Dover, NJ: Picatinny Arsenal,
    pp. 45–65.

Jacobs, Sigmund J., Liddiard Jr., Thomas P., and Drimmer, Bernard E. 1963. The shock-to-
    detonation transition in solid explosives. *Symposium (International) on Combustion*,
    9(1), 517–529. https://doi.org/10.1016/S0082-0784(63)80061-2

JASA. 1942. Proceedings: 103rd Annual Meeting: Biltmore Hotel, New York: Saturday,
    December 27, 1941. *Journal of the American Statistical Association*, 37(217),
    117–144. http://www.jstor.org/stable/2279439

Kirkwood, John G. 1935. Statistical mechanics of fluid mixtures. *The Journal of Chemical
    Physics*, 3(5), 300–313. https://doi.org/10.1063/1.1749657

Kirkwood, John G. 1939. Molecular distribution in liquids. *The Journal of Chemical
    Physics*, 7(10), 919–925. https://doi.org/10.1063/1.1750344

Kirkwood, John G. and Bethe, Hans. 1942. *The Pressure Wave Produced by an Underwater Explosion, Basic Propagation Theory, Part 1*. Tech. rept. OSRD #588 (May 15).

Kirkwood, John G. and Boggs, Elizabeth M. 1942a. Note on the theory of fusion. *The Journal of Chemical Physics*, 10(5), 307. https://doi.org/10.1063/1.1723729

Kirkwood, John G. and Boggs, Elizabeth M. 1942b. The radial distribution function in liquids. *The Journal of Chemical Physics*, 10(6), 394–402. https://doi.org/10.1063/1.1723737

Kirkwood, John G. and Monroe, Elizabeth. 1940. On the theory of fusion. *The Journal of Chemical Physics*, 8(10), 845–846. https://doi.org/10.1063/1.1750591

Kirkwood, John G. and Monroe, Elizabeth. 1941. Statistical mechanics of fusion. *The Journal of Chemical Physics*, 9(7), 514–526. https://doi.org/10.1063/1.1750949

Kirkwood, John G. and Montroll, Elliott W. 1942. *The Pressure Wave Produced by an Underwater Explosion II*. Tech. rept. OSRD #676 (July 1).

Kistiakowsky, George B. 1980. Reminiscences of wartime Los Alamos. In Badash, Lawrence, Hirschfelder, Joseph O., and Broida, Herbert P. (eds.), *Reminiscences of Los Alamos 1943–1945*. Dordrecht: D. Reidel Company, pp. 49–66.

Kistiakowsky, George B. 1982. Transcript of an oral history interview of George Kistiakowsky by Richard Rhodes in Cambridge, MA, January 15. Voices of the Manhattan Project. https://ahf.nuclearmuseum.org/voices/oral-histories/george-kistiakowskys-interview

Lassman, Thomas C. 2018. *Edward Condon's Cooperative Vision: Science, Industry, and Innovation in Modern America*. Pittsburgh, PA: University of Pittsburgh Press.

Lennard-Jones, John E. 1929. The electronic structure of some diatomic molecules. *Transactions of the Faraday Society*, 25, 668–686. https://doi.org/10.1039/TF9292500668

Lennard-Jones, John E. 1931. The nature of cohesion. *Nature*, 128, 462–463. https://doi.org/10.1038/128462a0

Lennard-Jones, John E. 2011. Transcript of an oral history interview of John Lennard-Jones by Patrick Charbonneau and Daphne Klotsa in Woodbridge, Suffolk, UK, May 21. Patrick Charbonneau Personal Collection.

Lennard-Jones, John E. and Devonshire, Albert F. 1937. Critical phenomena in gases: I. *Proceedings of the Royal Society of London Series A*, 163(912), 53–70. https://doi.org/10.1098/rspa.1937.0210

Linshitz, Henry. 1988. Interview of Henry Linshitz by Steven Heims, February 23. Niels Bohr Library & Archives. https://www.aip.org/history-programs/niels-bohr-library/oral-histories/5039

MacDougall, Duncan P. 1933. *The Production of Temperatures Below 1 Absolute by Means of the Demagnetization of* $Gd_2(SO_4)_3$. PhD Thesis. Berkeley, CA: University of California, Berkeley.

MacDougall, Duncan P. 1986. Transcript of an oral history interview of Duncan P. MacDougall by Lillian E. Hoddeson at Los Alamos, NM, April 9. LANL OH-116. Patrick Charbonneau Personal Collection.

MacGrayne, Sharon B. 1993. *Nobel Prize Women in Science: Their Lives, Struggles, and Momentous Discoveries*. Seacaucus, NJ: Carol Publishing Group.

Macrae, Norman. 1992. *John von Neumann: The Scientific Genius Who Pioneered the Modern Computer, Game Theory, Nuclear Deterrence and Much More*. New York: Pantheon Books.

Martin, Frederick J., Boggs, Elizabeth M., and Messerly, George H. 1943. *The Effect of Particle Size on the Detonation Velocity of Ammonium Picrate*. Tech. rept. OSRD # 1755 (August 31).

Martin, Joseph D. 2018. *Solid State Insurrection: How the Science of Substance Made American Physics Matter*. Pittsburgh, PA: University of Pittsburgh Press.

Martin, William T. and Hollcroft, Temple R. 1948. Second symposium in applied mathematics. *Bulletin of the American Mathematical Society*, 54, 842–844. https://doi.org/10.1090/S0002-9904-1948-09064-4

Mercer, Marc. 1993. Special Section: Elizabeth Boggs at 80. *Families [The New Jersey Developmental Disabilities Council]*, 2(2), 2–14.

Messerly, George H. 1938. *Thermal and Vapor Pressure Data for Tetramethylmethane from 13.22 K to the Boiling Point: The Entropy from the Raman Spectrum*. PhD thesis. State College, PA: Pennsylvania State College.

Miller, Claire C. 2018. Overlooked: Ada Lovelace 1815–1852. *The New York Times*, March 8. https://www.nytimes.com/interactive/2018/obituaries/overlooked-ada-lovelace.html

Monroe, Elizabeth. 1938. On the energies associated with the two-centre problem: I. General features of the energy surface. *Mathematical Proceedings of the Cambridge Philosophical Society*, 34(3), 375–381. https://doi.org/10.1017/S0305004100020314

Monroe, Elizabeth. 1939. *The Generalized Spheroidal Wave Equations and Their Application [to] Some Problems in Quantum Mechanics with Special Reference to the Two Center Problem*. PhD Thesis. Cambridge: Cambridge University.

Mott, Nevill F. 1955. John Edward Lennard-Jones, 1894-1954. *Biographical Memoirs of Fellows of the Royal Society*, 1, 174–184. https://doi.org/10.1098/rsbm.1955.0013

Murphy, Pamela M. 2023. Transcript of an oral history interview of Pamela M. Murphy by Patrick Charbonneau conducted online, February 2. Patrick Charbonneau Personal Collection.

Nature. 1955. Scientific civil service: Promotion of individual research workers. *Nature*, 172, 657–658. https://doi.org/10.1038/172657a0

Nethercott, Philip J. M. 1983. Obituary: Albert Frederick Devonshire (1911–1983). *Proceedings of the Bristol Naturalists' Society*, 42, 10–12. https://archive.org/details/proceedin404519811987bris

New York Times. 1938. 7 women winners of scholarship. *The New York Times*, March 27, 35. https://nyti.ms/4ai0Q2R

New York Times. 1941a. Troth is announced of Miss E. Monroe. *The New York Times*, June 22, SOCIETY: 2. https://nyti.ms/4cjBFPg

New York Times. 1941b. Elizabeth Monroe bride. *The New York Times*, September 21, SOCIETY: 40. https://nyti.ms/4c8Ul3U

New York Times. 1950. Philip Rollins, 81, authority on West. *The New York Times*, September 12, 26. https://nyti.ms/43GVtpV

New York Times. 1957. Mrs. P. A. Rollins, a clubwoman, 88. *The New York Times*, December 23, 22. https://nyti.ms/3v8SHPf

New York Times. 2000. Boggs, Jonathan David. *The New York Times*, March 3, A: 19. https://www.nytimes.com/2000/03/03/classified/paid-notice-deaths-boggs-jonathan-david.html

Padnani, Amisha and Bennett, Jessica. 2018. Overlooked. *The New York Times*, March 8. https://www.nytimes.com/interactive/2018/obituaries/overlooked.html

Parshall, Karen H. 2015. Training women in mathematical research: the first fifty years of Bryn Mawr College (1885–1935). *The Mathematical Intelligencer*, 37, 71–83. https://doi.org/10.1007/s00283-015-9540-2

Plant, Helen. 2005. Women scientists in British industry: technical library and information workers, c. 1918–1960. *Women's History Review*, 14(2), 301–322. https://doi.org/10.1080/09612020500200424

Prins, Jan A. and Petersen, Hendrik. 1936. Theoretical diffraction patterns corresponding to some simple types of molecular arrangement in liquids. *Physica*, 3(1), 147–153. https://doi.org/10.1016/S0031-8914(36)80218-3

Rayner-Canham, Marlene F. and Rayner-Canham, Geoff. 1998. *Women in Chemistry: Their Changing Roles from Alchemical Times to the Mid-twentieth Century*. Philadelphia: Chemical Heritage Foundation.

Rayner-Canham, Marlene F. and Rayner-Canham, Geoff. 2008. *Chemistry Was Their Life: Pioneering British Women Chemists, 1880–1949*. London: Imperial College Press.

Rhodes, Richard. 1986. *The Making of the Atomic Bomb*. New York: Simon & Schuster.

Rice, Stuart A. and Stillinger, Frank H. 1999. John Gamble Kirkwood 1907–1959. *Biographical Memoirs*, 77, 162–175. https://doi.org/10.17226/9681

Robinson, Tim. 2005. The Meccano set computers: a history of differential analyzers made from children's toys. *IEEE Control Systems Magazine*, 25(3), 74–83. https://doi.org/10.1109/MCS.2005.1432602

Rossiter, Margaret W. 1982. *Women Scientists in America: Struggles and Strategies to 1940*. Baltimore, MD: Johns Hopkins University Press.

Rowe, David E. 2021. *Emmy Noether: Mathematician Extraordinaire*. Cham: Springer.

Royall, C. Patrick, Charbonneau, Patrick, Dijkstra, Marjolein, et al. 2024. Colloidal hard spheres: triumphs, challenges and mysteries. *Reviews of Modern Physics*, 96, 045003. https://doi.org/10.1103/RevModPhys.96.045003

Saxon, Wolfgang. 1996. Dr. Elizabeth Monroe Boggs, 82, founder of group for retarded. *The New York Times*, January 30, B: 16. https://www.nytimes.com/1996/01/30/nyregion/dr-elizabeth-monroe-boggs-82-founder-of-group-for-retarded.html

Schwartz, John. 2020. Overlooked no more: Eunice Foote, climate scientist lost to history. *The New York Times*, April 27, D: 8. https://www.nytimes.com/2020/04/21/obituaries/eunice-foote-overlooked.html

Spitalnik, Deborah M. and Carroll, Robyn E. 2022. Transcript of an oral history interview of Deborah M. Spitalnik and Robyn E. Carroll by Patrick Charbonneau conducted online, October 18. Patrick Charbonneau Personal Collection.

Stoicheff, Boris P. 2002. *Gerhard Herzberg: An Illustrious Life in Science*. Montréal: McGill-Queens University Press.

Szasz, Ferenc M. 1998. James L. Tuck: scientific polymath and eternal optimist of the atomic west. In Hevly, Bruce W., and Findlay, John M. (eds.), *The Atomic West*. Seattle: University of Washington Press, pp. 136–156.

Tukey, John W. 1947. The January meeting on fluid dynamics. *Bulletin of the American Mathematical Society*, 53, 712–713. https://doi.org/10.1090/S0002-9904-1947-08832-7

Tullberg, Rita M. W. 1998. *Women at Cambridge*. Cambridge: Cambridge University Press.

Underhill, Gertrude. 1943. A collection of samplers. *The Bulletin of the Cleveland Museum of Art*, 30(1), 6–8. https://www.jstor.org/stable/25141035

van der Waerden, Bartel L. 1930. *Moderne Algebra. Teil 1*. Berlin: Springer-Verlag.

van der Waerden, Bartel L. 1997. Meine Göttinger Lehrjahre. *Mitteilungen der Deutschen Mathematiker-Vereinigung*, 5(2), 20–27. https://doi.org/10.1515/dmvm-1997-0208

Warner, John C. 1984. Transcript of an oral history interview of John C. Warner by John Heitmann at Gibsonia, PA, February 8. Chemical Heritage Foundation, Philadelphia, PA. Transcript 0044. https://digital.sciencehistory.org/works/rf55z890h

Westheimer, Frank H. 1979. Transcript of an oral history interview of Frank H. Westheimer by Leon Gortler at Harvard University, Cambridge, MA, January 4–5. Chemical Heritage Foundation, Philadelphia, PA. Transcript 0046. https://digital.sciencehistory.org/works/rb68xd255

Wilkes, Maurice V. 1939. Theoretical ionization curves for the E region. *Proceedings of the Physical Society*, 51(1), 138–146. https://doi.org/10.1088/0959-5309/51/1/315

Wilkes, Maurice V. 1940. The theory of reflexion of very long wireless waves from the ionosphere. *Proceedings of the Royal Society of London Series A*, 175(961), 143–163. https://doi.org/10.1098/rspa.1940.0050

Wilkes, Maurice V. 1985. *Memoirs of a Computer Pioneer*. Cambridge, MA: MIT Press.

Wykle, Helen. 2006. George Willis Pack: a name that will endure. D. H. Ramsey Library, University of North Carolina at Asheville. http://toto.lib.unca.edu/web_exhibits/WNC_pack/default_pack.htm

Zachary, G. Pascal and Bush, Vannevar. 2022. *The Essential Writings of Vannevar Bush*. New York: Columbia University Press.

8

Excelsior! John Wheeler, Katharine Way, and the Role of Women in the Exploration of the Microcosm

STEFANO FURLAN

> One keeps forgetting to go right down to the bottom.
> One does not put the question marks deep enough down.
>
> *Wittgenstein, Vermischte Bemerkungen*

8.1 Introduction

"Had we followed her lead, we might have thought of fission" (Wheeler, 2000, p. 23). These words are from none other than John Archibald Wheeler (1911–2008), reminiscing about the 1930s work of his very first PhD student, Katharine Way (1903–1995). To make sense of this rather stunning claim, Wheeler's vast corpus of published and unpublished writings is a precious but little explored source. If many have heard about Wheeler's research or his popularization of black holes, few have viewed him as an exceptional witness whose long life and career spanned almost a century. Although Way is not as famous today as many of Wheeler's other students, she did enjoy a distinguished career in nuclear physics and, if we are to believe Wheeler, she even reached the theoretical "quasi-discovery" of such a decisive phenomenon as fission. Section 8.2 offers a sketch of Way's life, based on available material; Section 8.3 takes a closer look at Way's early work and begins to context-ualize Wheeler's claim; Section 8.4 presents a critical reading of that claim within a broader scope, recognizing that Wheeler's texts are typically layered, and can therefore easily mislead readers. By examining his peculiar way of intertwining history, personal experience, and theoretical insights or guiding ideas, as well as his way of building narratives, we take a look at his view of the history and prehistory of nuclear fission and, as a sort of *fil rouge*, the role played by various women in it. These considerations place us in a better position to assess his late claim about Way.

The author wishes to express his gratitude to Patrick Charbonneau, Michelle Frank, Daniela Monaldi, Margriet van der Heijden, as well as Guido Bacciagaluppi, Alexander Blum, and the people of the American Philosophical Society Library, for discussions and comments.

8.2 Way, "Dear to the Community of Nuclear Physicists"

An important obstacle to investigating Way's life and work is the paucity of archival sources, as is unfortunately often the case when trying to reconstruct the stories and contributions of women scientists. The libraries of Duke University, where she spent a good part of her career after World War II, hold but a couple of slim folders, containing newspaper clippings that are not particularly relevant for the purpose of this study.[1] Wheeler's archives and recollections therefore become key sources, despite their historiographical problems. Although Wheeler and Way stayed in touch for almost six decades, Wheeler's archives contain only a few letters between the two, which they exchanged mainly on the occasion of academic celebrations. As we will see below, Wheeler's recollections can also be challenging to interpret. Before diving into this particular difficulty, however, some biographical context about Way is warranted. Previous authors have already gathered relevant biographical data about her (Martin et al., 1996; Howes and Herzenberg, 1999; Ware and Braukman, 2004). Here, we also assemble deeper insights into her early work on nuclear models, which are largely missing from the secondary literature.[2]

Katharine ("Kay") Jones Way was born in Sewickley, Pennsylvania, in 1903,[3] the second of three children from a relatively wealthy family that could afford to send her to private boarding schools first in Plainfield, New Jersey, and then in Greenwich, Connecticut. Her early life was marked by the loss of her mother, when she was 12. Shortly thereafter, her father, a lawyer, got remarried (Ware and Braukman, 2004, p. 670) to a practicing otolaryngologist, who acted as a role model of a working woman for Way. Given this family and educational background, and if we were to focus exclusively on Way's birth year, she would appear to belong to what John Slater (1900–1976) once called the lucky generation of physicists, in analogy with Malcolm Cowley's similar label for dos Passos, Faulkner, Fitzgerald, and Hemingway in literature (Schweber, 1990, p. 339). For Slater, that generation referred to physicists born to see in their prime, or contribute

---

[1] In these two thin folders of "biographical information" about her in the collections of the Rubenstein Library, at Duke University (Biographical Reference Collection, box 26; New Service Biographical Files, box 107), we can nonetheless read, for instance, that she actively supported "triangle women," meaning by that women working in the "Research Triangle" whose vertices are Durham (home of Duke University), Raleigh (home of North Carolina State University), and Chapel Hill. According to the page dedicated to Way on the website of the "Contributions of 20th-Century Women to Physics" (Keyes, 1997), she also served as the 1972–1973 president of the Durham Chapter of the American Association of University Women.

[2] The lack of archival sources is only a partial excuse for that, since her published work can be quite easily consulted.

[3] Second- or third-hand sources erroneously report 1902. Wheeler, born in July 1911, explicitly states that she was eight years his senior and that her 90th birthday was in 1993. This is further confirmed by a copy of her curriculum vitae preserved in Wheeler's archives (John A. Wheeler Papers, American Philosophical Society Library, Philadelphia, box 68). That discrepancy nevertheless illustrates the general neglect surrounding even the few straightforward pieces of information about Way's life.

to, the advent of quantum mechanics, typically through some *Lehrjahre* in Europe. However, if even Robert Oppenheimer, born in 1904, felt he had already arrived a bit too late to step on the main stage (Schweber, 2008, p. xi), Way's one-year chronological edge was not what it seemed. Her college education was indeed troubled for reasons that remain unclear, but which include illness and a change of curricular interests. Way first enrolled at Vassar College, in New York State, in 1920, but she had to interrupt her studies only two years later because of suspected tuberculosis (Ware and Braukman, 2004, p. 670). Following a two-year convalescence, she joined Barnard College, in New York City, for a couple of semesters (1924–1925). Still uncertain about choosing one field over another, however, she interrupted her studies once again. Only in 1929 did she begin to devote herself to mathematics at Columbia University, notably attending the lectures of Edward Kasner (relatively well known today for introducing the term "googol").[4] She attended classes there until 1934, meanwhile completing a BSc in 1932 (Ware and Braukman, 2004, p. 670).

It was during this time at Columbia that Way became intrigued by physics after reading – perhaps rather unusually for a future professional physicist – Bertrand Russell's *The ABC of Atoms* (Howes and Herzenberg, 1999, p. 42). Meeting Wheeler would do the rest: Way herself would later recall when she "was first seeing the world of physics through [Wheeler's] eyes."[5] It is not clear how they first met or why she moved for graduate work to the University of North Carolina at Chapel Hill.[6] What is certain is that, once Wheeler came back from Copenhagen in 1935, Way became the first graduate student of the young professor (eight years younger than her).[7] Her dissertation was dedicated to the "photoelectric cross section of the deuteron,"[8] which followed from her paper with the same title (Way, 1937). Even though she had fulfilled all PhD requirements by 1938, she held on to her graduate student status (Howes and Herzenberg, 1999, pp. 42–43), given the persistent unemployment after the Great Depression.[9] During that time, she also kept working on nuclear physics. Upon finally receiving a research fellowship at Bryn Mawr College, she officially completed her PhD and moved to the

---

[4] Kasner and Way, along with Erna Jonas and George Comenetz, also wrote a paper on centroidal polygons and groups (Kasner et al., 1934). That was Way's first paper.

[5] Way to Wheeler, April 27, 1983. (John A. Wheeler Papers, American Philosophical Society Library, Philadelphia, box 68.)

[6] Neither her obituary (Martin et al., 1996), nor the other referenced sources, nor her 1978 curriculum vitae kept in Wheeler's archives (John A. Wheeler Papers, American Philosophical Society Library, Philadelphia, box 68) provide any information about Way obtaining a MSc. Ware and Braukman (2004, p. 670) simply state that, after 1934, she "continued her education at the University of North Carolina."

[7] Wheeler's recollections (Wheeler, 2000, p. 150) do not seem to imply that they had met before.

[8] A copy – with the title indicated and dated 1938 – is preserved at the University of North Carolina (Wilson Library, North Carolina Collection, C378 UO2 1938 WAY).

[9] That did not prevent her from helping others in need: "While at graduate school in North Carolina, Way joined in efforts to provide food and clothing to textile workers who had lost their jobs as a result of strikes" (Ware and Braukman, 2004, p. 671). She kept her active concern for others also at an advanced age.

famous women's college for a year. She then obtained a teaching position at the University of Tennessee, where she stayed until 1942.

During the war, after hearing rumors about a nuclear project in Chicago, Way asked Wheeler if there was something she could do (Howes and Herzenberg, 1999, p. 42). She hence became part of the Manhattan Project as a member of the Metallurgical Laboratory, where Wheeler also worked (Wheeler, 2000, p. 150). There, she initially analyzed, together with Alvin Weinberg, the data from Fermi's first atomic piles to evaluate whether the flux of neutrons was sufficient to allow a self-sustaining chain reaction, and then, with Eugene Wigner, she developed the Way–Wigner estimate of the fission product decay (Way and Wigner, 1948).

Her involvement with wartime research left her with ethical concerns. In the summer 1945 she signed the Szilard petition opposing the use of atomic bombs against Japan. Right after the war, while pursuing nuclear research at Oak Ridge National Laboratory, she became publishing director for the Federation of the American Atomic Scientists,[10] and, along with Dexter Masters,[11] she co-edited *The New York Times* bestseller *One World or None: A Report to the Public on the Full Meaning of the Atomic Bomb*, in which Albert Einstein, Niels Bohr, Oppenheimer, Wigner, Arthur Compton, Leo Szilard, and others expressed concerns and offered to a larger public a problematization of the perspectives opened by the Nuclear Age (Masters and Way, 1946).[12] On the surface, Masters and Way were "invisible" editors, as the book contains no preface or comment signed by them. However, given Way's role as publishing director of the Federation and her endorsement of the Szilard petition, it seems safe to assume that she likely shared her views in the last section of the book, "Survival Is at Stake" (Masters and Way, 1946, pp. 215–220), signed by "The Federation of American (Atomic) Scientists" (Masters and Way, 1946, p. 215):

American scientists, acutely aware of their responsibilities for making known the full implications of their scientific developments, have been mobilizing ever since the end of the war. In October scientists from the atomic project joined forces, and in December the Federation of American Scientists, open to all scientists and engineers, was formed.

---

[10] The name was at first "Federation of Atomic Scientists" (founded in November 1945 by some scientists that had taken part in the Manhattan Project and considered it their duty to educate the public and promote nuclear disarmament), but the following year was changed simply to "Federation of American Scientists." See, for example, Smith (1965).

[11] Nephew of Edgar Lee Masters, Dexter Wright Masters (1909–1989) was actually a writer who, during the war, had served on the communication staff of the US Air Force and edited classified publications for research laboratories. Later, he would also write a novel, *The Accident*, about the final days of a physicist – inspired by the case of Louis Slotin – dying from radiation sickness (Masters, 1955).

[12] See also (Howes and Herzenberg, 1999). The reader is redirected to (Furlan, 2024) for some further contextual elements, such as Wheeler's own way of dealing with the reactions that the nuclear age elicited from the larger public – even if that does not necessarily mean (from what is known) that, especially in the postwar period, he and his former student had similar opinions.

Member associations across the country are encouraging qualified scientific and political discussion such as that presented in this book.

In the years that followed, according to Wheeler (2000, p. 150), "[s]he made the decision not to teach, electing instead to pursue research." In 1947, she started working at the National Bureau of Standards in Washington but, more generally, in the years after the war and for the rest of her career Way's main activity consisted of organizing the Nuclear Data Project,[13] which systematically gathered data about atomic and nuclear properties.[14] In 1964, she established a first journal of the project, *Nuclear Data Sheets*, which was followed in 1965 by another, *Atomic Data and Nuclear Data Tables*, to which Way's name is indissolubly associated as editor. That association, in particular, led Wheeler to call her "dear to the community of nuclear physicists for her own work in the field as well as for her contribution to theirs" (Wheeler, 1979, p. 254). From 1968 to the end of her career she was at Duke University, where Hertha Sponer (1895–1968) had taught (see Chapter 3 and Section 8.4).

Way's scientific *magnum opus* consisted of editing journals and volumes of nuclear data. As colleagues wrote in her obituary, "by her insistence on the critical evaluation of all published basic data and her ability to combine these data into as logical and self-consistent a set of nuclear structure properties as possible, Kay influenced an entire generation of evaluation experts and the presentation of data in physics literature" (Martin et al., 1996). If we open the two volumes she edited in 1973, *Atomic Data and Nuclear Reprints* (Way, 1973), in the short and sober preface we can read words that, in this case, were certainly written by her (Way, 1973, p. ix):

Since 1968 the journal Nuclear Data Tables has published four works on internal conversion which, taken together, present a complete set of values for the internal conversion coefficients. The collection of these four papers into a single volume with an additional table, Appendix B (to help find L-subshell ratios), provides a compact tool for the analysis of nuclear radiations. Such a tool is needed not only by nuclear spectroscopists in constructing decay schemes but also by health physicists in estimating dosages, nuclear engineers in calculating shielding requirements, and theorists in checking internal conversion theory itself."[15]

---

[13] The Nuclear Data Project was absorbed by the National Nuclear Data Center in the late 1990s.
[14] While an actual historiographical work about this project and Way's involvement is still missing, some useful notes can be found in (Tuli, 2008).
[15] In an unassuming tone, she goes on to explain in very simple terms what these efforts are about: "In a nuclear transition the energy released can be emitted as a photon, with $E = h\nu$, or can be transferred to an atomic electron which is ejected with an energy equal to $h\nu - B_i$ where B_i is the binding energy of the electron in the ith shell. The internal conversion coefficient is defined as the ratio of the probability of electron emission to the probability of gamma emission. This ratio depends, of course, on the shell from which the electron is ejected. Partial coefficients must be computed for each shell, and subshell, and summed to obtain the total conversion coefficient. Fortunately, the partial values decrease rapidly with increase in the principal quantum number so usually only the inner shells, K, L, and M, need be considered" (Way, 1973, p. ix).

This work, as is easy to guess, became an important reference for the experimental, theoretical, engineering, and even biomedical communities, as well as an opportunity to establish common standards among them. As Way writes without any flamboyance: "It is hoped that the present volume will make easier the comparison of experimental and theoretical values and thus lead to a deeper understanding of the conversion process" (Way, 1973, p. x). She played a similar role in enabling the coordination and cooperation of different communities and in organizing the consortium of universities doing research at the Oak Ridge facilities (Martin et al., 1996).

After decades of this kind of service, Way mentioned in a letter to Wheeler[16] that she was "planning to give up the journal [*Atomic Data and Nuclear Data Tables*] soon and work on a book on the ageing process." The last phase of her life was indeed characterized by an interest in health care for the elderly (Ware and Braukman, 2004, p. 671). Only a couple of years later, she retired from that role, but not before sending around a brief memoir titled "History of a Compilation Journal."[17]

From a brief exchange between Way and Wheeler at the end of April and beginning of May 1983,[18] we can surmise that she had a joyful retirement party at which she was presented with old pictures, notes, her thesis cover, and other mementos. (Wheeler too took part in the event and was responsible for many of these gifts.) It is not known what became of these materials, but from Way's April 27 letter[19] we do know that she felt she owed a debt of gratitude to Wheeler and the other participants, "one which I can repay only by trying to pass on the kindness and goodwill you showed to me. You gave me a wonderful shot of self-esteem to carry me forward into a less exhilarating future. I'll do my best to make good use of it." Nonetheless, in her words we can possibly sense a shadow, so to speak – that of farewell to scientific activities and, perhaps, of the loneliness of old age. (She never married.)

Way passed away in Chapel Hill in December 1995. Wheeler, in his autobiographical book published a few years later, recalls how, in 1993, when she turned ninety, "I visited her in her Chapel Hill retirement community, Carolina Meadows, where I found her as lively and opinionated as ever" (Wheeler, 2000, p. 150).[20]

---

[16] Way to Wheeler, September 2, 1980, John A. Wheeler Papers, American Philosophical Society Library, Philadelphia, box 68.

[17] Way to Wheeler, November 12, 1982, John A. Wheeler Papers, American Philosophical Society Library, Philadelphia, box 68.

[18] John A. Wheeler Papers, American Philosophical Society Library, Philadelphia, box 68.

[19] Way to Wheeler, April 27, 1983, John A. Wheeler Papers, American Philosophical Society Library, Philadelphia, box 68.

[20] It is unclear whether "opinionated" referred to some specific set of ideas or if it was merely a way to characterize her as a person of firm opinions (in the same page he described her as "not reluctant to express her definite opinions on people and events"). At first sight, a possible hint could be found in a letter to Wheeler dated April 27, 1983 (Way to Wheeler, April 27, 1983, John A. Wheeler Papers, American Philosophical Society Library, Philadelphia, box 68), in which she writes that her happy days in physics "were almost brought to a close in the McCarthy days from which I was rescued by that blessed chicken farmer" (she seems to be referring to an incident well known to those who knew her such as Wheeler, but of which no other trace has been

Having caught a glimpse of the relationship between Way and Wheeler from the latter's viewpoint, we may also quote here some delightful rhymes that she wrote in 1977, after the 65-year-old Wheeler retired from Princeton and moved to Texas:[21]

Dear John,
Here's some of what I learned from you
On how to work, what to do.
Before you start, make due reflections
On all exceptions, all corrections;
Never think a small chance nil
For what can happen surely will.
Then calculate with all your might
Make it right, make it tight.
Check by day, recheck by night
Phys. Rev. at last will show delight.

Now a toast to you and Janette
You've made this earth a better planette.[22]

Her affection is also displayed in her letter thanking Wheeler for his contribution to her own retirement party in 1983, recalling the "happy days [. . .] when I was first seeing the world of physics through your eyes."[23]

Wheeler, in turn, showed how highly he valued her role in the nuclear physics community when he, in 1978, along with Gertrude Scharff Goldhaber, Eugene Wigner, Alvin Weinberg, and others, nominated Way for the Tom W. Bonner Prize in Nuclear Physics from the American Physical Society (which, despite the strong supporting party, was neither awarded to her that year nor later).[24] Wheeler's letter, as generous as it is insightful, contains some telling statements:

I do not have to say anything about her unique contribution that is certainly worth the work of at least 20 people in analyzing nuclear data from a discriminating point of view so that it makes the lives of all of us so much more productive than would otherwise be the case.

---

found at present), but her troubles do not seem to have had a political or ideological origin. Further research suggests that what she was referring to is probably linked to the aftermath of an episode described in Libby (1979): in Chicago, as a friendly gesture, she had let a visitor from the Canadian Atomic Energy Program stay in her apartment, since she was out of town. This visitor was later convicted of treason in Canada and, consequently, Way too had a series of security-clearance difficulties. This is the context in which we need to place the remark by Ware and Braukman (2004, p. 30) that, during the McCarthy era, "[p]hysicist Katharine Way spoke out against the unfounded accusations sabotaging careers in her field, charges often rooted in the secret wartime conditions necessary to develop the atomic bomb." It remains unclear whether she "spoke out" before, during, or after the above-mentioned episode.

[21] The poem is part of an anthology of personal recollections, *A Family Gathering*, offered by many of Wheeler's former students and collaborators. Its index can be found in John A. Wheeler Papers, American Philosophical Society Library, Philadelphia, box 43.

[22] Janette (née Latourette Hegner) was Wheeler's wife.

[23] Way to Wheeler, April 27, 1983, John A. Wheeler Papers, American Philosophical Society Library, Philadelphia, box 68.

[24] In 1974, Way was included among the Distinguished Alumni of the University of North Carolina (Ware and Braukman, 2004, p. 671).

It has been said that she is responsible for the best storage and retrieval system in the whole of physics; and I agree completely. What I would want to add is something much closer to my own experience because after all she was the first graduate student I ever had and am very much aware of her contributions in exploring in the 1936–1938 period new pathways in nuclear physics, including her publications on the relevance on velocity-dependent forces in nuclear magnetic moments. In this connection, she analyzed the combined effect of rotation and high nuclear charge on the shape and magnetic moment of a rotating nucleus. She discovered that a heavy nucleus had no stable configuration if it rotated too fast, precursor – if we had only known it – of nuclear fission.

In wartime days I was close to her work for the metallurgical product of the Manhattan Project and can testify that her calculations of heating effects of the radioactive decay products of fission and her calculations of the effectiveness of control modes of special geometries were of direct value to the construction of the Hanford production reactors.[25]

This last passage brings us back to the crux of this chapter: nuclear fission. "If we had only known it" sounds here like an observation made in hindsight, without many regrets – markedly different from the later "Had we followed her lead . . ., " which instead suggests a concrete but neglected possibility, or even a responsibility. To understand this distinction more clearly, let us add a few elements to Wheeler's remark.

8.3 Way: On the Verge of Nuclear Fission?

At the 1938 New York meeting of the American Physical Society (Mehra and Rechenberg, 2001, p. 990), Way presented part of her graduate research that was not directly related to the topic of her dissertation. In a paper titled "Nuclear quadrupole and magnetic moments," she considered the cigar-like deformation of a spinning nucleus by trying different models, among which was the liquid-drop model. Written not long before the nomination letter above, Wheeler's recollection for the 1977 Minnesota symposium, "Nuclear Physics in Retrospect," at which eminent contributors to 1930s nuclear physics (those who were still alive, at least) gathered and discussed the developments of those years (Stuewer, 1979), provides additional insight about Way's 1938 paper (Wheeler, 1979, p. 266):[26]

---

[25] Wheeler to Henry Motz, August 15, 1978, John A. Wheeler Papers, American Philosophical Society Library, Philadelphia, box 68. The same folder also contains a copy of Gertrude Goldhaber's letter of nomination, from which we learn that she first proposed the nomination to Wheeler, who then enlarged the circle. In Alvin Weinberg's statement, also preserved in the same folder, we can read that "It was through her [Way's] foresight more than 30 years ago that the first nuclear information system was set up. Nuclear science could not have survived today had it not learned the lesson Kay Way taught: that rational handling of nuclear data is a key to a coherent and manageable nuclear science." Speaking of rational handling and planning, we may notice *en passant* that the motto of Barnard College is Ἑπόμενη τῷ λογισμῷ, somehow translatable as "Following the way of reason."

[26] As Robert Wilson said when he introduced Wheeler, "he knew the men, he was present at the moments. No one is better qualified to discuss this subject" (Stuewer, 1979, p. 215).

What about the rotation of heavier nuclei? Nothing was more obvious than to see what one could get out of the liquid drop model. The usefulness of this model had been demonstrated in Carl Friedrich von Weizsäcker's semiempirical formula for nuclear masses. Could one not go further and consider the implications of a surface tension for nuclear dynamics? And what was more natural than to go from statics to dynamics by the intermediate step of a steady rotation? Katherine [sic] Way and I asked each other what would be the order of magnitude of the magnetic moments one would expect for atomic nuclei on the liquid drop model. One day she came in and reported a difficulty. The equations gave no solution for the case of a sufficiently highly charged nucleus turning at a sufficiently great angular velocity. It was clear that one had to do in this case with some kind of instability. It took only 1939 and the discovery of Hahn and Strassmann to recognize the nature of the instability: nuclear fission. Why did we not go to the analysis of higher order terms in the deformation energy and predict fission in advance of its discovery? It was not any difficulty in the mathematics. It was a difficulty in the model. It failed to give the right magnitudes and the right trends for nuclear magnetic moments. [. . .] We were not in the domain in which a statistical or liquid drop model of rapidly rotating nuclei is possible.

This recollection is quite sober: in short, Wheeler hints that, in retrospect, they could have dedicated more attention to that instability and perhaps discovered nuclear fission a few months prior to the work of Otto Hahn, Fritz Strassmann, Lise Meitner, and Otto Frisch, but no such claim is made, and a sort of rationale is provided instead to explain why they did not think of going down that road.

A couple of decades later, however, in a series of interviews with Kenneth Ford which resulted in Wheeler's already quoted autobiographical book (Wheeler, 2000), the recollections take a more dramatic tone. The year is 1939. Bohr, accompanied by Rosenfeld, got some very confidential news about Meitner and Frisch's results on fission while he was departing for the US (Wheeler, 2000, p. 15):

By the time Bohr shook hands with me and the Fermis on the dock, he had a pretty good idea of a direction to go in to give a theoretical account of fission. That is what would occupy us intensely for the next few months. But at that shipside greeting, he said nothing about fission. In his characteristic way, he wanted to be sure that Meitner and Frisch got the credit they deserved before the news spread widely. Not even during his day with Fermi did Bohr breathe a word of the discovery.

Rosenfeld was accompanying Bohr but, "unaware of Bohr's concern about priority for Meitner and Frisch, spilled the beans to me on the train. I was excited. Here was a whole new mode of nuclear behavior that we had overlooked" (Wheeler, 2000, p. 15). When Bohr found out that Rosenfeld had spoken, he was apparently quite upset, but at that point the only thing left was to discuss the new physics and its implications (Wheeler, 2000, pp. 22–23):

To me, just as to Bohr, fission seemed immediately believable. I felt stupid not to have realized, several years before, that nuclei should be able to split. My student Katharine Way at the University of North Carolina had investigated magnetic properties of nuclei using the liquid-droplet model. Her equations had no solution when the nucleus rotated too fast. This told us that rapid spin could make a nucleus become unstable and fall apart. It would have been natural to ask ourselves whether there were other ways to make a nucleus come apart. Had we followed her lead, we might have thought of fission.

A general remark, first, in order to avoid easy misunderstandings. In the later narrative we may note a seeming change in emphasis on Meitner and Frisch, compared to "Hahn and Strassmann" in 1977, but this distinction is, in part, an artifact of the choice of quotes presented here, as well as a consequence of the markedly autobiographical tone of the second text. (Wheeler had a personal connection with Meitner and Frisch, but no personal connection with Hahn nor Strassmann, who even seem to disappear from the narrative.) As a matter of fact, as early as 1967, in the semi-popular *Physics Today*, Wheeler published, with Frisch himself, a history-inflected piece, in which Meitner's work is certainly not glossed over (Frisch and Wheeler, 1967), nor is it in the full text of the Minnesota talk (Wheeler, 1979, pp. 272–273; 299).[27] Furthermore, in a passage of the latter, Wheeler was clearly stressing that, *at the time of the events* (right after Bohr's Atlantic crossing), what there seemed to be to physicists like him was, on the one hand, an experimental result – that of "Hahn and Strassmann" – and, on the other, an interpretative "proposal," a "broad-brush picture" (Wheeler, 1979, pp. 272–273), that of Meitner and Frisch (who was also at the symposium). At least in Wheeler's intentions, that was not meant as a form of downplaying Meitner's role and, as we shall see in Section 8.4, he would even promote historiographical research on her work. As for Way, her absence from the 1967 paper seems understandable in a text of a few columns written with one of the *actual* discoverers of fission about its *actual* discovery.

The order of storytelling in Wheeler's recollections is rather important: in 1977, he soberly recalls Way's work, puts it in perspective among the issues they were pursuing at the time, and half a dozen pages later brings up Bohr's arrival with the news about fission.[28] In the later autobiographical book, when

---

[27] We will return to Wheeler's appreciation of Meitner in Section 8.4.

[28] Here is, for the sake of completeness, the 1977 text: "Across the stormy North Atlantic, unbeknownst to any of us, it brought the fateful find of fission from the Old World to the New: a few words spoken by Otto Robert Frisch to Niels Bohr right before he boarded the ship; a few words spoken by Bohr to me after his arrival. Coming with me to meet Bohr and his son Erik and Leon Rosenfeld were Enrico and Laura Fermi. They took father and son off for an overnight visit in New York before Bohr's three-month stay in Princeton, where he was to lecture on the quantum theory of measurement. Rosenfeld, his collaborator in the preparation of these lectures, I took on the train to Princeton and induced him to speak at the evening Journal Club on Hahn and Strassmann's discovery, still unknown in the U.S., on Meitner and Frisch's proposal that this was the breakup of a liquid droplet, and on Bohr's conclusion, worked out in conjunction with Rosenfeld on the voyage over, and written up four days later, that this new process fits in naturally with the compound nucleus theory of nuclear reactions. Rosenfeld's account created great excitement! It also greatly distressed Bohr when he heard about it. He felt obliged to protect Meitner and Frisch until their publication should have appeared (February 11, 1939); and to protect Frisch until his intended ionization chamber experiment at Copenhagen should have been done (January 16) and published (February 25, 1939). For that reason, he had said nothing about the discovery either at Columbia or to the Fermis. Bohr once arrived in Princeton, we set to work to go from Frisch and Meitner's broad-brush picture to a detailed analysis of the mechanism along the lines of the compound nucleus model and liquid drop model that Bohr – and I – had already

Wheeler mentions his student's work in the context of Bohr and Rosenfeld bringing the news from Europe, the effect is clearly different, in terms of pathos and personal involvement, as is the claim about the anticipation of fission. It would be perfectly reasonable to assume that, since many years had passed for Wheeler, the dimming of memory and the oral dimension of the interviews created a fluid and shifting set of recollections. For instance, "several years before" (Wheeler, 2000, p. 22) is certainly inaccurate: at most, it could have been a few months. However, an exploration of Wheeler's archives reveals that he and Ford did some extensive fact-checking between the interviews and the "autobiographical" book born out of them and that, for instance, Wheeler had among his papers[29] a work about the discovery of nuclear fission written by Spencer Weart (Weart, 1983), former director of the Center for History of Physics of the American Institute of Physics, whom he had known for a long time. It was actually Weart himself who sent Ford a copy of his historical reconstruction, which was certainly read by Wheeler.[30]

Two of Weart's claims are of particular interest here: that those who had come closer to theoretically identifying the sausage-like stretching of the nucleus (*Kernwurst*) leading to fission were Wheeler and his student Way[31] and, independently, Weizsäcker and his student Wilfried Wefelmeier; and that these hints were not followed because scientists working on the topic were thinking within a paradigm of sorts, with its blinker effect (Weart, 1983).[32] Even if Weart's mention of Way's work was not entirely independent of Wheeler's own recollections, the fact that Weart framed the latter in a broader context, supported by historical research, may have sounded like a full corroboration and may have thus prompted Wheeler to make "bolder" claims in his autobiographical book. As for the paradigm-like blindness, as we shall see particularly in Section 8.4, Wheeler does not seem totally satisfied with that kind of explanation – the question that almost haunted Wheeler's last years was: why were they all so "slow" to understand fission?

---

been expounding and applying. This work took not only the three months of Bohr's stay in Princeton but two additional months of finishing up until I could send it in for publication (June 28, 1939)" (Wheeler, 1979, pp. 272–273).

[29] John A. Wheeler Papers, American Philosophical Society Library, Philadelphia, box 208.

[30] Ford to Wheeler, March 3, 1994; Ford to Wheeler, March 5, 1994, which also contains what Ford had written to Weart (John A. Wheeler Papers, American Philosophical Society Library, Philadelphia, box 142).

[31] Her name is only mentioned in the reference given in a footnote (Weart, 1983, p. 130) – and from the other references that Weart lists, we can see that he too, for this at least, was depending on Wheeler's talk at the Minnesota symposium.

[32] More specifically, Weart argues that, from the beginning of the century on, the consideration of relatively small changes in nuclei – such as an alpha or beta particle – had been enough to explain what was discovered; furthermore, Gamow's successful theory of alpha decay discouraged one from thinking that a "piece" larger than an alpha particle could escape from the nucleus (Weart, 1983).

Apart from comparing Wheeler's recollections and their refraction amidst the work of others, we can also independently assess, at least in part, the soundness of his statements. As a matter of fact, Way, in spring 1939, did publish a paper based on the work she presented in 1938. "The liquid-drop model and nuclear moments," which appeared in the May 1939 issue of *Physical Review* (Way, 1939) and includes, as a sort of introduction, a brief mention – without reference – of the new results concerning fission. However, it has no further connections to the text corpus and, especially if someone already knows about Wheeler's story, it may look like it was added later.[33] The ensuing technical corpus of the paper could definitely have been written without knowing the recent results about fission. The conclusion, not flamboyant at all (consistent with Way's unassuming tone in her later writings, as we have seen in Section 8.2) finally comments on the lack of solutions within the framework of the liquid-drop model.

The overall impression is that there is no reason to contest what Wheeler says, especially in 1977 – but that does not mean that one has to naïvely embrace his later comments on that "missed opportunity" either. Yes, the lack of solutions could have suggested that an instability might exist, and that the nucleus could not remain intact, so to speak. It may have therefore provided a heuristic hint that something else was happening. This, in turn, along with other hints, may in principle have guided someone to nuclear fission. "Never think a small chance nil / For what can happen surely will," as Way wrote.[34] However, it would also have been reasonable to say: The liquid-drop model here no longer works, let us try something else. The alternatives, in a sense, were a choice between jumping out of the model or staying within its framework and taking the breakdown as a *physically* meaningful indication in the map of our exploration of nuclear regimes.

Nevertheless, the outcome may have not been a radical either–or scenario. Wheeler did not simply ignore Way's results and, somehow, he did try to keep them into account – albeit in negative terms. Because the picture of a turning body did not seem adequate to nuclear rotation, he considered other options in one of his papers with Edward Teller (Teller and Wheeler, 1938).[35] The predominant

---

[33] Way's manuscript was in fact received on March 25, 1939, that is after the events (and publications) recalled by Wheeler (Wheeler, 1979, p. 273).

[34] "A Family Gathering." John A. Wheeler Papers, American Philosophical Society Library, Philadelphia, box 43. See also Section 8.2. It is doubtful, however, that Way had this episode in mind, and, in any case, the tone of those lines certainly does not resonate with the feeling of regret expressed by Wheeler.

[35] Wheeler makes this link – if it may be called that – explicit at the Minnesota symposium, right after mentioning Way's results: "If turning as a rigid body did not give an acceptable picture for nuclear rotation, what model would? Edward Teller had ideas on this subject, and I had case examples, which we developed together in a paper on nuclear rotation. The main point was simple. The very low-lying rotational energy levels that one would have expected on the basis of rigid rotation would of necessity be "promoted" in energy for an elementary reason: A rotation, followed by a minor displacement of the particles in the system, would restore the set of particles to a configuration, identical up to a permutation, with the original configuration. Therefore, it was all important to know what kind of potential energy barrier had to be overcome to perform this slight displacement. We examined several idealized models. They showed how sensitive is the position of low-lying

difficulty, however, was that, in the largely unexplored realm of nuclear behavior, different models, emphasizing different traits, were also inducing people to think of the nucleus in different terms that were intuitively hard to combine.[36] And that is also why there is a big difference from the impression one can have in hindsight and the actual historical situation, with all its fog. Nonetheless, while there does not seem to be any trace that Way further developed those aspects of her work[37] (or any sign of regret that she did not), Wheeler apparently insisted on reproaching himself. In order to better understand that on multiple levels, and properly weigh his late claim about the "quasi-discovery" of fission, we have to broaden the horizons of our analysis.

8.4 Wheeler's Uses of the Past, Daring Conservatism, and the Role of "Heroines" in His Narratives

Over the years, Wheeler commented rather broadly about the role of women in the early phases of nuclear physics. He had personally met a few of them, and his reminiscences and descriptions may offer insights that help a reader to interpret and assess his particular claim about nuclear fission. However, it is necessary to first introduce a more general *caveat* about Wheeler's uses of the past. Readers of his papers may notice – beside his exuberant rhetoric and his rich use of metaphors and anecdotes from the past – some grand preludes, charged with historical pathos. That is quite idiosyncratic, but, in Wheeler's case, it is not solely ornament and rhetoric; it runs much deeper. In Wheeler's vast archival material, different facets of his engagement with the past can be identified and, for the sake of simplicity, categorized – as discussed at length elsewhere (Furlan, 2020, 2024) – according to the famous Nietzschean tripartition: an "antiquarian" effort, aiming at the preservation of the past for its own sake; a "monumental" use of history, in order to build grand narratives and spur collaborators or students in some directions of research, or at least to make

---

states of high angular momentum to the energy of the deformation. We concluded that it is reasonable to believe that certain low-lying levels are indeed rotational states" (Wheeler, 1979, pp. 266–267).

[36] For a technically detailed historical treatment one should also consider papers such as Wheeler (1937), but for our purposes here we can resort again to Wheeler's synthesis at the Minnesota symposium: "By that time the experimental evidence on nuclear spins and magnetic moments, so impressively correlated by Maria Mayer and by Haxel, Jensen, and Suess, had made it inescapably evident that nuclei show 'independent particle' properties. Niels Bohr and I were having to come to terms with The Great Accident of nuclear physics – the circumstances that the mean free path of particles in the nucleus is neither extremely short compared with nuclear dimensions (as assumed in the liquid drop picture) nor extremely long (as assumed in the earliest days of nuclear physics), but of an intermediate value. When we took this fact into account, we found that we could understand how a nucleus could at the same time show independent particle properties and yet behave in many ways as if it were a liquid drop" (Wheeler, 1979, p. 267). All this is clearly interesting also from the viewpoint of philosophy of science: see, for example, a few mentions in Massimi (2022, pp. 87–109).

[37] There was little reason, as they were explained within a few months, with all the implications that fission would have for life and research in the years that followed.

them feel part of a great and still-alive enterprise;[38] and a "critical" gaze, capable of selectively identifying analogous traits in past situations and re-functionalizing or transfiguring them in the present. By all accounts, Wheeler was very good at transmitting a kind of feeling of belonging to a tradition. Thanks to his teaching and his way of involving students, even in open research questions, physics no longer seemed something done mainly in another contin-ent during previous centuries, but a great collective effort they were part of (Wheeler, 2000, p. 84):[39]

> What of the girl or boy who reads about Newton and Maxwell and Bohr and Einstein and says, "I want to build on what they have built; I want to add to the sum of knowledge about the most basic laws of nature"? That child has gained somewhere, somehow, even more than average confidence in his or her own capabilities.

A similar sense of science as a great enterprise transpires from the grand narratives that Wheeler built in order to support and, in a sense, justify his own speculative ideas, still in lack of empirical corroboration or possibility of accurate confrontation.[40] In this sense, his monumental use of history was also entwined with a heuristic dimension. However, as shown at length by his promotion of historiographical research (as also illustrated below),[41] Wheeler, although he did not feel constrained by details in a historicist sense, had a genuine respect for the past. In other words, we are not dealing with a mere rhetorical strategy to better sell his ideas or to offer some pedigree to his own speculations.

Now, a question that we can ask, perhaps also in relation to Wheeler's military commitments such as the Matterhorn Project against what he perceived as the Soviet threat, or his relatively conservative political positions,[42] is whether those

---

[38] There is, of course, a whole Anglophone tradition of monumental history (intertwined with Nietzsche's reflections as well), from Carlyle to Emerson and beyond. The former, curiously enough if we want to stay close to nuclear physics, was studied by Enrico Fermi's wife, Laura Capon (Fermi, 1939). It would be interesting to evaluate the more or less mediated impact of this tradition on Wheeler, but that is not the purpose of the present work. We may however still add that also Niels Bohr's wife, Margrethe Nørlund, shortly after meeting her future husband, urged him to read Carlyle's *On Heroes and Hero Worship* (Aaserud and Heilbron, 2013, p. 155). In Bohr, possibly, Wheeler did not just find a hero for his own narratives.

[39] That is particularly underscored by Sergio Hojman (Misner, 2010, p. 20).

[40] It is quite easy to guess the relevance of all this for today's approaches in speculative fields such as quantum gravity and, in a sense, the considerations we will develop here should also make one cautious about the dangers of "philosophizing" on superficial or distorted historical grounds, as often happens with partisan accounts.

[41] By the 1950s, Wheeler played a crucial role in setting in motion the vast project *Sources for History of Quantum Physics*. See Furlan (2024) for a contextualization of his reasons and aims.

[42] The following passage by Leonard Susskind (a recollection of when he first met Wheeler in 1961) may serve as a good example to convey both a first impression and its overcoming: "The only way I can describe the man who greeted me is to say that he looked Republican. What the hell was I doing in this Wasp's nest of a university? Two hours later, I was completely enthralled. John was enthusiastically describing his vision of how space and time would become a wild, jittery, foamy world of quantum fluctuations when viewed through a tremendously powerful microscope. He told me that the most profound and exciting problem of physics was to unify Einstein's two great theories – General Relativity and Quantum Mechanics. He explained that only at the Planck distance would elementary particles reveal their true nature, and it would be all about geometry – quantum geometry. To a young aspiring physicist, the stuffy businessman exterior had morphed into an idealistic visionary. I wanted more than anything to follow this man into battle" (Susskind, 2008, p. 146).

narratives and monumental use of history were strictly "masculine" and gender-biased. Of course, Wheeler's uses of the past inevitably mirror the asymmetries that were part of that same past and the ways it was re-told – but more nuanced insights can also be gained by focusing on the female personalities in those grand narratives of his. Rather evocative pages are dedicated to them. We would not gain many insights by simply labeling (or de facto dismissing) Wheeler's narratives as "male tales," echoing a formula by Sharon Traweek (Traweek, 1988).[43] Aside from the fact (as suggested by the quote above) that Wheeler meant to provide inspiration to anyone, regardless of gender, we are rather going to examine his narratives to show a complex intertwining with the research he was pursuing at a given moment, or even, as we shall see, with his psychology and personal experience.[44] If we do not keep this in mind, the risk of misunderstanding him or remaining baffled by his unusual kind of remarks is high and, given his prestige, this could easily result in uncritically spreading some of his claims (such as, perhaps, the quasi-discovery of nuclear fission by a former student of his). It is thus useful to give a taste of what can be hidden beneath Wheeler's texts.

The first example of a woman scientist in Wheeler's historical narratives is an icon of the exploration of the microcosm: Marie Skłodowska Curie (1867–1934). If that choice is predictable, luckily Wheeler turns to something a bit less conventional already from the title: "Maria Skłodowska-Curie: Copernicus of the world of the small" (Wheeler, 1968). This is not one of those awkward expressions that may be found in some magazines and newspapers with improbable comparisons or juxtapositions – or not entirely, at least. When this piece was later included in an anthology of Wheeler's writings, the title was changed slightly: "Maria Skłodowska Curie and the world of the small," with the previous title assigned instead to the last of the sections into which the reprinted paper was divided (Wheeler, 1994). Delivered in Warsaw in October 1967, for the centenary of her birth (and thus obviously laudatory), it was written in a period that marked the beginning of a crucial turning point in the development of Wheeler's ideas. That is indeed the moment in which his geometrodynamical program – all the ambitions he had nurtured for the previous 15 years, while dusting off and developing

---

[43] This is not to deny that an analysis in terms of gender studies might offer additional perspectives or show that, after all, some of Wheeler's ideas were indeed biased, but that is simply not the purpose here.

[44] We could even suggest that Wheeler's case, rather than calling for easy labels, represents an opportunity to rethink our picture of science and scientific life against old and stiffened abstract schemes: his status as a key figure of twentieth-century physics is not contested, but the full reasons for that are somewhat vague and elusive. To fully grasp his role, we have to take into account a heuristic dimension (all the suggestions and guiding ideas he disseminated), a strong sense of conviviality in practicing science, and so on. If someone wants to contest a merely "resultatistic" picture of science, Wheeler's case is then a resource. By analogy, *mutatis mutandis*, Hermann Weyl's words about Emmy Noether, pronounced shortly after her passing away in 1935, may easily come to mind: "Her significance for algebra cannot be read entirely from her own papers; she had great stimulating power and many of her suggestions took final shape only in the works of her pupils or co-workers. And one cannot read the scope of her accomplishments from the individual results of her papers alone" (Weyl, 1981, p. 130).

general relativity – started showing its limitations and entering into a crisis, after which Wheeler came to question the very meaning of physical laws. The paper opens with a sort of curious, fatal (really with a sense of destiny), almost cinematographic parallel between Marie Curie's birth and first years of life and Maxwell's discoveries and future expectations (Wheeler, 1968, p. 1197):

The month of November 1867 that brought Maria Skłodowska into the world saw James Clerk Maxwell winning new insight into the laws of electromagnetism. Four years later, in the month of October, while Maria sat pouring water from one bottle to another and asking her childhood questions about how and why, Maxwell was delivering his introductory lecture on experimental physics at Cambridge University. After describing the new facilities and stressing the importance of experimental work for the young man and for society, Maxwell offers his vision of physics.

If referring to Maxwell is typical of narratives about the unification of physics – with some grounded reason, of course – Wheeler's evocation is nonetheless peculiar. Yes, he says, "Maxwell held fast to the long-term dream of an underlying unity. Yet to him 'unity' meant not so much one law as one substance" (Wheeler, 1968, p. 1197). By "substance" Wheeler was referring to a field ontology: an attempt that, *mutatis mutandis*, he himself was pursuing in that phase of his career, when he was trying to derive everything from the fields of general relativity. It does not really matter here whether such a characterization of Maxwell's vision is accurate or nuanced: the point is that it prepares the next step (almost a dialectical one), that is the passage from what Wheeler calls a "physics of substance" to a "physics of law." The key figure behind this shift is Marie Curie, he claims.[45]

What does this mean? In Wheeler's own words, "[t]he study of substance revealed law. Law in turn explained substance" – but here law does not mean any kind of law. It means something fundamental and as broad as a framework: "Three laws the great investigators gave us: the relativity principle, both special and general; the quantum principle; and electromagnetism" (Wheeler, 1968, p. 1197). In this sense, from the empirical study of substance we gained insights into some fundamentally overarching principles, but then it was these overarching principles that guided our understanding and exploration (I emphasize the heuristic aspect) of realms more and more distant from everyday experience. One has to trust these "laws," take them "seriously": "Electromagnetism, already discovered, was in effect rediscovered when at last it was taken seriously in the world of the small. To that end no one contributed more than Marie Curie" (Wheeler, 1968, p. 1197).[46] Now, this framing might just sound as an idiosyncratic take on the epistemology of science and the history of

---

[45] For Wheeler's way of thinking about and relying on well-established physical laws in those years, I redirect the reader to Blum and Furlan (2022) and Furlan (2022).

[46] A much later passage significantly insists again on that "seriously": "But the Maxwell equations held more. Forty years later, Einstein invented special relativity as a way – the only way – to assure the complete validity and self-consistency of Maxwell's equations. By taking them seriously and insisting on their universal validity,

physics; but it is fascinating how Wheeler, with much more eloquence than a brief quote reveals, is at the same time speaking, between the lines, about his own heuristic methodology and research program. When at the beginning of the 1950s he decided to dedicate himself to general relativity, he did not just intend to study another area of physics or tackle a specific question. He decided to take general relativity "seriously," that is try and squeeze everything possible from it,[47] without caring too much for all the confusing novelties of the particle zoo (that is, all the new unaccounted-for particles that were being found). With a half-playful nod to a slogan by Eisenhower, he gave this heuristic methodology a name: "daring conservatism" (Blum and Brill, 2020; Blum and Furlan, 2022, p. 284). This kind of attitude toward well-established principles, in general, and the reliance on the guide they can provide in the exploration and ordering of new phenomena remained for him a compass amid the confusion of possible alternatives – also because Wheeler (or so he claimed) had learned this modus operandi from one of his main inspiring figures, Bohr. We might even suggest that in this kind of methodological emphasis we can hear echoes of the allegedly "missed opportunity" in Way's work and nuclear fission, as we will better articulate: If only they had explored to the extreme what the model suggested! A daring conservative regret, we may say.

In 1967, however, the development of general relativity according to this methodology, as already mentioned, was starting to show some cracks – but, as a sort of compensation (or denial), it is interesting to notice how, in the paper dedicated to Marie Curie, Wheeler still raises the stakes and makes her not only a heroine of daring conservatism, but implicitly one member of a sort of Holy Trinity (the other two being Einstein and Bohr, occasionally replaced by Planck) which is mirrored by the three great "principles": relativity, quantum theory, and electromagnetism. In the midst of such considerations, there are also some lyrical passages that give life to the (quasi-)historical characters conjured up by Wheeler (Wheeler, 1968, p. 1198):

If the sad and lonely figure of Mme. Curie touches anyone's heart, and if he hears from those who knew her that she never smiled, let him read her works. A new step forward, by whomever made, captured her admiration. To her the search was one great enterprise; and all searchers, partners. How did she respond to the movement from a physics of substances towards a physics of law? She welcomed the new laws, she followed them, she preached them.

And then he lists her achievements. It should not sound surprising, now, that we spoke of a "monumental" use of history. Afterwards, Wheeler goes on with his

---

Einstein showed that space and time are relative concepts and that mass is energy! Hardly conclusions that a casual glance at Maxwell's equations would reveal" (Wheeler, 2000, p. 166).

[47] Wheeler truly meant to build everything out of the geometry of general relativity, but endowed with the Planck scale fluctuations of quantum foam, one of Wheeler's most suggestive creations. See again Blum and Furlan (2022) or Furlan (2022).

typical way of jumping back and forth between the past and the future, pointing out alleged anticipations of important results and, at the same time, on that basis, asking some questions about a future still to come – or better: about the future of physics as he was envisioning it.

What about the comparison with Copernicus? The reasons for this parallel between him and Curie, once we know the second part of the title of the speech (Copernicus of the world of the small), are actually easy to guess, besides the association to Poland, and Wheeler makes them explicit while evoking, in the conclusions, a sort of mental theater – or a black and white movie – in which the protagonists of the early Solvay conferences (among whom the no longer sad Madame Curie) take the floor again and give us hints on finding nature's secrets (Wheeler, 1968, p. 1200):

We see her in her later years, packing her suitcase with such happiness for a Solvay Congress, where she would walk and talk again with Lorentz, Planck, Einstein, Ehrenfest, and Bohr.

We see the magic circle and see Planck speaking. He repeats his great and familiar message. There is only one truly fundamental length in nature; a length free of all reference to the dimensions and rate of rotation of the planet on which we happen to live; free of any appeal to the complex properties of any solid, liquid, or gas; free of every reference to the mysterious properties of any elementary particle; what we call today the Planck length [. . .] and what we identify with the characteristic scale of the quantum fluctuations in the geometry of space.

The light shifts, the figures are regrouped, and Einstein is giving his famous account of the quantum fluctuations that pervade the electromagnetic field in every part of space, forerunner of modern quantum electrodynamics – the greatest triumph of theoretical physics since World War II – and happy guide to the meaning of quantum fluctuations in the geometry of space at the Planck scale of distances.

The Solvay Congress fades away, we are in an old shed in Paris, and we see a young woman working intently at her radium. She gave us the projectiles to penetrate a new world of small distances. She did more than anyone to open the door to 10^{-13} cm, as her countryman Copernicus did more than anyone to alert us to movement and meaning at the previously unimaginable distance of 10^{+13} cm. [. . .] Copernicus directed our gaze out to the domain of the unbelievably remote, and today we have come close to plumbing the greatest distances that we know how to conceive. The discoverer of radium by her life and work directs our gaze down to the world of the small. There many new decades of the distance scale still wait to spring into life and meaning, all the way from 10^{-16} cm to Planck's 10^{-33} cm. Marie Skłodowska Curie is our Copernicus in the still continuing voyage of exploration into the world of the unbelievably small.

This passage makes even more manifest the intertwinement with Wheeler's own quest, from the scale of relativistic cosmology to that of quantum foam, as well as with his own Blake-like gaze "to see a world in a grain of sand."[48] But, apart from illustrating

---

[48] Interestingly enough, Copernicus, within six years, would also become a character in Wheeler's grand narratives, the occasion being once again provided by an anniversary, his 500th birthday in 1973 (Wheeler,

the layeredness of these texts, this leads us to another story. Wheeler had not finished reflecting and talking about Curie: a decade later, at the previously mentioned 1977 Minnesota symposium (Stuewer, 1979), since the discussants included people like Wigner, Hans Bethe, Frisch, and Emilio Segrè, clearly Wheeler could not just go into "free storytelling mode," but he nevertheless made an unusual comment about her. Building upon the historical inquiries he had made to prepare the address for Curie's centenary, he told of his surprise at finding an anticipation of the conclusions he himself had later reached only thanks to the Polanyi–Wigner formula, so important in suggesting a way of considering nuclear behavior in Wheeler's own decisive work with Bohr on fission.[49] This signals to us how actively he was digging in and reflecting on the "prehistory" – or the prodromes – of the discovery of fission, and he was not afraid to present his thoughts in front of other renowned experts. This line of thinking did not merely represent for him a way of conveying pathos to a more or less generic audience (as it may have been in Warsaw) or invoking the rhetoric of the forerunner, evidently.

Wheeler never interacted personally with Curie. For the lyrical passages he could only rely on tales of people that had met her, on books, and on his own imagination. When he could add his own first-hand experience with other leading figures, however, his peculiar form of storytelling became even more effective. A later and easily guessable name in this respect is Lise Meitner, whom the young Wheeler met personally in the 1930s. He would also write a foreword to Patricia Rife's book, *Lise Meitner and the Dawn of the Nuclear Age*, published in 1999 (Rife, 1999). Actually, in the light of archival materials, it turns out that already in 1984, when Rife contacted Wheeler after finishing her PhD dissertation dedicated to Meitner,

---

1974). By that time, Wheeler had moved beyond his previous conception of physical laws and, once again, he would venture into bold parallelisms between "now" and "then" in order to provide new perspectives – it does not seem out of place to suggest that his meditations on the figure of Curie as "Copernicus of the small" prompted him to make a similar operation to give voice to his own ideas while celebrating Copernicus later (Blum and Furlan, 2022, p. 307).

[49] More in detail, here is the actual passage: "Years later, reading the works of Marie Skłodowska Curie in preparation for a celebration of her centenary, I was astonished to discover that she had thought about the breakup of a nucleus in just such terms. She referred to Rutherford and Soddy and their 1902 and still standard theory of radioactive decay. She recognized, as they did, that they were dealing with a process that transcended the understanding of the time. At the 1913 Solvay Congress she called attention anew to the mystery of the exponential law. She stressed the experimental evidence that an atom, if it had not yet decayed, had not aged at all, no matter how long it had lived. She proposed 'to look in the interior of the atom for the element of disorder necessary to explain the application of the law of chance.' She brought forward the suggestion of Debierne, first, that in the center of the atom there may exist a temperature much higher than the external temperature and second, that the mechanism involved may be identical with that of a monomolecular chemical reaction. She asks us to imagine 'a molecule which is moving about in the interior of a box endowed with a tiny hole.' Marie Skłodowska Curie goes on to say, 'When the molecule in the course of its motion meets the hole it leaves the box and the system is radically changed. If we have a great number of boxes each containing one molecule, and if the initial velocities and positions of the molecules are random, it may happen that the escape phenomenon is governed by the rule of chance, even though the constitution of the system itself is relatively simple.' Mme. Curie was in advance of her age. She put forward the right idea to describe nuclear fission during an epoch when she had to deal with the leakage of alpha-particles through a potential barrier!" (Wheeler, 1979, p. 280). Rather than merely dismissing this as a "teleological" historical remark, however, it may be worthwhile to relate it to Weart's aforesaid comments about the role of Gamow's model of alpha-decay (Weart, 1983, p. 102).

he strongly supported her initiative and gave her indications ("I'm <u>delighted</u> at your project," underlined by Wheeler himself).[50] The foreword to the book, then, was just a natural – but certainly telling – continuation of his following the genesis of the book.

What is striking in those few introductory paragraphs is the admiring character-ization that Wheeler offers of Meitner's "style": "Excelsior! Or, in everyday language, ever higher. That, in a word, says to me what I remember of Lise Meitner as unique. Identify a central and decidable issue of physics, devise a way to go after it, and pour on the attack" (Rife, 1999, p. ix).[51] It is not coincidental that this characterization is reminiscent of his description of Curie's resolute explor-ations of the world at lengths far removed from our everyday experience. However, it is also reminiscent, due to that sort of three-step description of Meitner's course of action, of what Wheeler used to call the "trilogy of action" of the "seekers of the larger view" (such as Einstein, Charles Darwin, James Hutton, and so on): "to rise to the larger view – and from that vantage point recognize what is missing, and fill the gap" (Wheeler, 1980, p. 185). Thus, once again, if we aptly connect Wheeler's passage to his other writings, we can find much more than what meets the eye – and in this case we can appreciate how the praise is indeed of the highest caliber. Curiously enough, at the end of that foreword Wheeler also mentions that some unspecified "postwar Hollywood filmmaker" intended to create a movie about Meitner, and that some of her friends had suggested to make Wheeler "her representative in negotiating the wording for her part in the film" (Rife, 1999, p. xii). But, in the end, she was not satisfied or comfortable with that project and withdrew from the initiative. At least, Wheeler adds in the same page, this opportunity offered him a last chance to see her and "appreciate anew the force of her character and her drive to 'Identify the Central Puzzle and Clear It Up'" (Rife, 1999, p. xii).

Once we speak of Meitner and Wheeler, we are again at the very heart of the discovery and understanding of nuclear fission, and we have already remarked in the previous section how her work was mentioned not only during the Minnesota symposium, but also in the article on fission that Wheeler published together with her nephew Frisch (Frisch and Wheeler, 1967). Meitner's role was a central part of the story, but what is important to stress here is again Wheeler's "methodological" emphasis. The *fil rouge* that, in his recollections, seems to link women to fission, or more generally to the early exploration of the microcosm, does not only run through results or historical facts: Wheeler repeatedly insists on method, broadly speaking.

---

[50] Wheeler to Rife, April 19, 1984, John A. Wheeler Papers, American Philosophical Society Library, Philadelphia, box 60.

[51] "Excelsior" in the title of this chapter comes from this passage; it is not a reference to Goethe's eternal feminine that draws us on high.

The latter aspect also shows up when, during the same symposium, he portrays another well-known physicist: Maria Goeppert Mayer (1906–1972), who collaborated in Baltimore with Karl Herzfeld, Wheeler's PhD advisor in the early 1930s. Once again, Wheeler's narrative tellingly resonates with his values, his way of conceiving research and teaching in physics, and perhaps something he wished he himself had had on some occasion (Wheeler, 1979, p. 225):[52]

Maria Goeppert-Mayer often, and from time to time her husband Joseph Mayer, crossed the lawn from chemistry to physics to participate in a seminar or consult Herzfeld. The higher administration had denied her any proper position. She did not teach. She nevertheless taught as effectively as anyone – by her questions and comments in seminars and colloquia. Her concern for whatever the issue was made her an example to us all. It is impossible to forget her quiet firmness, the rising emphasis at the end of her last sentence, and *her inability to leave a problem until it was clarified* (emphasis added).

Not long after the years evoked in this recollection, Wheeler, in 1934–1935, went to Copenhagen to work with Bohr, which was a life-changing experience, on both professional and personal levels. While he was there, he met many other well-known physicists, including Meitner, as already mentioned, and – he would add *en passant* in a later interview (Wheeler, 1988), but without special comments – Sponer. He also repeatedly spoke of the important role of Margrethe Bohr: actually, we might plausibly draw a parallel between what he says about the Bohrs' hospitality in Copenhagen (Wheeler, 2000, pp. 127–128) and what some of Wheeler's students remarked about his own welcoming house and the role of his wife Janette,

---

[52] Wheeler's later recollections about Goeppert Mayer are worth quoting as well, also as a sign of his attention to the condition of women in science: "Another theorist at Hopkins at that time was Maria Goeppert Mayer, later to win a Nobel prize for her work on the structure of nuclei. She and Herzfeld ran a seminar course in which a few of us sat around a table going chapter by chapter through the new German language book of Max Born and Pascual Jordan on quantum theory. It was an exciting way to learn the subject. Maria's husband, Joe Mayer, was a professor of chemistry at Hopkins. She came to Hopkins as a newlywed with a fresh Ph.D. from Göttingen in 1930, when she was just twenty-four. Suffering the fate of many women academics, she had no proper position, being treated as a guest, with little if any salary. Herzfeld had the good sense to recognize her ability and brought her into the workings of the department. Maria Mayer's situation got worse after the retirement of Joseph Ames as president of Hopkins in 1935. In the eyes of the new president, Isaiah Bowman, she suffered by being not only female but foreign. This made it easy for her husband, Joe, to accept a position at Columbia University in 1939. Even there, she was not appointed to a regular faculty position. She taught at Sarah Lawrence College while conducting her research at Columbia. During World War II, I had the chance to admire her abilities again when we worked together at the Met Lab in Chicago. After the war, while her husband, Joe, held a professorship in chemistry at the University of Chicago, she held half-time, nontenured research positions at the university and at Argonne National Laboratory. Not until 1960 did she get a professorship herself, when she and her husband moved to the University of California, San Diego. Her Nobel Prize was awarded in 1963. Despite the injustices she suffered for most of her professional life, she remained always cheerful and always vigorously active in theoretical physics. To her colleagues she was a valued full partner, whatever status she might be assigned by local administrations. A heavy smoker, Maria Mayer died in 1972 at the age of sixty-five" (Wheeler, 2000, pp. 96–97). She clearly crossed paths with Way, not just because of locations but also of research topics, but not much else seem to be known besides an acknowledgment by Goeppert Mayer: "The author is indebted to Dr Katherine [sic] Way, who pointed out the connection of the closed shells with neutron absorption cross sections" (Goeppert Mayer, 1948, p. 238).

whom Way's poem mentioned.[53] After that experience, the young Wheeler went back to the US and soon obtained a position at Chapel Hill, where he met Way.

However, there is still a blurry figure that would make her recurring appearance in Wheeler's European memories during his old age, regardless of the absence, in this case, of an element of personal reminiscence: Ida (née Tacke) Noddack (1896–1978). Her name is missing from the acts of the Minnesota symposium – not just from Wheeler's contribution, but from the analytical index of the entire volume. In Wheeler's case, and possibly in that of other participants, it was probably due to the lack of personal contacts with Noddack and to the relative oblivion in which she had fallen, but, in general, political factors may have been at play, too.[54] A decade earlier, however, Frisch, in his section of the joint article with Wheeler for *Physics Today* (Frisch and Wheeler, 1967, p. 48), had briefly mentioned her; Weart too discussed her work (Weart, 1983, p. 95), and that may have been one of the catalysts of Wheeler's "rediscovery" of her. In his autobiographical book she is inescapably present – but this time the "anticipation" of fission is not accompanied by a sense of regret, but almost a breath of relief (Wheeler, 2000, p. 15):

Even when the German chemist Ida Noddack suggested in 1934 that Fermi had in fact split the uranium nucleus, no one paid attention. It was, at the time, too radical a thought. (One can't help wondering whether Noddack's insight would have found a more receptive audience if it had come from a man instead of a woman.) In retrospect, the blindness of

---

[53] There would be no shortage of testimony, but I would like to offer here an exceptional and unexpected perspective, that of someone whose life was also intertwined with the long chain of events triggered by the understanding of nuclear fission. In Wheeler's archives we can find a letter by no less than Ya.B. Zel'dovich, dated February 18, 1986, congratulating Wheeler and his wife Janette on their golden marriage: "Every physicist knows John, it is an axiom. But I am happy to remember Tbilissi [sic] (Georgia, USSR), where I met you, Jeannette. Having seen you and spoken with you I understand better why is John so clever and good-humored and healthy as he is" (Zel'dovich to Wheeler, February 18, 1986. John A. Wheeler Papers, American Philosophical Society Library, Philadelphia, box 70). The wish of "many more happy years and everlasting peace," coming from a man who three decades earlier was one of the directors of the Soviet project to develop the H-bomb and addressed to one of the leaders of the US H-bomb project, is a precious document from a human point of view. Ten years earlier, as witnessed by another letter from Zel'dovich, his wife had suddenly died while on holiday on the Black Sea (Zel'dovich to Wheeler, September 9, 1976, John A. Wheeler Papers, American Philosophical Society Library, Philadelphia, box 70). Shortly before, Wheeler had sent them a "kind letter," "the very interesting Bargman [sic] book with" an "excellent article" by Wheeler himself (we can certainly infer that the volume was *Studies in Mathematical Physics: Essays in Honor of Valentine Bargmann*, which was indeed published in that year), and a present for Zel'dovich's wife, who "enjoyed" it "much." Then Zel'dovich, in his "broken English" (to borrow Hendrik Casimir's once famous expression among physicists), which here only adds to the humanness of the communication, goes on to say: "Our meeting in Warsaw and trip to Moscow were among the happy part of my life. I fear that never more I will feel myself the same as before." It continues, "I met your wife in Tbilisi many years ago. I remember exactly the nice Georgian garden and our conversation. It was a strong impression of integrity, moral courage and deep ethical principles, just what is needed from the wife of a scientist and mother of family. Please give her my best whishes [sic]" Last, it adds, "Be careful with her. Womens [sic] are often overstimating [sic] their forces and they whish [sic] not to worry us – and mens [sic] are so often egocentric!" (Zel'dovich to Wheeler, September 9, 1976, John A. Wheeler Papers, American Philosophical Society Library, Philadelphia, box 70). All this, besides the element of personal loss, clearly adds further depth to Zel'dovich's message 10 years later.

[54] See Magalhães Santos (2014), which underscores the role of the influential stories told by figures such as Fritz Paneth and Emilio Segrè, both of which had to leave Europe because of their Jewish ancestry, portraying the Noddacks (Ida and her husband Walter) as Nazi supporters and adding more episodic allegations of scientific dishonesty.

physicists and chemists to fission in the mid-1930s can be regarded as a blessing. Had scientists in Germany – and elsewhere – followed up on Ida Noddack's suggestion, it might well have been the Germans, not the Allies, who got the atomic bomb first. The history of the world could have been different.

In Wheeler's archives it is actually possible to see that, c. 1994 (that is, around the time of the interviews with Kenneth Ford), in a list of topics (the "Wheeler grab-bag") he was eager to discuss with students as an advisor, at the very first place there is "the mystery of Ida Noddack."[55] Even in a 1998 letter to Patricia Rife he encouraged her, since her project about Meitner was over, to investigate the "life of Ida Noddack and how Germany would have won the war if physicists Hahn and Fermi had believed her insights into fission."[56]

A psychological remark is now in order, since we are not suggesting that Wheeler was obsessed with some sort of Philip K. Dick-like Nazi dystopia or idly fantasizing about some alternative timeline, in which Way and he had acceler-ated the understanding of fission. Here we are rather dealing with a collection of closely interrelated events that, at once, represented for Wheeler the most important highlight of the first phase of his career and the moment when his investigations as a scientist could have a real impact on world history, in the midst of the horror of World War II, in general, and, on a more personal level, in connection with the traumatic death of a close and beloved family member. Wheeler's brother Joseph ("Joe") was killed between Florence and Bologna towards the end of 1944, and Wheeler received a postcard from him afterward with the invitation to "hurry up" with the nuclear project, hoping that it would put an end to the war.[57] (It is not coincidental that this is how Wheeler's autobiography opens.) The few months between his brother's death and the actual end of the war, plus – as Wheeler saw it – all the other human lives that were lost in those final phases, always remained for him a huge "What if . . . ?," which in a sense became an "attractor" (or fixed point) for regrets, counterfactuals, and other events he would have liked to see, or to be more sagacious to promote.[58]

Wheeler, also via his "methodological" remarks, seemed to somehow reproach himself for the lack of that daring conservative attitude that he would later display and even celebrate in the figure of Curie or Meitner – or, at least, if not a full-fledged

[55] John A. Wheeler Papers, American Philosophical Society Library, Philadelphia, box 189. To be precise, the typescript is not dated, but, in its text, there is an element that identifies 1994 as *terminus post quem*; keeping in mind that Wheeler was 83, however, it is safe to assume that he was not going to act as an advisor much later than that.

[56] Wheeler to Rife, October 6, 1998, John A. Wheeler Papers, American Philosophical Society Library, Philadelphia, box 157.

[57] See the dedication of Wheeler (2000), where the story is also narrated (pp. 18–19).

[58] There are even accounts of how, late in life, Wheeler, speaking in public on topics with some association with nuclear fission and thus (in his mind) also with all that, burst into tears and had to interrupt himself. I wish to thank Guido Bacciagaluppi and Paul Halpern for a couple of such testimonies.

daring conservatism, a similar kind of attitude: that of taking seriously one's theoretical tools and bringing them to their extreme consequences before throwing them away or laying them down. The missed leap from Way's results to fission may thus have become a matter of regret in hindsight, possibly also due to the personal feeling of having been almost touched in advance by history. Around all of this, at the end of the century, when Wheeler was approaching his nineties, there seems to be that soft light that shrouds distant memories when someone looks back over their long life. For sure, some contours are blurred and unclear, there are parallax effects, some spots are no longer in plain sight, but all of that, taken together, is definitely a lived experience, not a contrived and fanciful construction.

8.5 Conclusion

Within Wheeler's rich and eventful recollections, it is not irrelevant, neither factually nor symbolically, that he paid tribute in the form we have seen to Way's work. It would be easy to take his words at face value, repeat them authoritatively, give in to the temptation of sensationalism, and present Way as the "quasi-discoverer" of nuclear fission – perhaps another of the women who contributed to its understanding but who went unheard and fell into oblivion, at least temporarily. By weighing Wheeler's late claim, and by realizing that it needs to be tempered both due to the nature of his regret and to his peculiar way of expressing himself and using the past,[59] one may arguably recognize the description he made in the 1978 nomination of Way for the Bonner Prize as the most accurate and sober: "She discovered that a heavy nucleus had no stable configuration if it rotated too fast, precursor – if we had only known it – of nuclear fission."[60] All this, however, certainly does not diminish the value of Way's scientific activities and, as we have seen in that same nomination letter, the issue of fission is, after all, rather marginal. Perhaps more importantly, we realize that this was *Wheeler's* viewpoint: there does not seem to be, on the contrary, the slightest hint (even if the paucity of sources cannot be conclusive) that Way ever had similar regrets or looked at herself in that light – with good reason, if we look dispassionately at her published work. The

---

[59] Another regret, perhaps with some sort of *coquetterie* but also with a core of seriousness, was that, because in those days Bohr had to work with him on fission, Wheeler had in a sense deprived Einstein of the possibility of going on with his disputes with Bohr, and thus deprived the world of what could have come out of that discussion between his two revered mentors. See, for example, Wheeler (2006, p. 30): "But the news of nuclear fission, which Bohr received shortly before boarding his ship, changed the agenda. I like to say that I feel guilty for my role in preventing these discussions from taking place in anything like the extended and leisurely form that Bohr and Einstein had hoped for." In a sense, Wheeler would try to dialectically develop their conflicting stances in his own mind for the second half of the century.

[60] Hopefully, after this discussion of the sources, reference to Way's work, at least in the historical and philosophical literature, will not be made depending nonchalantly on a single paragraph in Mehra and Rechenberg's *The Historical Development of Quantum Theory* (Mehra and Rechenberg, 2001, pp. 990–991), which in turn depends (for that) on Wheeler's contribution to the 1977 Minnesota symposium.

Katharine Way who was "dear to the community of nuclear physicists" found another path, which, as we have sketched in this chapter, involved service for wider groups of people. There admittedly remain various gaps and open questions in her biography, but this contribution hopes to have provided some critical insights, to have drawn attention to unpublished archival material, and to stimulate further research on her.

References

Aaserud, Finn and Heilbron, John L. 2013. *Love, Literature, and the Quantum Atom: Niels Bohr's 1913 Trilogy Revisited.* Oxford: Oxford University Press.

Blum, Alexander S. and Brill, Dieter R. 2020. Tokyo Wheeler or the epistemic preconditions of the renaissance of relativity. In Blum, A.S., Lalli, R. and Renn, J. (eds.), *The Renaissance of General Relativity in Context.* Boston, MA: Birkhäuser, pp. 141–188.

Blum, Alexander S. and Furlan, Stefano 2022. How John Wheeler lost his faith in the law. In Ben-Menahem, Y. (ed.), *Rethinking the Concept of Laws of Nature.* Berlin: Springer, pp. 283–322. https://doi.org/10.1007/978-3-030-96775-8_11

Fermi, Laura 1939. *Thomas Carlyle.* Milan: Casa Editrice Giuseppe Principato.

Frisch, Otto and Wheeler, John A. 1967. The discovery of fission. *Physics Today*, 20(11), 43–52. https://doi.org/10.1063/1.3034021

Furlan, Stefano 2020. Einstein's mantle, Bohr's shadow: glimpses from Wheeler's Relativity Notebook III. In La Rana, A. and Rossi, P. (eds.), *Proceedings of the 39th SISFA Conference.* Pisa: Pisa University Press, pp. 221–227. https://doi.org/10.12871/978883339402233

Furlan, Stefano 2022. Pursuitworthiness between daring conservatism and procrastination: Wheeler and the path towards black holes. *Studies in History and Philosophy of Science*, 96, 174–185. https://doi.org/10.1016/j.shpsa.2022.10.001

Furlan, Stefano 2024. The smile of Mnemosyne: John Wheeler between the history of science and arts. *Scientia*, II(1), 1–35. https://doi.org/10.61010/2974-9433-202301-004

Goeppert Mayer, Maria 1948. On closed shells in nuclei. *Physical Review*, 74, 235–239. https://doi.org/10.1103/PhysRev.74.1547

Howes, Ruth H. and Herzenberg, Caroline L. 1999. *Their Day in the Sun: Women of the Manhattan Project.* Philadelphia, PA: Temple University Press.

Kasner, Edward, Jonas, Erna, Way, Katharine, and Comenetz, George 1934. Centroidal polygons and groups. *Scripta Mathematica*, 2, 131–138.

Keyes, Martha. 1997. Way, Katharine. Contributions of 20th-Century Women to Physics, UCLA Library. http://cwp.library.ucla.edu/Phase2/Way,_Katharine@862427327.html

Libby, Leona M. 1979. *The Uranium People.* New York: Charles Scribner's Sons.

Magalhães Santos, Gildo. 2014. A tale of oblivion: Ida Noddack and the "universal abundance" of matter. *Notes & Records of the Royal Society*, 68(4), 373–389. https://doi.org/10.1098/rsnr.2014.0009

Martin, Murray, Gove, Norwood, Gove, Ruth, Raman, Subramanian, and Merzbacher, Eugen 1996. Katharine Way. *Physics Today*, 49(12), 75. https://doi.org/10.1063/1.881582

Massimi, Michela 2022. *Perspectival Realism.* Oxford: Oxford University Press.

Masters, Dexter and Way, Katharine (eds.). 1946. *One World or None: A Report to the Public on the Full Meaning of the Atomic Bomb.* New York: McGraw-Hill.

Masters, Dexter 1955. *The Accident*. London: Cassell & Co.

Mehra, Jagdish and Rechenberg, Helmut 2001. *The Historical Development of Quantum Theory, Volume 6.2*. New York: Springer.

Misner, Charles W. 2010. John Wheeler and the recertification of general relativity as true physics. In Ciufolini, Ignazio and Matzner, Richard A. (eds.), *John Archibald Wheeler and General Relativity*. Dordrecht: Springer, pp. 9–27.

Rife, Patricia 1999. *Lise Meitner and the Dawn of the Nuclear Age*. Boston, MA: Birkhäuser.

Schweber, Silvan S. 1990. The young John Clarke Slater and the development of quantum chemistry. *Historical Studies in the Physical and Biological Sciences*, 20(2), 339–406. https://doi.org/10.2307/27757647

Schweber, Silvan S. 2008. *Einstein and Oppenheimer*. Cambridge, MA: Harvard University Press.

Smith, Alice K. 1965. *A Peril and a Hope: The Scientists' Movement in America, 1945–47*. Chicago, IL: Chicago University Press.

Stuewer, Roger H. (ed.). 1979. *Nuclear Physics in Retrospect*. Minneapolis, MN: University of Minnesota Press. https://doi.org/10.2307/27757647

Susskind, Leonard 2008. *The Black Hole War*. Boston, MA: Little, Brown, and Company.

Teller, Edward and Wheeler, John A. 1938. On the rotation of the atomic nucleus. *Physical Review*, 53, 778–789. https://doi.org/10.1103/PhysRev.53.778

Traweek, Sharon 1988. *Beamtimes and Lifetimes*. Cambridge, MA: Harvard University Press.

Tuli, Jagdish K. 2008. The Office of Nonproliferation & National Security. Talk at the workshop "Nuclear Structure and Decay Data: Theory and Evaluation," Trieste, 28 April–9 May. https://www.nds.iaea.org/workshops/smr1939/Lectures/J.K.Tuli/Lecture-JKT-Mod-1.pdf

Ware, Susan and Braukman, Stacy L. 2004. *Notable American Women: A Biographical Dictionary Completing the Twentieth Century*. Cambridge, MA: Belknap Press.

Way, Katharine 1937. Photoelectric cross section of the deuteron. *Physical Review*, 51(7), 552–556. https://doi.org/10.1103/PhysRev.51.552

Way, Katharine 1939. The liquid-drop model and nuclear moments. *Physical Review*, 55(10), 963–965. https://doi.org/10.1103/PhysRev.55.963

Way, Katharine (ed.). 1973. *Atomic and Nuclear Data Reprints* (2 vols.). New York: Academic Press.

Way, Katharine and Wigner, Eugene P. 1948. The rate of decay of fission products. *Physical Review*, 73(11), 1318–1330. https://doi.org/10.1103/PhysRev.73.1318

Weart, Spencer R. 1983. The discovery of fission and a nuclear physics paradigm. In Shea, William R. (ed.), *Otto Hahn and the Rise of Nuclear Physics*. Dordrecht: Reidel Publishing Company, pp. 91–133.

Weyl, Hermann 1981 [1935]. Memorial address. In Dick, A. (ed.), *Emmy Noether 1882–1935*. Boston, MA: Birkhäuser, pp. 112–152.

Wheeler, John A. 1937. Molecular viewpoints in nuclear structure. *Physical Review*, 52(11), 1083–1106.

Wheeler, John A. 1968. Maria Skłodowska Curie: Copernicus of the world of the small. *Science New Series*, 160(3833), 1197–1200.

Wheeler, John A. 1974. The universe as home for man. *American Scientist*, 62(6), 683–691.

Wheeler, John A. 1979. Some men and moments in the history of nuclear physics: The interplay of colleagues and motivations. In Stuewer, R. (ed.), *Nuclear Physics in Retrospect: Proceedings of a Symposium on the 1930s*. Minneapolis, MN: University of Minnesota Press, pp. 217–284.

Wheeler, John A. 1980. Einstein and other seekers of the larger view. In Goldsmith, M. (ed.), *Einstein: The First Hundred Years*. Oxford: Pergamon Press, pp. 183–193.

Wheeler, John A. 1988. Interview of John A. Wheeler by Finn Aaserud on May 4, 1988. Niels Bohr Library & Archives. https://www.aip.org/history-programs/niels-bohr-library/oral-histories/5063-1

Wheeler, John A. 1994. *At Home in the Universe*. College Park, MD: American Institute of Physics Publishing LLC.

Wheeler, John A. (with Ford, K.W.) 2000. *Geons, Black Holes, and Quantum Foam: A Life in Physics*. New York: W. W. Norton & Co.

Wheeler, John A. 2006. Mentor and sounding board. In Brockman, J. (ed.), *My Einstein*. New York: Pantheon Books, pp. 27–38.

9

Sonja Ashauer from São Paulo to Cambridge: A Journey to Quantum Electrodynamics

BARBRA MIGUELE AND IVÃ GURGEL

9.1 Introduction

Sonja Ashauer (1923–1948) was one of the first Brazilian women to work in theoretical physics and is understood to be the first Brazilian woman to earn a PhD in physics. Her pioneering trajectory has been the primary focus in the few secondary sources that exist about her, such as in the books, *Pioneiras da ciência no Brasil* (Melo and Rodrigues, 2006) and *Mulheres na física: casos históricos, panoramas e perspectivas* (Dantes and Chassot, 2015). This chapter extends beyond these sources by contextualizing her trajectory within the history of the early participation of Brazilian physicists in international collaborative research in quantum theory and nuclear physics.

Ashauer majored in physical sciences from the Faculdade de Filosofia, Ciências e Letras (FFCL) of the University of São Paulo (USP), both of which had been founded only six years prior to her starting her studies. At FFCL-USP, she joined the group of young researchers led by the Italian–Russian physicist Gleb Wataghin (1899–1986). That group included many enthusiastic young physicists, who stood out for their contributions to cosmic-ray physics and their lively discussions on cutting-edge topics in quantum theory. The bond between the members was so strong that Ashauer maintained ties with many of them even during her PhD years.

She attended the University of Cambridge in 1945–1947 and obtained a PhD under the supervision of Paul Dirac. There, she specifically worked on the problem of reformulating the equations of classical electrodynamics for a point particle interacting with its own field, as Dirac had first proposed in 1938.

In this chapter, we explore the connections between her experiences at FFCL-USP and Cambridge, particularly emphasizing the origins of her deep-seated interest in quantum electrodynamics and in the problems involving electrons interacting with their own field. Section 9.2 presents a brief overview of Ashauer's early life and family upbringing. Section 9.3 delves into her academic pursuits at FFCL-USP. It is subdivided into two subsections, the first providing a historical context for FFCL and its physics department before Ashauer's arrival, and the second exploring her scientific training and initial forays into theoretical physics. Moving forward, Section 9.4 is dedicated to her move to Cambridge and her relationship with Paul Dirac, as well as the challenges she encountered along the way. In Section 9.5, we explore Ashauer's theoretical research on quantum electrodynamics problems, first examining the scientific context, then reporting some early encounters with these problems that happened while she was still in the FFCL physics department before her move to Cambridge, and finally analyzing her own theoretical contributions. We end this section with her return to Brazil and her final year of life and conclude the chapter with some final remarks in Section 9.6.

9.2 Ashauer's Early Life

Sonja Ashauer was born in São Paulo, Brazil, on April 9, 1923, and died on August 21, 1948, at the age of 25, six months after defending her PhD thesis. She was the daughter of two German immigrants, Walter and Herta Ashauer.[1] Her father, an engineer specializing in the energy supply sector, contributed to the planning and construction of small hydroelectric plants in São Paulo; her mother worked as a housewife. Ashauer's sole sibling, Nils Ashauer (1937–2015), pursued a career in engineering, like their father.

The Ashauer family enjoyed a high socio-economic status. They lived in an upscale neighborhood of the city of São Paulo, Brooklin Paulista, in a house designed by Walter Ashauer himself. According to Nils Ashauer's recollections, their father's design included a laboratory where his children could learn science (see Figure 9.1).[2] It was fully equipped with a fume hood and essential tools, such as test tubes, scales, and other instruments.

In that laboratory, Sonja Ashauer conducted various experiments alongside her father, mainly using teaching kits he imported from Germany. They engaged in didactic experiments, including the hydrolysis of water. The room's many drawers were also used to store collections of natural artifacts like stones, woods, and leaves, which the young Sonja enjoyed gathering.

But the laboratory was not solely for experiments. It was also the room where Ashauer would often isolate herself for studying. Academic pursuits held significant value within the family, and hence, when she was studying, everyone in the

---

[1] The reasons why Ashauer's parents migrated to Brazil remain unknown.

[2] On May 8, 2015, Nils Ashauer was interviewed by Maria Amélia M. Dantes for the book, *Mulheres na física: casos históricos, panoramas e perspectivas*, which contains a chapter about Sonja Ashauer written by Dantes in collaboration with Walkiria Chassot (Dantes and Chassot, 2015). We were given access to the transcription of this interview by Chassot, but it was never published.

Figure 9.1 Sonja Ashauer studying in the laboratory of the Ashauer family's Brooklin house, c. 1940. Nils Ashauer's personal collection. Reproduced with permission.

house was expected to respect her focus. In fact, one of the few memories her brother retained of her was that he was forbidden from making noise when she went to the laboratory.[3]

Family encouragement for science was complemented with a good formal education. Ashauer did her primary schooling at the Vila Mariana German School

---

[3] When Sonja Ashauer died, Nils Ashauer was just 11 years old. Coexistence between the two siblings was limited, given that she was 14 years older. Throughout Nils' childhood, she was not a very present figure, as she was starting her career in physics, and was therefore living a routine of dedication to her studies and research. Added to this, she spent more than three years away in England. As a result, Nils did not have many memories of his sister.

(renamed Benjamin Constant School after World War II), and her secondary schooling at the Gymnásio do Estado de São Paulo, one of the oldest and most traditional schools in the city, founded in 1894. There, in addition to gaining a solid education in physics, chemistry, and mathematics, she learned English and French, thus making her quadrilingual (with Portuguese and German fluency).

In February and March 1940, at the age of 16, Ashauer took the exams to enter the physics course at FFCL-USP. She studied there for three years until she obtained her physics degree, and for one year more in 1943, to obtain a supplementary didactics degree.[4] But, instead of pursuing a career as a teacher in secondary schools, Ashauer joined the research group led by the Italian–Russian physicist Gleb Wataghin in FFCL physics department and then started her path towards theoretical physics research. The following sections more specifically explore the faculty and Wataghin's group.

9.3 Training at FFCL-USP

9.3.1 A New Place for Physics in Brazil: The Creation of FFCL-USP

USP was founded in 1934, after more than a decade of political engagement by the *O Estado* group, a collective of liberal intellectuals, journalists, and politicians linked to the newspaper *O Estado de São Paulo* – a very influential media organization in Brazil to this day – advocating for national political renewal through education. Their aim was to dispute the hegemony of the prevailing oligarchies that held political power nationwide and cultivate a new ruling elite grounded in science and intellectuality (Cardoso, 1982). Following political upheavals in Brazil in the early 1930s, including the ascent of Getúlio Vargas to power and the Constitutionalist Revolution of 1932 – a thwarted insurgency led by liberals from São Paulo who sought alternative means of regaining political influence – the *O Estado* group seized the opportunity to establish the university.

In 1933, seeking to foster ties with the state of São Paulo, Vargas appointed the newspaper's director, Armando Salles de Oliveira, as governor, and then USP began taking shape by absorbing existing professional and higher education institutes that already existed within the state of São Paulo, such as the Polytechnic School of Engineering, the Faculty of Medicine of São Paulo, the Law School of São Paulo, etc. In addition, a novel faculty was created, the Faculty of Philosophy, Sciences, and Letters, known under its Portuguese acronym as FFCL. This faculty was designed to be a central nucleus for USP (Schwartzman, 2001, ch. 5, p. 23), exclusively dedicated to the production and reproduction of "high culture and disinterested science" (Cardoso, 1982, p. 123).

---

[4] In order to enter a teaching career after graduating, FFCL students were required to complete an additional year of a didactics course. This was a path commonly taken by women who graduated from that institution.

To establish FFCL, the founders of USP deemed it crucial for its teaching staff to comprise foreign professors recruited through international "missions" to European universities (Wataghin, 1992; Petitjean, 1996), who could lead frontier scientific research programs and provide a Europe-inflected intellectual training.

For the physics department came the Italian–Russian theoretical physicist Gleb Wataghin, from the University of Turin, Italy. At first, Wataghin had no interest in coming to Brazil. In 1934, his career was still gaining momentum, and he feared that emigrating would isolate him from the European community. However, he was eventually convinced by his Italian colleagues. Wataghin was Russian[5] by birth, and this weighed heavily on his decision, given the fascist regime in Italy, as he later recalled (Wataghin, 2010, p. 11; author's translation from Portuguese):

There was fascism [in Italy]; I couldn't stay there. They also made me understand that it was difficult for me to get a full professorship in Italy [because he was not born Italian]. That it would be better to accept an offer, which was a generous one – at that time I received a salary of three thousand *réis*, which was a good salary. Then I went.

Wataghin only had a few years of research experience when he arrived at FFCL-USP, but he was already well acquainted with the community of theoretical physicists in Europe and had a solid scientific background (Videira and Bustamante, 1993, p. 269). Between the end of the 1920s and the beginning of the 1930s, he traveled to different parts of Europe and took part in the 1927 Volta Conference, held in Como, Italy, where he met Niels Bohr, Werner Heisenberg, Walter Heitler, Wolfgang Pauli, and Paul Dirac (Wataghin, 2010, p. 1).

Although Wataghin was mainly a theoretical physicist, at USP he was put in charge of creating the *complete* physics course. He was responsible for the chairs of general and experimental physics, and his mission was to create an experimental laboratory for the physics department (Wataghin, 2010, p. 12). Later, he also created the chair of theoretical and mathematical physics and the chair of celestial mechanics, which he took on as well.

During FFCL's first two years, not much research took place in the physics department. Wataghin was busy organizing the curriculum and teaching the very few physics and mathematics students, along with those from the Polytechnic School of Engineering. He nevertheless concurrently published several theoretical physics works on the properties of elementary particles (Wataghin, 1935a),

---

[5] Gleb Wataghin's nationality is sometimes given as Italian–Russian or as Italian–Ukrainian. He was born in the city of Birsula, on the outskirts of Cherson, Ukraine, lived and studied in Kiev, Ukraine, and in 1919 – still a teenager – he moved with his family to Italy where he acquired Italian citizenship. The matter of Wataghin's Ukrainian nationality is somewhat controversial. First, the Ukrainian territory in which he was born was part of the former Russian Empire. Second, he was the heir of noble Russians who strongly identified as Orthodox Russians (Silva, 2020, p. 210). Additionally, many of his contract documents in Brazil identified him as Russian (not Ukrainian), particularly during the war years when he could not claim his Italian nationality. Here, we therefore elect to refer to Wataghin's nationality as Italian–Russian.

Figure 9.2 Team of employees and professors from the Department of Physics at FFCL-USP, in the late 1930s. From left to right: Roberto Xavier de Oliveira (assistant professor), Maria (caretaker), Giuseppe Occhialini (visiting professor), Marcello Damy de Souza Santos (assistant professor), José (caretaker), Yolande Monteux (assistant professor), Abrahão de Moraes (assistant professor), Mario Schenberg (assistant professor), Gleb Wataghin (professor), and Francisco Bentivoglio Guidolin (laboratory technician). Historical Collection of the USP Physics Institute.

quantum electrodynamics theory (Wataghin, 1934a, 1934b, 1934c, 1935b, 1936a), and gravitational effects in quantum theory (Wataghin, 1936b). The research dynamics, however, underwent a significant shift in 1936, when Wataghin gained his first collaborators.

The first physicists to graduate from FFCL-USP were Marcelo Damy de Souza Santos, Mário Schenberg, and Paulus Aulus Pompéia. Damy[6] was the only one to graduate from the physics course, in 1936; Schenberg graduated the same year, but with a degree in mathematics (USP, 1938). Pompéia, by contrast, graduated as an electrical engineer from the Polytechnic School in 1935 and initiated collaborations with Wataghin and the physics department before subsequently earning a physics degree from FFCL (USP, 1939). In 1937, another graduate from the physics course joined the team, the French-born Brazilian Yolande Monteux, the first woman to graduate from a physics course in Brazil (USP, 1938) (Figure 9.2).

---

[6] Although Marcelo Damy de Souza Santos' last surname was Santos, he became mostly known as Damy.

In addition to these Brazilian students, in July 1937 another Italian physicist joined the department, assuming the chair of General and Experimental Physics: Giuseppe Occhialini. Invited by Wataghin, Occhialini came to Brazil, seeking refuge from the impending World War II and Italy's totalitarian regime due to his anti-fascist stance (Ribeiro de Andrade, 2006, p. 51; Gariboldi and Verzeroli, 2021, p. 134). At that time, he was already a well-recognized experimental physicist, especially thanks to his work in the early 1930s with Patrick Blackett's team at the Cavendish Laboratory in Cambridge, which led to the detection of the positron in cloud chambers.

Wataghin and Occhialini built an experimental laboratory focused mainly on cosmic-ray physics, which was still a relatively new branch of physics at that point. In addition, this research topic was an affordable choice for the physics department, as it did not require the use of expensive equipment and could be done using little laboratory space. Cosmic rays are indeed a natural and free source of elementary particles, and most of the instruments needed to study them could be simply crafted by the researchers themselves (Videira and Bustamante, 1993, p. 275; Wataghin, 2010, pp. 12–13). Only a few more complicated devices, such as the cloud chamber, required greater expertise and higher costs, but those still fit within the department's limited budget.

The research carried out by Wataghin and Occhiallini together with their Brazilian disciples produced significant results by the end of the 1930s. In 1939, Wataghin, Damy, and Pompéia detected "penetrating showers" of particles, a cascading effect of cosmic radiation, which occur when a primary cosmic ray interacts with particles in the Earth's atmosphere, thus leading to the creation of secondary particles in a cascading effect known as "showers."

This achievement led to two publications in *Physical Review*, one in 1940 (Wataghin et al., 1940) and the other in 1941 (Santos et al., 1941), and attracted the attention of foreign researchers, such as Arthur Compton. In 1941, Compton undertook a scientific and diplomatic expedition to South America,[7] during which he visited the physics department and even carried out research collaborations with the São Paulo group.[8] Compton's visit also resulted in the first Symposium on Cosmic Rays held in Brazil in 1941.[9]

This episode, like others that span the first years of the physics department at FFCL, sheds can shed light on another aspect of the academic unit: its international

---

[7] Freire and Silva (2014, 2019) argue that this expedition, financed by Nelson Rockefeller's Office of the Coordinator of Inter-American Affairs (OCIAA), did not occur solely for scientific purposes, but was also linked to US interests in expanding its cultural influence in South American countries during World War II, and hence to undermine the possibilities of them joining the Axis.

[8] Text about the University of Chicago scientific expedition that brought Arthur Compton to South America to study cosmic radiation. Historical Collection of the USP Physics Institute (http://acervo.if.usp.br/), IF-DF-VI -02–23–0000–00231–0.

[9] *Annals of the Symposium on Cosmic Rays*, published by the Brazilian Academy of Sciences. The symposium took place at the University of Brazil in Rio de Janeiro in 1941. Historical Collection of the USP Physics Institute (http://acervo.if.usp.br/), IF-DF-VI-02–26–0000–00261–0.

collaborations with physicists and institutions worldwide. In the historiography of the physics department, Wataghin is often credited as the person responsible for shaping that international aspect, a role emphasized in the works of Tavares et al. (2020a) and Silva (2020). While he remained in Brazil, Wataghin indeed actively maintained ties with scientists across different nations, enabling him to stay updated on the latest developments in physics. Moreover, Wataghin managed to get his students, after graduating, to visit institutions in Europe and the US, to do research internships and/ or postgraduate studies.

Establishing and maintaining networks with political and scientific figures from all over the world was a crucial part of Wataghin's scientific practice. For Silva (2020), it was a particular tactic of the intellectual, a necessary way of ensuring the advancement of his scientific work, and of the institutions and people with which he was associated. For Tavares, Bagdonas, and Videira (Tavares et al., 2020a), it was a constitutive element of Wataghin's identity and an ideal he strove to attain through concrete actions – such as international collaborations and exchanges – and to pass on to his students.

Either way, the fact is that Wataghin forged and sustained connections with scientists and scientific institutions worldwide, allowing him to keep up with state-of-the-art physics internationally while in Brazil, but also facilitating collaborative endeavors with his FFCL students. These networks not only enabled Wataghin to remain at the forefront of quantum physics but also to develop cutting-edge research programs for his students. In the same sense, it paved the way for some graduates to carry out exchanges with institutions in Europe and the US.

Damy went for postgraduate studies with William Lawrence Bragg at Cambridge University in 1938. Pompéia went to work with Arthur H. Compton at the University of Chicago in 1940. Schenberg was sent by Wataghin to Europe in 1938 to work with Paul Dirac in Cambridge. Schenberg, however, did not get to work with Dirac. He first went to Italy, where he was introduced to Enrico Fermi, and decided to work with him at the University of Rome instead of carrying on to England (Schönberg, 2010, p. 4; Coelho, 2020, p. 66). Schenberg also worked with Wolfgang Pauli at the University of Zurich and with Frédéric Joliot-Curie at the Collège de France in 1938–1939, before returning to Brazil. In the late 1940s, César Lattes was at the forefront of experiments carried out at Bristol that led to the discovery of the π meson. Ashauer followed a similar path as those physicists.

9.3.2 Ashauer at the Physics Department of FFCL

When Ashauer joined the department in 1940, an internationalist atmosphere had already set in. Wataghin, Occhialini, Schenberg, and Damy were teaching the physics classes. The latter two had just returned from Europe and resumed their activities as assistant professors (USP, 1953).

The physics department was a small space, with few students and few professors. The physics course, with two or three students per year, was the FFCL unit with the lowest enrollment (see USP, 1936, 1938, 1939, 1942), which resulted in a great closeness between participants and a constant sharing of ideas. It was in this atmosphere of closeness with professors and fellow research students that Ashauer started her training.

During her first two years at USP, Ashauer attended Damy's course in general and experimental physics, which spanned the initial four semesters of the physics degree. In 1940, the syllabus for the first year covered topics in metrology, mechanics, acoustics, and optics.[10] The course syllabus for the second year, in 1941, encompassed subjects like thermology, electricity and magnetism, and oscillations and waves.[11] During her final year, it was also under Damy's guidance that Ashauer enrolled in a "higher physics" course, the extension of general and experimental physics. Its syllabus included discussions on discharges in gasses and nuclear and atomic physics. Additionally, the course provided an introduction to cosmic-ray physics.[12]

Schenberg was Ashauer's professor in vector calculus during the first semester of 1940. Although the vector calculus course extended for a year, Schenberg only instructed it for one semester in 1940 since he planned to travel to Europe in the latter half of the year. Wataghin took over the instruction of Ashauer's class during the second semester. Additionally, Wataghin taught her rational mechanics, covering advanced mechanics topics, as well as mathematical and theoretical physics.[13] In that last course, Ashauer was introduced to differential equations in mathematical physics, covariance and contravariance, tensors, Maxwell's equations, relativity theory, and spinors and the Dirac equation.[14]

Ashauer distinguished herself as one of the top students in her class across all these courses, consistently earning the highest grades in most subjects, except for higher physics, where she still obtained a commendable score of 9/10.[15] In a 1942 statement, Damy reported that Ashauer stood out as one of the best students in his

---

[10] General and Experimental Physics program for the first-year class, from 1940. Historical Collection of the USP Physics Institute (http://acervo.if.usp.br/), IF-DF-III-01–11–0000–01114–0.

[11] General and Experimental Physics program for the second-year class, from 1941. Historical Collection of the USP Physics Institute (http://acervo.if.usp.br/), IF-DF-III-01–11–0000–01117–0.

[12] Higher Physics program for the third-year class, from 1941. Historical Collection of the USP Physics Institute (http://acervo.if.usp.br/), IF-DF-III-01–11–0000–01116–0.

[13] Rational Mechanics program for the second-year class, from 1941. Historical Collection of the USP Physics Institute (http://acervo.if.usp.br/), IF-DF-III-01–11–0000–01118–0

[14] Mathematical Physics program for the third-year class, from 1942. Historical Collection of the USP Physics Institute (http://acervo.if.usp.br/), IF-DF-III-01–11–0000–01119–0.

[15] Sonja Ashauer's undergraduate transcript. Available in Sonja Ashauer's Graduation Record in the IFUSP Academic Assistance Archive.

classes. For him, she always "showed a remarkably good understanding of physics problems" and "demonstrated complete precision and understanding of the things she was doing."[16]

Ashauer completed her Bachelor's degree in physics at the end of 1942, and she then remained in the physics department during 1943–1944, first, to spend an additional year taking the FFCL supplementary didactics course, and second, to undertake some research projects with her professors. Many FFCL students took the supplementary didactics course, especially female students, as there was a government incentive for college graduates to be hired as teachers in São Paulo state schools (Trigo, 1997, p. 68; Blay and Lang, 2004, pp. 51–52). Although Ashauer took this supplementary course, she never showed interest in being a schoolteacher, to the extent that her grades in the didactics course were lower than her grades in the physics courses.[17]

In a statement to the British Council's Research Studies Board, Wataghin wrote that Ashauer would be working with him from January 1943 until September 1944 on theoretical research into "various problems of quantum mechanics, in particular on the statistics of nuclei and elementary particles at extremely high temperatures."[18] In 1944, she indeed published a paper titled "On the quantum theory of the absorption coefficient" (Ashauer, 1944) in the *Annals of the Brazilian Academy of Sciences*, in which she used perturbation theory to calculate the absorption coefficients of an electron bound to an atomic nucleus after undergoing the photoelectric effect. This introductory work in her studies of quantum mechanics could be seen as a student's approach. Ashauer examined two types of state transitions, considering processes where the final state of the electron belongs to the continuous spectrum of energy eigenvalues relative to the nucleus's field: *bound–free* transitions, where the electron moves from an initial discrete energy state to a final state that belongs to a continuum spectrum, and *free–free* transitions, in which both the initial and the final state of the electron belong to the continuous spectrum.

In 1944, Ashauer was hired as Wataghin's first assistant to the Chair of Theoretical Physics and Mathematical Physics, a post she kept until her untimely death in August 1948 (USP, 1953). Working as a chair assistant was a fundamental part of the training of specialists graduating from FFCL, as this position enabled them to enter careers as researchers and academics. In the physics department,

---

[16] Statement by Marcello Damy about Sonja Ashauer written around 1943. Historical Collection of the USP Physics Institute (http://acervo.if.usp.br/), IF-DF-III-01–14–0000–00141–0.

[17] While in her physics course, Ashauer consistently received grades of 9 and 10; during her didactics course, her grades ranged from 5 to 8. This information is sourced from Sonja Ashauer's didactics course certificate, available in Sonja Ashauer's Graduation Record in the IFUSP Academic Assistance Archive.

[18] Statement by Gleb Wataghin to The Board of Research Studies of the British Council, March 18, 1946. Historical Collection of the USP Physics Institute (http://acervo.if.usp.br/), IF-DF-III-01–14–0000–00142–0.

occupying such a position meant not only the opportunity of being paid to conduct physics research, but also of joining Wataghin's research group.

At the beginning of her work as a chair assistant, Ashauer ventured into different studies linked to quantum theory and nuclear physics. Even though she was already a graduate, she was still in a training phase and needed to expand and deepen her knowledge of contemporary physics. Ever since Ashauer completed her Bachelor's degree, she and Wataghin had been expressing interest in her pursuing postgraduate studies in England. Her mentor therefore applied for a British Council's Research Studies Board postgraduate scholarship on her behalf.

While Ashauer worked as Wataghin's assistant, the work routine of the physics department underwent significant changes. In August 1942, Brazil entered World War II on the side of the Allies, which resulted in Wataghin, an Italian citizen, being replaced by Damy as department head. Although Wataghin remained a USP professor, he was unable to lead activities.

In 1943, also as a result of the country's entry into the war, an agreement was reached between the Brazilian Navy and the physics department to develop sonar for submarine detection. This project was led by Damy and had a significant impact on the department's activities. Some areas were closed for the exclusive use of Brazilian national defense research, in which Wataghin and other researchers did not participate.

Without certain laboratory resources at their disposal, experimental cosmic-ray research became constrained relative to what it had been (Tavares, 2017, p. 65; Silva, 2020, pp. 212–213). But those researchers not involved in Brazilian national defense efforts could still engage in theoretical endeavors. Within this group, among the professors, we find Wataghin and Schenberg – who at the time had just returned from his research trip in Europe. Occhialini was no longer present. Since he was Italian through and through – unlike Wataghin, who held dual nationality – Occhialini's visiting professor position was under threat as he was perceived to be a war enemy (Bustamante, 2024, p. 154; Ribeiro de Andrade, 2024, p. 157). He was forced to leave the department and took refuge in the Itatiaia National Park, on the border between the states of Rio de Janeiro and Minas Gerais, posing as a tourist guide.

Ashauer, Lattes, and José Leite Lopes formed the younger generation. Lattes, who joined the physics department a year after Ashauer, completed his physics degree in 1943. To this day, he is probably the most famous Brazilian physicist, thanks to his contributions to enhancing the nuclear emulsion method for particle detection with Cecil Powell's team at the University of Bristol around 1946–1947, which led to the detection of the π meson (Vieira and Videira, 2014; Tavares, 2017; Tavares et al., 2020b). Leite Lopes, a recent graduate from the Faculdade Nacional de Filosofia, came to the FFCL physics department on a scholarship to pursue an "improvement in physical sciences" course[19] with Schenberg and Wataghin. At that time, Ashauer had already graduated and had begun collaborating with Wataghin.[20]

Records of the scientific work conducted between 1942 to 1944 are scarce. International research trips did not occur during that period in the same way they had previously, resuming after the war in 1946. The exception was Ashauer, who went to Cambridge at the end of 1944. During this period, the department engaged in specific theoretical studies on quantum theory and nuclear physics, albeit with some constraints due to the limited circulation of physics publications. The US government, in particular, placed strict censorship on scientific journals outside the country in order to prevent technical information on nuclear fission from reaching enemy hands.[21]

As will be further discussed in Section 9.5, among the potential research avenues within the department during these years were topics related to theoretical and mathematical difficulties in quantum electrodynamics. This subject would later become a central interest of Ashauer, forming the basis of her PhD thesis under the guidance of Paul Dirac at the University of Cambridge.

9.4 Life in Cambridge

If at FFCL Ashauer learned to do research in close contact with professors and fellow research students, at Cambridge her experience was markedly different. Dirac, who was already a renowned theoretical physicist, showed no interest in constructing a school around himself or in training future generations.

When Ashauer arrived in England for her PhD, World War II had not yet ended. At the University of Cambridge and other British research centers, many researchers were still displaced from their usual activities to take part in the national war effort. Consequently, only a limited number of researchers were available, engaged in research unrelated to political and military interests.

In a letter sent to Wataghin shortly after settling in England,[22] Ashauer reported that there were only two theoretical physicists available at the university to teach classes and carry out mentoring work that year: Paul Dirac[23] and Allan Harris

---

[19] Report by Gleb Wataghin to the *Zerrenner Foundation*, on the academic activities of José Leite Lopes from April to August 1943, Historical Collection of the USP Physics Institute (http://acervo.if.usp.br/), IF-DF-III -02–21–0000–00211–0.

[20] Declaration from Gleb Wataghin to the Board of Research Studies of the British Council, March 18, 1946, Historical Collection of the USP Physics Institute (http://acervo.if.usp.br/), IF-DF-III-01–14–0000–00142–0.

[21] Business correspondence from J. E. Blackburn Jr., November 15, 1943. Historical Collection of the USP Physics Institute (http://acervo.if.usp.br/), IF-DF-IV-03–00–0000–00319–0.

[22] Ashauer to Wataghin, March 17, 1945, Historical Collection of the USP Physics Institute (http://acervo .if.usp.br/), IF-DF-I-02–00–0000–02117–0.

[23] Although, at first glance, it might sound as if Dirac's presence in Cambridge could be explained by his possible non-participation in any research linked to World War II, it is worth pointing out that Dirac did play a part, albeit small, in the British war effort. At the beginning of the 1940s, he collaborated occasionally as a consultant for

Wilson. In that same letter, she reported that only a few students were doing research in theoretical and mathematical physics. The classes she attended and the weekly colloquia organized by Dirac were therefore both rather empty.

The scene of a vacant Cambridge did not solely appear in Ashauer's letters. This scenario finds resonance in the memoirs of contemporaneous PhD students whose research was also supervised by Dirac, such as those of Sri Lankan physicists Christie Jayaratnam Eliezer and Subrahmaniyan Shanmugadhasan, and of Indian mathematician Harish-Chandra (see Harish-Chandra, 1987; Shanmugadhasan, 1987).[24]

Among physicists and historians of physics, the peculiar personality of Dirac, an introverted, taciturn man, is well known. Farmelo's detailed and empathetic biography (Farmelo, 2009) even suggests that Dirac may have been positioned somewhere on the autistic spectrum, which could potentially account for what might be perceived as his "limited" social skills compared with his genius for physics and mathematics.

What can be asserted about Dirac's personality is that numerous individuals who had the opportunity to know him or attend his classes recall him as a highly reserved individual, deeply involved in his theoretical research endeavors, but with limited teaching skills and with minimal commitment to his academic duties other than pure research.

Regarding PhD research supervision, Dirac had a relatively small number of graduate students. Helge Kragh reported that during his more than 50-year academic career he officially supervised fewer than 15 PhD students, most of them after 1945 (Kragh, 2003).[25] Ashauer was the only woman.

Despite his apparent willingness to supervise PhD research during the war years, he had little interest in doing so, and often adopted strategies to evade requests for guidance (Dalitz and Peierls, 1986, p. 155; Harish-Chandra, 1987, p. 35; Shanmugadhasan, 1987, p. 49; Kragh, 1990, pp. 253–254). In the first decades of his career, Dirac avoided collaborating as a supervisor in the research of young theoretical physics students, possibly because he did not want to be credited for the success of their project (Dalitz and Peierls, 1986).

This attitude was no different with Ashauer. In an essay in memory of Dirac, Subrahmaniyan Shanmugadhasan recalls Dirac's dodgy demeanor when he was

---

the British Government's Tube Alloys project, for developing a nuclear weapon (Kragh, 1990, p. 157; Farmelo, 2009, p. 321). However, his contributions were scarce and only lasted a short time, since "Dirac's heart remained in quantum mechanics" (Farmelo, 2009, p. 323).

[24] Eliezer began his PhD studies in 1941 and defended his thesis in 1946. Shanmugadhasan started his PhD studies in 1944 and defended his thesis in 1947. Harish-Chandra entered Cambridge a few months after Sonja Ashauer, also in 1945, and defended his thesis in 1947.

[25] Among whom were: Paul Weiss, Andrew Lees, H. R. Hulme, Fred Hoyle, Christie Jayaratnam Eliezer, Subrahmaniyan Shanmugadhasan, Harish-Chandra, Dennis Sciama, Richard Eden, R. J. N. Philips, H. J. D. Cole, and Sonja Ashauer.

approached by him and other colleagues of his, including Ashauer (Shanmugadhasan, 1987, p. 49):

Dirac was usually very reluctant to supervise research students. [. . .] Miss Ashauer told me that at their first meeting Dirac informed her that he did not have a problem for her but that Wilson had one ready and so perhaps they should go and see Wilson. It seems he repeated it on a later occasion. But both times, in her innocence, she did not react to Dirac's suggestion. Only later, after she knew more about Dirac's ways, did it occur to her that Dirac was probably trying to hand her over to Wilson.

Ashauer was not exceptional in being rejected by Dirac the first time she approached him, but her insistence on working with him surely was.

Many who sought out Dirac to supervise their research or to start collaborations gave up because of this attitude and decided to look for someone else (Kragh, 1990). It is therefore surprising that Ashauer saw him completely differently, even though she had experienced similar rejection. In a letter to Wataghin shortly after arriving in Cambridge, she described Dirac as "[. . .] an excellent supervisor – very accessible, always ready to help and very patient,"[26] a very different version from that of her other colleagues.

Apart from his apparent refusal to take on new supervisions, Dirac also did not spend much effort on his work as supervisor. Harish-Chandra commented on his experience being supervised by Dirac (Harish-Chandra, 1987, p. 35):

[. . .] my personal contacts with Dirac were rather infrequent. I went to his lectures but soon dropped out when I discovered that they were almost the same as his book. However, I did attend the weekly colloquium run by him. I found that he was very gentle and kind and yet rather aloof and distant. Therefore, I felt that I should not bother him too much and went to see him only about once each term.

Criticism of Dirac's low level of attention for teaching extends beyond his role as supervisor of PhD research. Dirac is also recognized as a physicist with little aptitude for collaborative work. He did not create a school around himself, nor did he seek to influence future generations of Cambridge physicists (Kragh, 1990). The relationship he maintained with his doctoral students was quite different from the collaborative relationship Ashauer knew from FFCL.

Perhaps due to her background, Ashauer decided she would not passively accept Dirac's attitude and carry on with her research alone. Rather than succumbing easily to her supervisor's distancing efforts, she actively and frequently sought meetings with him to discuss her research. This reliance on supervision developed gradually. As noted in Shanmugadhasan's memoirs, Ashauer maintained a closer connection with Dirac, consistently seeking his guidance whenever she felt any blockage or difficulty in her research (Shanmugadhasan, 1987, p. 51).

---

[26] Ashauer to Wataghin, March 17, 1945. See Note 22.

The relationship that Ashauer sought to develop with Dirac is also revealed in some of her letters with Wataghin. In 1947, during the last year of her doctorate, she was frustrated by the absence of Dirac, who had gone on sabbatical leave with his family as visiting professor at the Institute for Advanced Study in Princeton, New Jersey, US (Kragh, 1990, p. 180; Farmelo, 2009, p. 342). Dirac had then left all his Cambridge PhD students in the care of Nicholas Kemmer.[27] For Ashauer, the shift in guidance was meaningful, since she had placed a lot of trust in the way Dirac evaluated her work. As she wrote to Wataghin:

A short time ago Prof. Dirac told me that he has again been invited to lecture in Princeton, for a whole year this time. He is leaving Cambridge sometime in the summer (referred to here [northern hemisphere], of course). I am very disappointed about this, since he has only recently returned from America. When he goes I shall have to be placed under the supervision of somebody else, who can't possibly be interested in the kind of work I am doing. I am not yet in a condition to proceed far without any guidance. The main problem in research is, as Prof. Dirac aptly remarked, to find a problem to work on. It is all somewhat disheartening.[28]

In response to this letter, Wataghin informed her that he was initiating theoretical work on astrophysical problems, such as the distribution of atomic nuclei in the universe. He then suggested that if she wished to return to Brazil without her diploma, there would be theoretical work for her to do there. For him, Ashauer did not need to "give excessive importance to the degree," because what mattered was not the title, but the work.[29] Ashauer, however, declined the offer and continued with her research.

Despite feeling distressed about not having defined a problem to work on, she found an intriguing subject to explore during 1947: the non-physical solutions to the Lorentz–Dirac equation. This equation was proposed by Dirac during a moment of crisis in quantum electrodynamics theory, a topic into which we delve deeper in Section 9.5.

It is important to note that Ashauer's letter exchange with Wataghin did not primarily serve to provide updates on her life and Dirac's supervision; these topics were secondary. Its main purpose was to share reprints, articles, and other materials which she found in Cambridge with the FFCL physics department.

All the letters indeed included additional materials, typically copies of reprints of Dirac's works or other literature related to research in quantum mechanics and cosmic-ray physics. In one of these letters, Ashauer even sent a copy of the 1939

---

[27] Nicholas Kemmer (1911–1998) was a Russian–British physicist working in Cambridge. Throughout the first half of the 1940s, he had been working with the war effort on the Tube Alloys project. For this reason, until the end of the war, he did not participate in the academic life of Cambridge, only returning at the end of 1945, when he was transferred to the mathematics department, where Dirac and his supervisors worked.

[28] Ashauer to Wataghin, March 29, 1947, Historical Collection of the USP Physics Institute (http://acervo.if .usp.br/), IF-DF-I-02–00–0000–02157–0.

[29] Wataghin to Ashauer, May 8, 1947, Historical Collection of the USP Physics Institute (http://acervo.if.usp.br/), IF-DF-I-02–00–0000–02164–0.

edition of Dirac's textbook, *The Principles of Quantum Mechanics*, with the intention of integrating this material into course planning at FFCL.[30]

Ashauer, along with other physicists from the FFCL physics department, saw her stay in Cambridge not merely as an individual opportunity to gain experience abroad but as a necessary means of ensuring that scientific ideas and contributions from other countries could reach the department.

For Ashauer, integrating the department into the global physics community required a reciprocal exchange. It did not suffice to have foreign scientific works circulating in Brazil: it was also imperative to facilitate the circulation of Brazilian productions abroad. A demonstration of her commitment to this approach can be found in the conversation she reported having with Lajos Jànossy during the School of Theoretical Physics at the Dublin Institute for Advanced Studies in July 1945. She attended this school during her Cambridge summer vacation and wrote about it to Wataghin the following August.[31] At the school, she participated in presentations by Dirac, Erwin Schrödinger, Max Born, W. H. Peng, and Jànossy. Ashauer meticulously described each presentation to Wataghin, as they delved into recent research debates in quantum mechanics and cosmic-ray physics.[32]

In her report on Jànossy's presentation, she mentioned his discussion of cosmic-ray research and penetrating showers, which, as we have seen, had been discovered at FFCL-USP in 1939–1940. She pointed out that he referred to Wataghin's publications on the subject and the experiments carried out in Brazil. At the end of the presentation, Ashauer felt compelled to approach Jànossy and offer offprints from the Brazilian group, despite being only 22 years old and one of only two women at the event. She wrote:

Since he found the publications to be very brief, he wanted to know if I was aware of the details and if I could give him reprints, to which I had to disappoint him. Perhaps you could send him reprints directly (Physical Laboratories, Manchester University)? I also gave him a copy of "Symposium" [1941 Symposium on Cosmic Rays], which he was interested in because he's writing a book on cosmic rays. The other of the two copies I brought to England is in the Philosophical Library. If it's not too much trouble to send one or two more copies by ordinary post, I think I could distribute them, for example, to the University Library.[33]

One can see Ashauer's confidence in the São Paulo group and her dedicated efforts to disseminate the work produced by her colleagues in the FFCL physics department.

---

[30] Ashauer to Wataghin, June 26, 1947, Historical Collection of the USP Physics Institute (http://acervo.if.usp.br/), IF-DF-I-02–00–0000–02174–0.
[31] Ashauer to Wataghin, August 19, 1945, Historical Collection of the USP Physics Institute (http://acervo.if.usp .br/), IF-DF-I-02–00–0000–02125–0.
[32] Ashauer to Wataghin, August 19, 1945. See Note 31.
[33] Ashauer to Wataghin, August 19, 1945 (translated by the author from Portuguese). See Note 31.

9.5 Ashauer's Cambridge PhD

9.5.1 Scientific Context: The Divergence in the Self-energy of the Electron

Since 1927, when, following the fifth Solvay Conference, quantum theory was presented in its then most complete form, the challenge faced by theoretical physicists had been to extend and generalize the theory, aiming in particular to make it compatible with special relativity and Maxwell's electromagnetism. These endeavors led to the development of a new theoretical framework, quantum electrodynamics.

However, this program was not straightforward. From the late 1920s through the early 1940s, as new techniques for quantizing matter fields were proposed, difficulties, anomalies, and unsatisfactory mathematical results were encountered along the way. One of these challenges includes what have become known as "divergence" or "infinity" problems, in which physical observables that were expected to take a finite value, such as the zero-point energy and the self-energy of the electron, diverged within the formalism of the new theory.

The divergence in the self-energy of the electron – meaning the electron's interactions with its own electromagnetic field – was already a problem within special relativity. In classical theory, the electron is typically modeled as a tiny sphere with a finite-size radius. However, this approach poses a problem in special relativity, as if one considers the electron with finite dimensions and, therefore, a defined internal structure, the shape of the electron could appear different to observers in different inertial reference frames, violating the principle of invariance of the laws of physics under Lorentz transformations. Furthermore, when considering the electron's radius to be zero, the electron's self-energy, which is equivalent to e^2/r, tends to infinity, implying infinite mass and energy, which is physically unreasonable.

This issue remained unsolved within the field of relativity, often lingering in the background (Pais, 1986, pp. 370–372). However, the emergence of quantum mechanics in the 1920s meant that the classical model of an electron with finite size could no longer serve. As the wave–particle duality from quantum mechanics suggests that electrons exhibit both wave-like and particle-like characteristics, the radius of the electron must be zero within this framework. Then, the problem with the electron's divergent self-energy could no longer be ignored or placed in the background, as had happened in classical theory.

During the early 1930s, numerous attempts were made to resolve the difficulties arising from these divergences. But many of them resulted in unsatisfactory solutions, led to new anomalies, or fell short of eliminating the infinities (Rueger, 1992; Miller, 1994, pp. 41–83).

Nowadays, quantum electrodynamics is a well-established and successful theory of modern physics. The renormalization techniques proposed in 1947 by Richard

Feynman, Julian Schwinger, and Sin-Itiro Tomonaga adjusted the parameters of the theory to eliminate its divergences (Rueger, 1992, pp. 330–337; Schweber, 1994). But until this solution was formulated, several attempts to fix the formalism were advanced during a period of theoretical crisis that lasted almost two decades.

One early approach, often attributed to Bohr, relied on the correspondence principle, employing a semi-classical treatment of radiation problems where quantization was applied selectively to either matter or the radiation field. This method, considered by some as "ad hoc," faced criticism for its inconsistency with the formalism of quantum electrodynamics. Moreover, in certain scenarios, it still resulted in divergent outcomes. Heisenberg proposed another approach, speculating on the existence of a fundamental length of the order of the classical electron radius. His assumption also relied on the correspondence principle, tied to Bohr's applicability limits for quantum electrodynamics theory. But unlike Bohr, Heisenberg assumed this theory could only describe processes in which the length r (associated with an energy hc/r or a momentum h/r) is greater than the fundamental length. This approach also faced criticism for postulating parameters for a yet non-existent theory. Dirac and Peierls pursued a different path, developing subtraction techniques to handle the divergences. This method, involving subtracting divergent terms, raised concerns about its arbitrariness, and lacked clarity in terms of its physical validity, failing to address the underlying issues and provide a deeper understanding of the physical processes (Rueger, 1992, pp. 322–323; Miller, 1994, pp. 58–60).

This series of unsuccessful endeavors to resolve problems with the theory resulted in widespread criticism of quantum electrodynamics in the first half of the 1930s (Kojevnikov, 2002; Rueger, 1992; Schweber, 1994). Pauli was among the early figures to abandon attempts at resolving its theoretical issues. Heisenberg made multiple attempts, at times exploring the subtraction techniques, at times assuming infinities might possess physical significance. He, as others, never abandoned the belief that quantum electrodynamics was to be considered a provisional theory that would be replaced by a future revolutionary world scheme (Bokulich, 2004, pp. 386–393; 2008, pp. 49–72). However, if at the beginning of the decade there was a collective anticipation that the challenges posed by the divergence problem would be resolved through the introduction of an entirely new conceptual framework in physics – a sort of "revolutionary" change – as the years passed, this revolutionary dream dissipated, giving rise to a prevailing sense of pessimism regarding the development of quantum electrodynamics.

Most theoretical physicists ceased to contribute new formulations of the theory by the second half of the 1930s and redirected their efforts toward other issues that appeared more promising. Dirac himself refrained from making new attempts for a few years (Kojevnikov, 2002, p. 242) until a novel idea emerged in 1938. We will direct our attention to this particular idea, which was explored in depth by Ashauer during her PhD studies.

Dirac believed that the problem of divergences in the self-energy of the electron was not solely a problem of quantum electrodynamics but had its origins in classical electrodynamics (Dirac, 1938). As he stated (Dirac, 1938, p. 149):

One may think that this difficulty will be solved only by a better understanding of the structure of the electron according to quantum laws. However, it seems more reasonable to suppose that the electron is too simple a thing for the question of the laws governing its structure to arise, and thus quantum mechanics should not be needed for the solution of the difficulty. Some new physical idea is now required, an idea which should be intelligible both in the classical theory and in the quantum theory, and our easiest path of approach to it is to keep within the confines of classical theory.

Dirac viewed all physical theories as open theories. He considered that even the best-established physical theories could undergo revisions, modifications, and corrections (Bokulich, 2004, p. 378; 2008, pp. 49–51). Classical and quantum theories were, as he saw it, different parts for the same theoretical framework; one was an extension of the other. For this reason, he suggested that the solution to the divergence problem should involve improving the classical theory for the movement of charged particles.

By treating the electron as a point charge, disregarding its internal structure, and considering the radiation field on it as the difference between the field of outgoing radiation leaving the neighborhood of the electron (its field of radiation reaction) and the incoming field incident on the electron, Dirac derived Eq. (9.1). This equation is known as the Lorentz–Dirac equation for the motion of the electron, due to its similarities with Lorentz's classical equations of motion. Dirac, however, expanded upon Lorentz's formulation by incorporating terms related to external fields, radiation reaction forces, and corrections aimed at avoiding divergences:

$$m\frac{\mathrm{d}v_\mu}{\mathrm{d}\tau} - \frac{2}{3}e^2\frac{\mathrm{d}^2v_\mu}{\mathrm{d}\tau^2} - \frac{2}{3}e^2\left(\frac{\mathrm{d}v}{\mathrm{d}t},\frac{\mathrm{d}v}{\mathrm{d}t}\right)v_\mu = ev_\nu F^\nu_{\mu_{ext}} \tag{9.1}$$

Here, m is the electron's mass, e is its charge $F^\nu_{\mu_{ext}}$ is the force due to the incident field on the electron, v_μ is the four-velocity, and τ is the proper time.

For the simplest case of a free electron, with no external field, Dirac obtained:

$$a\frac{\mathrm{d}v_\mu}{\mathrm{d}\tau} - a\frac{\mathrm{d}^2v_\mu}{\mathrm{d}\tau^2} - \left(\frac{\mathrm{d}v}{\mathrm{d}t},\frac{\mathrm{d}v}{\mathrm{d}t}\right)v_\mu = 0 \tag{9.2}$$

with $a = \frac{3m}{2e^2}$. Dirac found two types of solutions for Eq. (9.2). In the first, referred to by Dirac as the "physical solution," velocity results in a constant.

The second, referred to as a "non-physical solution," involves hyperbolic functions of the electron's proper time. The non-physical solution predicts that the electron velocity would rapidly approach the speed of light, and it would do so even in the absence of an external field. This means that the electron velocity would exceed the speed of light even before information about the external field reaches the electron. This outcome was taken as unacceptable due to its implications for the principle of causality and is, therefore, discarded from the pool of viable solutions.

Dirac's paper, though elegant, introduced new theoretical challenges that required resolution, such as understanding and modifying the non-physical solutions. The problem of infinities ceased to be an issue for theoretical physicists only after renormalization techniques emerged in 1947. However, Dirac was dissatisfied with these techniques until the end of his life (Kragh, 1990, pp. 165–188; Kojevnikov, 2002, pp. 249–250; Farmelo, 2009, pp. 334–338), and never abandoned the belief that a new scheme should be introduced to better address the divergence problems.

During the late 1930s and early 1940s, interest in alternative formulations that attempted to resolve the difficulties with infinities in quantum electrodynamics waned significantly. Dirac's conceptualization of physical theories found little resonance among his peers; the kind of effort he put into the reformulation in classical theories did not spark the creation of a new school of thought. Even for Dirac himself, the approach posed challenges. However, it did resonate with a select group of theoretical physicists, including Ashauer and other theoretical physicists of FFCL physics department.

9.5.2 The Studies on the Divergence Problem within the FFCL Physics Department

In 1943, within the FFCL physics department, the theoretical physicists excluded from research for Brazilian national defense engaged in studies that included Dirac's proposal on the modification of classical equations of motion for a point electron. During this period, Ashauer was completing her supplementary didactics course and initiated her theoretical research alongside Schenberg and Wataghin. Concurrently, Leite Lopes was pursuing an "improvement course in physical sciences,"[34] a kind of temporary contract for graduate physicists, where he received a scholarship to attend classes in theoretical physics with Wataghin and classes in classical and celestial mechanics with Schenberg, in addition to conducting research and publishing his findings. During that year, he studied with Schenberg the theory of radiation emitted by a point electron.

---

[34] Report by Gleb Wataghin to the *Zerrenner Foundation*, on the academic activities of José Leite Lopes from April to August 1943. See Note 19.

Leite Lopes and Schenberg's studies resulted in two papers published in *Physical Review* in 1945 – one written by Schenberg alone and the other a joint effort – both critical of Dirac's approach and of the techniques he used to arrive at the Lorentz–Dirac equation. In the first, Schenberg used the Klein–Gordon equation to arrive at the same equations of motion as Dirac, without resorting to the method of subtracting infinities as Dirac had done in his paper (Schönberg, 1945). In the second, they discussed the radiation field of an electron defined by Dirac in the 1938 paper (Dirac, 1938), arguing that it should be corrected to give the radiation losses (Lopes and Schönberg, 1945).

These works were developed while Ashauer was already a young researcher in the physics department and was carrying out research with Wataghin and Schenberg. Nevertheless, it appears that she did not engage with the criticisms raised by Leite Lopes and Schenberg. Perhaps she did not delve deeply into this proposition nor develop strong opinions on the problem surrounding the reformulation of classical equations for the motion of the electron until initiating her doctoral studies at Cambridge.

As was discussed in the previous section, during her first years at Cambridge, Ashauer struggled searching for a problem to investigate. It was only later that she settled on the issues elaborated in her thesis. Unlike her peers from the FFCL physics department, Ashauer strongly embraced Dirac's assumptions and dedicated her doctoral research to exploring and interpreting the non-physical consequences of his 1938 paper.

9.5.3 *Ashauer's Contribution in Her PhD Thesis*

The work of a theoretical and mathematical physicist frequently involves abstraction and speculations that may lack counterparts in the real world. Their task consists of pushing the boundaries of established theories and axioms, probing unexplored realms within the physical-mathematical domain even when these theoretical endeavors do not culminate in revolutionary breakthroughs.

This was the case for numerous formulations crafted during the 1930s, aimed at resolving the infinities within quantum electrodynamics. It was also the case with Ashuaer's PhD work at Cambridge, where she explored and extrapolated the pictures of the "physical-mathematical world" that were conjured up by the non-physical solutions of the Lorentz–Dirac equation (Ashauer, 1947, p. 1):

Some of these difficulties [of the infinity self-energy which appear in the theory of the interaction of an electron with an electromagnetic field] arise only in the quantum theory, as a result of the quantization of the field, whilst others have their classical counterpart. It seems reasonable to look for ways of eliminating the latter difficulties in the classical theory first, since if this classical theory can be put into Hamiltonian form we are able to pass at once to a quantum theory that is free from the corresponding divergences.

As stated above, in the case of a free electron without an incident field, the Lorentz–Dirac equation yields solutions of the type:

$$\begin{cases} v_\mu = \sin h \, (Ae^{a\tau} + B), & \mu \neq 0 \\ v_0 = \cos h \, (Ae^{a\tau} + B), & \mu = 0 \end{cases} \tag{9.3}$$

where $a = \frac{3m}{2e^2}$, τ is the electron proper time, and A and B are arbitrary constants of integration. By these solutions, when $A \neq 0$, the velocity of the electron increases and tends toward the velocity of light as $\tau \to \infty$. In other words, a light signal emitted from the origin at the time $x_0 = 0$ never catches up with the accelerating electron, a phenomenon termed self-accelerating motion.

In 1938, Dirac proposed a universal final boundary condition on the electron acceleration to eliminate these non-physical solutions, according to which the electron should pass to a state of motion without acceleration once the incident field ceases to act.

Ashauer recognized the limitations of this condition, as it implied that for the electron to lose acceleration after the passage of an external field, it must have possessed an acceleration of such magnitude before the pulse's arrival as to exactly cancel the effect of the pulse, almost as if it anticipated the pulse's arrival. Once again, this was an assumption that challenged the principle of causality. Therefore, Ashauer opposed the use of this additional condition. Her aim was to build a physical picture of the self-accelerating electron when no boundary condition is imposed on its solutions (Ashauer, 1947).

To do so, she solved Eq. (9.3) by changing the independent variable from τ to $\varphi = Ae^{a\tau} + B$, and then graphically examined the field energy distribution in the electron's neighborhood when it had already attained a velocity close to the speed of light. She showed that as the electron accelerated, the bulk of the energy of the field produced by it was concentrated in a small region close to the electron, forming an energy-disk. In this picture, she observed that the disk flattened out as time increased, and that the energy density within the disk-area diminished more rapidly than time when time increased.

Although results showed an energy rate that continually increased over time, Ashauer deemed this examination as physically plausible, according to the following consideration (Ashauer, 1947, p. 55):

Although the rate of energy radiated by the electron goes on increasing with time, yet the bulk of this energy does not get very far from the electron since the velocity of the latter is almost that of light. The electron thus carries its field with it. The large negative acceleration energy of the electron is in this sense compensated for by a large positive field energy close to it in the form of a disk of limited dimensions, so that the electron as a whole behaves like a small disk travelling with a large velocity.

She thus arrived at a portrayal of the electron as an entity with confined dimensions and with all its mass concentrated within its radius. In a sense, one can say that her extensions of Dirac's equation for the classical electron's motion yielded somewhat ironic conclusions, given that Dirac's original work aimed to avoid assumptions about the structure of the electron. In addition, this outcome also introduced challenges in transitioning to a quantum description of the self-accelerated electron.

Although Ashauer's PhD thesis does not yield many tangible results for the challenges of her time, it sheds light on the issues tackled by theoretical physics researchers in the 1940s. It underscores the complexity inherent to the theoretical landscape at the intersection of classical and quantum physics.

In February 1948, she successfully defended her PhD thesis. Upon returning to Brazil shortly thereafter, Ashauer resumed her position as Wataghin's assistant to the Chair of Theoretical Physics and Mathematical Physics at the FFCL physics department. Limited information exists regarding any new projects she may have undertaken during this period. Tragically, in August of that year, her scientific career came to an abrupt halt. At just 25 years old, Ashauer contracted pneumonia and passed away, leaving this world behind with only a handful of research publications: her thesis and a few derived papers.

9.6 Final Remarks

In this chapter, we delved into the scientific path of Sonja Ashauer, a Brazilian woman who ventured into the field of quantum electrodynamics theory in the early days of its development. Graduating in physical sciences from FFCL-USP in 1942, Ashauer entered a scene shaped by the experimental research on cosmic-ray physics of the physics department led by Wataghin. Yet even amid this experimental fervor, Ashauer was able to start carving a path in theoretical physics, with the mentorship and guidance of Gleb Wataghin.

By examining her scientific path and encounters, one may reach new ways of evaluating the practices and networks that characterized the research at the FFCL physics department. She was one of the first physicists in the department to venture into theoretical physics rather than experimental cosmic-ray physics, during a period of profound uncertainty and crisis in theoretical physics worldwide, given the ongoing divergence problems in quantum electrodynamics.

While there is limited evidence of her involvement in cosmic-ray physics research within the physics department, this does not diminish her dedication to the department in other ways. Notably, during her PhD at Cambridge, she played a crucial role in contributing to and upholding the internationalization practices learned under Wataghin, actively sharing papers both from Cambridge to São Paulo and from São Paulo to Cambridge.

Throughout her PhD studies at Cambridge, Ashauer struggled to adapt to a research environment that felt more secluded than the collaborative and enthusiastic atmosphere she had experienced at FFCL. Working with Dirac presented a different dynamic, given his taciturn and reserved personality, initially displaying an unwillingness to guide her. However, this did not prevent her from persisting in working under Dirac's supervision, eventually developing a student–professor bond with him.

Ashauer's early struggles at Cambridge involved searching for a research problem. However, she later embraced Dirac's 1938 program of reformulating classical equations for the motion of point particles, an approach aimed to solve divergences on the electron's self-energy within quantum electrodynamics. In her PhD thesis, she explored and extrapolated the non-physical solutions of the equations proposed by Dirac in 1938 (Dirac, 1938).

However, her life was tragically brief, as she passed away only a few months after defending her PhD upon her return to Brazil. Despite the brevity of her career, Ashauer's scientific journey offers more than mere insights into her individual contributions.

One cannot help but wonder about the potential directions her career might have taken had fate not intervened. Would Ashauer have responded positively when, shortly afterward, publications on renormalization techniques arrived in Brazil? What would she have thought of these methods? Would she have persisted in her research in quantum electrodynamics, or would she have shifted her focus to other topics, such as the astrophysics problems suggested by Wataghin, or maybe cosmic-ray research? And would she have remained at USP as an assistant professor, or would she have applied for a full professor position elsewhere? These questions must remain unanswered, leaving us to contemplate the incalculable possibilities and durability of Ashauer's scientific efforts.

References

Ashauer, Sonja. 1947. *Problems on Electrons and Electromagnetic Radiation*. PhD Thesis. Cambridge: University of Cambridge.

Ashauer, Sonja. 1944. Sobre a teoria quântica do coeficiente de absorção. *Anais da Academia Brasileira de Ciências*, XVI, 245–254.

Blay, Eva A. and Lang, Alice B. da S. G. 2004. *Mulheres na USP: horizontes que se abrem.* São Paulo: Associação Editorial Humanitas.

Bokulich, Alisa. 2008. *Reexamining the Quantum-Classical Relation: Beyond Reductionism and Pluralism.* Cambridge: Cambridge University Press.

Bokulich, Alisa. 2004. Open or closed? Dirac, Heisenberg, and the relation between classical and quantum mechanics. *Studies in History and Philosophy of Science Part B: Studies in History and Philosophy of Modern Physics*, 35(3), 377–396. https://doi.org/10.1016/j.shpsb.2003.11.002

Bustamante, Martha C. 2024. Giuseppe Occhialini and the history of cosmic-ray physics in the 1930s: from Florence to Cambridge. In Gariboldi, Leonardo, Gervasi, Massimo, Sironi, Giorgio, Treves, Aldo, and Tucci, Pasquale (eds.), *The Scientific Legacy of Beppo Occhialini: Formative Years and the Return to Italy in 1950*, 2nd ed. Cham: Springer, pp. 141–155.

Cardoso, Irene de A. R. 1982. *A universidade da comunhão paulista (o projeto de criação da Universidade de São Paulo).* São Paulo: Autores Associados/Cortez Editora.

Coelho, Alexander B. 2020. Mário Schenberg na rede científica transnacional de Gleb Wataghin: a primeira geração de físicos brasileiros. *Em Construção: arquivos de epistemologia histórica e estudos de ciência*, 7, 55–78. https://doi.org/10.12957/emconstrucao.2020.47821

Dalitz, Richard H. and Peierls, Rudolf. 1986. Paul Adrien Maurice Dirac. 8 August 1902–20 October 1984. *Biographical Memoirs of Fellows of the Royal Society*, 32, 138–185. https://doi.org/10.1098/rsbm.1986.0006

Dantes, Maria A. M. and Chassot, Walkiria C. F. 2015. Sonja Ashauer (1923–1948). In Saitovitch, Elisa M. B., Funchal, Renata Z., Barbosa, Marcia C. B., Pinho, Suani T. R. de, and Santana, Ademir E. de (eds.), *Mulheres na física: casos históricos, panorama e perspectivas*. São Paulo: Livraria da Física, pp. 95–109.

Dirac, Paul A. M. 1938. Classical theory of radiating electrons. *Proceedings of the Royal Society of London. Series A. Mathematical and Physical Sciences*, 167(929), 148–169. https://doi.org/10.1098/rspa.1938.0124

Farmelo, Graham. 2009. *The Strangest Man: The Hidden Life of Paul Dirac, Mystic of the Atom.* New York: Basic Books.

Freire, Olival Jr. and Silva, Indianara. 2014. Diplomacia e ciência no contexto da Segunda Guerra Mundial: a viagem de Arthur Compton ao Brasil em 1941. *Revista Brasileira de História*, 34(67), 181–201. https://doi.org/10.1590/S0102-01882014000100009

Gariboldi, Leonardo and Verzeroli, Mattia. 2021. Beppo Occhialini in Brazil between physics and politics. In *Proceedings of the 40th Annual Conference of the Società italiana degli storici della fisica e dell'astronomia*. Pisa: Pisa University Press, pp. 133–139.

Harish-Chandra 1987. My association with Professor Dirac. In Kursunoglu, Behram N. and Wigner, Eugene P. (eds.), *Paul Adrien Maurice Dirac: Reminiscences about a Great Physicist*. Cambridge: Cambridge University Press, pp. 34–36.

Kojevnikov, Alexei. 2002. Dirac's quantum electrodynamics. In Balashov, Yuri and Vizgin, Vladimir (eds.), *Einstein Studies in Russia*. Boston, MA: Birkhäuser, pp. 229–259.

Kragh, Helge. 1990. *Dirac: A Scientific Biography.* New York: Cambridge University Press.

Kragh, Helge. 2003. Paul Dirac: the purest soul in an atomic age. In Knox, Kevin C. and Noakes, Richard (eds.), *From Newton to Hawking: A History of Cambridge University's Lucasian Professors of Mathematics*. Cambridge: Cambridge University Press, pp. 387–424.

Lopes, José L. and Schönberg, Mario. 1945. The radiation field of a point electron. *Physical Review*, 67(3–4), 122–123. https://doi.org/10.1103/PhysRev.67.122.2

Melo, Hildete P. de and Rodrigues, Lígia M. C. S. 2006. *Pioneiras da ciência no Brasil*. São Paulo: Brazillian Society for the Progress of Science (SBPC).

Miller, Arthur I. 1994. *Early Quantum Electrodynamics: A Source Book*. Cambridge: Cambridge University Press.

Pais, Abraham. 1986. *Inward Bound: Of Matter and Forces in the Physical World*. New York: Oxford University Press.

Petitjean, Patrick. 1996. Entre ciência e diplomacia: a organização da influência científica Francesa na América Latina, 1900–1940. In Hamburger, Amélia I., Dantes, Maria A. M., Paty, Michel and Petitjean, Patrick (eds.), *A ciência nas relações Brasil–França (1850–1950)*. São Paulo: Edusp/Fapesp, pp. 89–120.

Ribeiro de Andrade, Ana M. 2024. Occhialini's trajectory in Latin America. In Gariboldi, Leonardo, Gervasi, Massimo, Sironi, Giorgio, Treves, Aldo, and Tucci, Pasquale (eds.), *The Scientific Legacy of Beppo Occhialini: Formative Years and the Return to Italy in 1950*, 2nd ed. Cham: Springer, pp. 157–174.

Rueger, Alexander. 1992. Attitudes towards infinities: responses to anomalies in quantum electrodynamics, 1927–1947. *Historical Studies in the Physical and Biological Sciences*, 22(2), 309–337. https://doi.org/10.2307/27757684

Santos, Marcello D. de S., Pompeia, Paulus A., and Wataghin, Gleb. 1941. Showers of penetrating particles. *Physical Review*, 59(11), 902–903. https://doi.org/10.1103/PhysRev.59.902.2

Schönberg, Mario. 1945. Classical theory of the point electron. *Physical Review*, 67(3–4), 122–122. https://doi.org/10.1103/PhysRev.67.122

Schönberg, Mario. 2010. Mario Schenberg, transcript of an interview by Carla Costa and Tjerk Franken at Recife-PE, Brazil, in 1978. Rio de Janeiro: Center for Research and Documentation on Contemporary History of Brazil (CPDOC).

Schwartzman, Simon. 2001. *Um espaço para a ciência: a formação da comunidade científica no Brasil*. Brasília: Ministério de Ciência e Tecnologia: Conselho Nacional de Desenvolvimiento Científico e Tecnológico: Centro de Estudos Estratégicos.

Schweber, Silvan S. 1994. *QED and the Men Who Made It: Dyson, Feynman, Schwinger, and Tomonaga*. Princeton: Princeton University Press.

Shanmugadhasan, Subrahmaniyan. 1987. Dirac as a research supervisor and other remembrances. In Taylor, J. G. (ed.), *Tributes to Paul Dirac*. Bristol: Adam Hilger, pp. 48–57.

Silva, Luciana V. S. da 2020. *Ciência, universidade e diplomacia científica: a trajetória brasileira de Gleb Vassilievich Wataghin (1934–1971)*. PhD Thesis. São Paulo: Faculty of Education, University of São Paulo. https://doi.org/10.11606/T.48.2020.tde-29092020-165017

Tavares, Heráclio D. 2017. *Estilo de pensamento em física nuclear e de partículas no Brasil (1934–1975): César Lattes entre raios cósmicos e aceleradores*. PhD Thesis, Rio de Janeiro: HCTE, Federal University of Rio de Janeiro.

Tavares, Heráclio D., Bagdonas, Alexandre, and Videira, Antonio A. P. 2020a. Transnationalism as scientific identity: Gleb Wataghin and Brazilian physics, 1934–1949. *Historical Studies in the Natural Sciences*, 50(3), 248–301. https://doi.org/10.1525/hsns.2020.50.3.248

Tavares, Heráclio D., Gurgel, Ivã, and Videira, Antonio A. P. 2020b. César Lattes e as técnicas de produção e detecção de mésons: a prática científica como objeto histórico.

Revista Brasileira de Ensino de Física, 42, e20200330. https://doi.org/10.1590/1806-9126-RBEF-2020-0330

Trigo, Maria H. B. (1997). *Espaços e tempos vividos: estudo sobre os códigos de sociabilidade e relação de gênero na Faculdade de Filosofia da USP (1934–1970)*. PhD Thesis. São Paulo: Department of Sociology, University of São Paulo,

USP. 1936. *Annuario da Universidade de São Paulo: 1934–1935*. São Paulo: Official Press of the State of São Paulo.

USP. 1938. *Anuario da Universidade de São Paulo: 1936–1937*. São Paulo: Printing Company of "Revistas dos Tribunais."

USP. 1939. *Anuário da Universidade de São Paulo: 1938–1939*. São Paulo: Edition of the University of São Paulo.

USP. 1942. *Anuário da Universidade de São Paulo: 1940–1941*. São Paulo: Edition of the University of São Paulo.

USP. 1953. *Anuário da Faculdade de Filosofia, Ciências e Letras (Universidade de Sõa Paulo): 1939–1949*. São Paulo: Publications Section of the University of São Paulo.

Videira, Antonio A. P. and Bustamante, Martha C. 1993. Gleb Wataghin en la universidad de São Paulo: un momento culminante de la ciência brasileña. *Quipu*, 10(3), 263–284.

Vieira, Cássio L. and Videira, Antonio A. P. 2014. Carried by history: Cesar Lattes, nuclear emulsions, and the discovery of the pi-meson'. *Physics in Perspective*, 16(1), 3–36. https://doi.org/10.1007/s00016-014-0128-6

Wataghin, Gleb. 1934a. Bemerkung über die Selbstenergie der Elektronen. *Zeitschrift für Physik*, 88, 92–98. https://doi.org/10.1007/BF01352311

Wataghin, Gleb. 1934b. Sull'elettrodinamica relativistica e sull'irraggiamento nell'urto degli elettroni veloci. *Il Nuovo Cimento*, 11, 635–647. https://doi.org/10.1007/BF02960430

Wataghin, Gleb. 1934c. Über die relativistische Quanten-Elektrodynamik und die Ausstrahlung bei Stößen sehr energiereicher Elektronen. *Zeitschrift für Physik*, 92, 547–560. https://doi.org/10.1007/BF01339359

Wataghin, Gleb. 1935a. Sobre as propriedades das partículas elementares. *Annaes da Academia Brasileira de Sciencias*, 7(3), 273–276.

Wataghin, Gleb. 1935b. Sulle relazioni di commutazione nell'elettrodinamica quantistica. *Il Nuovo Cimento*, 12, 290–283. https://doi.org/10.1007/BF02961199

Wataghin, Gleb. 1936a. Sul movimento oscilatorio dell'elettrone secondo la teoria di Dirac. *La Ricerca Scientifica*, Anno VII, 2(5–6), 333–334.

Wataghin, Gleb. 1936b. Sulle forze d'inerzia secondo la teoria quantistica della gravitazione. *La Ricerca Scientifica*, Anno VII, 2(5–6), 341.

Wataghin, Gleb. 2010. Gleb Wataghin, transcript of an interview by Cylon Eudóxio Silva at Campinas-SP, Brazil, in 1975. Rio de Janeiro: Center for Research and Documentation on Contemporary History of Brazil (CPDOC).

Wataghin, Gleb, Santos, Marcello D. de S., and Pompeia, Paulus A. 1940. Simultaneous penetrating particles in the cosmic radiation. II. *Physical Review*, 57(4), 339–339. https://doi.org/10.1103/PhysRev.57.339

Wataghin, Lucia. 1992. Fundação da Faculdade de Filosofia, Ciências e Letras da Universidade de São Paulo: a contribuição dos professores italianos. *Revista do Instituto de Estudos Brasileiros*, 151–173. https://doi.org/10.11606/issn.2316-901X.v0i34p151-173

10

Untangling Entanglement History: Early Quantum Contributions of Chien-Shiung Wu

MICHELLE FRANK

Science is not static but ever growing and dynamic. It involves not just the addition of new information, but the continuous revision of old knowledge. From a flat earth to a round globe, from classical Newtonian Mechanics to Quantum Mechanics, there are many illustrations. It is the courage to doubt what has long been established and the incessant search for its verification and proof that pushes the wheel of science forward.[1]

Chien Shiung Wu
AAUW Achievement Award Acceptance Speech, 1959
& Nishina Memorial Lecture 1983

10.1 Introduction

In October 2022, the Nobel Prize Committee in Stockholm, Sweden, released a public statement championing experimental innovation. The three physicists recognized that year – John Clauser, Alain Aspect, and Anton Zeilinger – contributed to the world's understanding of quantum entanglement over a span of decades. Each devised increasingly sophisticated and innovative "Bell tests" to probe whether hidden

Portions of this work were originally published by *Scientific American* in Spring, 2023. Sincere thanks are due to my WiHQP co-editors, Daniela Monaldi, Patrick Charbonneau, and Margriet van der Heijden. I cannot imagine a more inspiring group of partners. Thank you also to anonymous reviewers for thoughtful suggestions as this chapter developed. I'm grateful to the Consortium for History of Science, Technology and Medicine; the Center for History of Physics at the American Institute of Physics; the Sloan Foundation and the Leon Levy Center for Biography for fellowship and funding support. Special thanks to Joanna Behrman for championing this research during its early stages and to mentors from the Biography and Memoir Program at the CUNY Graduate Center who shared guidance and advice. Elise Crull and Charles Liu gave generously of their time and provided important suggestions. Colleagues at the 2023 AIP Early Career Conference at the Niels Bohr Institute offered feedback as the project deepened. Thank you to the extraordinary Mina Rees Library team, archivists at Columbia University's Rare Book and Manuscript Library, the Berkeley Bancroft Library, AIP's Niels Bohr Library & Archives, the Nobel Archives, the Institute for Advanced Study, and the CalTech Archives and Special Collections, all of whom offered valuable research support. Thank you also to David Kaiser for making time to discuss the finer points of entanglement history, and for pointing me to additional materials. I am very grateful to Abhay Pasupathy, Andrew Millis, Morgan May, and the Columbia University Physics faculty for providing insight about Wu's years at Columbia. Most of all, my heartfelt gratitude to George Sterman and also to Chia-Hsiung Tze for their tireless guidance, scientific and historical insights, and encouragement along the way. Any errors that remain are my own.

[1] C. S. Wu Papers. Rare Book and Manuscript Library, Columbia University. Box 9, Folder 25. Published in Wu (2008).

variables might explain the bizarre behavior of subatomic particles that seem deeply connected, even at a distance. Working separately, each of the three men had improved on his predecessors' experimental designs, with dazzling results (NobelPrize.org, 2022).

By 2022, entanglement had been part of the global zeitgeist for close to a century. It also had become a trusted component of applied and fundamental research, due in significant part to work that took place between the 1970s and the present. Today, the deeply counter-intuitive phenomenon we call entanglement is studied and applied in laboratories worldwide to develop future quantum sensors, quantum computers, and more.

When the 2022 physics prizes were announced, though, one aspect of entanglement's history was missing from public conversation. Nobel press releases, award ceremony speeches, lectures, and news media often celebrate the prize recipients' predecessors. But in 2022, there was almost no public mention of a Chinese American experimentalist whose research had been described earlier as the first documented observation of quantum entanglement in a laboratory.

That same experimentalist is perhaps best known for proving that parity conservation was violated for weakly interacting particles, leading to the 1957 Nobel Prize for two other men. In 1957, a year when the Nobel Committee could have added a third laureate but chose not to do so, Chen Ning Yang and Tsung-Dao Lee became the first Chinese physicists in history to receive the world's most prestigious scientific prize. Many believe that the missing laureate in 1957 was Chien-Shiung Wu.[2]

Unlike other stories of women's omission, Lee and Yang warmly supported Wu.[3] Until the end of her life, she fiercely supported her junior colleagues too. She would later declare, "[M]ost truly fine and creative scientists are so obsessed and busy in their own pursuit of research that they do not waste time on irrelevant or petty jealousy."[4]

To this day, though, there is lively debate about whether Wu was improperly passed over for the 1957 Nobel Prize. At the height of Wu's career, anti-Asian sentiment in the West held many Asian scientists back. American culture in the post-war era placed a higher premium on women's relationships than on their scientific rigor. Like a prism, the 1957 Nobel Prize to Yang and Lee separated out

---

[2] A word about names: In China, the surname (or family name) is typically stated first, and the given name second. This contrasts with the western practice of placing the given name first, and the family name second. When referring to well-known historical figures from Chinese history, this text uses the Chinese naming convention of placing surname first (e.g., Yuan Shikai, Hu Shih, Sun Yat-sen). Where the individual in question spent the majority of their adult life in the US, the western naming convention is followed.

[3] Familiar examples include Jocelyn Bell Burnell, after she reported her findings on quasars, and the notorious tension between Rosalind Franklin and her competitors, Watson and Crick.

[4] C. S. (Chien-Shiung) Wu Papers, 1945. Box 9, Folder 20. Dina Moche oral history interview questions, including responsive notes from Wu.

these elements of identity like bands of light, rendering the impact of gender more visible.

Although Wu is best known for her parity experiment, this chapter focuses on Wu's quantum foundations research before and after her 1956 discovery. It examines Wu's contribution to and omission from the dominant narrative of entanglement history. The chapter also addresses cultural and gendered expectations of the 1950s and beyond, inviting readers to consider how these factors combined to obscure the history of Wu's scientific contributions long after the 1957 Nobel Prize in Physics. Wu may be regarded as a paradigmatic case of intersectional disadvantage; she also has a record of extraordinary accomplishments. She was one of the mid-twentieth century's most influential female physicists, but recognition has centered largely on her parity experiment, which has at times overshadowed other aspects of her groundbreaking work. Often, she found herself the only woman of color in a sea of white men (see Figure 10.1). In recent years, she has re-emerged as an icon for diversity, equity, and inclusion in STEM (science, technology, engineering, and mathematics).

Figure 10.1 Wu at a particle physics conference in Israel, 1967 (seated in front row). American Association of Physics Teachers (AAPT), courtesy of AIP Emilio Segrè Visual Archives.

Rather than allowing Wu's entanglement contributions to quietly drop away from the historical record, recent and current work has begun to bring forward Wu's significance for quantum foundations history. Here, I discuss the writings of Indianara Silva and Francisco Javier (F. J.) Duarte against the backdrop of recent work by David Kaiser, among others (Duarte, 2012; Silva, 2019, 2022; Kaiser, 2022).

The chapter begins with a brief biographical sketch (Section 10.2). It then offers an introduction to Wu's 1949 photon experiment (Section 10.3) and outlines how her later return to quantum foundations research became entangled with work of other experimentalists (Section 10.4). The chapter considers the societal backdrop against which Wu's work transpired (Section 10.5), the criticisms of Wu's entanglement experiments and available rebuttals (Section 10.6), and how her contributions to the understanding of entanglement have begun to resurface (Section 10.7). Closing reflections (Section 10.8) invite additional inquiry about the production of historical scientific narratives and public recognition. By revisiting the existing scholarship on the history of entanglement, this chapter hopefully will inspire further research about what Wu may have had in mind as early as 1949 when she began to investigate John Wheeler's proposed test for "pair theory," and later when she took up the topic again in earnest.

10.2 Formative Years

Wu was born in 1912, in the same year as the New Republic of China.[5] She was raised in a river town 20 miles outside Shanghai, and she grew up in the cross-currents of Chinese nationalism, China's New Culture Movement, and the student-led May Fourth Movement emerging in the wake of the Treaty of Versailles. These transformative social and political pressures contributed to the overhaul of trad-itional Chinese society during her formative years.[6]

Relatively few records exist about Wu's mother, but we know that her father, Wu Zhong-Yi, was a revolutionary and a supporter of equality between girls and boys (Zhu, 2001, pp. 46, 53, and 65).[7] To celebrate the arrival of his new daughter and the

---

[5] Most historians agree today that Wu was born in 1912, the same year as the end of the Qing Dynasty. Whether due to differences between the Chinese lunar calendar and the Gregorian calendar, or to bureaucratic record-keeping errors, Chien-Shiung Wu's birthdate can be found listed in the US as April 29, May 13, May 29, May 31, 1911, 1912, and even 1914, across a variety of US immigration records, secondary sources, and related public documents.

[6] The basic facts of Wu's life and science have been documented in multiple secondary sources supporting this section. See, for example, Lubkin (1971), Bertsch McGrayne (1998), Zhu (2001), Wang (2008), Benczer-Koller (2009), Hammond (2010), Hargittai (2012), Jiang and Wong (2014), Calvin (2017), Dihal (2018), Yomtov (2018), Hamish (2020), Johnston (2020), Yuan (2021), AAUW (n.d), and US National Park Service (n.d.).

[7] The first gift that Zhong-Yi gave his wife was a set of textbooks so that she could continue her education (Zhu 2001, pp. 46, 53, and 65).

end of dynastic rule, Zhong-Yi hosted a party, a practice that traditionally had been reserved for the birth of sons (Zhu, 2001, pp. 43–44). He announced his new daughter, and he gave her what would have been considered a distinctly masculine name. The syllables of that name (Chien-Shiung) translate into English as "Strong Hero" (Zhu, 2001, pp. 28 and 44; Wang, 2008, p. 363). In his daughter's honor, Zhong-Yi announced his plan to open the region's first elementary school for girls (McGrayne, 1998, p. 256; Zhu, 2001, p. 46).

As a child, Wu attended the elementary school that her father had founded. Then her family faced a crisis. There were no local schools for girls beyond the elementary years. Her father consulted the family matriarch who agreed that her 11-year-old great-granddaughter should be sent to study 50 miles away from home at the Suzhou Women's Normal School (Zhu, 2001, pp. 55–57). From that point forward, Wu's studies always led her away from home.

In Suzhou, Wu rose swiftly to the head of her class. A year after graduation, she enrolled at National Central University in Nanjing where she was, again, a top student. Afterward, Wu spent a year as a teaching assistant at Zhejiang University, famously referred to as "the Cambridge of the East" by China scholar, Joseph Needham (University of Cambridge, 2009; PR Newswire, 2022). Next, she moved to a research assistantship at Academia Sinica, west of Shanghai, where she apprenticed under another woman of physics: Gu Jing-Wei, focused on X-ray crystallography (Benzcer-Koller, 2009, p. 5; Jiang and Wong, 2014). Gu would become an important mentor to Wu. Then in 1936, at the age of 24, having reached the limit of what China could offer to her in terms of physics training, Wu boarded the *SS Hoover* headed for the US.

Political movements in China had called for science and democracy throughout Wu's life, and for a new generation of scholars who could strengthen the nation (Zhu, 2001, p. 62; Wang 2008, p. 364; Tu, 2012, p. 76). Wu personified the values of her upbringing, with every reason to imagine that she would bring her advanced training back to the nation of her birth. She had no way to know it at the time, but she would never see most of her family again.[8]

The ship Wu sailed on, the *SS Hoover*, made landfall on August 26, 1936, in San Francisco (Zhu, 2001, p. 43). Wu had been accepted to the University of Michigan, but before traveling further she toured the University of California, Berkeley (UC Berkeley) physics department. She saw Ernest Lawrence's cyclotron, and she was introduced to another Berkeley physics student who had Americanized his name to Luke Yuan. He was the grandson of China's first

---

[8] Wu's family cautioned her against returning home during the civil war in China after World War II. The ensuing political and economic instability, and eventually the Cultural Revolution, as well as the US policies of restricting re-entry for Chinese Americans who left the country, all created obstacles that interfered with Wu's ability to visit her family before their deaths. Her father died of natural causes shortly after the 1957 Nobel Prize. Her youngest brother committed suicide during the Cultural Revolution (Zhu, 2001, p. 52).

president Yuan Shikai. When Yuan, China's erstwhile new leader, moved to consolidate power and suppress his opponents, Wu's revolutionary father had tried to topple him (Zhu, 2001, p. 43).

Writers and historians remain undecided as to whether Wu's view of the Berkeley cyclotron or of the handsome young physicist played a greater role in her next move. Perhaps there was another explanation altogether. Rumors of gender discrimination at the University of Michigan troubled Wu; she was told that the Michigan student union building did not permit women to enter through the front door (University of Michigan, n.d.).[9] Though we may never know the exact reason, Wu sought out the Berkeley physics department's director, Raymond Birge. Birge approved her admission to UC Berkeley, despite severe overcrowding in the physics department, and despite the fact that the semester had already begun (Birge, 1966, p. 4; McGrayne, 1998, pp. 260–261; Zhu, 2001, pp. 71–72; Benczer-Koller, 2009, p. 5).[10]

At Berkeley, Ernest Lawrence became Wu's official advisor. Emilio Segrè became a crucial mentor. J. Robert Oppenheimer became a teacher and a friend. Wu, once again, quickly rose to the top of her class, but this time, she did so under extraordinary pressure. Less than a year after her arrival in the US, Japan invaded China. The following December, the infamous Rape of Nanjing shocked the world. Japanese soldiers ransacked China's capital, raping and murdering countless civilians. Wu had completed her university studies in Nanjing only a few years earlier. The escalation of World War II then cut off communication between the US and much of China. Wu would receive virtually no news of her family for eight years (Zhu, 2001, p. 135).

Meanwhile, West Coast anxiety about Asian immigration escalated to a fever pitch. From as early as 1882, Chinese immigration to the US had been severely restricted, first by the Chinese Exclusion Act, and then by a series of laws extending those restrictions, including the Scott Act, the Geary Act, the McCreary Act, and the 1924 Johnson–Reed Act (Saelee, 2024). Narrowly tailored exceptions selectively allowed Chinese students to study in the US, and under those exceptions Wu had been allowed to enroll in a PhD program. Nonetheless, immigration officials investigated Wu's student status and raised questions about Wu's qualifications for graduate work.[11]

---

[9] Another explanation was offered in 1971: When speaking to an interviewer for Smithsonian Magazine, Wu said that she based her decision to transfer on the high number of Chinese students enrolled at Michigan, because she wanted a more complete immersion experience in American culture (Lubkin, 1971).

[10] Though Wu had learned three foreign languages (German, Japanese, and English), Birge would later report that he believed Wu spoke no English when she arrived. (Birge, 1966, p. 43; Zhu, 2001, pp. 71–72).

[11] See, for example, Department of Justice. Immigration and Naturalization Service. San Francisco District Office. Wu Chien Shiung – Case Number: 36611/012–22 – Ship of Arrival: President Hoover, 08/26/1936. Series: Immigration Arrival Investigation Case Files, 1884–1944, 1884.

Under these unfavorable conditions, Wu completed her PhD in four years. Her dissertation research on the fission products of uranium was so sophisticated and sensitive that it would be requested by the US military and embargoed until the end of World War II. Her doctoral thesis had two sections. The first section on bremsstrahlung, or braking radiation, in beta decay, was titled "The Internal and External Continuous X-Rays Excited by the Beta-Particles of Phosphorus 32." The second, and more militarily sensitive section was titled "Some Fission Products of Uranium." It focused on the decay products of the xenon chain and grew from a collaborative project with Emilio Segrè in 1939.

Despite the significance of her research, Wu had difficulty finding a job after graduation. Anti-Asian sentiment grew more severe in the US, evidenced most distinctly by Japanese internment. The American political climate impacted Chinese Americans too, despite the fact that China was a political ally. Equally, if not more significant, at the time of Wu's graduation, almost none of the top research universities in the US had ever hired a woman professor to their physics faculty (McGrayne, 1998, p. 265; Benczer-Koller, 2009, p. 7).[12]

After graduation, without a secure faculty position, and without having been able to contact family for almost half a decade, Wu worked as a research assistant in Lawrence's Radiation Laboratory. Berkeley's acting comptroller wrote to warn that Wu's employment would be approved only on a temporary basis. Less than a year later, he wrote again, saying "Miss Wu is not eligible for employment," and "immediate steps should be taken to dismiss this employee from your staff" (Zhu, 2001, pp. 429–430).

Lawrence came to Wu's defense, writing letters to US immigration officials about the importance of Wu's work to the war effort.[13]

By 1942, when Oppenheimer moved his work to Los Alamos, Wu had moved east for a teaching position at Smith College. After two years of research assistantships, Smith's employment offer must have been welcome, but it was not the only incentive for Wu's cross-country move: A quarter of a century after Wu's father's military division had tried to unseat Yuan Shikai, Yuan's grandson would take Zhong-Yi's daughter's hand in marriage (Zhu, 2001, p. 43). Instead of a wedding hosted by family, Robert Millikan (Luke's advisor and at the time one of the biggest names in physics), hosted the bride and groom. Soon after, Wu's new husband

---

[12] By 1936, Hertha Sponer had become the first female faculty member on the physics faculty at Duke University despite active opposition from Robert Millikan at Caltech. Raymond Birge wrote to Sponer to offer his support, implicitly acknowledging that the role might fall short of being fully satisfactory but implying that she should be grateful to have found any position at all. (See Chapter 3 in this volume.)

[13] E. O. Lawrence, letter to the Commissioner of the Immigration and Naturalization Service of the Department of Justice of the United States, November 6, 1940. In Lawrence Files. Ernest Orlando Lawrence Files, "Correspondence and Papers," call number 72/117/c, carton 18, folder 36, "General files – Wu, Chien-Shiung," Bancroft Library, University of California, Berkeley.

accepted a position with RCA Corporation based at Princeton University.[14] Smith College was much closer to Luke than Berkeley.

After a backyard wedding at Millikan's home, and a cross-country move, Wu began to settle in at Smith. It was an uneasy fit. Smith was a teaching college with limited resources for research. Wu was an experimentalist. The following year, with supportive letters from Lawrence, Wu became the first woman hired to the Princeton University physics faculty as an instructor (McGrayne, 1998, p. 266; Zhu, 2001, p. 431; Calvin, 2017, pp. 2–29; 4–9)[15] A few months later, the Manhattan Project recruited her. Beginning in March of 1944, she would play a quiet, crucial, and perhaps conflicted role in the development of the atomic bomb (Yuan, 2021).

In New York City, Wu's research for the Manhattan Project focused on gaseous diffusion of uranium to generate fissionable material for the US's new weapon, and she worked on improved radiation detection devices to provide more precise measurements than previously had been possible (Rossiter, 1995, p. 5; Wang, 2008, p. 366; Kam et al., 2024, p. 30). Meanwhile, Wu continued to navigate repeated inquiries and threats of deportation for years.[16]

After the end of World War II, Wu received an offer for a full-time research position at Columbia University. She accepted readily. She would remain a member of the Columbia faculty for the rest of her career.

10.3 Entangling Photons

Quantum entanglement emerges from rigorous mathematics and physics, but it also gives romantic poets a run for their money. Abner Shimony famously called entanglement "passion at a distance" (Howard, 2009, p. 3). It is not hard to imagine that a physicist separated from her family by international conflict might find the phenomenon of interest.

Conventional wisdom about the history of entanglement, in its broadest brush-strokes, goes a little something like this: Early quantum researchers of the twentieth century, from Niels Bohr to Werner Heisenberg, to Erwin Schrödinger, to Albert Einstein, were fascinated, mystified, and unsettled by the uncanny relationships between subatomic particles separated by great (relative) distances from one

---

[14] Millikan to Murray, August 14, 1942; Murray to Millikan, August 17, 1942; Yuan to Millikan, October 6, 1942; Millikan to Yuan, October 13, 1942. Robert Andrews Millikan Collection. Series IV, Millikan Correspondence; Folder 23.21. California Institute of Technology Archives and Special Collections.

[15] Historian, Scott Calvin, notes a publication by Princeton suggesting that women did occasionally provide instruction to graduate students; however, the publication does not specify whether such instruction took place within the physics department (Calvin, 2017, citing Princeton University, n.d.)

[16] Department of Justice. Immigration and Naturalization Service. San Francisco District Office. Wu, Chien Shiung – Case Number: 36611/012–22 – Ship of Arrival: President Hoover, 08/26/1936, Series: Immigration Arrival Investigation Case Files, 1884–1944, 1884.

another. Einstein complained about "spooky action at a distance," suggesting that it showed where quantum mechanics remained incomplete. Schrödinger agreed, along with his cat. The Einstein–Podolsky–Rosen (EPR) paper emerged in 1935, largely drafted by Podolsky (Einstein et al., 1935).[17] Schrödinger coined the term "entanglement" to describe the mystifying links between spatially separated particles. And then, in the turmoil of World War II, all eyes turned toward nuclear physics and weapons of mass destruction. During and after the war, scientists who felt curious enough to raise ongoing questions about quantum foundations were told to "shut up and calculate" (Freire, 2006; Kaiser, 2012, p. 1, p. 20; Whitaker, 2016, p. 279). Quantum foundations research shifted to a back burner for decades, disparaged as philosophy rather than "real science" until the emergence of John Stewart Bell.[18]

Then, according to the dominant narrative, in 1964, Bell published "On the Einstein Podolsky Rosen paradox" in *Physics Physique Физика* (Bell, 1964), and a few years later, John Clauser found a copy of Bell's theorem in the Columbia University library. Clauser designed an experiment to test Bell's theorem in partnership with Stuart Freedman, and the pair published their findings in 1972. Alain Aspect took Clauser's work further, adding new and improved technologies between 1976 and 1981, changing detector settings during the flight of the photons in his new experiments. Anton Zeilinger followed with quantum random number generators and cosmic Bell tests, making it less and less likely that hidden variables could account for the strange relationships between distant particles. Quantum entanglement research, as a topic for experimentalists, had been reborn. Finally, in 2022, the Nobel Committee recognized all three men's pioneering research, establishing the bona fides of quantum entanglement once and for all, and trumpeting the demise of hidden-variables theory.

The conventional account includes beautiful discoveries, innovative designs, mechanical insights, and brilliant accomplishments by tremendous scientific minds and experimentalists whom the world is right to revere.

And yet . . . the standard account relegates some important details to the margins. Wu is one example.

***

---

[17] In the *Stanford Encyclopedia* (Fine, 2017), Fine writes: " . . . [T]he cumbersome machinery employed in the EPR paper makes it difficult to see what is central. It distracts from rather than focuses on the issues. That was Einstein's complaint about Podolsky's text in his June 19, 1935 letter to Schrödinger. Schrödinger responded on July 13 reporting reactions to EPR that vindicate Einstein's concerns. With reference to EPR he wrote: 'I am now having fun and taking your note to its source to provoke the most diverse, clever people: London, Teller, Born, Pauli, Szilard, Weyl. The best response so far is from Pauli who at least admits that the use of the word "state" ["Zustand"] for the psi-function is quite disreputable. What I have so far seen by way of published reactions is less witty. . . . It is as if one person said, "It is bitter cold in Chicago"; and another answered, "That is a fallacy, it is very hot in Florida"' [Fine, 1996, p. 74]."

[18] For a much more thorough treatment of this history, see, for example, Horne et al. (1991), Freire (2006), Kaiser (2012), and Whitaker (2016).

By November of 1949, none of the 2022 physics laureates had finished grade school yet. That winter, Wu and her Columbia University graduate student, Irving Shaknov, began work on a new experiment. Physicists had understood since 1932 that when an electron and a positron collide, they annihilate, releasing two photons that fly apart in separate directions (Anderson, 1932; Süvegh and Marek, 2011, p. 1463).[19] In 1946, the American theorist John Wheeler proposed an experimental test to demonstrate that the resulting pair of photons would be orthogonally polarized relative to one another (Wheeler, 1946).[20] Wu and Shaknov wanted to try Wheeler's proposed test of pair theory.

The Wu–Shaknov experiment required a supply of positrons, which could be created in a basement laboratory with a new version of the device Wu saw when she first arrived at Berkeley – the cyclotron. Lawrence had invented the cyclotron; while Wu was still a graduate student at Berkley, it had earned her advisor the 1939 Nobel Prize in Physics (NobelPrize.org, n.d.). Not much later, Columbia University built its own cyclotron with a multi-ton magnet so huge that, according to university folklore, administrators had to blast a hole in an exterior wall and recruit the football team to maneuver the giant blocks of iron into Pupin Hall, home of the physics department.[21] In 1949, 13 stories below Wu's university office in Pupin, the Columbia cyclotron bombarded a sheet of copper with deuterons, generating an unstable isotope, Cu-64, as a source of positrons (Wu and Shaknov, 1950).[22]

By the time Wu and Shaknov began their search for conclusive evidence of pair theory, two other teams had already attempted similar experiments. The results of the first team had a high margin of error and were considered to be insufficiently reliable (Bleuler and Bradt, 1948; Wu and Shaknov, 1950). A second team obtained results where the polarization asymmetry ratios were consistently lower than the theorists' predictions (Pryce and Ward, 1947; Snyder et al., 1948; Hanna, 1948; Wu and Shaknov, 1950; Duarte, 2012, pp. 311–318). In the winter of 1949, Wu had a slightly new design with more sophisticated detectors than her predecessors (Wu and Shaknov, 1950; Silva, 2022). Wu had become well known for her extreme

---

[19] "In the case of two-photon annihilation, the energy conservation law has one more serious consequence on photons. Both should have an energy of 0.511 MeV. This energy is so uniquely characteristic to electron–positron annihilation that Anderson (1932) proved the existence of positrons by the detection of this annihilation radiation" (Süvegh and Marek, 2011, p. 1463).

[20] Pair theory has been traced to Paul Dirac's work on electron–positron pairs (Duarte, 2012, p. 312). Duarte writes, "The pair theory that Wheeler refers to is the theory of electron-positron pairs to which Dirac [Dirac, 1930] was an important contributor."

[21] Author's interview with Abhay Pasupathy, Professor of Physics at Columbia University, October 28, 2022. Dr. Pasupathy works in Wu's former office, confirmed its location, and shared the story of the Columbia cyclotron, later confirmed by email.

[22] Wu's comments in *Physical Review* included a thank you to the "cyclotron group" for their help in creating the Cu-64 sample. ("We also wish to thank the cyclotron group for preparing the Cu-64 source . . . ") (Wu and Shaknov, 1950).

precision and experimental acumen, even more so after she decisively proved Enrico Fermi's theory of beta decay on the heels of a decade worth of unsuccessful attempts by others (McGrayne, 1998, p. 265; Wang, 2008, p. 366; Jiang and Wong, 2014, pp. 107–108). Given her track record, the new photon experiment was worth pursuing.

Wu and Shaknov packed their copper isotope into a tiny aluminum capsule, eight millimeters long. This arrangement produced annihilation radiation, which then passed through a hole in the middle of a lead block to select gamma rays that were oriented in opposite directions (Wu and Shaknov, 1950; Silva, 2022, p. 738). Wu and Shaknov tracked the gamma rays with newly developed scintillation counters – an improvement to their predecessors' use of Geiger counters – and they spent 30 continuous hours counting the tiny flashes.[23] With the improved detectors, they wrote, ". . . there will be an increase in the coincidence counting rate of one hundred times," relative to the results of their predecessors (Wu and Shaknov, 1950).

Wu and Shaknov published their findings on New Year's Day, 1950, as a one-page letter in *Physical Review* (Wu and Shaknov, 1950). Their experiment had set out to prove that Wheeler's test worked for pair theory, and it had succeeded. But perhaps even more significantly, their work suggested that pairs of photons from annihilation were polarized at right angles to one another consistently, despite the quantum uncertainties in their individual polarizations, as if somehow connected even at a distance. In other words, the Wu–Shaknov experiment quietly offered the first documented evidence of entanglement between particles of light (Duarte, 2012; Horne et al., 1991, pp. 356–72; Zeilinger, 1999, pp. 288–97).

And then . . . the 1949 revelation about entangled photons passed through public awareness, and across time, like a neutral particle might pass through empty space. Shaknov lost his life on a data collecting mission for the Navy when his plane was shot down during the Korean War a few years after the experiment (Center for Naval Analysis, n.d.). Wu's photons moved into obscurity. For over 70 years, the 1949 observation would be hidden in plain sight.

10.4 Entangled Timelines

Hidden in plain sight, but with one particularly significant exception. Like Einstein, another physicist expected a more reasonable explanation for entanglement – one that might not be so spooky. An explanation that could be attributed to hidden variables.

That physicist was David Bohm, and he would think and write about hidden variables for years (e.g. Bohm, 1952; Bohm and Aharanov, 1957; see also Horgan,

---

[23] For a fuller description of the apparatus, see Silva (2022); see also Kam et al. (2024).

1993; Freire, 2009, pp. 577–616, describing Bohm's scholarship). In 1957, Bohm and his graduate student, Yakir Aharonov, wrote about how photon research could harness the famous EPR paradox to reveal hidden variables. "[T]here has been done an experiment which, as we shall see, tests essentially for this point, but in a more indirect way," Bohm and Aharanov wrote (Bohm and Aharanov, 1957, p. 1074). They pointed to the Wu–Shaknov experiment.

In 2022, historian of science, Indianara Silva, pointed out that it was precisely *this* 1957 paper by Bohm and Aharanov that inspired new work by John Stewart Bell (Silva, 2022). Bohm had every reason to know of Wu's stellar reputation. He linked his by-then controversial argument about hidden variables to Wu's well-regarded experimental acumen. He acknowledged Wu in a footnote. Like Wu, Bohm had been a PhD student at Berkeley. Although he was a few years junior to Wu, both scientists had studied under Oppenheimer, and both had worked in Lawrence's prestigious radiation laboratory.

Silva was an early pioneer in excavating Wu's hidden entanglement history.[24] Silva highlights how Bohm's 1957 article inspired Bell, who proposed that the number of quantum coincidences between particles could be predicted and counted. Like Bohm, Bell wrote about entanglement for much of his career, but Bell's theoretical work on hidden variables took a quantum leap after Bohm and Aharanov's 1957 paper (Bell, 1964; Whitaker, 2016, pp. 120–122, 155–157, 200–202). In his now-famous article in *Physics Physique Физика*, Bell discussed the 1957 Bohm–Aharanov paper which had referenced Wu's experiment, and he launched what would come to be known as Bell's inequality (Bell, 1964; see also, Bell, 1981; Kaiser, 2012).

Wu inspired Bohm. Bohm inspired Bell. Bell inspired Clauser to design a new experiment, one Clauser hoped would prove Bell right, showing that hidden variables were real.[25]

*

Fast forward to Columbia University as the radical 1960s eased into the 1970s. Almost two decades had passed since Wu and Shaknov's *Physical Review* article (Wu and Shaknov, 1950). Quantum interpretational questions had begun to move from the margins back to center stage (Freire, 2005; Kaiser, 2012). Wu revisited her earlier 1949 experiment. There was a new graduate student to train: Leonard Kasday. Silva notes that by then, Wu was far more professionally secure and addressed questions about quantum mechanics directly.[26] Even so, Wu favored

---

[24] Shortly before this volume went to press, Chon-Fai Kam, Da Hsuan Feng, and Cheng-Ning Zhang published an article in *Physics Today,* adding valuable new details to this history (Kam et al., 2024).

[25] For a fuller treatment of this chronology, see Silva (2022).

[26] After 1958, Wu held a full-time tenured professor position at Columbia.

traditional quantum entanglement interpretations, and not Bohm's hidden-variables theory (Silva, 2022).

Kasday, working under Wu's guidance, improved on the 1949 Wu–Shaknov experiment by refining the original design two decades after Wu and Shaknov had first demonstrated pair theory. There were at least two significant changes. First, rather than using aluminum, Kasday followed the examples of others who had repeated the Wu–Shaknov experiment, selecting thin plastic scintillation scatterers. Second, he changed the geometric design, and he implemented a four-fold coincidence detection model to avoid accidental events (Kasday, 1972).

In the opening pages of his thesis, Kasday makes it quite clear that the idea for his central experiment came directly from Wu. "I would like to express sincere appreciation of Professor C. S. Wu for suggesting this research ...," he wrote (Kasday, 1972). Kasday first presented his experimental results in 1971 at the Varenna Summer School – the International School of Physics "Enrico Fermi" held near Lake Como in Italy – and he published his talk in the ensuing proceedings (Silva, 2022, p. 746; Kasday, 1971). His thesis was deposited the following year (Kasday, 1972).

Of course, 1972 was the very same year that Clauser's watershed experiment with Freedman would be published too.

Historian of physics, David Kaiser, makes it clear that Clauser had begun thinking about Bell's theorem as early as 1967. In *How the Hippies Saved Physics*, we learn that Clauser was rebuffed by his own dissertation advisor when he proposed to work on Bell's theorem for his thesis (Kaiser, 2012, p. 44). Things might have gone differently if Clauser had worked under Wu. Instead, Clauser would wait until after graduation before submitting an abstract to the annual meeting of the American Physical Society (APS) outlining his experimental ideas (Clauser, 1969; Freire, 2005). By then, two more years had passed. The 1969 abstract caught Abner Shimony's eye; his own graduate student, Michael Horne, was interested in fundamental physics too (Kaiser, 2012, p. 44; Freire, 2005, p. 590). Clauser's 1969 bulletin for the APS sparked a collaboration, leading to what would come to be known as the CHSH paper (after the names of the authors, Clauser, Horne, Shimony, and Richard Holt).

Clauser, Horne, Shimony, and Holt critiqued the 1949 Wu–Shaknov experiment (Clauser et al., 1969). Kasday's improvements followed in response to the 1969 CHSH critiques, which are discussed in more detail in Section 10.6.

Then, on April 3, 1972, in their "Experimental test of local hidden-variable theories," Clauser and Freedman went further, including a footnote that critiqued the new experiment too – this one conducted by Kasday, Jack Ullman and Wu ("KUW"). In their footnote 6, Clauser and Freedman referred to Kasday's 1970 bulletin from two years earlier and to a paper "pending publication" (Freedman and Clauser, 1972).[27]

Two years later, in February 1974, Kasday, Ullman, and Wu would submit results from their collaborative experiment testing EPR to *Il Nuovo Cimento*. The authors waived the opportunity to review proofs to speed up publication. While they waited, in May 1974, Wu shared Kasday's findings in a letter to Stuart Freedman (Silva, 2022, p. 747). She wrote, in pertinent part:

... we carried out the investigation on the Angular Correlation of Compton-scattered Annihilation Photons with much improved precision and accuracy and also made an attempt to relate its interpretation with hidden variables. I am taking the liberty of sending you a copy of our paper (just being sent out for publication now) in thinking that you might be interested in our interpretation.[28]

The KUW article reached the public in February 1975 (Kasday et al., 1975).

It would be fascinating to uncover whether Wu was or was not aware of Clauser's 1969 APS abstract (Clauser et al., 1969) when she first suggested that her own graduate student conduct an experiment testing EPR. The crucial point, though, is not when Wu learned of Clauser's ambitions; rather, Kasday's thesis hints at Wu's own view of her research from decades earlier. Clearly, Kasday viewed the Wu–Shaknov experiment as having been directly relevant to EPR. Presumably, his impressions were shaped through dialogue with Wu, his advisor. In his thesis, Kasday wrote (Kasday, 1972, pp. 7–8):

[I]t was J. S. Bell (Bell 64, 70) who, in 1964, showed that a suitable experiment would rule out Einstein's complete theories – if the experimental results agreed with the quantum predictions. It is shown in this chapter that, if certain reasonable assumptions are permitted, *then a more extensive version of the Wu–Shaknov measurement is a suitable experiment.* Before Bell's work was published, the Wu–Shaknov experiment had been already repeated .... Wu and Shaknov had analyzed the linear polarizations by Compton scattering in aluminum (emphasis added).

He continued a few pages later (Kasday 1972, p. 31):

Our results, therefore, are consistent with the work of *Wu and Shaknov (Wu 50), who were the first to show good quantitative agreement between quantum theory and experiment* (emphasis added).

Similarly, the 1975 Kasday–Ullmann–Wu publication in *Il Nuovo Cimento* referred back to the 1949 Wu–Shaknov experiment. The 1975 team pointed to the 1957 Bohm–Aharanov paper that relied on Wu–Shaknov to bolster their claim, writing (Kasday et al., 1975, p. 636):

---

[27] Presumably they were pointing to Kasday's thesis. It is also possible they anticipated the Kasday-Ulmann-Wu publication of 1975.

[28] Wu to Freedman. May 16, 1974. C. S. Wu Papers. Rare Book and Manuscript Library, Columbia University.

Using aluminum scatterers and anthracene detectors, *Wu and SHAKNOV (4) measured this* *Φ-dependence as early as* 1950 and found good agreement with theory ... However, the inefficient detectors available at the time required collection of events over a wide range of scattering angles and thus large corrections for geometrical effects . . .. *Our aim was to test* *the predictions of quantum mechanics* for this distribution over a range of scattering angles with as few uncertainties about normalization or geometrical corrections as possible.

The reason why it is worth-while to lavish so much attention on this measurement is *because it is often referred to in discussions of "hidden variable" theories in quantum* *mechanics* (emphasis added).

No records have yet come to light revealing what Wu's private thoughts may have been, but Kasday's acknowledgment and the substance of his thesis suggest that by at least the 1970s, if not earlier, Wu viewed her new graduate student's research as a continuation of work she, herself, had initiated in 1949 (Kasday, 1972, pp. 7–8 and 30). Moreover, Wu's letter to Freedman and the effort to speed up publication in *Il Nuovo Cimento* suggest a race for priority.

One must be cautious, of course, about suggesting Wu had quantum foundational questions in mind as early as 1949 when she and Shaknov first set out to test pair theory. After all, in the first sentence of their *Physical Review* letter, Wu and Shaknov explained that Wheeler's "Polyelectrons" had inspired their work, and the word "entanglement" never appeared in the one-page publication. For that matter, whether Wheeler thought about entanglement when he first published "Polyelectrons" in 1946 remains debatable. Wheeler scholar Stefano Furlan notes that Wheeler had never been one to emphasize "entanglement as the quintessential aspect of quantum mechanics." He has pointed out that, similarly, "in the ... intellectual autobiography [Wheeler] published at the end of the century, *Geons, Black Holes, and Quantum* *Foam*, the word "entanglement" never appears."[29] Kaiser might agree. In a 2022 interview, he said "I think he [Wheeler] was just trying to think about antimatter as a new kind of conceptual plaything to be perfectly honest. And certainly not thinking about, you know, entanglement or EPR . . ."[30]

And yet, Furlan also notes that Wheeler included "Polyelectrons" in his 1983 anthology, *Quantum Theory and Measurement*. There, the chapter called "Polyelectrons" appears as the very first entry in a section titled "'Hidden Variables' Versus 'Phenomenon' and Complementarity." This placement invites "Polyelectrons" to function as an on-ramp to work by quantum dissidents like Bohm (writing about the EPR paradox); Bell; Clauser, Horn, Shimony, and Holt; among others. Clearly, by the 1980s, Wheeler connected his own work (Wheeler, 1946) with quantum foundations, even if in hindsight.[31]

---

[29] Stefano Furlan. Email correspondence with the author, July 16, 2023. (Dissertation forthcoming.)
[30] David Kaiser. Zoom interview with author. November 2, 2022.
[31] Wheeler was known for his retrospective reflections, as Stefano Furlan points out in Chapter 8.

By 2012, historian of physics F.J. Duarte drew a similar connection, calling Wheeler's work on pair theory "the essence of entanglement" (Duarte, 2012, pp. 311–318). Duarte also relied on the Wu–Shaknov experiment to bolster his own claim that Maurice H. L. Pryce and John Clive Ward deserved greater recognition as quantum pioneers.[32] He wrote (Duarte, 2012, p. 316):

Beyond the EPR implications, and Bohm and Aharanov's interpretation, the [Wu–Shaknov] experiment seeded from Dirac's ideas ..., discussed by Wheeler [Wheeler, 1946], illustrated and described with the correct quantum physics by Pryce and Ward [Pryce and Ward, 1947] had all the ingredients of quantum entanglement as mentioned by Dalitz ... [Dalitz, 2000].

Duarte continued,

Clauser et al. [1969] ... argued that the WS experiment (Wu 1950) did not produce "evidence against local hidden-variable theories" due to their 'use of Compton polarimeters [Clauser et al., 1969].' *In their post Bell paper, Wu and colleagues had responded that "even though a Compton experiment cannot rule out hidden-variable theories, it can provide strong evidence against them* [Kasday et al., 1975]" (emphasis added).

Duarte's step-by-step analysis was built to emphasize Pryce and Ward's importance. But the implicit premise in Duarte's argument was that Wu and Shaknov revealed the quantum foundational implications of their own work in 1949–1950 when they cited Pryce and Ward.[33]

Given Wu's and Shaknov's direct nod to Pryce and Ward in their 1950 publication, Wu's later inclusion of Pryce and Ward's work in the 1975 KUW paper, and the interpretation of later historians like Duarte and Silva, who have brought Wu's connection with quantum foundations into sharper focus, it's not unreasonable to ask whether Wu and Shaknov may have perceived the quantum connection too. If so, it's worth asking again why Wu did not mention entanglement in her 1950 letter.[34] On the one hand, it is entirely possible that in 1949 she might not have had

---

[32] Pryce and Ward (1947) and Snyder et al. (1948) had been inspired by Wheeler's work to develop a theoretical framework for the experiment Wu and Shaknov later conducted in 1949. For our purposes, what is most relevant here is Duarte's reliance on Wu–Shaknov to support his argument: Duarte made a point of noting that Wu and Shaknov (1950) cited Pryce and Ward in their letter to *Physical Review*. The assumption that Wu and Shaknov were, indeed, the first to document entangled particles in the laboratory, in other words, was baked securely into Duarte's claim. Duarte had studied under Ward.

[33] Duarte also noted a publication by C. N. Yang, interpreting Wheeler's 1946 work along similar lines, writing, "It should also be pointed out that Yang (Yang 1950) wrote a famous paper on the quantum selection rules governing the disintegration of a particle into two photons that also refers to Wheeler (Wheeler 1946) plus Pryce and Ward (Pryce 1947) in regard to a proposed 'experimental verification' of the phenomenon" (Duarte, 2012, p. 315, citing Yang, 1950). Relatedly, the physicist Da Hsuan Feng recounted that at a meeting with Professor Chen-Ning Yang in Yang's home in Beijing, Yang told him that Wu was the first to experimentally demonstrate entanglement. Emails from Da Hsuan Feng to author, May 22, 2024; February 3, 2024; author's interview with Da Hsuan Feng, February 8, 2022.

[34] Of note, Clauser and Shimony would later state that the Wu–Shaknov experiment: "... determined the parity of the ground state of *positronium* by a method suggested by Wheeler that consisted of measuring the polarisation correlation of gamma rays produced by positronium annihilation. The photons Compton-scattered, and two-photon coincidences were observed as a function of azimuthal scattering angles, alpha and beta. Two relative

entanglement in mind. On the other hand, she and Shaknov may have been hesitant to discuss evidence of entanglement because, at the time, quantum-foundations work had been stigmatized as junk science. The stigma persisted for decades, as Olival Freire and Kaiser make abundantly clear in their leading work on this topic (Kaiser, 2012, 2022; Freire, 2019). As Kaiser would later express, "back then, the idea of using an experiment to prove or disprove theories about quantum physics or to test for local hidden variables was 'not even an inkling' for most physicists" (Frank, 2023). Instead, researchers who explored questions about entanglement often disguised their research because backlash could stymie a promising career (Clauser, 2002; Kaiser, 2012, 2022; Freire, 2019).

Eventually, as we now know, the quantum foundations tide would change. The irony of course, as Freire and Kaiser and others have pointed out, is that when Clauser and Freedman (then a Berkeley PhD student) conducted their own test of Bell's theorem and published their results in 1972, Clauser had wanted to prove hidden variables were real. Instead, he *disproved* the existence of hidden variables and demonstrated entanglement with even greater certainty. Clauser had counted coincidences, much as Bell suggested, but he found a far greater number of coincidences than hidden variables could explain. Clauser's work prompted Alain Aspect and Anton Zeilinger's later experiments, which closed lingering loopholes and further supported quantum mechanical interpretations of entanglement. Together, all three men's experiments led to the 2022 Nobel Prize in Physics (NobelPrize.org, 2022).

Wu, on the other hand, seems already to have held the view that hidden-variables theory would fail when she supervised Kasday. Indeed, in 1974 when she reflected on Kasday's initial results, she had written that the results "should certainly quiet those proponents of hidden variables" (Silva, 2022, p. 746).

In 1975, Bell personally invited Wu to participate in a small quantum foundations conference scheduled at the Centro di Cultura Scientifica Ettore Majorana in Erice, Italy, focusing on "experiments on the basic notions of quantum mechanics and, more particularly, people interested in *doing* such experiments" (emphasis in original).[35]

Wu politely declined the invitation.

---

angles 0 and pi/2 were employed. From the ratio of these two coincidence rates they were *able to infer that the parity of the ground state is negative*. Bohm and Aharonov (1957), with different motivations, showed that these data are explained by quantum mechanics ... but are incompatible with the assumption that the ensemble of photon pairs can be described by a mixture of states, each of which is a product of two single-photon polarisation states. *They therefore concluded that the data of Wu and Shaknov confirm the existence of states of two-particle systems which are 'non-separable', even though the particles are spatially remote from each other* (see appendix 1)" (Clauser and Shimony, 1978, p. 1914, emphasis added). Of course, the 1950 Wu–Shaknov letter was as silent about the "parity of the ground state" of "positronium" as it was silent about "entanglement" and quantum foundations.

[35] Letter from John Bell to C. S. Wu, 1975. C. S. Wu Papers. Rare Book and Manuscript Library, Columbia University.

10.5 Parity Redux

Looking back today, we can speculate about a number of reasons Wu might have chosen to turn down Bell's invitation. One possible explanation is practical and simple. In short: she was very busy.

In the early 1970s, a series of policy changes and eventually new legislation in the US, brought the Civil Rights Act of 1964 to university settings. Columbia University, like many universities, found itself under the microscope. In 1965, an executive order prohibited federal contracts with any institution that discriminated in employment on the basis of race, color, national origin, or religion. By 1967, a new executive order extended these policies to address discrimination based on sex, and in January 1969, the US Department of Health, Education and Welfare began monitoring universities to evaluate their compliance with the new rules. During this time period, the government sought data about hiring and promotion practices from Columbia University,[36] where Wu had been serving on the faculty since the 1940s.

Columbia was viewed as having stubbornly delayed (for over 30 months) in providing responsive documents to address the government's concerns about inequitable hiring and promotion practices.[37] Time lags exacerbated an already tense situation, and what began as a low-stakes compliance investigation became a high-stakes academic funding emergency. By November 1971, the US government's Office of Civil Rights took steps toward preventing Columbia University from receiving further federal grants or contracts.[38]

With substantial government cash-flow suddenly at risk, Columbia found itself in crisis. University leadership called on Wu. The Executive Vice President tapped her, along with several other faculty members, to serve on an Emergency Advisory Committee to make recommendations on the status of women and affirmative action.[39]

Columbia leaders pointed out that at similarly situated universities, committees like Wu's had been viewed as evidence of the university's bona fides, perhaps justifying extensions of time to provide the government with a response. At Columbia, Wu's new group was expected to screen and analyze internal documents

---

[36] Statement to the Columbia University Senate, Dr. William McGill, February 25, 1972. C. S. Wu Papers. Box 6, Folder 19. Rare Book and Manuscript Library, Columbia University.

[37] "Where We Are and How We Got Here" memo with handwritten corrections. C. S. Wu Papers. Box 6, Folder 28. Rare Book and Manuscript Library, Columbia University.

[38] Statement to the Columbia University Senate, Dr. William McGill, February 25, 1972. C. S. Wu Papers. Box 6, Folder 19. Rare Book and Manuscript Library, Columbia University; "Where We Are and How We Got Here" memo. C. S. Wu Papers. Box 6, Folder 28. Rare Book and Manuscript Library, Columbia University.

[39] Different committee names were used at different times, including the "Emergency Advisory Committee," the "Affirmative Action Faculty Advisory Committee," and the "Columbia University Commission on the Status of Women." C. S. Wu Papers. See, for example, Box 6, Folders 28–31. Rare Book and Manuscript Library, Columbia University.

containing years' worth of hiring and promotion data and to make detailed recommendations to the university for policy improvements.

In other words, in addition to her own scientific research and teaching responsibilities in the early 1970s, Wu was expected to get the messy government investigation cleaned up.[40] In the process of doing so, it would come to light that Wu's own salary had lagged behind her male contemporaries for years. Despite her outstanding scientific reputation, Wu had not advanced at Columbia as quickly as her male peers (McGrayne, 1998; Rossiter, 2012, p. 25; Jiang and Wong, 2014, p. 111; Calvin, 2017, p. 4–10). In this complicated context – as the leader of her employer's affirmative action committee, and also as the long-standing recipient of unequal pay and inequitable promotion practices – Wu brought diplomacy, savvy, and political acumen to bear.

From the start, Wu's committee ran into repeated document production delays that rivaled the government's difficulties. Despite a series of delicately worded memos to colleagues and supervisors, the requested documents were not forthcoming.[41] Lawsuits were a "strategy of last resort," as Rossiter makes clear in her analysis of the time period (Rossiter, 2012), but eventually, Wu and her committee met privately with the American Civil Liberties Union (ACLU) at her apartment in Manhattan. ACLU attorneys prepared a complaint to file with the New York City Commission on Human Rights.[42] Wu prepared yet another carefully crafted letter to the university leadership, politely suggesting that without the requested materials and an opportunity to review the faculty's recommendations, her committee would have no alternative but to disband.[43]

The subtle threat must have brought the university up short. Documents came to light showing persistent disparities. Wu, herself, was one example. By 1956, she had been working at Columbia for more than a decade, but she was still only an associate professor. Colleagues had been hired and promoted on all sides while she taught, mentored, and continued her research. Men who were professionally junior to Wu on the physics faculty earned more than she did (Rossiter, 2012, p. 25;

---

[40] See, for example, William Theodore deBary to C. S. Wu, April 10, 1972. C. S. Wu Papers. Box 6, Folder 31. Rare Book and Manuscript Library, Columbia University; "Where We Are and How We Got Here" memo. C. S. Wu Papers. Box 6, Folder 28. Rare Book and Manuscript Library, Columbia University; C. S. Wu to President William McGill, May 30, 1972. C. S. Wu Papers. Box 6, Folder 31. Rare Book and Manuscript Library, Columbia University.

[41] See, for example, C. S. Wu et. al. to William J. McGill, University President, December 2, 1971; February 4, 1972; February 23, 1972; April 4, 1972; April 8, 1972. C. S. Wu Papers. Box 6, Folder 31. Rare Book and Manuscript Library, Columbia University.

[42] Untitled memo labeled "Participants in April 3rd Meeting at Dr. Wu's Home, From Steering Committee Re: "Attached Charges" and attachments, including Draft Complaint by Commission on Human Rights. C. S. Wu Papers, Box 6, Folder 26.

[43] See, for example, C. S. Wu et. al. to William J. McGill, University President, December 2, 1971; February 4, 1972; February 23, 1972; April 4, 1972; April 8, 1972. C. S. Wu Papers. Box 6, Folder 31. Rare Book and Manuscript Library, Columbia University.

Calvin, 2017, pp. 4–10). In the midst of the governmental funding debacle over equal pay, it could hardly be more apt that the law of physics that Wu had challenged in 1956–1957 had been called the "principle of parity."

Along with a handful of other women and faculty of color, Wu's compensation was adjusted, and the university's hiring and promotion practices were revised.[44] Thereafter, Columbia faculty framed the episode as a success story and a partnership, one in which the University's new policy had helped the Office of Civil Rights to craft Higher Education Guidelines Executive Order 11246 – a tool which would come to be used as a reference guide for universities across the nation.[45]

*

It had been a long road. Wu was brilliant, but by as early as the 1950s, some of her professional relationships had begun to fray (Zhu, 2001, p. 273). Her partners were unfailingly polite and collaborative when face-to-face, but privately some complained about her input. Colleagues tossed off gendered and ethnic slurs. Some of her students expressed resentment about her tinkering, her seemingly tireless help to make everything just a little bit better. Seemingly, her dedication exhausted those around her. Colleagues and competitors joked about her tireless work ethic. They referred to her as a "slave driver" and as the "Dragon Lady" (Zhu, 2001, p. 320, p. 448, p. 452). When that lost its appeal, they criticized her parenting (McGrayne, 1998, p. 270). In person, they often addressed her directly as "Madame Wu," ostensibly, to convey respect. But as almost any English-speaking woman can attest, the word "Madame" is layered with complexity; the word hints simultaneously at prostitution, orientalism, and power. "If it is not too late, I would appreciate very much if you would delete the title Madame Wu," she wrote to an editor in 1974 at *Physical Review*. "At Columbia, my title is Professor Wu."[46] "Avoid using Mme" she noted in 1981 in hand-written corrections to Paul Forman's "Fall of Parity" exhibit materials.[47] The moniker proved nearly impossible to shake.

Wu's diligence had been celebrated during her youth; early after her arrival in the US, her newcomer status charmed graduate mentors and classmates, but even they expressed their appreciation in gendered and racialized terms. News media and faculty advisors focused on her appearance. "Petite Chinese girl shows research in atom smashing," declared the *San Francicso Tribune* (1941). Segrè wrote about Wu's beauty in his autobiography (Segrè, 1993). Lawrence penned recommendation letters

---

[44] Additional details of the governmental inquiry into Columbia's practices can be found in C. S. Wu Papers, Rare Book and Manuscript Library, Columbia University, Box 6.
[45] "Where We Are and How We Got Here" memo. C. S. Wu Papers, Box 6, Folder 28.
[46] C. S. Wu to George L. Trigg, Editor, *Physical Review*, February 27, 1974. C. S. Wu Papers. Box 1, Folder 3. Rare Book and Manuscript Library, Columbia University.
[47] C. S. Wu Papers. Box 3, Folder 11. Rare Book and Manuscript Library, Columbia University.

on her behalf that described Wu as, "altogether a decorative addition to any laboratory" (Zhu, 2001, p. 101).[48]

By the time Bell invited Wu to the conference on quantum foundations, she was serving as APS President and might have had little energy left. Moreover, her position on the hidden-variables debate was already clear. We will never know how Wu might have contributed to future foundations research if she had attended the Erice conference. What we do know is that her absence did not prevent others from continuing to critique earlier Wu experiments. Section 10.6 analyzes those claims.

Before moving further, it is worthwhile to note the clear advances in the designs of each successive Bell test that followed. The Wolf Prize in 2010 and the Nobel Prize in 2022 confirm subsequent experiments' innovation and importance. In recognizing later innovative designs, though, beginning with Clauser–Freedman in 1969, and continuing through Aspect's switching polarizers during the flight of the photons and Zeilinger's cosmic Bell tests, it seems only natural that history would celebrate and highlight a foundational experiment too. Instead, and despite her experiments being cited as a springboard by some of the very same researchers who would go on to become Nobel laureates, Wu's name would not rise to wider public attention in connection with entanglement experiments for decades. The situation seems nearly to jump off the page as an illustration of Rossiter's work on chronic undervaluation of women's scientific contributions (Rossiter, 1993).[49]

10.6 Critiquing Wu

As early as 1969, Clauser, Horne, Shimony, and Holt, later Clauser and Freedman, and still later Clauser and Shimony, distinguished Clauser's experimental approach from Wu's. They asserted in 1969, in 1972, and again in 1978 that Wu's experimental designs could not conclusively disprove a hidden-variables theory.[50]

First, Clauser, Horne, Shimony, and Holt argued that the angles Wu and Shaknov studied in their 1949 experiment made it impossible to distinguish between hidden-variable versus quantum-mechanical theories; specifically, if polarization measurements were taken at relative angles of 0 and 90 degrees (including multiples of 90 degrees, such as 180 and 270 as in the Wu–Shaknov experiment) then the predicted measurement outcomes would be identical for quantum mechanics and for local hidden-variable models, leaving questions about the basis for entanglement

---

[48] See also letter from Ernest H. Wakefield to Chien-Shiung Wu. October 14, 1974. C. S. Wu Papers, Rare Book and Manuscript Library, Columbia University (referring to Wu as "Dear Cover Girl" and "my favorite female scientist" when she was in her early 60s).

[49] Rossiter coined the term, "The Matilda effect" in her later work on women's scientific invisibility (Rossiter, 1993).

[50] Notably, Wu's own publication with Kasday and Ullman in 1975 raised a similar concern, but the authors explained why they believed their results to be relevant nonetheless (Kasday et al., 1975).

unresolved. With this critique in mind, Kasday had refined the original 1949 geometric design and measurement angles.

From the very beginning, though, because there were no tools that could effectively measure the polarizations of high-energy photons, both of Wu's teams had used the workaround suggested by Wheeler and calculated by Pryce and Ward: Compton scattering (Wheeler, 1946; Pryce and Ward, 1947; Kasday, 1972). By measuring correlations in the scattering angles of the Compton-scattered particles that had been bumped by the photons in question, the experimentalists drew inferences about photon polarization. Like any other measuring device, those secondary particles were a step removed from the photons themselves.

As noted above, Kasday and his team members wrote up their results three times: first in 1971; next in Kasday's thesis of 1972; and a third time for *Il Nuovo Cimento* in 1975. In the 1975 publication, Kasday, Ullman, and Wu hedged their claim ever so slightly. They wrote (Kasday et al., 1975, p. 637):

> We will show . . . that this experiment is not ideal for testing Bell's theorem, but it does make any theory that satisfies Bell's theorem and reproduces our results look quite artificial.

Clauser et al. did not completely agree. They extended their original critique to the new experiment, arguing that the lack of sufficiently sophisticated measuring apparatus for high-energy photons rendered proxy measurements through Compton scattering insufficient.[51] They argued that Kasday, Ullman, and Wu's use of the Klein–Nishina formula to compute the relation between the results of Compton scattering and hypothetically ideal polarization analyzers introduced a logical flaw into the analysis; they challenged the auxiliary assumptions of the KUW research that connected Compton scattering results with true and accurate measurements (Kasday, 1972, p. 40), and they claimed that such assumptions should be understood to discredit the KUW conclusions (Freedman and Clauser, 1972; Clauser and Shimony, 1978).[52]

In short, because Compton scattering measures the angle of secondary particles rather than the photons directly, the Clauser team viewed the KUW measurement as a step too far removed. Later, Clauser and Shimony also argued that the mathematical formulae justifying Compton scattering analysis, itself, presupposed a quantum

---

[51] Notably, Kasday had raised the same concern in his thesis directly (Kasday, 1972, pp. 20–21).

[52] As noted by Clauser et al., Kasday, Ullman, and Wu's auxiliary assumptions were:

" . . .

(i) in principle, ideal linear polarisers can be constructed for high-energy photons;
(ii) the results, which would be obtained in an experiment using ideal analysers, and those obtained in a Compton scattering experiment, are correctly related by quantum theory."

Kasday devoted attention to this question in his thesis too, with an appendix titled, "Experimental Evidence for the Validity of the Quantum Relations Between Compton and Ideal Polarization Analysis" (Kasday, 1972, p. 80).

state. "Combining quantum mechanics and a general local realistic theory," they wrote, "results in a fatally incorrect handling of both theories" (Clauser and Shimony, 1978).

Kasday had addressed a similar concern in his own publication, but he also proposed a proof suggesting that the problem might be considered a red herring. In recent years, the Clauser teams' critique has come under criticism by others too, albeit obliquely. Duarte, for example, suggested in 2022 that a critique of Compton scattering in experiments may have been misplaced (Duarte, 2022, pp. 12–2):

To question the claim of the scattering experiments as the first observations of quantum entanglement is to question the impact of all other physics experiments that rely on more than one direct and immediate observation to deduce a valid result, [...]

He continued (Duarte 2022, pp. 12–13):

As long as no observational attempts are made to intrude in the measurement process, consisting of the incidence of the entangled gamma ray quanta and the scattering of the radiation towards the detector, there is no pragmatic motive to doubt the validity of the measurements from the scattering experiments.

Kasday, Ullman, and Wu had also raised concerns with the Clauser–Freedman assumptions. Clauser's team was upfront about their own assumption that all photons emerging from a polarizer would have the same detection probability, independent of the polarizer's orientation (Clauser et al., 1969 1969; Clauser and Shimony, 1978, p. 1904). Clauser et al. also assumed that for every pair of emissions, the probability of a count with the polarizer in place would be less than or equal to the corresponding probability with the polarizer removed (Freedman and Clauser, 1972; Clauser and Shimony, 1978). Kasday, Ullman, and Wu suggested that the Clauser–Freedman assumptions raised non-trivial concerns.

In 1978, Clauser and Shimony reviewed the experimental landscape and the collection of assumptions, including their own. They concluded, "Despite our caution concerning the CHSH and CH [Clauser and Horne, 1974] assumptions, we regard the experimental refutation which relies upon them to be compelling."[53]

---

[53] Clauser and his co-authors refer to their own assumptions as "mild" and view other experimentalists' assumptions as more significant (Clauser et al., 1969; Freedman and Clauser 1972; Clauser and Shimony, 1978; see also Kaiser, 2022, p. 1296). They wrote (Clauser and Shimony, 1978, p. 1916): "The experimental data of KUW are in good quantitative agreement with the quantum-mechanical predictions. This experiment is less decisive, in our opinion, as a refutation of the family of local realistic theories than are the cascade-photon experiments discussed in Section 5, because it relies upon assumptions which are considerably stronger than the assumption needed by the latter. If assumptions (i) and (ii) are not made, then a local hidden-variables model can be constructed (Horne 1970, Bell 1971 (see Kasday 1971)) which yields the same predictions for the experiment as those by quantum mechanics." But they also acknowledged, "This consideration, by itself, is not a fully sufficient reason to prefer the cascade-photon experiments, since a local hidden-variables model, albeit a much more artificial one, also exists which yields quantum-mechanical predictions for those experiments."

Years later, though, Michael Horne (of CHSH and CH) would acknowledge in an oral history with the American Institute of Physics (AIP) that even the cascades which Clauser's team had favored were insufficient to settle the hidden-variables question (Horne, 2002).

10.7 Recovering a Quantum Legacy

In 2022, Silva published a chapter in the *Oxford Handbook of the History of Quantum Foundations*, which aimed to dispel any remaining doubt about the crucial role the 1949 Wu–Shaknov experiment and its successors had played in later entanglement research (Silva, 2022). In 2022, Silva built on and refined arguments that she had begun to advance a few years earlier (Maia Filho and Silva, 2019). She carefully tracked the history of theory and experiment, demonstrating that when Wu and Shaknov made the first precise measurement of pair theory in 1949, they also became the first experimentalists to document entanglement between photons, inspiring decades of later research in quantum foundations. She identified a string of publications by physicists and historians who acknowledged the importance of the 1949 Wu–Shaknov observation. The list began with Bohm in 1957 and continued through Whitaker, Duarte, and even Zeilinger, one of the three 2022 Nobel laureates who had written in 1999 that "an earlier experiment by Wu and Shaknov (1950) had demonstrated the existence of spatially separated entangled states" (Zeilinger, 1999, p. 293; Whitaker, 2016, pp. 120–122, 155–157, 200–202; Silva, 2022, pp. 737, 742, and 749).[54] Silva acknowledged Clauser, Freedman, and Shimony's critique of Wu's use of Compton scattering, but she also firmly established Wu's priority in documenting entangled photons in the laboratory.

Thus, although the 1949 Wu–Shaknov experiment may not have been designed to test the *reasons* for entanglement, Silva's work made it increasingly clear that the Wu–Shaknov test was the first experiment to document entangled photons. Of course, Wu's later work in the 1970s was pivotal as well. Duarte's comments about the validity of Compton scattering measurements offer additional credibility (Duarte, 2022, pp. 12-2–12-3).

In 2023, Columbia University released Kasday's thesis electronically for the first time, and an accompanying blog post commented on the work and its limitations. "The pieces were all there, but the technology at the time was not," said Ana Asenjo-Garcia, a quantum physicist and professor at Columbia University (Neff, 2023).

---

[54] At the same time, Zeilinger bracketed his claim with a caveat pointing out that the Wu–Shaknov experiment had failed to collect data about other than orthogonal measurement directions, and therefore could not have distinguished between hidden variables versus quantum-mechanics interpretations.

Looking back today, Wu and her graduate students' use of Compton scattering might be understood as introducing a loophole into entanglement research. So, too, might Clauser and Freedman's early designs, which established equipment settings prior to the flight of the photons. In the *Oxford Handbook,* (Freire 2022; the same publication where Silva's work on Wu appears), Kaiser discusses the "locality loophole" afflicting Clauser, Freedman, and subsequent experimentalists, pointing out that components of experimental apparatus could conceivably communicate with one another (Kaiser, 2022). Aspect's later design set out to address the locality loophole by changing the polarizer angles while the photons were in flight; he managed to partially address the locality loophole, but he nonetheless faced a different problem – namely, the setting selections were not entirely random. In other words, the locality loophole persisted, albeit in a narrower form. Fifteen years later, Zeilinger's use of quantum random number generators moved another step toward closing the locality loophole, but it still left open a "fair-sampling loophole;" the new tests still only measured about 5% of all photon pairs. It remained conceivable that the selection process somehow introduced an invisible bias or preference for measuring precisely those photons that would discredit a hidden-variables interpretation. After locality and random sampling concerns were more fully addressed, a further loophole still lingered, as Kaiser explains. Referred to as the "freedom of choice" loophole, this theory posits that a hypothetically common cause in the distant past might affect the experimental parameters (including, for example, the detector settings in a Bell test) even in the absence of direct communication between disparate parts of the experimental apparatus. Such a common cause might exist at great distance in the overlapping areas of the light cones for the experimental settings, even if that common cause remained utterly undetectable to contemporary observers (Kaiser, 2022, pp. 349 and 361).

In other words, although Clauser et al. (1969), Freedman and Clauser (1972), and Clauser and Shimony (1978) critiqued Wu's design, each of the Nobel laureates who were celebrated in 2022 for their Bell tests overtly acknowledged that they relied on designs that also invite critique; in each of their experimental apparatus, so-called loopholes left open one or more possibilities by which hidden-variables theory might survive.

With each improvement in the experimental design, from Wu–Shaknov to Kasday–Ullman–Wu, to Clauser–Freedman, to Aspect et al., to Zeilinger et al., not to mention other experimentalists between and beyond, designs and apparatus steadily improved, and the likelihood of exceptional caveats diminished, to the point where the risk of missing a hidden-variables scenario has become vanishingly small. And yet, even the most advanced experiments to date – namely, Zeilinger's "cosmic Bell tests," which rely on light from stars which are hundreds of light years away from one another – cannot fully eliminate hidden-variable possibilities, even if such hypotheses must rely on retrocausality and enormous, backward-facing light cone claims to keep the contrarian prospect of hidden variables on the table.

10.8 Concluding Remarks: Connection and Perception

In his 1957 Nobel lecture, Yang directly addressed how crucial Wu's parity experiment had been, making a bold statement before the committee and assembled guests that the results were due to Wu's team's courage and skill (Yang, 1958). Lee would later plead with the Nobel Committee to recognize Wu's work.[55] Oppenheimer publicly hinted that Wu should have shared the 1957 prize at a celebratory banquet at the Institute for Advanced Study.[56] Segrè called the discovery of the non-conservation of parity "probably the major development of physics after the war" (Segrè, 1967) and he nominated Wu for the Nobel in 1965 and 1968.[57]

In the years that followed, other scientists criticized Wu's exclusion from the highest court of scientific achievement too. In 1987, physicists sought joint support from Academia Sinica in Taipei and the Chinese Academy of Science in Beijing to re-nominate Wu.[58] In 1991, Douglas Hofstadter, the author of *Gödel, Escher, Bach*, recruited 30 scientists to write letters to the Nobel Committee recommending Wu for the physics prize (Zhu, 2001, pp. 347–351). In 2018, 1600 researchers invoked Wu's name in an open letter to CERN challenging current-day sexism in physics. "[T]here are at least four women whose work is relevant for particle physics who are widely viewed as having deserved the Nobel prize but who did not receive it, in some cases even though their male colleagues did," the letter said. Wu's name appeared at the top of that list (see Grant, 2018; Frank, 2023).

Wu would go on to become the first woman to receive the Comstock Prize in Physics in 1964 from the National Academy of Sciences, the first female president of the APS in 1975, the first physicist to receive the prestigious Wolf Prize in 1978, and in 1990 the first living physicist to have an asteroid named in her honor.[59] Her work at Columbia pushed open doors to science and academia in the West for women and scientists of color. In China, she is revered. In 2021, the US Postal

[55] T. D. Lee nominating C.S. Wu, January 15, 1964. The Nobel Archives, The Royal Swedish Academy of Sciences, Stockholm. See also, T. D. Lee to Nobel Committee, January 15, 1964. C. S. Wu Papers. Box 3, Folder 1. Rare Book and Manuscript Library, Columbia University.
[56] Institute for Advanced Study. Director's Office. General Files. Box 48. Folders 9, 10, and 11.
[57] Emilio Segrè nominating C. S. Wu, January 11, 1965, and January 5, 1968. The Nobel Archives, The Royal Swedish Academy of Sciences, Stockholm.
[58] Wheeler to Lundquist, February 6, 1987. John Wheeler Papers. American Philosophical Society; Tuan to Lundquist, September 3, 1986. John Wheeler Papers, American Philosophical Society.
[59] For additional listings of many honors Wu received, see, for example, Benzcer-Koller (2009, p. 15), AAUW (n. d.), and Thomas (2022).

Service (USPS) released a Forever stamp with Wu's portrait (USPS News, 2020, 2021). Today Wu's parity experiment is understood as an early step on the path to what would become the Standard Model of particle physics, and it points toward possible answers about why matter exists in our universe at all (Johnston, 2020; Keating, 2021).

Wu's early entanglement work, however, remained largely unknown to the broader public for decades. During those decades, beginning in earnest almost 20 years after Wu and Shaknov's letter to *Physical Review* (Wu and Shaknov, 1950), the work of Clauser, Aspect, and Zeilinger took center stage. Experiment by experiment, these physicists, and their teams, closed loophole after loophole, and they ruled out one alternative explanation after another, until finally the Nobel Committee determined that quantum entanglement was the only explanation left standing.

In the history of quantum physics, incompleteness has never been a mark of irrelevance. Loopholes and incremental steps always have been part and parcel of the arc of scientific progress. Although Wu could not have received a Nobel Prize posthumously, the 2022 physics awards celebrated a set of connected experiments building upon the foundations of research that had predated and inspired them – indeed as almost all great scientific discoveries have done for centuries. Today, as a result of the collaborative work of historians and physicists alike, Wu's early entangled photons are finally coming to light. The Wu–Shaknov experiment, and the subsequent research of Kasday, Ullmann, and Wu can and should be celebrated as a crucial foundation, and as a springboard to the experimental discoveries that followed.

References

Anderson, Carl. 1932. Energies of cosmic-ray particles, *Physical* Review, 41(4), 405–21, https://doi.org/10.1103/PhysRev.41.405

AAUW (American Association of University Women). n.d. Chien-Shiung Wu overlooked for Nobel Prize. https://www.aauw.org/resources/faces-of-aauw/chien-shiung-wu-overlooked-for-nobel-prize

Bell, John S. 1981. Bertlemann's socks and the nature of reality. *Journal de Physique Colloques* 42(C2), C2-41–C2-62. https://doi.org/10.1051/jphyscol:1981202

Bell, John. S. 1964. On the Einstein Podolsky Rosen paradox. *Physics Physique Физика*, 1(3), 195–200. https://doi.org/10.1103/PhysicsPhysiqueFizika.1.195

Benczer-Koller, Noemie. 2009. *Chien-Shiung Wu 1912–1997: A Biographical Memoir.* Washington, DC: National Academy of Science.

Birge, Raymond T. 1966. *History of the Physics Department v.4 (1932–42)*. Berkeley, CA: University of California.

Bleuler, Ernst and Bradt, Helmut. L. 1948. Correlation between the states of polarization of the two quanta of annihilation radiation, *Physical Review*, 73(11), 1398–1398. https://doi.org/10.1103/PhysRev.73.1398

Bohm, David 1952. A suggested interpretation of the quantum theory in terms of "hidden" variables. I. *Physical Review*, 85, 166. https://doi.org/10.1103/PhysRev.85.166

Bohm, David and Aharonov, Yakir. 1957. Discussion of experimental proof for the paradox of Einstein, Rosen, and Podolsky. *Physical Review*, 108(4), 1070–1076. https://doi.org/10.1103/PhysRev.108.1070

Calvin, Scott. 2017. *Beyond Curie: Four Women in Physics and Their Remarkable Discoveries, 1903 to 1963.* http://iopscience.iop.org/book/978-1-6817-4645-6

Center for Naval Analysis. Irving Shaknov: a singular life. https://www.cna.org/about-us/research/history/irving-shaknov-a-singular-life

Clauser, John. 2002. Interview of John Clauser by Joan Bromberg in Walnut Creek, California, May 20, 21, and 23, 2002. Niels Bohr Library & Archives. https://www.aip.org/history-programs/niels-bohr-library/oral-histories/25096.

Clauser, John F. and Horne, Michael A. 1974. Experimental consequences of objective local theories. *Physics Review D*, 10(2), 526–535. https://doi.org/10.1103/PhysRevD.10.526

Clauser, John. F. and Shimony, Abner. 1978. Bell's theorem. Experimental tests and implications. *Reports on Progress in Physics*, 41(12), 1881–1927. https://doi.org/10.1088/0034-4885/41/12/002

Clauser, John F., Horne, Michael A., Shimony, Abner, and Holt, Richard A. 1969. Proposed experiment to test local hidden-variable theories. *Physical Review Letters*, 23(15), 880–884. https://doi.org/10.1103/PhysRevLett.23.880

Dalitz, Richard. H and Duarte Francisco, J. 2000. John Clive Ward. *Physics Today*, 53(10), 99–100. https://doi.org/10.1063/1.1325207

Dihal, Kanta. 2018. Where state politics meets gender politics: Chien-Shiung Wu and the Manhattan Project. https://www.ladyscience.com/where-state-politics-meets-gender-politics-chienshiung-wu-and-the-manhattan-project/no40

Dirac, Paul A. M. 1930. On the annihilation of electrons and protons. *Cambridge Philosophical Society*, 26, 361–375.

Duarte, Franciso J. 2012. The origin of quantum entanglement experiments based on polarization measurements. *European Physical Journal H*, 37, 311–318. https://doi.org/10.1140/epjh/e2012-20047-y

Duarte, Franciso J. 2022. *Fundamentals of Quantum Entanglement*, 2nd ed. Bristol: IOP Publishing.

Einstein, Albert, Podolsky, Boris, and Rosen, Nathan. 1935. Can quantum-mechanical description of physical reality be considered complete? *Physical Review*, 47(10), 777–80. https://doi.org/10.1103/PhysRev.47.777

Fine, Arthur. 1996. *The Shaky Game: Einstein, Realism and the Quantum Theory*, 2nd ed, Chicago. IL: University of Chicago Press.

Fine, Arthur. 2017. The Einstein–Podolsky–Rosen argument in quantum theory. *Stanford Encyclopedia of Philosophy Archive*. https://plato.stanford.edu/entries/qt-epr

Frank, Michelle. 2023. The little-known origin story behind the 2022 Nobel Prize in Physics. *Scientific American*, April 1. https://www.scientificamerican.com/article/the-little-known-origin-story-behind-the-2022-nobel-prize-in-physics

Freedman, Stuart J. and Clauser, John F. 1972. Experimental test of local hidden-variable theories. *Physical Review Letters*, 28(14), 938–941. https://doi.org/10.1103/PhysRevLett.28.938

Freire, Olival Jr. 2005. Science and exile: David Bohm, the Cold War, and a new interpretation of quantum mechanics. *Historical Studies in the Physical and Biological Sciences*, 36(1), 1–34. https://doi.org/10.1525/hsps.2005.36.1.1

Freire, Olival Jr. 2006. Philosophy enters the optics laboratory: Bell's theorem and its first experimental tests (1965–1982). *Studies in History and Philosophy of Science Part B: Studies in History and Philosophy of Modern Physics*, 37(4), 577–616. https://doi.org/10.1016/j.shpsb.2005.12.003

Freire, Olival Jr. 2009. Quantum dissidents: research on the foundations of quantum theory circa 1970. *Studies in History and Philosophy of Science Part B: Studies in History and Philosophy of Modern Physics*, 40(4), pp. 280–289. https://doi.org/10.1016/j.shpsb.2009.09.002

Freire, Olival Jr. 2019. *David Bohm: A Life Dedicated to Understanding the Quantum World*. Springer International Publishing.

Freire Olival Jr. (ed.) 2022. *Oxford Handbook of the History of Quantum Interpretations*. Oxford Handbooks. Oxford: Oxford University Press.

Grant, Andrew. 2018. High-energy-physics community condemns sexist talk at CERN. *Physics Today*, October 5. https://doi.org/10.1063/PT.6.2.20181005a

Hammond, Richard. 2010. *Chien-Shiung Wu: Pioneering Nuclear Physicist*. New York: Chelsea House.

Hanna, R.C. 1948. Polarization of annihilation radiation, *Nature*, 162(4113), 332–332. https://doi.org/10.1038/162332a0

Hargittai, Magdolna. 2012. Credit where credit's due? *Physics World*, September, 13. https://physicsworld.com/a/credit-where-credits-due

Horgan, John. 1993. Last words of a quantum heretic: just before he died, the physicist David Bohm explained his life-long quest to understand the universe and how he developed his alternative view of the quantum world. *New Scientist*, February 27. https://www.newscientist.com/article/mg13718624-400

Horne, Michael. 2002. Interview of Michael Horne by Joan Bromberg in Dorchester, United Kingdom, September 12, 2002. Niels Bohr Library & Archives. https://www.aip.org/history-programs/niels-bohr-library/oral-histories/34331

Horne, Michael, Shimony, Abner, and Zeilinger, Anton. 1991. Down-conversion photon pairs: a new chapter in the history of quantum mechanical entanglement. *Quantum Coherence*, 356–372. https://doi.org/10.1142/9789814439251_0031

Howard, Don. 2009. Passion at a distance. In Myrvold, Wayne C. and Christian, Joy (eds.), *Quantum Reality, Relativistic Causality, and Closing the Epistemic Circle: Essays in Honour of Abner Shimony*, Dordrecht: Springer, pp. 3–11.

Jiang, Caijian and Wong, Tang-Fong. 2014. *Madame Wu Chien-Shiung: The First Lady of Physics Research*. World Scientific Publishing.

Johnston, Hamish. 2020. Overlooked for the Nobel: Chien-Shiung Wu. *Physics World*, October 2. https://physicsworld.com/a/overlooked-for-the-nobel-chien-shiung-wu

Kam, Chon-Fai, Zhang, Cheng-Ning, and Feng, Da Hsuan 2024. Chien-Shiung Wu's trailblazing experiments in particle physics. *Physics Today*, December 1 https://digital.physicstoday.org/physicstoday/library/page/december_2024/AD

Kaiser, David. 2012. *How the Hippies Saved Physics: Science, Counterculture, and the Quantum Revival*. New York: W.W. Norton.

Kaiser, David. 2022. Tackling loopholes in experimental tests of Bell's inequality. In Freire Olival Jr. (ed.), *Oxford Handbook of the History of Quantum Interpretations*. Oxford: Oxford University Press, 339–370.

Kasday, Leonard R., Ullman, Jack D. and Wu, Chien-Shiung. 1975. Angular correlation of Compton-scattered annihilation photons and hidden variables. *Il Nuovo Cimento B (1971–1996)*, 25(2), 633–661. https://doi.org/10.1007/BF02724742

Kasday, Leonard R. 1972. *The Distribution of Compton Scattered Annihilation Photons, and the Einstein–Podolsky–Rosen Argument*. PhD Thesis. New York: Columbia University. https://doi.org/10.7916/5q2y-3494

Kasday, Leonard R. 1971. Experimental test of quantum predictions for widely separated photons. Proceedings of the International School of Physics "Enrico Fermi," Lake Como, Italy, pp. 195–210.

Keating, Brian. 2021. Polarization peregrinations: using cosmic lampposts to find keys to nature's laws. *Simons Foundation*, January 14. https://www.simonsfoundation.org/event/polarization-peregrinations-using-cosmic-lampposts-to-find-keys-to-natures-laws

Lubkin, Gloria. 1971. The charming First Lady of experimental physics. *Smithsonian*, 52–57. "Wu, Chien-Shiung (1912–1997)."

Maia Filho, Angevaldo Menezes and Silva, Indianara. 2019. O experimento WS de 1950 e as suas implicações para a segunda revolução da mecânica quântica. *Revista Brasileira de Ensino de Física*, 41(2). https://doi.org/10.1590/1806-9126-rbef-2018-0182

Maia Filho, Angevaldo Menezes and Silva, Indianara. 2019. A trajetória de Chien Shiung Wu e a sua contribuição à Física. *Caderno Brasileiro de Ensino de Física*, 36(1), 135–157.

McGrayne, Sharon Bertsch. 1998. *Nobel Prize Women in Science: Their Lives, Struggles, and Momentous Discoveries*, 2nd ed. Washington, DC: Joseph Henry Press.

Neff, Ellen. 2023. A quantum thesis comes to light. *Columbia Quantum Initiative*, October 2. https://quantum.columbia.edu/news/quantum-thesis-comes-light

NobelPrize.org. 2022. Nobel Prize in Physics, 2022. Nobel Prize Outreach AB 2024. www.nobelprize.org/prizes/physics/2022/summary

NobelPrize.org. n.d. The Nobel Prize in Physics 1939. https://www.nobelprize.org/prizes/physics/1939/lawrence/facts

PR Newswire, 2022. "Cambridge of the East" shines: global-minded Zhejiang University plays a growing role in pushing SDGs. *PR Newswire*, November 9, https://www.prnewswire.com/news-releases/cambridge-of-the-east-shines-global-minded-zhejiang-university-plays-a-growing-role-in-pushing-sdgs-301689240.html

Princeton University. n.d. Princetoniana women. https://princetoniana.princeton.edu/history/women

Pryce, Maurice and Ward, John C. 1947. Angular correlation effects with annihilation radiation. *Nature*, 160, 435. https://doi.org/10.1038/160435a0

Rossiter, Margaret W. 1993. The ~~Matthew~~ Matilda effect in science. *Social Studies of Science*, 23(2), 325–341.

Rossiter, Margaret W. 1995. *Women Scientists in America: Before Affirmative Action, 1940–1972*. Baltimore: Johns Hopkins University Press.

Rossiter, Margaret W. 2012. *Women Scientists in America: Forging a New World since 1972*. Baltimore, MD: Johns Hopkins University Press.

Saelee, Mike. Chinese Exclusion Act: topics in chronicling America. Introduction. https://guides.loc.gov/chronicling-america-chinese-exclusion-act/introduction

San Francisco Tribune. 1941. Petite Chinese girl shows research in atom smashing. *San Francisco Tribune*, April 25.

Segrè, Emilio. 1993. *A Mind Always in Motion: The Autobiography of Emilio Segrè*. Berkeley, CA: University of California Press.

Segrè, Emilio. 1967. Interview of Emilio Segré by Charles Weiner, in Segré's home, Lafayette, California, February 13, 1967. Niels Bohr Library & Archives. https://www.aip.org/history-programs/niels-bohr-library/oral-histories/4876

Silva, Indianara. 2022. Chien-Shiung Wu's contributions to experimental philosophy. In Freire, Olival Jr. (ed.), *The Oxford Handbook of the History of Quantum Interpretations*. Oxford: Oxford University Press.

Snyder, Hartland, Pasternack, Simon, and Hornbostel, John. 1948. Angular correlation of scattered annihilation radiation. *Physical Review*, 73, 440.

Süvegh, Károly and Marek, Tamas. 2011. Positron annihilation spectroscopies. In Vértes, Attila, Nagy, Sándor, Klencsár, Zoltan, Lovas, Rezsö G., Rösch, Frank. (eds.), *Handbook of Nuclear Chemistry*. Boston, MA: Springer.

Thomas, Heather. 2022. Dr. Chien-Shiung Wu: Premier nuclear physicist. *The Library of Congress Blogs*, May 17. https://blogs.loc.gov/headlinesandheroes/2022/05/dr-chien-shiung-wu-premier-nuclear-physicist

Tu, Weiming. 2012. Confucian spirituality in contemporary China. In Yang, Fenggang and Tamney, Joseph B. (eds.), *Confucianism and Spiritual Traditions in Modern China and Beyond*, Leiden: Brill, pp. 75–96.

University of Cambridge. 2009. Students from "Cambridge of the East" take part in exchange. https://www.cam.ac.uk/news/students-from-cambridge-of-the-east-take-part-in-exchange

University of Michigan. n.d. Michigan Union Opens. University of Michigan Student Life, University Unions History. https://uunions.umich.edu/history

US National Park Service. n.d. Dr. Chien-Shiung Wu, The First Lady of Physics https://www.nps.gov/people/dr-chien-shiung-wu-the-first-lady-of-physics.htm

USPS News. 2020. Hello, 2021 – U.S. Postal Service announces upcoming stamps. *USPS News*, November 17, 2020. https://about.usps.com/newsroom/national-releases/2020/1117-usps-announces-upcoming-stamps.htm

USPS News. 2021. Nuclear physicist Chien-Shiung Wu to be honored on a forever stamp. *USPS News*, February 1. https://about.usps.com/newsroom/national-releases/2021/0201ma-nuclear-physicist-chien-shiung-wu-to-be-honored-on-forever-stamp.htm

Wang, Zuoyue. 2008. Chien-Shiung Wu. In Koertge, Noretta (ed.), *New Dictionary of Scientific Biography*. Detroit, MI: Charles Scribner's Sons/Thomson Gale, pp 363–368.

Wheeler, John A. 1946. Polyelectrons. *Annals of the New York Academy of Sciences*, XLVIII, 219–238. https://catalog.hathitrust.org/Record/001993508

Whitaker, Andrew. 2016. *John Stewart Bell and Twentieth Century Physics: Vision and Integrity*. Oxford: Oxford University Press.

Wu, Chien-Shiung and Shaknov, Irving. 1950. The angular correlation of scattered annihilation radiation. *Physical Review*, 77(1), 136–136. https://doi.org/10.1103/PhysRev.77.136

Wu, Chien-Shiung. 2008. The discovery of the parity violation in weak interactions and its recent developments. *Lecture Notes in Physics*, 746, 43–70. https://doi.org/10.1007/978-4-431-77056-5_4

Yang, Chen Ning. 1958. Law of parity conservation and other symmetry laws. *Science*, 127(3298), 565–569. https://doi.org/10.1126/science.127.3298.565

Yang, Chen Ning. 1950. Selection rules for the dematerialization of a particle into two photons, *Physical Review*, 77(2), 242–45, https://doi.org/10.1103/PhysRev.77.242

Yomtov, Nelson. 2018. *Chien-Shiung Wu: Nuclear Physicist. Women in Science.* Minneapolis, MN: Essential Library, an imprint of Abdo Publishing.

Yuan, Jada. 2021. Discovering Dr. Wu, *Washington Post*, December 13. https://www .washingtonpost.com/lifestyle/2021/12/13/chien-shiung-wu-biography-physics-grandmother

Zeilinger, Anton. 1999. Experiment and the foundations of quantum physics. *Reviews of Modern Physics*, 71(2), S288–S297. https://doi.org/10.1103/RevModPhys.71.S288

Zhu, Yuelin. 2001. *Chien-Shiung Wu: An Intellectual Biography.* PhD Thesis. Boston, MA: Harvard University.

11

From Quantum Physics to Ethics: Grete Hermann on Heisenberg's Cut

ANDREA REICHENBERGER

11.1 Introduction

In recent years, Grete Hermann's (1901–1984; Figure 11.1) contributions to the philosophy of quantum physics have increasingly become the focus of scholarly attention. For example, Brigitte Falkenburg speaks of her "late appraisal" (Falkenburg, 2021); Mara Beller honors her in the book *Quantum Dialogue* (Beller, 1999); and Louisa Gilder describes the conversations between Werner Heisenberg, Carl Friedrich von Weizsäcker. and Hermann in her popular book *The Age of Entanglement* (Guilder, 2009). Of particular note are also the anthology on Hermann, edited by Guido Bacciagaluppi and Elise Crull (Crull and Bacciagaluppi, 2016), and the German-language edition by Kay Herrmann (Herrmann, 2019). Both works give an impressive overview of Hermann's life and work. The German edition further includes many of Hermann's important writings on physics, natural philosophy, and epistemology as well as excerpts from her correspondence (Reichenberger, 2020).

Considering the state of research, one point is, however, striking. Almost all published scholarship on Hermann deals either biographically and historically with her political activism (e.g. Hansen-Schaberg, 2016) – from her anti-fascist resistance, her participation in the reconstruction of the Federal Republic of Germany in the postwar period, up to her educational policy – or philosophically with her

I would like to express my deepest gratitude to Patrick Charbonneau, Michelle Frank, Margriet van der Heijden, and Daniela Monaldi for their highly professional editing support and enormously helpful feedback. I am very thankful to the inspiring working group Women in the History of Quantum Physics for letting me be part of this incredible network. I would also like to thank the librarians of the Archive of the Friedrich Ebert Foundation in Bonn/Bad Godesberg and the stuff of many other archives that I have consulted. Without their friendly and obliging assistance and help, my archival work and thus my contribution would not have been possible. I had the pleasure of working as a junior research group leader at the Department of Mathematics at the University of Siegen and most recently at the Deutsches Museum at the Technical University Munich, where I not only received financial support, but was also able to work with great and impressively competent colleagues. Translations are from the author, unless otherwise specified. Funding source: Deutsche Forschungsgemeinschaft Award Identifier / Grant number: RE 2929/4-1

Figure 11.1 Portrait of Grete Hermann c. 1932. © Archive of the Friedrich Ebert
Foundation Bonn/Bad Godesberg (AdsD). Photo collection. Sign. 6/FOTA029916.

systematic interpretation of quantum mechanics (e.g. Cuffaro, 2022). This chapter
aims to build a bridge between the two facets by considering her contributions to
quantum theory and ethics. In particular, Hermann's previously unknown contribution to the Third Zürich Talks in 1951 (see Section 11.7), which outlines mutual
influences between philosophical debates, ethical issues, and politics, is presented
as an appendix to the chapter. More specifically, after a brief biographical sketch
(Section 11.2), her views of Johann von Neumann's proof (Section 11.3) and on
Werner Heisenberg's cut (Section 11.4) are considered. Section 11.5 takes a look at
a previously unknown lecture that Hermann gave at the Third Zurich Talks in 1951.
It broadens and deepens our perspective on Hermann's critical reflection on the
philosophical foundations of quantum mechanics. In a concluding outlook
(Section 11.6), some lessons are drawn from today's perspective for rediscovering
Hermann's work.

11.2 Grete Hermann: A Biographical Sketch

Hermann was Emmy Noether's first doctoral student in Göttingen, graduating in
1925 after defending a dissertation on the theory of polynomial ideals (Hermann,
1926). That same year, on December 10, she passed the examination for the
teaching profession at secondary schools (first state examination) with mathematics
and physics as the main subjects and philosophical propaedeutics as an additional

one.[1] Her examiner in philosophy was Leonhard Nelson, who had become *Privatdozent* in Göttingen with the (eminent) support of David Hilbert. The following January, Hermann became Nelson's private assistant and worked with him on a volume about ethics and pedagogy.

Nelson was a charismatic figure, an ethical socialist, and both an anti-democrat and an anti-Marxist. He proposed a new kind of socialism based on a neo-Kantian theory of values and virtues in the tradition of Jakob Friedrich Fries (1773–1843), a strong liberal and unionist. Nelson founded the New Friesian School, the Jakob Friedrich Fries Society, the Philosophical-Political Academy, and – together with mathematician and educationalist Minna Specht – the International Socialist Fighting League (Internationaler Sozialistischer Kampfbund, ISK), a successor to the International Socialist Youth League. The neo-Kantianism of the Nelson School was politically and ethically motivated. Philosophy was understood as a way of life. Political activism and philosophical publishing practices went hand in hand.

After Nelson's untimely death in 1927, Specht, an educator, teacher, and socialist, became director of the Philosophical-Political Academy. Hermann closely collaborated with her and supported her work. Together, they decided to posthumously publish Nelson's work and to continue the pedagogical and philosophical reform that he had initiated. It was therefore no coincidence that Hermann later published her two most important works on the philosophy of quantum mechanics in the series of treatises from the Friesian School started by Nelson, *Neue Folge der Abhandlungen der Fries'schen Schule* (Hermann, 1935, 1937).

In September 1934, Hermann participated at the Eighth International Congress of Philosophy in Prague. She actively contributed to a number of discussions on specific topics, including "L'importance de l'analyse logique pour la connaissance" (The importance of logical analysis for knowledge), "Point de vue descriptif et point de vue normatif dans les sciences sociales" (Descriptive and normative perspectives in the social sciences), and "La mission de la philosophie dans notre temps" (The mission of philosophy in our time). At the same place, the First International Congress on the Unity of Science was held as a satellite to the main event.

Two years later, June 21 to 26, 1936, the Second International Congress on the Unity of Science took place in Copenhagen. Hermann was once again among the participants. Her review of the lecture of philosopher Moritz Schlick, one of the best-known representatives of the Vienna Circle, was published in the journal *Erkenntnis* later that year (Hermann, 1936)[2]. In June 1936, Hermann – along with

---

[1] Inge Hansen-Schaberg, an expert in educational science and women's studies in exile, wrote excellent contributions to the biography of Hermann, for example, Hansen-Schaberg (2012, 2016). Unfortunately, there is no up-to-date biography that comprehensively addresses Hermann's ethical, political, and scientific work in its systematic and historical context.

[2] The lecture was delivered by philosopher Philipp Frank; see Schlick (1936). For Frank's own lecture see Frank (1936).

fellow German philosophers Eduard May and Thilo Vogel – was awarded the Richard Avenarius Prize by the Academy of Sciences of Saxony, in Leipzig, for her work at the intersection of quantum physics and philosophy. September 5 to 7, 1936, at the invitation of Hermann, a small group of scientists and philosophers, including Carl Friedrich von Weizsäcker, Paul Bernays, Otto Meyerhof, and Adolf Kratzer, met at the Kaiser Wilhelm Institute for Medical Research at the University of Heidelberg to discuss the philosophical implications of the latest developments in quantum mechanics. Unfortunately, Heisenberg, Heinrich Scholz, and Walter Dubislav canceled at short notice. In addition, Moritz Schlick, who had agreed to speak, was murdered on June 22, 1936. (For further reading and more details on the Heidelberg conference, see Reichenberger (2024).)

Between 1931 and 1936, Hermann was highly active in the fight against National Socialism, playing a leading role in the ISK. By then, the ISK operated throughout Germany and had numerous contacts abroad, including in Switzerland and in China. In 1931, Willi Eichler moved the headquarters of the ISK from Göttingen to Berlin, and most of the members followed him there. Other members initially continued to teach at the Walkemühle school near Kassell, which was founded by Specht in 1924. When the Walkemühle school was seized and occupied by the Sturmabteilung[3] in March 1933, Specht succeeded in reopening the school in Østrupgaard, near Faaborg on the island of Funen in Denmark. This would not have been possible without the helping hands and the organizational support of Gustav Heckmann and Hermann. When the resistance work against National Socialism in Berlin became too dangerous, some of the ISK members emigrated to Paris under the leadership of Eichler. Hellmuth von Rauschenplat, alias Fritz Eberhard, took over the leadership of the ISK in Germany. The members who remained in the country spread their resistance work across German cities and towns. Hermann stayed in Berlin for a short time, but then accepted Heisenberg's invitation to come to Leipzig. (Heisenberg probably knew nothing of Hermann's close involvement in the resistance movement at the time.) Between 1934 and 1936 Hermann worked and lived on Østrupgaard.

Shortly before emigrating to London in 1938, Hermann contracted a marriage of convenience to a British man, which gave her the protection of local citizenship.[4] She joined the German Educational Reconstruction Committee, a project in Great Britain to plan and prepare a reorganization of the education system in postwar Germany (Klafki, 2019).

After Hermann's return to Germany in 1946, she joined the Social Democratic Party (Sozialdemokratische Partei Deutschlands, SPD), helped to establish the

---

[3] The Sturmabteilung (or Storm Troopers, SA) was the paramilitary wing of the Nazi Party.
[4] From then on, her official name was Grete Henry-Hermann, but this chapter refers to her as Grete Hermann throughout.

Education and Science Union, and dedicated all her efforts to education and teaching practices. From 1949 until her retirement in 1966, Hermann held a full professorship at Bremen's Pädagogische Hochschule. She also played a decisive role in UNESCO's educational policy in postwar Germany, in the West German Campaign against Atomic Death (*Kampf dem Atomtod*), and behind the scenes of the SPD's Godesberg Program.

From 1961 to 1978, Hermann was also the chairwoman of the Philosophical-Political Academy.[5] She supervised the publication of Nelson's works and was co-editor of the international journal of analytic philosophy, *Ratio*. Hermann's last public lecture was in 1981 at a Kant Congress of the Friedrich Ebert Foundation in Bonn (Herrmann and Neißer, 2023), where she spoke about Nelson and the foundations of liberal socialism.

Hermann's political activism is usually treated separately from and independently of her contributions to the philosophical interpretation of quantum mechanics. Léna Soler and Alexander Schnell (Soler and Schnell, 2001), as well as Giulia Paparo and co-workers (Paparo, 2012) did emphasize the importance of the intellectual background: Hermann's motivation to deal with quantum mechanics was the Neo-Friesian philosophy as propagated by Nelson. However, only Paparo linked Grete Hermann's views on politics and ethics to her interpretation of quantum physics when she wrote: "Finally, Hermann draws some ethical parallels from her analysis of the situation in physics: in the same way quantum mechanics shows us the impossibility of absolute knowledge, in ethics, it is possible to look for self-determination free from absolute values"(Paparo, 2012, p. 65).

For Grete Hermann, the method of philosophizing was the Socratic dialogue. The consistent application of the Socratic dialogue as a method of rational argumentation according to given rules was the hallmark of the Nelson circle. It was therefore a consequent step for Hermann to seek the dialogue with Heisenberg and other physicists in order to discuss the relationship between quantum mechanics and philosophy. A key question was whether Kant's philosophy could still be upheld in view of these scientific results in physics. For Grete Hermann, however, a moral lesson was more important: The ideal of absolute certainty and universal truth should be abandoned.

Hermann's concept of critical philosophy was committed to the principal revisability of knowledge that allowed her to criticize any politicization of science and, conversely, any scientification of politics and ethics. It is therefore worthwhile to link Hermann's discussion of quantum mechanics much more closely with her extraordinary life and her other philosophical works on politics, ethics, and law.

---

[5] The "Philosophisch-Politische Akademie" (PPA) is a registered non-profit association in Bonn. When the Nazis seized power in 1933, the PPA was abandoned and re-established in 1949. Minna Specht once again became the chairwoman. She was succeeded by Grete Hermann from 1961 until 1978.

This chapter cannot address these issues comprehensively. However, we will use an example to show that Hermann's interest in quantum theory was sparked by her criticism of reductionist scientism. Just as sociology cannot be reduced to biology, cultural evolution to biological evolution, or ethics to legal positivism, neither can human freedom of will and freedom to act be justified and explained by quantum-mechanical indeterminism.

11.3 Grete Hermann on Johann von Neumann's Proof

In his book on the mathematical foundations of quantum mechanics (Von Neumann, 1932), Johann von Neumann proposed an axiomatic formulation of quantum theory in terms of the Hilbert space formalism and the statistical interpretation based on Max Born's rule which allows one to calculate the probabilities for different eigenvalues of a certain observable. According to the standard reading, the book became famous because Von Neumann wanted to prove that there can be no theories with hidden variables. But this interpretation requires a critical modification, for Von Neumann intended to prove a weaker assertion, namely that there can be no theory of hidden variables with dispersion-free states that simultaneously and consistently assigns unique values to all observables of a quantum system (Corry, 2024).

A considerable number of historical studies have argued that it would be anachronistic to read Von Neumann against the background of John Bell's critique and the ensuing debate about no-go theorems, for example, Rédei (1996), Stöltzner (2001), Acuña (2021), and Mitsch (2022). Von Neumann himself considered his axiomatized quantum mechanics provisional insofar as it was based on the physical and mathematical results that were available at the time when he wrote the book (Golub and Lamoreaux, 2024). In sum, Von Neumann's axiomatic approach consisted of three steps:

(1) the analogous formal presentation of matrix mechanics and wave mechanics as an eigenvalue problem;
(2) the presentation of quantum mechanics as a theory of Hilbert space;
(3) the representation of quantum-mechanical observables as Hermitian operators.

Von Neumann introduced the Hilbert space formalism and demonstrated the equivalence of matrix and wave mechanics by showing that they were two different instantiations of Hilbert space formalism, "one being the space of square-summable sequences, the other being the space of square integrable functions," in the words of a recent historical treatment (Duncan and Janssen, 2023, p. 642). In addition, Von Neumann introduced a set of five assumptions, from which he derived that there were no dispersion-free states. The first three assumptions (axioms **A–C**) refer to relations of observables and quantum-mechanical states, and are much less contro-versial than the last two assumptions – axioms **D** and **E** – which describe properties that an expectation value functional should have (Von Neumann, 1932, p. 163 ff.). The fifth postulate, concerning the linearity of the expectation values of observa-bles, has been the subject of most discussion. In the course of this discussion, the question of whether Hermann had actually discovered that Von Neumann's proof was circular increasingly took center stage; see, for example, (Mermin and Schack, 2018).

In a 1935 essay on the natural-philosophical foundations of quantum mechanics, Hermann had indeed criticized Von Neumann's proof. According to her, his proof stands and falls with the assumption that the expectation value of a sum of physical quantities is equal to the sum of the expectation values of both quantities; in German: "Der Erwartungswert einer Summe physikalischer Größen ist gleich der Summe der Erwartungswerte beider Größen. Mit dieser Voraussetzung steht und fällt der NEUMANNsche Beweis" (Hermann, 1935, p. 223). At this point, it is worth taking a closer look at Hermann's original argument (Hermann, 1935, p. 224):[6]

The interpretation of the expression $(R\varphi, \varphi)$ is decisive for the entire proof. As the formalism shows, the approach that the expression indicates the expected value of the quantity $\mathfrak{R}$ on systems in the state φ essentially comes down to the same thing as the probability interpret-ation of the wave functions. The considerations linked to this interpretation can therefore be transferred here without further ado. Except for the proof of the impossibility of new determinants, which is to be provided here first, the expression $(R\varphi, \varphi)$ may denote the expectation value of the $\mathfrak{R}$ -measurements, but only for ensembles (*Scharen*) composed of physical systems in the state φ. However, for $(R\varphi, \varphi)$ to be applicable on any subset of such ensembles as well, one must leave open whether the expected value is the same in all such subsets that can be created on the basis of new characteristics. If this is left open, however, one can no longer conclude from the summation rule applicable to $(R\varphi, \varphi)$ that in these subsets, too, the expected value of the sum of physical quantities is equal to the sum of their expected values. This, however, eliminates a necessary step in NEUMANN's proof. If, on the other hand, one – with NEUMANN – does not reveal this step, one has tacitly included in the interpretation the unproven assumption that the elements of a set of physical systems characterized by φ cannot have any distinguishing features on which the result of the $\mathfrak{R}$-measurement depends. The impossibility of such features, however, is precisely the assertion to be proven. Thus, the proof runs into a circular argument.[7]

---

[6] Hermann adopted Neumann's notation of systematically distinguishing between physical quantities using capitals like $\mathfrak{R}$ and quantum-mechanical operators non-bold capitals like R (Dieks, 2017). Also, note that Hermann, who was a socialist, does not include *von* Neumann's aristocratic particle in her writing.

[7] Original: "Die Interpretation des Ausdrucks $(R\varphi, \varphi)$ ist hierbei entscheidend für den ganzen Beweis. Der Ansatz, daß er den Erwartungswert der Größe $\mathfrak{R}$ an Systemen im Zustand φ angibt, kommt – wie der Formalismus zeigt – im wesentlichen auf dasselbe heraus wie die Wahrscheinlichkeitsdeutung der Wellenfunktionen. Die an diese Deutung angeknüpften Überlegungen lassen sich daher hier ohne weiteres übertragen: Bis auf den Beweis der Unmöglichkeit neuer Bestimmungsstücke, der hier erst erbracht werden soll, darf der Ausdruck $(R\varphi, \varphi)$ den Erwartungswert der $\mathfrak{R}$-Messungen nur für solche Scharen physikalischer Systeme bezeichnen, denen die aber

Much later, Max Jammer rediscovered Hermann's critique while he was working on his book *The Philosophy of Quantum Mechanics* (Jammer, 1974). Jammer objected that Grete Hermann's critique was not justified because Von Neumann used the case of pure quantum states to motivate the axiomatic requirement of linearity, not to derive it (Mermin and Schack, 2018).

In line with Jammer, Dennis Dieks argued that "Grete Hermann's work, although interesting both historically and philosophically, did not succeed in coming to grips with von Neumann's proof either. Interestingly, Hermann's technical objections are – in their detail – equally misdirected" (Dieks, 2017, p. 137). Dieks referred to Hermann's works from 1933[8] and 1935 (Hermann, 1935), without apparently having read the correspondence between her and Jammer. He could then have nuanced and contextualized many of his critical points against Hermann since, in this correspondence, Hermann herself nuanced her previous works.

For example, in her letter to Jammer on April 11, 1968 (Figure 11.2), she underlined that the assumption of linear additivity of expectation values for hypothetical dispersion-free states would be a decisive weakness of Von Neumann's proof, if that proof were based on the assumption that $\langle \varphi | R \varphi \rangle$ represented the average value of measurements in any ensemble whose elements agreed with each other not only with respect to φ but also with respect to arbitrary as yet undiscovered quantities. However, she also admitted (Herrmann, 2019, p. 607):

Well, if my accusation of circular reasoning is based on a misunderstanding, then, of course, I am gladly retracting it. However, with or without a vicious circle, Neumann's approach does not help when it comes to the subject I was dealing with in my discussion of quantum mechanics at the time. I was provoked to my attack [against Neumann] at the time in Leipzig by repeated references from various interlocutors that Neumann had proven that "hidden parameters" that I had repeatedly thrown into the debate are mathematically impossible. However, I had not restricted the question about hidden variables to the condition "as far as quantum mechanical states are concerned," beyond which Neumann does not go![9]

---

auch nur die Bedingung auferlegt ist, sich im Zustand φ zu befinden; er muß es hingegen, um anwendbar zu bleiben, offen lassen, ob dieser Erwartungswert auch in allen Teilmengen solcher Scharen, die aus diesen auf Grund irgend welcher neuer Merkmale herausgegriffen werden, derselbe ist. Läßt man das aber offen, dann kann man aus der für (Rφ, φ) geltenden Summationsregel nicht mehr folgern, daß auch in diesen Teilmengen der Erwartungswert der Summe physikalischer Größen gleich der Summe ihrer Erwartungswerte ist. Damit aber entfällt ein notwendiger Schritt im NEUMANNschen Beweis. Gibt man dagegen – mit NEUMANN – diesen Schritt nicht preis, dann hat man in die Interpretation stillschweigend die unbewiesene Voraussetzung aufgenommen, daß es an den Elementen einer Schar durch φ charakterisierter physikalischer Systeme keine unterscheidenden Merkmale geben könne, von denen das Ergebnis der ℜ-Messung abhängt. Die Unmöglichkeit solcher Merkmale ist aber gerade die zu beweisende Behauptung. Der Beweis läuft somit auf einen Zirkel hinaus."

[8] Churchill/DRAC 3/11. The Papers of Professor Paul Dirac, Churchill Archives Centre, Churchill College, Cambridge. See Hermann (2019).

[9] Original: "Nun, wenn mein Zirkelschluss-Vorwurf auf einem Missverständnis beruht, dann nehme ich ihn natürlich gern zurück. Allerdings: ob mit oder ohne Zirkelschluss, für das Thema meiner damaligen Auseinandersetzung mit der Quantenmechanik helfen Neumanns Ausführungen mir nicht weiter. Provoziert zu meinem Angriff wurde ich damals in Leipzig durch wiederkehrende Hinweise verschiedener Gesprächspartner, Neumann habe die von mir immer wieder in die Debatte geworfenen 'verborgenen Parameter' als mathematisch

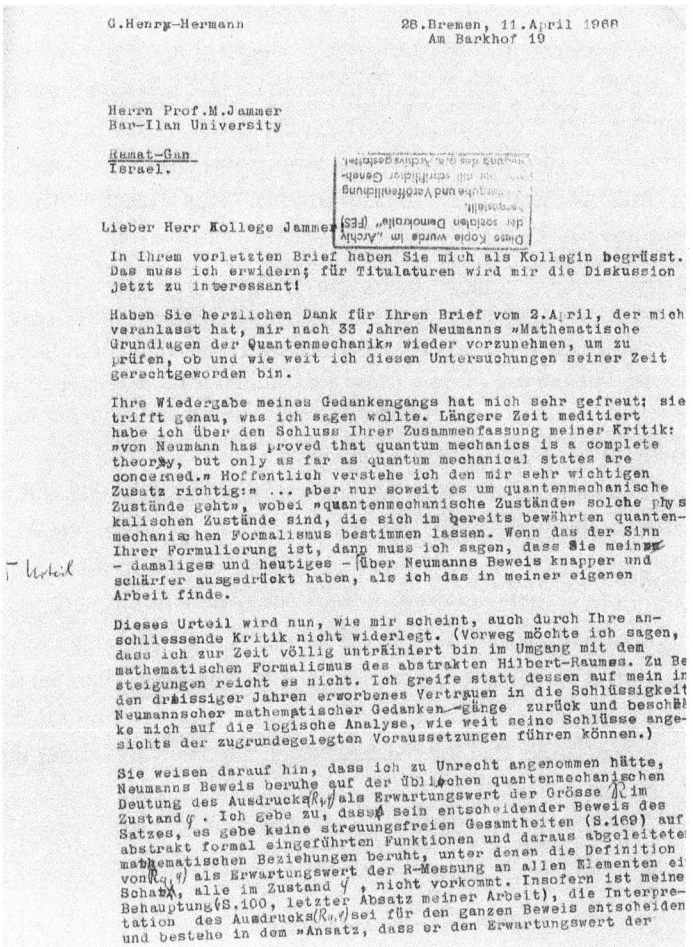

Figure 11.2 Letter from Grete Hermann to Max Jammer dated April 11, 1968. Archive of the Friedrich Ebert Foundation Bonn/Bad Godesberg (AdsD). Sign. 1/ GHAJ000010, No. 54.

Thus, Hermann clearly recognized what Dieks was underlining: Von Neumann's axiomatization was embedded in the formal setting of self-adjoint operators in a Hilbert space and based on Born's statistical interpretation of quantum mechanics. She did not intend to criticize this point, nor did she intend to make a plea for hidden-variables research. The main point of her criticism was that Von Neumann drew an overly hasty conclusion from his result. Von Neumann claimed that the

unmöglich erwiesen. Ich hatte die Frage nach ihnen aber eben nicht an die Bedingung gebunden 'as far as quantum mechanical states are concerned,' über die Neumann nicht hinausgeht!"

statistical character of quantum mechanics necessarily required the "decision ... against causality, because all ensembles have dispersions, even the homogeneous" (Von Neumann, 1932, p. 171).[10] Hermann considered this conclusion to be overly hasty, because it failed to distinguish between indeterminism, predictability, and causality. She argued that the theory of quantum mechanics, asserting the unpredictability of precise results in measurements, was faced with the following dilemma (Hermann, 1935, p. 226):

Either quantum theory itself designates the causes that completely determine these measurement results – but how does it then want to prevent the researcher from determining these causes in individual cases and predicting the measurement result from them? Or quantum theory does not designate these causes – how does it then want to exclude the possibility of future discoveries of these causes without arbitrarily anticipating the exploration of as yet unknown areas of nature?[11]

Hermann's point of criticism was that the statistical nature of quantum mechanics does not necessarily imply the causality or acausality of quantum processes. What Von Neumann's theorem really shows is that the standard quantum formalism is complete insofar as it cannot be transformed into a hidden-variable theory within Hilbert space. In general, any axiomatization of a theory should best be understood as provisional, practical, and tailored to theory development and exploration. The meaning of axioms, however, is determined in terms of content. Only after an interpretation has taken place can one prove, for example, that the quantum-mechanical probabilities defined as such fulfill the axioms, for example, by assigning the probability amplitudes to projection operators. Axiomatization is context-dependent and contingent.

11.4 Grete Hermann on Werner Heisenberg's Cut

The fateful years of laying the foundational stones of quantum mechanics have been told many times – see, for example, Mehra (1987), Schirrmacher (2005), and Renn (2013). From a historical perspective it is remarkable that both Erwin Schrödinger and Heisenberg claimed as an advantage of their own approach that it was more *anschaulich* (intuitive). In 1927, Heisenberg presented his paper "Über den anschaulichen Inhalt der quantentheoretischen Kinematik und Mechanik" (On the intuitive content of quantum theoretical kinematics and mechanics) (Heisenberg, 1927). This work, which was a response to Schrödinger, is considered

---

[10] Original: "Damit ist im Rahmen unserer Bedingung die Entscheidung gefallen, und zwar gegen die Kausalität: denn alle Gesamtheiten streuen, auch die einheitlichen."
[11] Original: "Entweder sie nennt selber die Ursachen, die diese Messungsergebnisse vollständig bestimmen – wie aber will sie es dann dem Forscher verwehren, diese Ursachen im einzelnen Fall zu bestimmen und aus ihnen das Messungsergebnis vorauszuberechnen? Oder sie nennt diese Ursachen nicht – wie will sie dann, ohne willkürlich der Erforschung noch unbekannter Naturgebiete vorzugreifen, die Möglichkeit künftiger Entdeckungen dieser Ursachen ausschließen?"

to be the birth of the uncertainty principle. It states "that canonically conjugate observables can be measured simultaneously only with the constraint that the product of their mean errors should be no less than a limit set by Planck's constant" (Ozawa, 2015, p. 2006).

Heisenberg declared at the beginning of his paper, "We believe we understand the physical content of a theory when we can see its qualitative experimental consequences in all simple cases and when at the same time we have checked that the application of the theory never contains inner contradictions."[12] Heisenberg proceeded with a thought experiment. It is called "Heisenberg's microscope" because it is about the measurement of the position of an electron by a microscope. Heisenberg pointed out that the position of an electron can be determined only up to an uncertainty Δx that depends on θ and the wavelength λ of the incoming light. The measurement of the electron's position is more accurate the shorter the wavelength of the light used to illuminate the electron is and further reasoning then led to his "inequality relation."[13] Mathematically, this inequality relation, later called "uncertainty principle," describes a fundamental limit for the product of the accuracy of certain related pairs of measurements of a quantum system, such as position and momentum. Such pairs of variables are known as complementary variables or canonically conjugate variables.

The mathematical formalism of the uncertainty principle was worked out later in the wake of the discussion on Heisenberg's papers from 1925 and 1927. Against this background Heisenberg's thought experiment caused a lot of confusion and trouble because it seemed to provide an explanation of the uncertainty principle based on classical optics, which turned out to be highly problematic. Bohr, among others, criticized that the uncertainty relations were not a consequence of the limited accuracy of the measuring instruments (Murdoch, 1987). Quantum probability and randomness, he stressed, are irreducible and intrinsic features of quantum mechanics.

In response to such criticism, Heisenberg introduced another thought experiment, in which he introduced a hypothetical cut in a system consisting of a quantum object and a measuring apparatus (Heisenberg, 1934). Heisenberg proposed to divide such a system into two parts: one, the quantum object to be measured; the other, the measuring apparatus. According to Heisenberg a quantum state is disturbed by any kind of measurement and consequently changes, because a measurement is always accompanied by an uncontrollable change of state.

---

[12] Original: "Eine physikalische Theorie glauben wir dann anschaulich zu verstehen, wenn wir uns in allen einfachen Fällen die experimentellen Konsequenzen dieser Theorie qualitativ denken können, und wenn wir gleichzeitig erkannt haben, daß die Anwendung der Theorie niemals innere Widersprüche enthält" (Heisenberg, 1927, p.172).

[13] A sketch of Heisenberg's thought experiment is given, for example, by Wick (1995, p. 34 ff.).

For Heisenberg, the object–instrument divide was a clear tool to limit the applicability of classical notions – like position or momentum – to microphysical phenomena. Bohr, among others, understood the "cut" as a feature of complementarity between the observed system and the observer. Bohr had introduced the concept of complementarity in his lecture at the International Congress of Physics, held in the Italian city of Como in 1927. To quote the famous closing sentence of his Como lecture (Bohr, 1928, p. 590):

I hope, however, that the idea of complementarity is suited to characterize the situation, which bears a deep-going analogy to the general difficulty in the formation of human ideas, inherent in the distinction between subject and object.

Bohr formulated the concept of complementarity qualitatively and in an interpretative variety of meanings, thus sparking many controversies and misunderstandings (Favrholdt, 1999). However, a certain reminiscence of the Kantian terminology is always present (Kaiser, 1992). Alluding to Kant who maintained that "we can cognize of things a priori only what we ourselves have put into them" (Kant, 1998, version B, p. xviii), in Bohr's view, the behavior of atomic and subatomic objects could not be separated from the measuring instruments that constitute the context in which the measured objects behave.

Hermann, who took up Bohr's considerations, made a very clear distinction between three different concepts of complementarity that Bohr used in relation to three different applications in quantum theory (Hermann, 1935). First, complementarity referred to the wave model and the particle model, often called the continuous and discontinuous pictures of atomic phenomena. Second, complementarity referred to the assignment of physical variables, namely to a couple of conjugate variables like position and momentum. Third, complementarity referred to the relationship between causality and space–time localization as mutually exclusive but complementary.

All three variants of complementarity had, according to Hermann, one thing in common: the quantum-mechanical characterization of a state is dependent on the measurement context. This kind of context-dependence became obvious through Heisenberg's cut. It allowed one to argue: (i) for the impossibility of any extension of quantum theory by hidden variables, (ii) for a specific kind of retrodictive causality, and (iii) against Nelson's theory of free will.

By argument (i), Hermann concluded that the quantum-mechanical formalism excludes the question of new variables to be discovered (Hermann, 1935). Any extension of quantum theory would lead to a causal overdetermination and by that conflict with the hitherto existing results (Henry-Hermann, 1948), because it was precisely the impossibility to uniquely determine Heisenberg's cut which, according to Hermann, would forbid one to extend quantum theory. The cut is, to a certain extent, left to the physicist's decision and is not clearly predetermined by the natural phenomenon under investigation nor objectively restricted to a specific spatio-temporal location of the process under investigation.

Argument (i) is closely linked to argument (ii). By distinguishing between causality and predictability, Hermann aimed to explain why quantum theory precluded predictability without excluding a subsequent or after-the-fact reconstruction of the causes of the particular outcome. Thus, a limitation on predictability as codified in the probability distribution would not necessarily result in a violation of causality. That meant that quantum mechanics "does not claim the existence of acausal processes. On the contrary, the unavoidable indeterminacy in the prediction of atomic processes arises precisely from the proof that at no point in the process itself is anything causally indeterminate and can therefore be determined by future discoveries if necessary," as she later argued (see Section 11.7).

Hermann used the term "overdetermination" to argue against what she considered to be a mistaken notion of causality, namely to presuppose that every kind of a causal relationship could be represented by a causal chain, that is, a linear ordered sequence of events linking the causes with its effects. The idea that causality is a cause-and-effect chain and that a violation of causality occurs when the chain is interrupted is a metaphor from everyday language. This also applies to the concept of a causal leap. Whether this picture is adequate to understand the nature of causality and its possible violation in quantum mechanics is questionable. If we assume the principle of complementarity and interpret it as a principle of contextuality, then it is in any case misleading to assume a picture of causality as a single causal chain that embraces every aspect of a process. This does not mean the refutation of causality, but it only implies that the notion of causality on which Laplace's determinism[14] is based must be modified. So much for Hermann's reflections on causality – for further reading see Crull (2022) and Cuffaro (2022). Her considerations on causality leave many questions unanswered, which Hermann does not address in detail, and which concern, among others, the concept of time in quantum physics.

In argument (iii), Hermann argued that Heisenberg's cut uncovered a fundamental mistake Nelson made when analyzing human action in analogy to

---

[14] An equation of motion in classical mechanics is a second-order ordinary differential equation, that is, a function of the position $\mathbf{r}$ of the object, its velocity (the first time derivative of $\mathbf{r}$, $\mathbf{v} = \frac{d\mathbf{r}}{dt}$, and its acceleration (the second derivative of $\mathbf{r}$, $\mathbf{a} = \frac{d^2\mathbf{r}}{dt^2}$), and time t. According to the standard reading this means that if we know the positions and velocities of all particles at a certain instant, and all the forces that are present, we can uniquely determine all (past, present, and future) states of the system. This statement is disputable from both a systematic and historical point of view even within classical mechanics. From a systematic point of view, the differential equations alone are not sufficient to explain the concept of time and the "split" between past, present, and future. Differentiable functions are continuous. Insofar as there is no split at all, the distinction between past, present, and future becomes meaningless. From a historical point of view it is noteworthy that Pierre-Simon Laplace himself, in his *Essai philosophique sur les probabilités* (Laplace, 1814 [2009]), founded his idea of determinism metaphysically, on the basis of the principle of sufficient reason that presupposes an intelligence with perfect knowledge and perfect calculating capacities, that is, a god-like entity referred to as the "eternal geometer."

physical processes, in order to find a solution to the apparent conflict between natural necessity and human's freedom and responsibility. For Nelson, ethics was a practical science. He distinguished it from theoretical science, that is to say, natural science. While the latter is based on natural laws, ethics is based on moral laws. While natural science explores cause-and-effect relationships, ethics deals with reason and values. Determinism, understood from the perspective of natural science, means causal determinism: every "effect" necessarily has to have a "cause." In ethics one would, by analogy, say that everything happens for a reason. However, even if it is the case that everything has a reason or a cause, it does not follow that reasoning could be reduced to causes, or vice versa. But that would be precisely the case if, according to Nelson, we turned ethics into natural science or vice versa. In the entry for Nelson in the fifth volume of the *Encyclopedia of Philosophy*, Hermann pointedly summarized Nelson's position (Henry-Hermann, 1967, p. 466):

Ethical standards, according to Nelson, are valid for human action in nature and are therefore directly relevant to two apparently mutually exclusive forms of legality: the theoretical form, according to which everything that happens in nature (including human behavior) is determined by natural laws working through the existing powers, and the practical one, which presents the human will with duties that can either be violated and ignored or become man's purpose. Within the framework of the critique, Nelson thoroughly examined the question of how man's freedom could be reconciled with this natural law.

Hermann considers the reason given by Nelson as to why indeterminism is not a condition of the possibility of free will to be mistaken. First, Nelson's analogy between "causally determined" and "determined by reason" fails. The analogy is not misguided because effects have causes and, analogously, consequences have reasons, but because Nelson assumed that both realms were completely determined. The assumption of completeness, however, is a metaphysical one that can only be justified by the existence of an omniscient and outstanding observer. Second, Nelson erroneously conflates determinism with causality. The assumption that "determinism equals causality" as well as that "indeterminism equals acausality" is misleading because it ignores in both cases the many faces and meanings of determinism and causality in very different contexts. Third, it is disputable to attribute to human beings the ability to liberate their own actions from chance. Chance always determines human life and actions, even if not completely.

At this point, Hermann brings Heisenberg's cut into play. We make a cut, so to speak, in which we move from the previous endeavor to another perspective, or another way of looking at things. The transfer of this mode of expression is justified in particular by the fact that it is not possible to identify objectively the cut in the process at which it must be made. It is not acausality, but contextuality which carries the Laplacian determinism of classical mechanics to the point of absurdity.

Analogous to the "insight that every physical description of nature is bound to the question on which it is based and cannot grasp its object from all sides that are essential to it," we have to respect that in everyday life "we remain bound to different questions and ways of looking at things, which limit each other in their application, but also complement each other" (Henry-Hermann, 2019, p. 401). If we ask for the function of free will as a premise for law and the rule of law which is based on moral autonomy, then we are operating in a different context than that of quantum-mechanical theory formation and its foundation. Hermann formulated the following thought experiment to make this point clear (Henry-Hermann, 2019, p. 401):

Let us therefore assume that there is no reason why a radium atom decays right now and not in 100 years, or why a quantum of light hits a photographic plate at this point and not at another. This would certainly not mean that the radium atom had the freedom of ethical decision to determine when it wanted to decay, or that the light quantum chose the place of its impact on the basis of ethical considerations. Human freedom, i.e., the freedom of choice, has by no means been proven possible by the acausality of atomic processes. For the people whose ethical decision we are asking about are not atoms or light quanta. And, on the other hand, quantum physical objects do not make ethical decisions.[15]

Hermann emphasizes that social interactions between people and quantum interactions are the topics of two different research areas. We do not assume that quanta are able to make ethical decisions and act morally. We attribute this ability to people as a right and a duty. Anyone who claimed that free will could be verified or falsified by the results of physical research would turn physics into physicalism. Such criticism is of course older than quantum physics. We do not need quantum physics to understand "that we are able to see and understand the world not only through physical-causal glasses" (Henry-Hermann, 2019, p. 406).[16] Just to give a well-known example: gravity is the cause of a stone falling to the ground, but it is not the reason why a person took their own life by jumping from a high-rise building. For this act, the responsible party is not gravity, but the person who made the decision.

One might object to Hermann's argument that complementarity and contextuality in the quantum world have nothing to do with the context dependence of our

---

[15] Original: "Nehmen wir also einmal an, es gäbe keine Ursache dafür, daß ein Radiumatom gerade jetzt und nicht etwa erst in 100 Jahren zerfällt, oder dafür, daß ein Lichtquant eine photographische Platte gerade an dieser und nicht an einer anderen Stelle trifft. Das hieße gewiß nicht, daß das Radiumatom die Freiheit ethischer Entscheidung besäße, zu bestimmen, wann es zerfallen wolle, oder daß das Lichtquant sich auf Grund ethischer Erwägungen den Ort seines Auftreffens aussuchte. Die menschliche Entscheidungsfreiheit ist durch die Akausalität atomarer Prozesse noch keineswegs als möglich erwiesen. Denn die Menschen, nach deren ethischer Entscheidung wir fragen, sind keine Atome oder Lichtquanten. Und diese mikrophysikalischen Partikel fällen keine ethischen Entscheidungen."
[16] "Die Einsicht, daß jede physikalische Naturbeschreibung an die ihr zugrundeliegende Fragestellung gebunden ist und ihren Gegenstand nicht von allen überhaupt für ihn wesentlichen Seiten erfassen kann, gibt den Weg wieder frei für die alte Überzeugung, daß wir die Welt nicht nur durch die physikalisch-kausale Brille betrachten können."

knowledge. That is correct. However, Hermann did not claim that one follows from the other. On the contrary, Heisenberg's cut, understood metaphorically, reminds us and suggests that we should differentiate here.

11.5 Widening Perspectives

The archival collection of the Swiss philosopher Magdalena Aebi (1898–1980) in the Zentralbibliothek Zürich contains a handwritten manuscript by Hermann dated August 1, 1951 (see Section 11.7). At first glance, this previously unknown manuscript only reproduces what she has published elsewhere. But it is worth reading more carefully. In this text, Hermann takes a remarkably harsh line with Kant. She states that "we can no longer share Kant's conviction that critical philosophy must succeed in completely liberating the rational moment of knowledge from the empirical and grasping it a priori in synthetic judgements."[17]

What is important, however, is less the content than the context. Right at the beginning, Hermann referred to Paul Bernays, who opened the Third Zürich Talks. Bernays had addressed the relationship between theory and experience with a new perspective that placed the empirical and the rational side of the genesis and foundation of knowledge on top of each other in a dialectic, and open-ended process (Bernays, 1952). Bernays concluded that the development of exact science had disavowed Kant's doctrine of synthetic knowledge a priori.

Hermann agreed with Bernays that it was not appropriate to oppose intuitiveness and conceptualization, experience and rationality in such a dualistic and static way as was done in Kantian philosophy. She used Heisenberg's cut in order to consolidate situatedness and perspectivity contrary to an all-encompassing worldview.

In 1948, Bernays had invited Wolfgang Pauli to edit a special issue on Bohr's complementary principle in *Dialectica*, a journal founded in 1946 by Bernays, along with Ferdinand Gonseth, a Swiss mathematician and philosopher, and Gaston Bachelard. In their opening remarks on the first issue, the editors presented their international journal as "a sustained philosophical effort" in order to keep science "on the horizons of human values" (Bachelard et al., 1947). Shortly after Pauli's return from Princeton to Zürich in 1946, he had become a member of the consulting board of the journal.

Since the late 1920s, Gonseth had been developing an approach to the foundations of mathematics and the philosophy of science in terms of an open philosophy, which was decisively ethically motivated. In 1938, he organized with the support of the International Institute of Intellectual Cooperation (a forerunner of the

---

[17] Original: "Ja wir können, im Licht dieser Erfahrung, die Kantische Überzeugung nicht mehr teilen, wonach es der kritischen Philosophie gelingen müsse, das rationale Moment der Erkenntnis vollständig vom Empirischen zu befreien und in synthetischen Urteilen a priori zu fassen."

UNESCO) the first "Entretiens de Zurich." From this meeting up to Gonseth's death in 1975, Bernays and Gonseth enjoyed an intellectual friendship. After World War II, UNESCO took over the patronage of the second meeting. It took place 10 years after the first, April 19 to 22, 1948, and had for topic the idea of dialectics.

The International Union of Logic, Methodology and Philosophy of Science, which was founded by Gonseth and Bernays together with Karl Dürr and Karl Popper in 1947, also organized four other Zürich Talks, the last of which took place in 1958. The topics were: the principle of duality (theory and experience), foundations and possible applications of probability theory and statistics, the status of a philosophy of science, and, finally, the Sixth Zürich Talk was dedicated to the question of whether and to what extent thinking machines can contribute to human knowledge.

In the special issue on the concept of complementary, Pauli wrote a detailed editorial.[18] The common editorial intention of the special issue was to extend the concept of complementarity beyond physics and to make it viable for an open philosophy. This led to heated discussions that were not just about conceptual issues. In particular, the link drawn between complementarity and dialectics suggested a Marxist underpinning of complementarity that was too political for Bernays, among others. No less controversial was the interpretation of the principle of complementarity from a Kantian perspective. Against this historical background, Hermann presented her contribution to the Third Zürich talk.

In the audience was the then graduate student Alexander Israel Wittenberg (1926–1965). Hermann's talk impressed him so much that he decided to make her broadening of the perspective on Heisenberg's cut the central idea of his doctoral dissertation on concept formation in the foundations of mathematics (Wittenberg, 1957). In 1957, Wittenberg completed his doctorate at the ETH Zürich under the guidance of Ferdinand Gonseth and Paul Bernays.[19] Wittenberg

---

[18] In this issue, Albert Einstein published "Quantum mechanics and reality" (Einstein, 1948). Other authors were, among others, Niels Bohr, Heisenberg, and Louis de Broglie.

[19] The following biographical outline can be found on the website of the Alexander Wittenberg Fonds (F0747), York University Archives & Special Collections (CTASC): "Wittenberg was born in Berlin in 1926 to a family of Russian Jewish immigrants. The family escaped Germany immediately after the 1933 Nazi rise to power and found refuge in neighboring France. In 1942 the Wittenberg family was forced to flee once again, this time to Switzerland. Although uprooted, Wittenberg continued his education and in 1957 completed his doctorate at ETH Zürich, under the guidance of Gonseth and Bernays. During the post-war years Wittenberg taught mathematics at several Swiss high schools, developed an interest in mathematical education and started his own family after marrying Marlyse Wittenberg, née Marx. In 1956 Wittenberg accepted the role of associate professor at Université Laval in Québec City and hence relocated to Canada together with his young family. In 1963 he moved to Toronto after being invited to join the newly established York University as a professor in the mathematics department. Proficient in German, French and English, he published his research in all three languages, altogether authoring five books and more than thirty articles, reviews and public addresses. Wittenberg was also an active participant in various contemporary debates regarding educational policies in North America and Europe."

then presented a new approach to the foundations of mathematics, which might be called "contextualism." It was accompanied by philosophical arguments for the epistemological nature and practices of proving theorems within conceptual frameworks. We always use certain concepts that are up for discussion in a given investigation, but reserve the right to rely on certain other concepts for the purposes of this investigation and to use them as if they were given to us. We make cuts, so to speak, within our conceptuality.

Bernays was very impressed by Wittenberg's work and saw his own views partially confirmed in it. Bernays recommended Wittenberg's dissertation to Kurt Gödel in his correspondence (Feferman et al., 2003). In the end, Gödel's view on Wittenberg turned out negative, maybe because of Wittenberg's criticism of Platonism. Gödel also seems not to have understood Wittenberg's intention, namely to provide an open and tolerant foundation for mathematics, hand in hand with efforts to democratize mathematics and science education (Wittenberg, 1968). This goal became a broad political and public concern in Wittenberg's academic life. The same applies to Hermann. Efforts in the special issue of *Dialectica* to extend Bohr's principle of complementarity beyond physics to philosophy were also committed to this concern. Bernays wrote a programmatic article on this subject with explicit reference to Hermann (Bernays, 1948).

Like Hermann, Bernays drew an analogy between the idea of complementarity and the Friesian concept of "splitting of truth." Ernst Friedrich Apelt, in a lecture course on natural philosophy at the University of Jena in 1842–1843 referred to Fries's principle of splitting the truth (*Gesetz der Spaltung der Wahrheit*) (Apelt, 1906):

Human knowledge is not like a flat surface that can be seen completely and at a glance from some high vantage point; rather, it resembles a hilly landscape, of which one must gradually piece together a complete picture from partial views. There are several heights, several viewpoints one above the other, each of which offers a different view and where one thing is sometimes revealed and sometimes concealed.[20]

It is beyond the scope of this chapter to go into more detail about how Hermann combined Apelt's concept of splitting the truth with the concept of complementarity to argue for the contextuality of ethical norms and values in order to find a middle ground between relativism and universalism. In any case, from this point of view, the notion of complementarity was no longer closely correlated to

---

[20] Original: "Die menschliche Erkenntnis gleicht nicht einer ebenen Fläche, die man von irgendeinem hohen Standpunkt herab vollständig und mit einem Blick übersehen könnte; sondern sie gleicht vielmehr einem Hügelland, von dem man sich nur nach und nach ein vollständiges Bild aus teilweisen Ansichten zusammensetzen muß. Es gibt mehrere Höhen, mehrere Standpunkte übereinander, von denen jeder einen anderen Anblick darbietet und wo sich bald das eine zeigt und bald wieder verbirgt."

quantum mechanics, which was developed later. It referred to the context-dependence of the concept of truth against the background of neo-Friesian philosophy.

11.6 Conclusion and Outlook

What are the benefits of integrating women's contributions and hidden figures into the history of quantum mechanics? The aim of this chapter on Grete Hermann has been to provide an exemplary answer to this question and, at the same time, to open up a new perspective on her interpretation of quantum physics.

By including women's contributions, we first obtain a more accurate picture of the history and philosophy of quantum mechanics. This constructive expansion deepens our knowledge of history and provides a justified criticism of a male-dominated history of heroes. But the example of Hermann provides even more. It corrects the philosophical-historical view of the neo-positivist dominance over the early understanding of quantum mechanics. In fact, the discussions in the 1930s were much more diverse than suggested by analytical philosophy in the second half of the twentieth century.

Second, a methodological integration of women's contributions builds a bridge between two much-discussed approaches to the history of physics from a feminist perspective, namely between a history of physics that concentrates on concepts and theories and a history of physics that reflects social, political, and economic aspects.

Mara Beller has further developed this approach to a "dialogical history of science" in her excellent book *Quantum Dialogue* (Beller, 1999). This work can be read as an attempt to redefine the social history of natural science by overcoming the contrast between the rational-cognitive and socially determined aspects of scientific development. Feminist epistemologies have analyzed the interrelatedness of knowledge practices as entanglements between the knower and the known for a long time. Objectivity as a "view from nowhere" has been questioned and replaced by "situated knowledges" (Haraway, 1988) in the need for ongoing conversation and dialogue. This fits Hermann's understanding of critical philosophy as a social practice surprisingly well. Science not only generates knowledge, but scientific research and knowledge construction are also socially constituted.

At this point, however, one should not forget that, in quantum mechanics, complementarity and contextuality mean something very specific. The specific technical meaning of both terms is lost if they are transferred and extended to other knowledge contexts and circumstances. Contextuality understood as the situatedness of our knowledge, and complementarity as a kind of splitting the truth cannot be derived and justified from quantum mechanics. This would amount to an unfounded confusion of research object and research subject and would fail to recognize that the methods with which quantum physics investigates quantum objects are different from those with which sociology investigates the social world and reality. We might learn this lesson from Grete Hermann's own interpretation of quantum physics.

11.7 Appendix: Contribution to the Third Zürich Dialogue, April 20, 1951

The archival collection of the Swiss philosopher Magdalena Aebi (1898–1980) in the Zentralbibliothek Zürich contains a handwritten manuscript by Grete Hermann with an introductory letter from Magdalena Aebi, Zürichbergstrasse 144, Zürich, dated August 1, 1951 (typewritten with handwritten signature). Here is a transcription, preceded by a translation, of the draft sent on August, 29, 1951.

Bernays assumes that the development of exact science has disavowed Kant's doctrine of synthetic knowledge a priori; he only holds onto the recognition of a rational factor in the knowledge of nature, the content of which, however, cannot be grasped in philosophical terms. It seems to me that too much is being abandoned here. The path of modern physics from the available experimental results to theory should be examined carefully to see what of Kant's basic assumptions has been discarded and what remains. I would like to illustrate this with the example of the law of causation. Kant's assertion that the conviction of the universal causality of natural events cannot be refuted at all by experience remains correct: because experience is incomplete, it is physically reasonable and possible for every physical process for which sufficient causes cannot yet be demonstrated to search for new data and connections in the realm of the hitherto unknown that causally determine the observable quantities of the process in question.

This consideration seems to contradict Heisenberg's uncertainty relations. Where is the mistake? The uncertainty relations make the exact prediction of atomic processes impossible. If one understands this proof to mean that "acausal" processes have been discovered here, then the consequence of Kant's reasoning is unavoidable, according to which the search for previously unknown causes in the area of as yet unknown atomic relationships is a physically meaningful task. In this interpretation, the limits of predictability could be regarded as provisional limits of physical knowledge, and the proof that they are caused by the atomic process itself would be impossible.

However, the derivation of the indeterminacy relations from the results of dualism experiments does in fact rule out the possibility of arriving at more precise predictions through the discovery of new physical data or causal relationships than the indeterminacy relations allow. The indeterminacy in the quantum-mechanical prediction occurs exclusively at the point of the "cut," that is, where the physicist moves from the quantum-mechanical calculation of the process under consideration to the classical description of the apparatus he is using. Such a transition becomes necessary because atomic systems, depending on the type of observations made on them, exhibit wave or corpuscular characteristics. They can neither be unambiguously determined according to the model of a propagating wave nor according to the moving corpuscle. The quantum-mechanical formalism makes it possible to summarize the quantities observed in the experiment and their correlations. But for the prediction of future observations, it must be translated back into the classical terms that correspond to these models.

At this point lies the unavoidable cut in the representation of physical processes. However, its position is left to the physicist's decision within a certain range and is not objectively bound to a specific spatio-temporal location of the process under investigation. If it is a matter of investigating atomic processes with the aid of a microscope and a photographic plate, the physicist can either place the section in his calculations at the point where the light is reflected by the atomic particles and thus thrown into the microscope, or at the point where the light blackens the photographic plate after passing through the microscope. At none of these points does it make sense to search for new physical data or correlations that have a causal effect on the observable process and enable a prediction that goes beyond the indeterminacy relations. This is because the physicist can move the cut away from each of them in his calculation of the process and thus gain a description that contains no causal indeterminacy. Any addition to his theory would therefore lead to a causal overdetermination and thus to a contradiction with quantum mechanics.

Quantum mechanics therefore does not claim the existence of acausal processes. On the contrary, the unavoidable indeterminacy in the precalculation of atomic processes results precisely from the proof that at no point in the process itself is anything causally indeterminate and can therefore be determined by future discoveries if necessary. The limits of prediction that have been proven here are not compatible with the classical version of the causal principle. It is characterized by the idea of Laplace's demon, according to which complete knowledge of the physical determinants of natural events at a point in time and knowledge of all the laws of nature must be sufficient to mathematically calculate every past and every future event. However, this idea did not fail because of the discovery of gaps in the causal connection of natural events, but because of the even more curious discovery that the atomic processes in quantum mechanics cannot be represented in an objective model that is independent of the physicist's question and experiment. The physicist must switch to different descriptions of the process depending on his question and the nature of his observation, and accept indeterminacies in this transition. Classical physics and the classical version of the causal principle tacitly contain the presupposition that physics gives us an objective, uniform model of natural events.

This presupposition has only become visible as such through quantum mechanics and at the same time has been proven to be false. The causal principle must be purged of it. We have thus had the experience that an apparently universally valid principle that cannot be refuted by experience has shown itself to be linked to an unconsciously made and untenable presupposition. Such experiences can be repeated. Indeed, in the light of this experience, we can no longer share Kant's conviction that critical philosophy must succeed in completely liberating the rational moment of knowledge from the empirical and grasping it a priori in synthetic judgments. However, purifying a principle of previously unrecognized presuppositions does not mean refuting it outright. Every such purification process should therefore be linked to the proof of which tacit and untenable presuppositions have been uncovered here and which of the classical considerations remain unchallenged. I am convinced that with such a procedure – for example, also in relation to the criticism of the classical ideas of space and time that is expressed by the theory of relativity – far more will be preserved and, on the other hand, far more profound new perspectives will open up than the popularizing presentations of modern physics would have us believe.

Original: "Bernays nimmt an, daß die Entwicklung der exakten Wissenschaft die Kantische Lehre von den synthetischen Erkenntnissen a priori desavouiert habe; er hält nur fest an der Anerkennung eines rationalen Faktors in der Naturerkenntnis, dessen Gehalt aber in philosophischen Begriffen nicht erfaßt werden könne. Mir scheint, daß damit zu viel fallen gelassen wird. Man sollte den Weg der modernen Physik von den vorliegenden experimentellen Ergebnissen zur Theorie genau daraufhin untersuchen, was von den Grundannahmen Kants desavouiert worden ist und was stehen bleibt. Ich möchte das am Beispiel des Kausalgesetzes erläutern. Kants Behauptung, daß die Überzeugung von der durchgängigen Kausalität des Naturgeschehens durch Erfahrung überhaupt nicht widerleget werden könne, bleibt richtig: Da die Erfahrung unabgeschlossen ist, ist es bei jedem physikalischen Vorgang, für den sich hinreichende Ursachen bisher nicht aufweisen lassen, physikalisch sinnvoll und möglich, im Bereich des bisher Unbekannten nach neuen Daten und Zusammenhängen zu suchen, die die beobachtbaren Größen des betreffenden Vorgangs kausal determinieren.

Diese Überlegung scheint mit den Heisenbergschen Unschärferelationen im Widerspruch zu stehen. Wo liegt der Fehler? Die Unbestimmtheitsrelationen machen die exakte Vorausberechnung atomarer Vorgänge unmöglich. Versteht man diesen Nachweis so, daß hier "akausale" Vorgänge entdeckt seien, dann ist die Konsequenz der Kantischen Überlegung unvermeidlich, wonach das Aufsuchen der bisher nicht bekannten Ursachen im Bereich noch unbekannter atomarer Zusammenhänge eine physikalisch sinnvolle Aufgabe ist. Die Schranken der Vorausberechenbarkeit könnten also gerade bei dieser Deutung nun aber als vorläufige Grenzen der physikalischen Erkenntnis angesehen werden, und der Beweis, daß sie durch den atomaren Vorgang selber bedingt sind, wäre unmöglich.

Nun schließt aber die Ableitung der Unbestimmtheitsrelationen aus den Ergebnissen der Dualismus-Experimente in der Tat die Möglichkeit aus, durch die Entdeckung neuer physikalischer Daten oder kausaler Zusammenhänge zu exakten Vorausberechnungen zu gelangen, als die Unbestimmtheitsrelationen sie zulassen. Die Unbestimmtheit in der quantenmechanischen Voraussage tritt nämlich ausschließlich an der Stelle des "Schnitts" ein, d.h. da, wo Physiker von der quantenmechanischen Berechnung des behandelten Vorgangs zur klassischen Beschreibung der von ihm benutzten Apparatur übergeht. Ein solcher Übergang wird notwendig, weil atomare Systeme, je nach der Art der an ihnen angestellten Beobachtungen, Wellencharakter oder Korpuskelmerkmale aufweisen. Sie lassen sich weder eindeutig nach dem Modell einer sich ausbreitenden Welle noch nach dem bewegten Korpuskel bestimmen. Der quantenmechanische Formalismus ermöglicht zwar eine Zusammenfassung der im Experiment festgestellten Größen und ihrer Zusammenhänge. Aber er muß für die Vorhersage künftiger Beobachtungen wieder in die klassischen Begriffe, wie sie diesen Modellen entsprechen, zurückübersetzt werden.

An dieser Stelle liegt der unvermeidliche Schnitt in der Darstellung physikalischer Vorgänge. Seine Lage ist aber in einem gewissen Bereich der Entscheidung des Physikers überlassen und nicht objektiv an eine bestimmte raum-zeitliche Stelle des untersuchten Vorgangs gebunden. Handelt es sich etwa um die Untersuchung atomarer Vorgänge mit Hilfe eines Mikroskops und einer photographischen Platte, so kann der Physiker in seinen Berechnungen den Schnitt entweder an die Stelle legen, wo das Licht an den atomaren Partikeln reflektiert und damit in das Mikroskop geworfen wird, oder dorthin, wo das Licht nach seinem Durchgang durch das Mikroskop die photographische Platte schwärzt. An keiner dieser Stellen hat es Sinn, nach neuen physikalischen Daten oder Zusammenhängen zu suchen, die auf den beobachtbaren Vorgang kausal einwirken und eine über die Unbestimmtheitsrelationen hinausgehnde Vorausberechnung ermöglichen. Denn der Physiker kann in seiner Berechnung des Vorgangs den Schnitt von jeder von ihnen fortverlegen und damit für seine Beschreibung gewinnen, die keine kausale Unbestimmtheit enthält. Jeder Ergänzung seiner Theorie würde also zu einer kausalen Überbestimmung und damit zum Widerspruch mit der Quantenmechanik führen.

Die Quantenmechanik behauptet also nicht das Vorliegen akausaler Vorgänge. Die unvermeidbare Unbestimmtheit in der Vorausberechnung atomarer Vorgänge ergibt sich im Gegenteil gerade aus dem Nachweis, daß an keiner Stelle des Vorgangs selber noch etwas kausal unbestimmt sei und daher gegebenenfalls durch künftige Entdeckungen bestimmt werden könne. Mit der klassischen Fassung des Kausalprinzips sind die hiermit erwiesenen Schranken der Vorausberechnung nicht vereinbar. Sie ist charakterisiert durch die Idee des Laplaceschen Dämons, wonach die vollständige Kenntnis der physikalischen Bestimmungen des Naturgeschehens in einem Zeitpunkt und die Kenntnis aller Naturgesetze hinreichen müsse, jedes vergangene und jedes künftige Geschehen mathematisch zu errechnen. Diese Idee ist aber nicht gescheitert an der Entdeckung von Lücken im Kausalzusammenhang des Naturgeschehens, sondern an der noch merkwürdigeren Entdeckung, daß die atomaren Prozesse in der Quantenmechanik nicht in einem objektiven von Fragestellung und Experiment des Physikers unabhängigen Modell dargestellt werden können. Der Physiker muß nach seiner Fragestellung und der Art seiner Beobachtung, zu verschiedenen Beschreibungen des Vorgangs übergehen und bei diesem Übergang Unbestimmtheiten in Kauf nehmen. Die klassische Physik und die klassische Fassung des Kausalprinzips enthalten stillschweigend die Voraussetzung, daß die Physik uns ein objektives, einheitliches Modell des Naturgeschehens gebe. Diese Voraussetzung ist erst durch die Quantenmechanik als solche sichtbar geworden und zugleich als falsch erwiesen. Das Kausalprinzip muß von ihr gereinigt werden. Wir haben also die Erfahrung gemacht, daß ein anscheinend allgemeingültiger und durch Erfahrung nicht widerlegbarer Grundsatz sich verknüpft gezeigt hat mit einer unbewußt gemachten und unhaltbaren Voraussetzung. Solche Erfahrungen können sich wiederholen. Ja wir können, im Licht dieser Erfahrung, die Kantische Überzeugung nicht mehr teilen, wonach es der kritischen Philosophie gelingen müsse, das rationale Moment der Erkenntnis vollständig vom Empirischen zu befreien und in synthetischen Urteilen a priori zu fassen. Einen Grundsatz von bisher unerkannten Voraussetzungen zu reinigen, heißt aber nicht, ihn schlechterdings zu widerlegen. Jeder solcher Bereinigungsprozess sollte daher mit dem Nachweis verbunden werden, welche stillschweigende und nicht haltbare Voraussetzungen hier aufgedeckt worden ist und was von den klassischen Überlegungen unangefochten stehen bleibt. Ich bin davon überzeugt, daß bei einem solchen Vorgehen – z.B. auch gegenüber der von der Relativitätstheorie angemeldeten Kritik an den klassischen Vorstellungen von Raum und Zeit – weit mehr erhalten bleibt und andererseits auch weit tiefgehende neue Ausblicke sich eröffnen, als die popularisierenden Darstellungen der modernen Physik uns glauben machen wollen.''

References

Acuña, Pablo. 2021. Must hidden variables theories be contextual? Kochen & Specker meet von Neumann and Gleason. *European Journal for Philosophy of Science*, 11, 41. https://doi.org/10.1007/s13194-021-00347-8

Apelt, Ernst Friedrich. 1906. *Über Begriff und Aufgabe der Naturphilosophie*. Göttingen: Vandenhoeck & Ruprecht, vol. 1.

Bachelard, Gaston, Bernays, Paul, and Gonseth, Ferdinand. 1947. Editorial/Éditorial/ Geleitwort. *Dialectica*, 1(2), 5–10. https://www.jstor.org/stable/42963795

Beller, Mara. 1999. *Quantum Dialogue. The Making of a Revolution*. Chicago, IL: University of Chicago Press.

Bernays, Paul. 1948. Über die Ausdehnung des Begriffes der Komplementarität auf die Philosophie. *Synthese*, 7(1/2), 66–70. https://www.jstor.org/stable/20114017

Bernays, Paul. 1952. Dritte Gespräche von Zürich. Ansprache. *Dialectica*, 6(22), 130–136. https://www.jstor.org/stable/42964030

Bohr, Niels. 1928. The quantum postulate and the recent development of atomic theory. *Nature*, 121, 580–590. https://doi.org/10.1038/121580a0

Corry, Leo. 2024. Von Neumann and impossibility, from Gödel to EDVAC. In Knudsen, Toke and Carter, Jessica (eds.), *Mastering the History of Pure and Applied Mathematics*. Berlin: De Gruyter, pp. 75–100.

Crull, Elise. 2022. Grete Hermann's interpretation of quantum mechanics. In Freire, Oliver Jr. (ed.), *The Oxford Handbook of the History of Quantum Interpretations*. Oxford: Oxford University Press, pp. 567–586.

Crull, Elise and Bacciagaluppi, Guido (eds.). 2016. *Grete Hermann: Between Physics and Philosophy*. Dordrecht: Springer.

Cuffaro, Michael E. 2022. Grete Hermann, quantum mechanics, and the evolution of Kantian philosophy. In Peijnenburg, Jeanne, and Verhaegh, Sander (eds.), *Women in the History of Analytic Philosophy*. Cham: Springer, pp. 114–145.

Dieks, Dennis. 2017. Von Neumann's impossibility proof: mathematics in the service of rhetorics. *Studies in History and Philosophy of Science Part B: Studies in History and Philosophy of Modern Physics*, 60, 136–148. https://doi.org/10.1016/j.shpsb.2017.01.008

Duncan, Anthony and Janssen, Michel. 2023. Von Neumann's Hilbert space formalism. In *Constructing Quantum Mechanics, Volume 2: The Arch, 1923–1927*. Oxford: Oxford University Press, pp. 642–670.

Einstein, Albert. 1948. Quanten-Mechanik und Wirklichkeit. *Dialectica*, 2, 320–324. https://doi.org/10.1111/j.1746-8361.1948.tb00704.x

Falkenburg, Brigitte. 2021. Grete Hermann's philosophy of quantum mechanics: a late appraisal. *Hopos. The Journal of the International Society for the History of Philosophy of Science*, 11(1), 201–210. https://doi.org/10.1086/712935

Favrholdt, David. 1999. *Niels Bohr: Collected Works. Volume 10: Complementarity Beyond Physics (1928–1962)*. Amsterdam: Elsevier.

Feferman, Solomon, Dawson, John W., Goldfrab, Warren, Parsons, Charles, and Sieg, Wilfried (eds.). 2003. *Kurt Gödel: Collected Works*. Oxford: Oxford University Press.

Frank, Philipp. 1936. Philosophische Deutungen und Mißdeutungen der Quantentheorie. *Erkenntnis*, 6, 303–317. https://www.jstor.org/stable/20011825

Golub, Robert and Lamoreaux, Steven K. 2024. A retrospective review of von Neumann's analysis of hidden variables in quantum mechanics. *Academia Quantum*, 1(1). https://doi.org/10.20935/AcadQuant7311

Guilder, Louisa. 2009. *The Age of Entanglement*. New York: Vintage Books, Random House.

Hansen-Schaberg, Inge. 2012. Prof. Dr. Grete Henry-Hermann (1901–1984). *Akten-Einsicht (Zeitschrift für Museum und Bildung*, volume 74). Berlin: LIT Verlag, pp. 104–122.

Hansen-Schaberg, Inge. 2016. A biographical sketch of Prof. Dr Grete Henry-Hermann. In Crull, Elise and Bacciagaluppi, Guido (eds.). *Grete Hermann: Between Physics and Philosophy*. Berlin: Springer, pp. 3–16.

Haraway, Donna. 1988. Situated knowledges: the science question in feminism and the privilege of partial perspective. *Feminist Studies*, 14(3), 575–599. https://doi.org/10.2307/3178066

Heisenberg, Werner. 1927. Über den anschaulichen Inhalt der quantentheoretischen Kinematik und Mechanik. *Zeitschrift für Physik*, 43, 172–198. https://doi.org/10.1007/BF01397280

Heisenberg, Werner. 1934. Wandlungen der Grundlagen der exakten Naturwissenschaft in jüngster Zeit. *Naturwissenschaften*, 22, 669–675. https://doi.org/210.1007/BF01500104

Henry-Hermann, Grete. 1948. Die Kausalität in der Physik. *Studium Generale*, 1(6), 375–383. https://doi.org/10.1007/978-3-658-16241-2_19

Henry-Hermann, Grete. 1967. Nelson, Leonard. In Edwards, Paul (ed.), *The Encyclopedia of Philosophy, Volume 5*. New York: Macmillan, pp. 463–467.

Henry-Hermann, Grete. 2019. Ethik und Naturwissenschaften. Ein Beitrag zu der von Carlo Schmid angeregten Diskussion über die Voraussetzungen eines sozialistischen Programms [1950]. In Herrmann, Kay (ed.), *Grete Henry-Hermann: Philosophie – Mathematik – Quantenmechanik*. Wiesbaden: Springer VS, pp. 399–407.

Hermann, Grete. 1926. Die Frage der endlich vielen Schritte in der Theorie der Polynomideale. *Mathematische Annalen*, 95, 736–788. https://doi.org/10.1007/BF01206635

Hermann, Grete. 1935. Die naturphilosophischen Grundlagen der Quantenmechanik. *Abhandlungen der Fries'schen Schule, Neue Folge*, 6(2), 69–152. (Reprinted in Herrmann (2019, pp. 204–258).)

Hermann, Grete. 1936. Zum Vortrag Schlicks. *Erkenntnis*, 6(1), 342–343. (Reprinted in Herrmann (2019, pp. 273–274).)

Hermann, Grete. 1937. Über die Grundlagen physikalischer Aussagen in den älteren und den modernen Theorien. *Abhandlungen der Fries'schen Schule, Neue Folge*, 6(3/4), 309–398. (Reprinted in Herrmann (2019, pp. 275–334).

Hermann, Grete. 2019. Determinismus und Quantenmechanik [1933]. In Herrmann, Kay (ed.), *Grete Henry-Hermann: Philosophie – Mathematik – Quantenmechanik*. Wiesbaden: Springer VS, pp. 155–203.

Herrmann, Kay (ed.). 2019. *Grete Henry-Hermann: Philosophie – Mathematik – Quantenmechanik*. Wiesbaden: Springer VS.

Herrmann, Kay and Neißer, Barbara. 2023. Leonard Nelson und die Grundlagen des freiheitlichen Sozialismus. In Herrmann, Kay and Neißer, Barbara (eds.), *Grete Henry-Hermann: Sittlichkeit und Vernunft. Frauen in Philosophie und Wissenschaft*. Wiesbaden: Springer VS, pp. 195–200.

Jammer, Max. 1974. *The Philosophy of Quantum Mechanics*. New York: Wiley & Sons.

Kaiser, David. 1992. More roots of complementarity: Kantian aspects and influences. *Studies in History and Philosophy of Science Part A*, 23(2), 213–239. https://doi.org/10.1016/0039-3681(92)90033-3

Kant, Immanuel. 1998. *Critique of Pure Reason* (P. Guyer and A. Wood, eds.). Cambridge: Cambridge University Press

Klafki, Wolfgang. 2019. Vernunft – Erziehung – Demokratie. In Braun, Karl-Heinz, Stübig, Frauke, and Stübig, Heinz (eds.), *Allgemeinerziehungswissenschaft. Systematische und historische Abhandlungen*. Wiesbaden: Springer, pp. 155–176.

Laplace, Pierre Simon. 1814 [2009]. *Essai philosophique sur les probabilités*. 5th ed. Cambridge: Cambridge University Press.

Mehra, Jagdish. 1987. Niels Bohr's discussions with Albert Einstein, Werner Heisenberg, and Erwin Schrödinger: the origins of the principles of uncertainty and complementarity. *Foundations of Physics*, 17(5), 461–506. https://doi.org/10.1007/bf01559698

Mermin, David and Schack, Rüdiger. 2018. Homer nodded: Von Neumann's surprising oversight. *Foundations of Physics*, 48(9), 1007–1020. https://doi.org/10.1007/s10701-018-0197-5

Mitsch, Chris. 2022. Hilbert-style axiomatic completion: on Von Neumann and hidden variables in quantum mechanics. *Studies in History and Philosophy of Science*, 95, 84–95. https://doi.org/10.1016/j.shpsa.2022.06.016

Murdoch, Dugald. 1987. *Niels Bohr's Philosophy of Physics*. New York: Cambridge University Press.

Ozawa, Masanao. 2015. Heisenberg's original derivation of the uncertainty principle and its universally valid reformulations. *Current Science*, 109, 2006–2016. https://www.jstor.org/stable/24906690

Paparo, Giulia. 2012. *Grete Hermann: Mathematician, Philosopher and Physicist*. M. Phil. Thesis. Utrecht: University of Utrecht. https://studenttheses.uu.nl/handle/20.500.12932/45560

Rédei, Miklos. 1996. Why John von Neumann did not like the Hilbert space formalism of quantum mechanics (and what he liked instead). *Studies in History and Philosophy of Science Part B: Studies in History and Philosophy of Modern Physics*, 27(4), 493–510. https://doi.org/10.1016/S1355-2198(96)00017-2

Reichenberger, Andrea. 2020. *Grete Hermann: Between Physics and Philosophy*. Edited by Elise Crull and Guido Bacciagaluppi; *Grete Henry-Hermann: Philosophie – Mathematik – Quantenmechanik*. Edited by Kay Herrmann [Book review]. *Mathematical Intelligencer*, 42, 80–82. https://doi.org/10.1007/s00283-020-09978-w

Reichenberger, Andrea. 2024. Philosophie und Physik: Zu Grete Hermanns neukantianischen Interpretation der Quantenmechanik. In Herrmann, Kay and Schwitzer, Boris (eds.), *Kantisches Denken in der Tradition von Jakob Friedrich Fries und Leonard Nelson im 20. Jahrhundert: Wirkungen und Aktualität*. Stuttgart: Springer/Metzler, pp. 271–295.

Renn, Jürgen. 2013. Schrödinger and the genesis of wave mechanics. In Reiter, Wolfgang L. and Yngvason, Jakob (eds.), *Erwin Schrödinger: 50 Years After*. Zürich: European Mathematical Society Publishing House, pp. 9–36.

Schirrmacher, Arne. 2005. *Dreier Männer Arbeit in der frühen Bundesrepublik: Max Born, Werner Heisenberg und Pascual Jordan als politische Grenzgänger*. Berlin: MP, preprint 296. https://www.mpiwg-berlin.mpg.de/sites/default/files/Preprints/P296.pdf

Schlick, Moritz. 1936. Quantentheorie und Erkennbarkeit der Natur. *Erkenntnis*, 6, 17–26. https://www.jstor.org/stable/20011826

Soler, Léna and Schnell, Alexander. 2001. Grete Henry-Hermanns Beiträge zur Philosophie der Physik. In Miller, Susanne and Müller, Helmut (eds.), *In der Spannung zwischen Naturwissenschaft, Pädagogik und Politik. Zum 100. Geburtstag von Grete Henry-Hermann*. Bonn: Philosophisch-Politische Akademie, pp. 16–27.

Stoeltzner, Michael. 2001. Bell, Bohm, and von Neumann: some philosophical inequalities concerning no-go theorems and the axiomatic method. In Placek, Tomasz and Butterfield, Jeremy (eds.), *Non-Locality and Modality*. Dordrecht: Kluwer Academic Publishers, pp. 37–58.

Von Neumann, Johann. 1932. *Mathematische Grundlagen der Quantenmechanik*. Berlin: Springer.

Wick, David. 1995. *The Infamous Boundary: Seven Decades of Controversy in Quantum Physics*. Boston, MA: Birkhäuser.

Wittenberg, Alexander. 1957. *Vom Denken in Begriffen. Mathematik als Experiment des reinen Denkens*. Basel: Birkhäuser.

Wittenberg, Alexander. 1968. Priorities and responsibilities in the reform of mathematical education. An essay in educational meta-theory. *Dialectica*, 22(1), 58–74. https://www.jstor.org/stable/42964723

12

Women Take the Lead: A Physics Laboratory Under the Dictatorship in Portugal, 1940s–1960s

ANA SIMÕES AND MARIA PAULA DIOGO

12.1 Introduction

This chapter places itself at the confluence of three marginalized streams in historical narratives concerning quantum physics and quantum mechanics: women scientists, groups in peripheral regions and countries not at the forefront of scientific developments, and standard, non-epoch-making contributions.[1] It specifically examines the contributions to quantum physics made by the Portuguese physicist Lídia Salgueiro (1917–2009) and a team of women researchers at the Laboratory of Physics of the Faculty of Sciences of the University of Lisbon (Laboratory of Physics for short), roughly from the 1940s to the 1960s, during the right-wing dictatorial regime known as Estado Novo (New State).

Between 1929 and 1947, in an academic environment still centered on teaching and therefore uncongenial to scientific research, the Laboratory of Physics was able to assert itself as a successful research school in atomic and nuclear physics. Particularly notable is that under the joint leadership of the physicists Cirilo Soares (1883–1950) and Manuel Valadares (1904–1982) one in three staff members was a woman, a ratio only previously attained in Paris at the Marie Curie Laboratory, with which Valadares entertained close connections, and at the Viennese Institute for Radium Research.

We thank the editors of this volume for the invitation to participate and for their comments and suggestions throughout the writing of this chapter. We also thank anonymous referees. We thank João Ferreira for discussions concerning his parents, Lídia Salgueiro and José Francisco Gomes Ferreira. This work has been supported by the Foundation for Science and Technology (FCT) under project UIDB/00286/2020. All translations from Portuguese to English are our own.

[1] There is an extensive bibliography on science and gender, mostly concerning well-known women scientists working at mainstream places and associated with epoch-making contributions, but also offering historiographical reflections on how to address science and gender issues. See, for example, Rossiter (1982, 1999), Schiebinger (1987), Keller (1995), Kohlstedt (1995, 1999), Kohlstedt and Longino (1997), Govoni (2009), and Oertzen et al. (2013). A classic book on women in radioactivity is Rayner-Canham and Rayner-Canham (1997). Specific bibliographic references on the topic of this chapter include Boudia (2001), Rentetzi (2004, 2007), and Pigeard-Micault (2013, 2018).

The Laboratory of Physics was dismantled in 1947 following a political purge by the Estado Novo, which affected many institutions, including the teaching staff of the universities, and led to the expulsion of Valadares. Despite his affiliation with the then-illegal communist party and his connections with the Joliot-Curies, our earlier work has shown that this expulsion was not strictly political. In a Portuguese academic context unreceptive to scientific research, discussions within the Faculty of Sciences on its role, and on scientific policy generally, were crucial. They generated a hostile academic environment, which most immediately led to the dissolution of the Laboratory of Physics (Gaspar and Simões, 2011, pp. 334–343), after which the Spanish right-wing physicist Julio Palacios (1891–1970) was appointed to direct the remaining group of researchers.

While Palacios opted for a research agenda focused on electrochemistry, thus severing any ties with Valadares, Lídia Salgueiro and affiliated women researchers at the Laboratory of Physics managed to secure their own research following the lines previously set up by Valadares. Moreover, and perhaps unexpectedly given Palacios' political leanings and the gender labor policies during the Estado Novo dictatorship, the percentage of women more than doubled between 1957 and 1966. This group of women then successfully extended their research into quantum physics to the study of radiation emitted at the atomic and nuclear levels, with a particular emphasis on X-ray spectroscopy.

In the first section of this chapter we briefly discuss the academic background of the Laboratory of Physics from the perspective of the organization of higher education and the changing position of women in Portugal; the second section tackles the emergence and development of a research school at the Laboratory of Physics from 1929 to 1947, in terms of its scientific agenda, relations to Portuguese funding agencies, researchers' mobility, connections with foreign laboratories, as well as the contributions of women researchers; the third section examines how, in the unfavorable period succeeding the downfall of the research school headed by Soares and Valadares, the Laboratory of Physics was reorganized and became a hub for women to the extent that they took the lead in various ways; we close the chapter with some concluding remarks.

12.2 From a Male-Dominated Polytechnic School to a Female-Friendly Faculty of Sciences

The Lisbon Polytechnic School was founded in 1837, when Portugal was ruled by a constitutional monarchy, to provide propaedeutic training for students aiming at military engineering careers, and at enrolling in medical-surgical schools. Consequently, despite the absence of any legal restrictions on its attendance by women, the Polytechnic School was fundamentally a male-dominated educational institution (Carolino and Mota, 2013; Simões et al., 2013, pp. 131–145).

In a country with an extremely high rate of illiteracy (1878, 82,4%; 1890, 79%; 1900, 77%; 1911, 75,1%) (Grácio, 1971) and but a single university throughout the nineteenth century, women's access to education only began in the last years of the monarchy and was boosted as part of the "education for all" trend embraced by the republican movement. The objective of female high-school education was to enable the privileged fraction of young women from the middle and upper classes to better fulfill their roles as future housewives and mothers.

The establishment of the republic in 1910 brought education issues to the center of the political sphere. For the republican ideology, education of the broad population was a *sine qua non* condition to create the new republican citizen, a believer in science as a substitute for religious faith. Although republican educational efforts were mainly aimed at basic education, higher education also featured on the republic's agenda. The change in higher education largely focused on restructuring the Portuguese university system by creating new universities, faculties, and courses. It is in this context of change that the University of Lisbon was created in 1911. Its Faculty of Sciences was installed at the premises of the former Polytechnic School, from which it also inherited technical and scientific staff, and scientific organizational structure.

In 1926, a military coup put an end to the republican regime. During the 16 years of the regime, the situation had been slowly deteriorating politically, socially, and economically due to a combination of internal struggles within the Republican Party, ruinous loans taken on by the Portuguese state mainly to support the construction of infrastructures in Portugal's mainland and its African colonies (Portugal declared bankruptcy in 1892), and the instability created during and in the aftermath of World War I. The ensuing military dictatorship led to the recruitment of António de Oliveira Salazar as finance minister in March 1928, who managed to fully reverse Portugal's unbalanced economic situation within a year. To achieve this, Salazar postponed public works and other state investments, privatized the public sector, introduced steep tax hikes, and froze salaries nationwide.

However, some of the projects from previous governments were implemented (Simões and Diogo, 2022). Among them stood the establishment of the Board for National Education (Junta de Educação Nacional, 1929–1936), an institution promoting scientific research and dependent on the Ministry of Public Instruction. With a very limited budget, its leading team granted funds to laboratories and libraries, and individual scholarships to members of the scientific community (Lopes, 2017).[2] The Board for National Education was later renamed Institute for High Culture (Instituto para a/de Alta Cultura, 1936–1952/1952–1976), a sign of the growing intervention of

---

[2] Decree-Law 1941, April 11, 1936.

Salazar's authoritarian state, which took advantage of the advance of fascism in Europe – Franco's dictatorship in Spain, German Nazism, and Mussolini's fascism – to impose its own fascist-like dictatorship. (Salazar became prime minister in 1932 and established an enduring right-wing dictatorship that lasted until 1974.)

It is in the political context of the first republic and then of Salazar's dictatorship that the scientific, educational, and students' profile of the Faculty of Sciences must be understood. Even though the Polytechnic School was a male-dominated institution, 31 female students enrolled from 1880 to 1911. This reflected a wider European societal context, which called for institutions for women's education following their rising entry into the job market. It also resulted from the spread within Portugal of republican ideology, which gained momentum and eventually opened the way to a change from the monarchical to a republican regime in 1910. Unlike its predecessor, the Faculty of Sciences, established right after the onset of the republic in 1911, revealed itself early on to be a female-friendly institution.

However, even the republican female-friendly profile had its limits. Women were allowed to access positions in the public sector, as they were considered to have equal intelligence and capacity for work as men, but managerial and initiative-taking roles were "naturally reserved for men."[3] The use of the epithet "naturally" clearly referred to the assumption of an unquestionable "scientific" canon based on biology – women are physically weaker and must be protected by men – and extended to the social realm, through the protection of women in their capacity as mothers and wives, crucially responsible for the family well-being. During the Estado Novo dictatorship, this social normativity became a legal norm through laws that assigned unequal and hierarchical roles to men and women. This was evident in the form of prohibitions and limitations on certain professions, as well as instances of wage and occupational discrimination.[4] Women were barred from pursuing careers in diplomacy, judicial magistracy, and leadership roles in local administration. Primary school teachers faced restrictions on their right to marry, requiring authorization from the Ministry of National Education, and hospital nurses and air hostesses were not allowed to marry. Until 1967, married women needed their husband's permission to enter employment contracts – the so-called "marital power" – with husbands retaining the power to cancel contracts without their wives' consent. Within the industrial sector and various collective labor agreements, women were often barred from engaging in specialized professional roles, being confined to undervalued and unskilled tasks. So, except for the lower-income classes, most women in the job market were either unmarried or widowed.

---

[3] Decree-Law 4676, July 19, 1918.
[4] Decree-Law 14535, October 31, 1927, ratified by the ministerial order of September 15, 1934.

The profile of the female population attending the Faculty of Sciences varied over time: three decades after its creation, in the academic year 1941–1942, 27% of its students were women as well as more than one in six (18%) of its teaching staff. In its first decade of existence, five women and nine men were granted undergraduate degrees; no data exists for the 1920s, but in the 1930s, 93 of the 316 graduates were women (22%); and in the 1940s, 254 out of 518 were women (51%) (Simões et al., 2013, pp. 137–139).

The professional trajectory of science graduates – both men and women – remains to be scrutinized. It is possible to state, however, that a substantial number of the women graduates who entered professional life became teachers in female high schools, while a much smaller number embarked on scientific research. (Their preparation during the course of their studies left a lot to be desired.)

At the Faculty of Sciences, both curricular organization and course content were quite conservative, and only from the end of the 1940s onwards were plans for the revision and subsequent reform of courses insistently discussed in the sessions of the School Council composed of all its full professors. Therefore, changes were slow. Some teaching staff, among whom stood the physicist Valadares, took upon themselves the task of organizing, in the late 1930s, a series of extra-curricular advanced courses (Gil, 2003; Gaspar, 2009, pp. 81–86). Others also promoted, over the years, free courses covering advanced subjects, such as modern physics and radioactivity,[5] regularly inviting foreign scientists to lecture on cutting-edge subjects.[6]

Gradually, courses were updated and the training of the few who ventured into a doctorate improved. While in the 1920s only one doctorate in science was awarded by the University of Lisbon, in the 1930s seven doctorates in science were awarded, and in the 1940s, eight, including for the first time three to women (Moura, 1972): Seomara da Costa Primo (1895–1986) in biological sciences, in 1942; Lídia Salgueiro and Marieta da Silveira (1917–2004), both in physico-chemical sciences, in 1945 and 1946, respectively. Even within this small group of women, the two scientific cultures that fought for hegemony in the Faculty of Sciences are discernible: the one preserving the canonical tradition of self-education and encyclopedism, embodied by Seomara da Costa Primo, and the

---

[5] Historical Archives of the Museum of Sciences of the University of Lisbon, AHMUL-MUHNAC, Lv. 1441. *Atas do Conselho Escolar da Faculdade de Ciências. Livro nº 7. Abril 1933 a 28 Julho 1944*. Session June 22, 1934; *Livro nº 9. 1952 a 1963*, p. 49v. Session October 22, 1957.

[6] Historical Archives of the Museum of Sciences of the University of Lisbon AHMUL-MUHNAC, Lv. 1441. *Atas do Conselho Escolar da Faculdade de Ciências. Livro nº 7. Abril 1933 a 28 Julho 1944*. Session of October 16, 1934: the biochemist Carl Neuberg and the chemist Georges Urbain visit the Faculty of Sciences, p. 23; Session November 9, 1942: the physicist Edoardo Amaldi will deliver a series of conferences on Nuclear Physics and on Neutrons, p. 152. AHMUL-MUHNAC, Lv. 1442. *Atas do Conselho Escolar da Faculdade de Ciências. Livro nº 8. 18 Novembro 1944 a 1952*, Session of December 18, 1948: Julio Palacios suggests that the physicist Julio Garrido should deliver lessons about "La estrutura atomica de los cristales," p. 79v.

emerging one, defending the affirmation of a new scientific ethos based on original research work, personified both by Lídia Salgueiro and Marieta da Silveira. It is no coincidence that Lídia Salgueiro and Marieta da Silveira were supervised by senior researchers of the Laboratory of Physics, the group who pushed forward at the Faculty of Sciences a new scientific ethos centered on experimental physics research.

12.3 Fighting for Research: The Exemplary Case of the Laboratory of Physics

From 1929 onwards, under the leadership of Soares, its director, helped by his collaborator, Valadares, the Laboratory of Physics designed a scientific agenda capable of ensuring the necessary funds for its development via the Board for National Education. The Board awarded scholarships for training abroad to the teaching assistants Herculano Amorim Ferreira (1895–1974), Amaro Monteiro (1898–1979), Francisco Mendes (1907–1975), Teles Antunes (1905–1965), Aurélio Marques da Silva (1905–1965), Armando Gibert (1914–1985), and Valadares (Gaspar and Simões, 2011, pp. 318–320).

Valadares' first scholarship (1929) – to specialize in radon radiotherapy, critical to qualify for a physics job at the Portuguese Institute of Oncology – was followed by another scholarship at Marie Curie's Laboratory in Paris (from 1930 to 1933) to work under the supervision of Marie Curie on radioactivity, which was then quickly becoming nuclear physics (Hughes, 2003, pp. 368). In late 1933, he was awarded a PhD in physics for his work on spectroscopy by crystal diffraction of γ radiation (Valadares, 1933). During his stay in Paris he worked with Salomon Rosenblum, with whom he became good friends. Later, after Valadares' expulsion in 1947, he would get a job at the Laboratoire de l'Aimant Permanent of the Centre National de la Recherche Scientifique, directed by Rosenblum in Paris (Salgueiro and Carvalho, 2001, p. 72).

In 1933, upon his return to the Laboratory of Physics, Valadares began a research program on radioactivity and X-ray spectroscopy, at first improvising and re-using old equipment. Later, in 1936, he was supported by modest funds from the recently created Institute for High Culture, but had to abandon radioactivity and concentrate on X-ray spectroscopy to adjust to the stringent budget conditions. As a result, he failed at discovering the element with atomic number 85, of the radium family, identified by Horia Hulubei and Yvette Cauchois at the Laboratoire Curie in July 1939 (Gaspar and Simões, 2011, p. 332).[7] Due to budget constraints, a material culture based on the re-utilization of pieces of old equipment found in different places emerged, as often is typical in peripheral contexts. From the late 1940s onwards, this bricolage culture was adopted by Lídia Salgueiro and her women's research team.

---

[7] The discovery of element 85 is usually wrongly attributed to Corson, MacKenzie, and Segrè. See Scerri (2013).

Valadares assembled his own equipment consisting basically of an X-ray tube with a changeable anti-cathode, a vacuum system, and a high-tension coil. The anti-cathode, consisting of the metal to be studied, was bombarded with electrons, and the resulting X-ray diffraction spectrum was detected by a photographic device. His paper reporting results on the satellite lines Lα of lead was published in 1938 (Valadares and Mendes, 1938).

Two years later, in 1940, the Institute for High Culture created "Centers for Studies" to circumvent the perils associated with granting scholarships to young researchers to travel to other European countries during the war years. The new centers were research units endowed with funds that afforded researchers the means, albeit limited, to conduct research and training in Portugal. One of the first such units was precisely the Centre for Studies in Physics at the Laboratory of Physics where Valadares asserted himself as the scientific leader of a research school on X-ray spectroscopy, radioactivity, and nuclear physics.[8] Collaborating in leadership functions were Marques da Silva and Gibert.

Breaking with the all-male composition of the Laboratory of Physics until then, from 1942 to 1946 the group included several women. Besides Valadares and Marques da Silva (Gibert was abroad), Teles Antunes, Mendes, Carlos Braga (1899–1982), José Sarmento (1899–1986), and Luiz Rivoir Alvarez, a collaborator of Palacios, the women were Glaphyra Vieira (1912–1995), the first woman teaching assistant to be hired in the Laboratory of Physics (Vieira, 1989),[9] Lídia Salgueiro, Marieta da Silveira, Maria Valentina Saraiva, and Judite Pereira. The group included not only staff members but also visiting scholars from other universities and young women PhD students to be supervised toward their doctoral degrees. Braga, Sarmento, and Lídia Salgueiro were supervised by Valadares and Marieta da Silveira was supervised by Marques da Silva (Gaspar, 2018, p. 44).

Valadares first met Lídia Salgueiro in 1941, when he was a member of the oral examination jury of the last course she had to pass to complete her undergraduate degree in physico-chemical sciences. At the end of the exam, Valadares invited her to enroll in research under his supervision. She knew he had been supervised by Marie Curie, whose scientific trajectory impressed her much, having just read her biography (Curie, 1937). But although attracted to research, her plans were different. She wanted to become a high-school teacher, as did most women science graduates, to guarantee a stable economic situation for herself, her sister, and her long-widowed mother, who

---

[8] For a discussion on the characteristics of Valadares' research school in terms of leadership, main themes, methodological approaches, training of new generations, publication outlets, national and international collaborations and networking, see Gaspar (2009, pp.152–181) and Gaspar and Simões (2011).

[9] Biographical details, Maria da Conceição Abreu e Paula Contenças, "Glaphyra Silva Vieira: a primeira mulher assistente no Laboratório de Física da Ciências ULisboa," Ciências ULisboa (February 11, 2024) at https://ciencias.ulisboa.pt/pt/noticia/11-02-2024/glaphyra-silva-vieira-a-primeira-mulher-assistente-no-laboratorio-de-fisica.

had taken pains to grant her children a proper education. Therefore, she prepared for the entrance exams for the internship granting the high-school teacher diploma. Coincidentally, Soares was a member of her examination jury. He invited her to become a teaching assistant at the Laboratory of Physics. She hesitated once again, but in the end opted for a university career, attracted by the prospect of both doing research and teaching (Carvalho, 2011, pp. 61–64).

Lídia Salgueiro's thesis focused on the γ-radiation spectrum originating from the radioactive transmutation RaD (lead-210) → RaE (bismuth-210). She tinkered with precarious materials and instruments to assemble her experimental apparatus. She set up a Bragg spectrometer with a rotating crystal by adapting an X-ray spectrometer belonging to the Laboratory of Chemistry, in which a clockwork mechanism, eventually replaced by an electric motor, controlled the movement of the crystal. The radioactive source was a set of five old radon tubes given by the Portuguese Institute of Oncology. Photographic detection was used to detect the gamma rays; however, it required long film exposures (up to a month), and film development was a demanding task hampering the pace of research (Carvalho, 2011, p. 64). This experimental setup became the core of her PhD, which was awarded in December 1945 (Salgueiro, 1945) (Figure 12.1).

Figure 12.1 Rotating crystal spectrometer adapted for the study of γ-radiation. Lídia Salgueiro, "Espectro gama dos derivados de vida longa do Radão" (PhD thesis, University of Lisbon, 1945), printed photograph I.

Besides working under difficult material conditions, wartime restrictions hindered scientific communication. When student and supervisor heard that the French physicist Marcel Frilley was working on the same topic, the two secured priority by publishing a note in the recently created journal *Portugaliae Physica* (Salgueiro, 1944; Carvalho, 2011, p. 65; Gaspar and Simões, 2011, p. 330).

In 1942, one year after Lídia Salgueiro joined Valadares' research team, Marieta da Silveira also joined the group. Like Lídia Salgueiro, she had completed high school, enrolled at the Faculty of Sciences, and obtained her undergraduate degree in the physico-chemical sciences in 1941. She also aimed at preparing for the entrance exams to the internship granting the high-school teacher diploma, but during her internship she was contacted by the director of the Laboratory of Chemistry to become a teaching assistant, and she accepted. However, she realized that no research was actually taking place at the Laboratory of Chemistry, unlike at the Laboratory of Physics. Having Marie Curie as a role model, and eager to do research in radioactivity, she took the liberty of asking Soares to accept her in the Center for Studies in Physics. For her dissertation, she studied γ-radiation emitted by uranium and its descendants using the method of absorption (Silveira, 1946; Brotas et al., 2011, p. 51; Gaspar and Simões, 2011, p. 330).

In later years, many striking parallels showed up again between the painstaking careers of Lídia Salgueiro and Marieta da Silveira. Their careers both unfolded at the Faculty of Sciences where they stayed until their retirement, one at the Laboratory of Physics, the other at the Laboratory of Chemistry, becoming full professors only decades after obtaining their doctorates, Lídia Salgueiro in 1970 and Marieta da Silveira in 1979. Lídia Salgueiro became the connecting link between the Soares–Valadares era and the subsequent one, in which she played a leading role. In contrast to their lifelong careers, those of the other women researchers were not as extended either for personal, family, or professional reasons.[10]

The international recognition of the Laboratory of Physics can concretely be assessed from Robert Beyer's 1949 book, *Foundations of Nuclear Physics*, a compilation of 13 facsimile articles considered central to the emergence of nuclear physics as a subdiscipline, followed by an extensive bibliography on nuclear physics that Beyer considered to be reasonably complete for the period ending in 1947. That bibliography includes several papers by Lídia Salgueiro, Marieta da Silveira, Braga, Gibert, Valadares, and Marques da Silva (Beyer, 1949; Gaspar and Simões, 2011, p. 333). Lídia Salgueiro and Marieta da Silveira's works are cited in relation to chapters by Enrico Fermi and George Gamow on topics of γ-radiation and radioactivity, and among scientists such as Ernest Rutherford, Erwin Schrödinger, or Lisa Meitner.

---

[10] For the contributions of Glaphyra Vieira, Maria Valentina Saraiva, and Judite Pereira, see Gaspar (2009, pp. 174–175).

Despite being in the periphery of Europe, the Laboratory of Physics was part of a growing network of European laboratories working on radioactivity. Its connection to other laboratories stemmed from researchers' mobility and the legacy of Marie Curie's laboratory, where Valadares had witnessed an environment characterized by a strong female presence (Boudia, 2001; Pigeard-Micault, 2013, 2018). Valadares' laboratory also came to attract a significant number of women, just like Marie Curie's laboratory in Paris and the Viennese Institute for Radium Research. The Viennese institute represented an especially significant model, as it was headed by Stephan Meyer, a supportive and politically committed man who played an active role in creating opportunities for women in radioactive research (Rentetzi, 2004; Pigeard-Micault, 2013). In Paris, women researchers made up 25–30% of the total between 1906 and 1934, and in Vienna their share amounted to 38% in 1934 (Rentetzi, 2004, p. 361). In the Laboratory of Physics, from 1929 to 1947, one in three staff members were women. Valadares' experience at Curie's laboratory, added to his leftist political leanings, might have predisposed him to accept women as collaborators on an equal footing with male colleagues.

In the context of a short parliamentary election period in October–November 1945, Salazar being forced to simulate democratic elections to please the Allies following the end of World War II, different scholars, including Valadares, expressed their political opinions in the press, on the radio, or at rallies. Surprised by the positive response to the opposition's proposals, Salazar's regime quickly regained absolute control of political life, heavily repressing those who had stood out during the election campaign. University professors were punished for their public interventions: First, scholarships were suspended, and then, in 1947, 21 faculty members from various universities were dismissed. Valadares was among them, along with Marques da Silva and Gibert.

However, Valadares' dismissal cannot be seen as the result of a strictly political clash only. Political retaliation was reinforced by internal disagreements over scientific policies and the relevance of original research work and a new experimental research culture. The conflict was exacerbated by Valadares' affiliation with the then-illegal communist party and by the fact that many members of the Scientific Council of the Faculty of Sciences were supporters of, or sympathetic to, the regime and unwilling to defy it (Gaspar and Simões, 2011, pp. 338–341). Following the expulsions, Lídia Salgueiro and Marieta da Silveira, together with other colleagues, signed a petition asking for the reintegration of those expelled. They handed it to the director of the Faculty of Sciences. Unsurprisingly, the petition was unsuccessful, and the Institute for High Culture instead immediately canceled their scholarships (Carvalho, 2011, p. 65; Brotas et al., 2011, p. 54).

Reacting to these dramatic events, Soares handed in his resignation. The Soares–Valadares research school gathered at the Laboratory of Physics ended abruptly. Helped by the Joliot-Curies, Valadares went into exile in Paris. Still in 1947, he became *chargé de recherche* and then *maître de recherche* at Rosenblum's laboratory. He was promoted to *directeur de recherche* in 1957 and succeeded Rosenblum following his death in 1959 as director of the Laboratoire de Spectrometrie Nucléaire et de Spectrometrie de Masse.

12.4 Reorganizing Research: A New Leadership and an Unexpected Hub for Women

Following Soares's resignation, the Spanish right-wing physicist and researcher Palacios, formerly the director of the X-ray section of the Spanish National Institute of Physics and Chemistry, who had been formerly in contact with the Laboratory of Physics, was appointed by the Institute for High Culture as the new director of the Center for Studies in Physics. In 1948, right after the dismantling of the Laboratory of Physics and the end of the Soares–Valadares era, four women stayed at the Laboratory working in atomic and nuclear physics: Lídia Salgueiro, Marieta da Silveira, Glaphyra Vieira, and Maria Helena Blanc de Sousa (1921–2011) (Gaspar, 2018, p. 52; Vieira, 1989, pp. 33–34; Blanc de Sousa, 1989, pp. 21–22). In 1950, only Lídia Salgueiro and Marieta da Silveira remained, and in 1951 Lídia Salgueiro was left by herself, as Marieta da Silveira had moved to the Laboratory of Chemistry, which could benefit more from her expertise with radioactive materials.

In the position of director, which Palacios held until 1956, he recruited collaborators in electrochemistry, his new field of research, possibly to eliminate any troublesome scientific links to an uncomfortable past. Despite this change, Palacios allowed freedom of research to Lídia Salgueiro, who took upon herself the task of continuing the scientific agenda previously led by Valadares. This was a clear, obvious, and brave choice. In her unpublished autobiographical notes, she is adamant: "I decided to do everything possible to continue the work begun by Valadares, especially in the field of X-rays, despite the difficulties that such a task imposed. I think my ambition was realized" (Carvalho, 2011, p. 66). Valadares was away but not absent. He continued to advise from a distance, sending materials when needed and accommodating visiting researchers from the Laboratory of Physics if necessary. Although Lídia Salgueiro was not the director of the Laboratory of Physics and the Center for Studies in Physics, she was the de facto scientific leader of the research line with a scientific past that most wanted to obliterate. Palacios recognized her scientific merit, and even managed to get her a scholarship from the Institute for High Culture as a technical assistant to replace her canceled research fellowship. But Lídia Salgueiro declined, despite the convenience of a complement to her income, considering this offer to be a downgrade of her status as a researcher and faculty member (Carvalho, 2011, p. 66).

In 1951, Lídia Salgueiro (pictured in Figure 12.2) was then left with just one (male) teaching assistant, PhD student and future collaborator, José Francisco

Figure 12.2 Portrait of Lídia Salgueiro (1951). Reproduced with permission of João Pedro Salgueiro Gomes Ferreira.

Gomes Ferreira (1923–1982), who was hired by Palacios in 1949 but who nevertheless embraced research in atomic and nuclear physics. Their initial hierarchical relationship developed into a scientific partnership and a lifelong union. They married in 1953 (Figure 12.3).[11] Lídia Salgueiro was 35 years old, six years older than him. Their first child was born in 1958, after infertility exams while in Scotland on a research stay and upon suggestion of their scientific advisor Marion Ross (Carvalho, 2011, p. 67). Their second child was born in 1959. From 1951 to 1954 they were the sole researchers continuing Valadares' research agenda.

In 1952, Palacios was invited to become the first director of the newly founded Center for Nuclear Energy Studies, based at the Portuguese Institute of Oncology. One suspects that political reasons were behind this singular choice of localization

[11] In reminiscence notes, Lídia Salgueiro's very good friend, the mathematician and geophysicist Maria Augusta Perez Fernandes tells how she and Lídia Salgueiro toured Spain by train in 1948. Lídia Salgueiro suggested that they should invite Gomes Ferreira, then in his last year of the undergraduate degree in physico-chemical sciences, to escort them. According to Maria Augusta Perez, it was during this trip that "their mutual enchantment began" (Fernandes, 1989, p. 26).

Figure 12.3 Portrait of the couple Salgueiro-Gomes Ferreira at their wedding (1953). Reproduced with permission of João Pedro Salgueiro Gomes Ferreira.

of the Center for Nuclear Energy Studies outside academia, contrary to similar centers associated with the launch of research into nuclear energy, all of which were installed within science faculties. For the government it was mandatory to suppress any links to a past connected with Valadares, whose activities in Paris associated with the World Peace Council, a movement of Soviet tendency, left no doubt as to his political leanings and connections with other leftist scientists such as Frédéric Joliot-Curie. His activities in Paris were closely followed by the Portuguese political police PIDE.[12] Although Palacios allowed Lídia Salgueiro to extend Valadares'

---

[12] National Archives of Torre do Tombo, ANTT Archives, "Ficha de Informação," PIDE-DGS, Manuel José Nogueira Valadares, SR 229/47 NT2592, pp. 289–291.

research lines, it is nevertheless clear that the Laboratory of Physics was stigmatized during the early Cold War period. Lídia Salgueiro was sure that the letters
from Valadares were all opened, and their contents checked by the political police
(Carvalho, 2011, p. 66).

Palacios' permissive attitude vis-à-vis Lídia Salgueiro might have resulted from
multiple factors, from his acknowledgment of Valadares' scientific expertise to the
recognition of her scientific merit, dedication and resilience, and the realization that
the leadership of a woman was not a real menace in a male-dominated environment
(Bourdieu, 1998). He also knew that his growing political as well as academic and
institutional influence facilitated his ability to hire collaborators whom he diverted
to the Center for Nuclear Energy Studies (Palacios, 1958, pp. 12–13).[13] By contrast,
research conditions at the Laboratory of Physics and at its affiliated Center for
Physical Studies were much less attractive than at the more powerful Center for
Nuclear Energy Studies, so that Lídia Salgueiro's hiring capacity was not
facilitated.

At the beginning of his Paris exile, in 1949, Valadares resumed connection with
his former laboratory, where Lídia Salgueiro pushed forward his research agenda,
which she appropriated as her own. He sent Lídia Salgueiro photographic plates
with imprints of α-particles emitted by a strong radioactive source of ThC (bismuth-
211) to foster work in nuclear physics. The technique of photographic emulsions
had been developed in the 1920s and 1930s mostly by Marietta Blau and Hertha
Wambacher, two women at the Viennese Radio Institute (Rentetzi, 2004, p. 378;
Galison, 1997, pp. 160–210).[14] Lídia Salgueiro and her colleague Glaphyra Vieira
studied the α-radiation with an improvised microscope set up using old equipment
(Salgueiro and Vieira, 1952).[15] Work also continued along Valadares' former
research lines, and Lídia Salgueiro finally managed to publish the results of their
joint research started in 1946 on X- and gamma-ray spectroscopy under vacuum
(Valadares and Salgueiro, 1949). For once, and counteracting their bricolage
material culture practice, they also used equipment which Valadares had in Paris
and which he considered "to be the first in the world to enable gamma-ray
spectroscopy of a crystal in a vacuum" (Valadares, 1950, p. 100). In the early
1950s, Lídia Salgueiro turned to X-ray spectroscopy again, using Valadares's
former equipment. Work was taxing as it required regular adjustments to the
instruments in order to maintain experimental conditions during the long film
exposures needed for the photographic method. The vacuum system also regularly

---

[13] From 1952 to 1958, a team of eight men and one woman, recently graduated from the Faculty of Sciences, were
selected to enroll in radioisotopes research in nuclear physics at the Portuguese Institute of Oncology.
[14] After 1938, Cecil Powell of Bristol University applied nuclear photographic plates to the study of cosmic
radiation.
[15] In addition, she published the paper with Maria Helena Blanc de Sousa (Salgueiro and Blanc de Sousa, 1951).

broke (Ramos and Marques, 1989, p. 42). Despite the possibility of using new equipment acquired by Palacios – a high-cost multipurpose Philips X-ray apparatus – which was also useful for atomic physics by means of photon bombardment of metal screens, Lídia Salgueiro refused Palacios' help. She found the old equipment advantageous, notwithstanding the exacting and tiresome conditions, probably because, on the one hand, she was more familiar with it and, on the other hand, she wanted to avoid closer (unwanted and undesirable) contact with Palacios.[16]

In 1956, Gomes Ferreira, who had received his PhD in 1955 under the informal supervision of Valadares (Gomes Ferreira, 1954), and Lídia Salgueiro began a year-long stay at the Department of Natural Philosophy of the University of Edinburgh, headed by Norman Feather, to specialize in the technique of nuclear photographic plates. Their aim was to study the particles emitted during radioactive transmutations. The couple studied the transmutation thorium-229 → radium-225 to establish its complete decay scheme. They analyzed the tracks of both α-particles and electrons to observe cascading phenomena, that is the emission of two electrons stemming from internal conversion of energy associated with an α-particle. However, their work stagnated upon their return to Portugal in mid-1957 due to its complexity and the paucity of equipment.[17] An adequate microscope could not be acquired before 1959, and it took until 1961 to publish the paper titled "Contribution to the study of the disintegration scheme of thorium-229, with nuclear plates," reporting work started in Edinburgh. For this work, Lídia Salgueiro and Gomes Ferreira were awarded the Artur Malheiros prize of the Academy of Sciences of Lisbon.[18]

Back in Lisbon, they were surprised to find a new academic and research scenario. In December 1956, Palacios resigned as director of the Center for Physical Studies to focus exclusively on the more promising prospects offered by the Center for Nuclear Energy Studies located at the Portuguese Institute of Oncology. He was succeeded by Monteiro, a former colleague of Valadares from the late 1920s (to whom we already referred), who had recently been promoted to a full professorship. Free from the influence of Palacios, the Center for Physical Studies at the Laboratory of Physics entered a more relaxed and less competitive period, concentrating wholeheartedly on atomic and nuclear physics research.

Lídia Salgueiro and Gomes Ferreira conducted research and supervised teaching assistants at the Laboratory of Physics on an equal footing, in the sense that

---

[16] The old Valadares's equipment was finally replaced by a Beaudouin apparatus in 1959.

[17] Instituto Camões Historical Archives, Lídia Coelho Salgueiro, Box 0358, File 12, and Francisco Vitorino Gomes Ferreira, Box 3239, File 2.

[18] Instituto Camões Historical Archives, Box 3251, File 3, Report of scientific activity during 1962, of Centro de Estudos de Física anexo à FCL, December 19, 1962. Contribuição para o Estudo, com placas nucleares, do esquema de desintegração do [229]Th. Prémio Artur Malheiros da Academia das Ciências de Lisboa (Ciências Físicas e Químicas), 1961.

scientific research formerly directed solely by the senior researcher Lídia Salgueiro was now shared with the junior researcher Gomes Ferreira.[19] Gomes Ferreira soon combined research and teaching activities with scientific management positions. He became a full professor in 1966, to be followed by positions at the Faculty of Sciences as Director of the Center for Physics Studies at the Laboratory of Physics, Director of the Instituto Geofísico Infante D. Luiz, Deputy Director of the Faculty of Sciences, and, finally, Vice-Rector of the University of Lisbon.

Unlike her husband, who had a canonical male academic career, Lídia Salgueiro did not harbor academic ambitions – she only became a full professor in 1970, in belated recognition of her scientific skills and dedication to the Laboratory of Physics and the Faculty of Sciences. She was a behind-the-scenes researcher, whose scientific guidance and relevance were core to the group, but who preferred to be almost invisible to the outside world, except for the relationship she maintained with Valadares.[20] The Salgueiro–Gomes Ferreira partnership enacted, in the scientific realm, complementary personal characteristics and expected gendered roles, in the sense of female self-effacement and male visibility, but not in the sense of an asymmetrical division of scientific labor, an area in which often Lídia Salgueiro dominated.

Lídia Salgueiro's focus was first and foremost on training new generations of women scientists in a laboratory context. Upon the couple's return from Edinburgh, in May 1957, the woman physicist Otilde Costa (1926–1999), a teaching assistant at the Faculty of Sciences of the University of Porto joined the Laboratory of Physics to work on spectroscopy with the Philips equipment for photonic bombardment. She was awarded a PhD in 1959 (Costa, 1959) and stayed with the group until 1966 on successive scholarships by the Institute for High Culture.[21] In 1959, Teresa Gonçalves-Ramos (1935–2006), teaching assistant at the Faculty of Sciences of the University of Lisbon, began work on the transmutation scheme of radium-226. She was granted a PhD in 1963 (Gonçalves, 1963). In both instances, and in subsequent work, they relied on the ability to read particle tracks in nuclear plates, an area in which Lídia Salgueiro and Gomes Ferreira had specialized since their stay in Edinburgh.

To calculate the energy of the particles, tracks imprinted on the plates had to be studied with the aid of an adequate microscope. This undertaking required a huge number of hours of observation to measure the horizontal and vertical projections of tens of thousands of particles' trajectories in the emulsion (Ramos and Marques,

[19] The role of couples in science has been addressed in Abir-Am and Outram (1987), Pycior et al. (1996), and Lykknes et al. (2012).

[20] Even the collection of stamps that Lídia Salgueiro put together at the end of her life on the history of physics was suggested by Valadares (Carvalho, 2011, p.70).

[21] See "Maria Otilde Barbosa Pereira da Costa," Memória ScientÍfica Pioneiras na FCUP (2023) at https://www.fc .up.pt/memoriascientifica/pnfcup-maria-otilde-barbosa-pereira-costa

1989, p. 43). Taxing researchers with this task was specific to the Portuguese group. In other groups, including Cecil Powell's in Bristol and Joaquín Catalá de Alemany's in Valencia, the researchers' burden was reduced by hiring women as technical assistants, either without scientific training, as in the case of the "Bristol scanning girls," or by giving them adequate scientific training as in the Valencia case, with some of the women "microscopists" eventually defending PhD theses in physics (Gaspar, 2018, pp. 57 and 284). But in all cases this routine, tiring, and time-consuming work, requiring constant discipline, thoroughness, and attention, still depended on women. The literature on science and gender has pointed out that these characteristics, being regarded as feminine, were associated with female researchers in different research fields, including astronomy and crystallography.[22]

After 1965, the technique of using nuclear plates was abandoned at the Laboratory of Physics due to the inability to hire technical assistants and the impossibility of grounding work solely on the few researchers available in the group (Gaspar, 2018, pp. 283–284). Research continued on topics addressed since the Salgueiro–Gomes Ferreira stay in Edinburgh, focusing on atomic physics using the Philips equipment formerly acquired to study fluorescence spectra by photonic bombardment, and the Beaudouin installation to study spectra obtained by electron bombardment. They explored the atomic-level yields of elements of atomic numbers between 73 and 92 (Gomes Ferreira et al., 1965; Salgueiro et al., 1965). They also counted on the collaboration of new women researchers: Inês Gonçalves-Marques (1938–2017), Maria Amália Campos, and Maria de Lurdes Tavares.

The Laboratory of Physics ended up acting as a hub for those more ambitious female students who sought to sidestep the otherwise lower career expectations compared with their male colleagues.[23] The possibility of carrying out scientific research in a small group, led by a woman, in the family-like environment provided and stimulated by the Salgueiro–Gomes Ferreira couple, offered to many, despite its difficulties, a more seductive professional choice than teaching at the secondary-school level.

The domestic-like atmosphere that Lídia Salgueiro created around her turned the Laboratory of Physics into a "second home" for everyone, with tea-time breaks to socialize and to lighten the long hours of laboratory practice.[24] This routine, less competitive and more fraternal (Ramos and Marques, 1989, p. 41), translated into

---

[22] Examples are the group of Edward C. Pickering, the director of Harvard Observatory, between 1877 and 1919, who hired numerous women to process the enormous amount of astronomical data collected, and who became known as Pickering's "harem" or "calculators" (see Chapter 1); the group of crystallographers led by John D. Bernal who, at Birkbeck College, London, used X-ray diffraction to identify the structure of various crystalline substances of biological interest; and the extensive calculations of NASA's space programs. See Rossiter (1982, 1999).

[23] In the Portuguese academic system, salaries just depended on academic position, never on gender.

[24] For the socially constructed notion of women as caretakers by opposition to characteristics of so-called male leadership see Koven and Michel (1993).

joint participation in research tasks and joint authorship of articles. In a way, the Laboratory of Physics offered a safe space for women to develop their research careers which, although possibly academically less ambitious, offered, in any case, a new possibility for female affirmation in the scientific world.

The Salgueiro–Gomes Ferreira's era at Laboratory of Physics thus embraced Valadares' vision, including fostering the recruitment of men and women on an equal basis. In the Soares–Valadares era, one in three staff members were women; under the Salgueiro–Gomes Ferreira leadership the previous gender profile was not only replicated but even considerably amplified during the decade following their return from Edinburgh (1957–1966). In a small-sized group of less than 10 researchers at its peak, women often outnumbered men, reaching a maximum imbalance between 1960 and 1966, when for one to three male researchers there existed four to six female researchers: Lídia Salgueiro, Otilde Costa, Teresa Gonçalves-Ramos as well as Inês Gonçalves-Marques, Lurdes Tavares, and Amália Campos (Gaspar, 2018, pp. 52, 58, and 284–287).

Several factors may account for the fact that this gender pattern in the Laboratory of Physics would reverse in the following decade. They include the restructuring and modernization of academic university curricula in physics and chemistry and government's increasing support, both concerning human and material resources, for fundamental research in academic contexts. Research careers in physics in academic contexts then became more attractive to men, despite the competition provided by the State Laboratory of Nuclear Physics and Engineering, which since 1961 attracted the most talented male undergraduates in physics and engineering to the Portuguese nuclear program (Gaspar, 2018, p. 287).[25]

The allegedly feminine characteristics (patience, endurance, discipline, attention) required for scanning nuclear emulsions may partially account for women's strong presence in the Salgueiro–Gomes Ferreira group, with the unlikely twist that they did not play secondary roles of technical assistants but were instead active researchers. However, specific context-dependent explanations, relying on the social, institutional, and political dimensions of Portuguese society which impacted the Laboratory of Physics, must be considered, since many of the laboratories in the same or similar research fields in Europe – except for the Marie Curie Laboratory and the Viennese Radio Institute – did not achieve such a high proportion of women researchers. Concerning the political context, the explanation usually given to

---

[25] Additionally, with the onset of the colonial war in Africa in 1961, which lasted until the Carnation Revolution of 1974, and the establishment of democracy in Portugal, young men who entered as researchers in the State Laboratory (the same was true for other general directorates of the Board) were allowed to serve part of the mandatory military service in the Laboratory, usually the time comprised between the end of the six-month instruction period and either the start of a commission in Africa or, if they were not appointed to any commission, the end of the two-year period of mandatory military service. Informal written information provided by João Duarte Cunha, research assistant in the Laboratory of Physics in the period 1968–1975 to Maria da Conceição Abreu and Ana Simões.

account for the role of women in the Viennese Radio Institute is grounded in a unique constellation of progressive politics and support from politically alert personalities and offers useful hints for a comparative assessment (Rentetzi, 2004; Rentetzi, 2007). While the Portuguese political dictatorial context and Palacios' political allegiances were at the antipodes of progressive politics, Valadares' influence extended across the two periods of the Laboratory of Physics under analysis, thus linking leftist politics and the Marie Curie Laboratory with the situation in Lisbon.

For certain, the impact of female presence in the Laboratory of Physics both during the Soares–Valadares and the Salgueiro–Gomes Ferreira eras, together with the singular role played by Lídia Salgueiro shaped the experimental practice and material culture of the group and influenced the arduous process of affirmation of a new culture of experimental research in physics at the Faculty of Sciences.[26]

12.5 Concluding Remarks

Most literature on gender and science addresses the importance of women scientists, usually focusing on women who have clearly influenced their fields, that is, women who, like their male counterparts, are internationally renowned. This slanted vision of the scientific community is built from the perspective of the so-called centers, and for a long time it presented itself as the only way of looking at the production and circulation of knowledge. In this framework, the so-called peripheral areas – European and non-European – were considered as only capable of reproducing and mimicking what was done in the centers. The increasingly large bibliography and scholarship dedicated to centers and peripheries shows how much this image, resulting from a specific lens of analysis, lacks depth and diversity (Gavroglu et al., 2008; Diogo et al., 2016; Diogo and Simões, in press). By changing the lens, the peripheries reveal a very different agency grounded on the multiple local strategies for producing knowledge and fostering communities of scientists who are very far from the passivity formerly attributed to them.

The same happens when one turns the gaze to the role of women in peripheral countries. The questions to be asked should not only be about whether their contributions were relevant in international terms, but also about their role and contributions in building their respective national scientific research communities. In the Portuguese case, the political framework has also to be considered when assessing the role of women in science. In Portugal, Salazar's dictatorship

---

[26] Furthermore, the long-term impact of the predominance of women in the periods under analysis in this chapter on the high percentage of female teaching and research staff at the Department of Physics of the Faculty of Sciences of the University of Lisbon created in the 1980s, which succeeded the Laboratory of Physics, remains to be analyzed.

combined political repression with a view concerning women's education and professional career that supported one of Salazar's ideological pillars – the praise of the traditional family based on a patriarchal model, where women should stay home to take care of their children, with their individual freedom legally dependent on the authorization of their husbands.

These historiographic and historical contexts, together with a *longue durée* perspective encompassing both the national landscape of higher education and the international networks connecting radioactive laboratory spaces across Europe, frame the conclusions of this chapter on the contributions to quantum physics at the national and international levels made by Lídia Salgueiro and a team of women researchers at the Laboratory of Physics.

National and institutional contexts, political circumstances, personal idiosyncrasies, and historical contingencies are thus the main variables that shape the case study presented in this chapter. Together they account for the singularity of the Laboratory of Physics during the Soares–Valadares era, in terms of the development of a research school in an academic context still unreceptive to scientific research, but resonating with practices appropriated from abroad and put in place in demanding financial and material circumstances. In such a context, women researchers were nurtured, perhaps following the model of the Marie Curie Laboratory in Paris in which Valadares was trained.

The second phase of the Laboratory of Physics, following the 1947 political dismissals, is even more intriguing. Although the new director, the right-wing physicist Palacios purposefully diverted research from the former scientific agenda of leftist Valadares, he nonetheless opted not to interfere with Lídia Salgueiro's scientific agenda. He probably considered that restrictive financial and material conditions would sufficiently hinder her team's work, so that a woman would not represent a real threat to his status and choices. In addition, Salazar's government saw no use for pure science in the field of experimental atomic and nuclear physics, despite Salazar's awareness that nuclear energy applications were becoming vital. Because both the regime's and Palacio's attentions were diverted to other scientific topics and research spaces, first Lídia Salgueiro, then the Salgueiro–Gomes Ferreira couple managed to attract women researchers who, starting out with lower career expectations than their male colleagues,[27] were perhaps less concerned about entering a less competitive research group, even though it was under the suspicion of being close to the political opposition to Salazar. Despite surveillance from the political police, there was a certain freedom of action, which enabled Lídia Salgueiro to continue the tradition inherited from Valadares, both scientifically

---

[27] As Palacio himself recognized, most women who graduated ended up as high-school teachers. Instituto Camões Historical Archives, Box 3250, File 3, letter from Julio Palacios to Presidente of IAC, October 10, 1948, p. 1.

(in terms of research, networking, and supervision from a distance) and gender-wise, consolidating a very specific research niche at the Laboratory of Physics. Dedicated to experimental atomic and nuclear physics, then already part of "normal science" (Kuhn, 1967), the Laboratory of Physics contributed to a survey of the spectra of radiation emitted by various radioactive elements and a taxonomic mapping of the microscopic world.

Political circumstances, scientific and networking choices, and gender issues are inextricably entangled in the case study addressed in this chapter. Focusing on gender issues has enabled us not only to identify similarities and differences in accounting for the changing contexts of the Laboratory of Physics, but also for its specificities vis-à-vis other European laboratories. The Laboratory of Physics was a female research space, in the sense that unlike other institutions it included mostly women researchers. It is impossible to avoid speculating whether the Laboratory of Physics' female profile was precisely what provided the room for maneuvering under Salazar's dictatorship, and for pursuing research in atomic and nuclear physics along Valadares' lines. Although there were women opponents to the regime who played a strong political role in the public sphere, we do not know the political allegiances of the women who gravitated around the Laboratory of Physics, except for Marieta da Silveira's communist inclinations (Brotas et al., 2011, p. 57). In any case, women scientists did not have a strong voice in academia, and were perceived essentially as a workforce, not as dangerous revolutionaries fighting for their rights. It is precisely this invisibility that allowed their work to flourish (Kuchinskaya, 2014). Although not at the forefront of scientific developments, their work remained critical to sustain research, showing that the relevance of a specific group can be asserted through "normal science." Less glamorous than other endeavors, perhaps, their work was nevertheless an integral part of the many avenues asserting experimental atomic and nuclear physics on a global scale.

References

Abir-Am, Pnina G. and Outram, Dorinda. 1987. *Uneasy Careers and Intimate Lives: 1789–1979*. New Brunswick, NJ: Rutgers University Press.

Beyer, Robert T. 1949. *Foundations of Nuclear Physics*. New York: Dover.

Blanc de Sousa, Maria Helena. 1989. Uma velha amiga. In *Jubileu de José Gomes Ferreira, Prof. Catedrático de Física da F.C.L.* Lisbon: Faculty of Sciences of the University of Lisbon, pp. 21–22.

Boudia, Soraya. 2001. *Marie Curie et son laboratoire: sciences et industrie de la radioactivité en France*. Paris: Éditions des archives contemporaines.

Bourdieu, Pierre. 1998. *La domination masculine*. Paris: Éditions du Seuil. English version: Bourdieu, Pierre. 2002. *Masculine Domination*, Palo Alto, CA: Stanford University Press.

Brotas, Manuela, Viegas, Francisca, and Maia, Elisa. 2011. Marieta Amélia da Silveira (1917–2004). A professora que não se esquece. In Simões, Ana (ed.), *Novas memórias de professores cientistas*. Lisbon: Faculty of Sciences of the University of Lisbon, pp. 50–59.

Carolino, Luís Miguel, Mota, Teresa Salomé, eds. 2013. The polytechnic experience in the nineteenth-century Iberian Peninsula. *HoST*, 7. https://www.johost.eu/vol7_spring_2013/vol7.htm

Carvalho, Luísa. 2011. Lídia Salgueiro (1917–2009). Fragmentos de uma vida: infância, percurso, paixões, o fim (excerpts from unpublished autobiographical notes). In Simões, Ana (ed.), *Novas memórias de professores cientistas*. Lisbon: Faculty of Sciences of the University of Lisbon, pp. 60–71.

Costa, Maria Otilde Barbosa Pereira Da. 1959. *Contribuição para o estudo das probabilidades relativas de ionização dos elementos de número atómico elevado*. PhD Thesis. Porto: University of Porto.

Curie, Eve. 1937. *Madame Curie: A Biography*. Paris: Hachette Books.

Diogo, Maria Paula, Gavroglu, Kostas, and Simões, Ana, eds. 2016. STEP matters: historiographical considerations. *Technology & Culture*, 57(4), 926–997. https://doi.org/10.1353/tech.2016.0112

Diogo, Maria Paula and Simões, Ana. in press. Center and periphery as historiographical categories. The art of fugue. Saraiva, Tiago, Schaffer, Dagmar, Bray, Francesca, and Valleriani, Matteo (eds.), *Cambridge History of Technology. Technology. Meaning and Methods, Volume 1*. Cambridge: Cambridge University Press.

Fernandes, Maria Augusta Perez. 1989. Reminiscências. In *Jubileu de José Gomes Ferreira, Prof. Catedrático de Física da F.C.L.* Lisbon: Faculty of Sciences of the University of Lisbon, pp. 23–28.

Galison, Peter. 1997. *Image and Logic. A Material Culture of Microphysics*. Chicago, IL: University of Chicago Press.

Gaspar, Júlia. 2009. *A investigação no Laboratório de Física da Universidade de Lisboa*. Lisbon: CIUHCT.

Gaspar, Júlia. 2018. *Percursos da física e da energia nucleares na capital portuguesa. ciência, poder e política 1947–1973*. Lisbon: Edições Colibri–CIUHCT.

Gaspar, Júlia and Simões, Ana. 2011. Physics on the periphery: a research school at the University of Lisbon under Salazar's dictatorship. *Historical Studies in the Natural Sciences*, 41(3), 303–343. https://doi.org/10.1525/hsns.2011.41.3.303

Gavroglu, Kostas, Patiniotis, Manolis, Nieto-Galan, Agustí, et al. 2008. Science and technology in the European periphery: some historiographical reflections. *History of Science*, 46(2), 153–175. https://doi.org/10.1177/007327530804600202

Gil, Fernando Bragança. 2003. Núcleo de Matemática, Física e Química: uma contribuição efémera para o movimento científico português. *Boletim da Sociedade Portuguesa de Matemática* 49, 77–92.

Gomes Ferreira, J. 1954. *Contribuição para o estudo da intensidade das bandas satélites da risca lα de elementos de número atómico compreendido entre 73 e 92*. PhD Thesis, Lisbon: University of Lisbon. (Published in *Revista da Faculdade de Ciências*, 2nd series B, 3, 65–140.)

Gomes Ferreira, J., Costa, M. O., Gonçalves, M. I., and Salgueiro, L. 1965. Le rendement de transition de Coster-Krönig LI→LIII des éléments de nombre atomique compris entre 73 et 92. *Journal de Physique*, 26, 5–8. https://doi.org/10.1051/jphys:019650026010500

Gonçalves, Maria Teresa. 1963. *Contribuição para o estudo do espectro de electrões de conversão interna emitidos na desintegração do 226 Ra*. PhD Thesis, Lisbon: University of Lisbon. (Published in *Revista da Faculdade de Ciências*, 5–62.)

Govoni, Paola. 2009. Donne in un mondo senza donne. Le Studentesse delle Facoltá Scientifiche in Italia (1877–2005). *Quaderni Storici*, 1, 213–246.

Grácio, Rui. 1971. Ensino primário e analfabetismo. In Serrão, Joel (ed.), *Dicionário história de Portugal*, Lisbon: Iniciativas Editoriais, vol. 3, p. 51.

Hughes, Jeff. 2003. Radioactivity and nuclear physics. In Nye, Mary Jo (ed.), *The Modern Physical and Mathematical Sciences, the Cambridge History of Science, Volume 5*. Cambridge: Cambridge University Press, pp. 350–374.

Keller, Evelyn F. 1995. Gender and science: origin, history, and politics. *Osiris*, 10, 26–38. https://doi.org/10.1086/368741

Kohlstedt, Sally G. 1995. Women in the history of science: an ambiguous place. *Osiris*, 10, 39–58. https://doi.org/10.1086/368742

Kohlstedt, Sally G. and Longino, Helen (eds.). 1997. *Women, Gender and Science: New Directions*. Chicago, IL: Chicago University Press.

Kohlstedt, Sally G. (ed.). 1999. *History of Women in the Sciences. Readings from Isis*. Chicago, IL: Chicago University Press.

Koven, Seth and Michel, Sonya 1993. *Mothers of a New World: Maternalist Politics and the Origins of Welfare States*. New York: Routledge.

Kuchinskaya, Olga. 2014. *The Politics of Invisibility: Public Knowledge about Radiation Health Effects after Chernobyl*. Cambridge, MA: MIT Press.

Kuhn, Thomas S., Heilbron, John L., Forman, Paul, and Allen, Lini. 1967. *Sources for History of Quantum Physics. An Inventory and Report*. Philadelphia, PN: The American Philosophical Society.

Lopes, Quintino. 2017. *A europeização de Portugal entre guerras. A junta de educação nacional e a investigação científica*. Lisbon: Caleidoscópio.

Lykknes, Annette, Opitz, Donald, and van Tiggelen, Brigitte (eds.). 2012. *For Better or For Worse? Collaborative Couples in the Sciences*. Basel: Birkäuser.

Moura, Maria José Sabino de (ed.). 1972. *Doutoramentos na Universidade de Lisboa 1911–1971*. Lisbon: University of Lisbon.

Oertzen, Christine von, Rentetzi, Maria, and Watkins, Elizabeth S. 2013. Finding science in surprising places: gender and the geography of scientific knowledge. *Centaurus*, 55 (2), 73–80. https://doi.org/10.1111/1600-0498.12018

Palacios, Julio. 1958. Actividade e Planos de Trabalho, Centro de Estudos de Física da Comissão de Estudos de Energia Nuclear. In 1ª Reunião dos Técnicos Portugueses de Energia Nuclear (1st Meeting of Portuguese Nuclear Energy Technicians). Lisbon.

Pigeard-Micault, Nathalie. 2013. *Les femmes du laboratoire de Marie Curie*. Paris: Glyphe.

Pigeard-Micault, Nathalie. 2018. Marie Curie et les femmes de son laboratoire. In Bréchemier, Dominique and Laval-Turpin, Nicole (eds.), *Femmes de sciences. Quelles conquêtes? Quelle reconnaissance?* Jacou: Éditions L'Harmattan, pp. 109–124.

Pycior, Helena M., Slack, Nancy G., and Abir-Am, Pnina G. (eds.). 1996. *Creative Couples in the Sciences*. New Brunswick, NJ: Rutgers University Press.

Ramos, M. Teresa and Marques, M. Inês. 1989. Trinta anos devotados à ciência. In *Jubileu de José Gomes Ferreira, Prof. Catedrático de Física da F.C.L.* Lisbon: Faculty of Sciences of the University of Lisbon, pp. 41–44.

Rayner-Canham, Marelene F. and Rayner-Canham, Geoffrey (eds.). 1997. *A Devotion to Their Science. Pioneer Women of Radioactivity*. Philadelphia, PA: Chemical Heritage Foundation. https://doi.org/10.1515/9780773566583

Rentetzi, Maria. 2004. Gender, politics, and radioactivity research in interwar Vienna: the case of the Institute for Radium Research. *Isis*, 95, 359–393. https://doi.org/10.1086/428960

Rentetzi, Maria. 2007. *Trafficking Materials and Gendered Experimental Practices. Radium Research in Early 20th Century Vienna*. New York: Columbia University Press.

Rossiter, Margaret W. 1982. *Women Scientists in America. Struggles and Strategies*. Baltimore, MD: Johns Hopkins University Press.

Rossiter, Margaret W. 1999. "Women's work" in science, 1880–1910. In Kohlstedt, Sally G. (ed.), *History of Women in the Sciences. Readings from Isis*. Chicago, IL: Chicago University Press, pp. 381–398.

Salgueiro, Lídia. 1944. Spectographie du rayonnement γ émis par le dépot actif à evolution lente du radon. *Portugaliae Physica* 1, 67–72.

Salgueiro, Lídia. 1945. *Espectro gama dos derivados de vida longa do radão*. PhD Thesis. Lisbon: University of Lisbon.

Salgueiro, Lídia and Blanc de Sousa, Maria Helena. 1951. Influence de la tension d'excitation sur les satellites des raies $L\beta_2$ de l'or, du plomb et du bismuth. *Portugaliae Physica* 3(2), 95.

Salgueiro, Lídia and Vieira, Glaphyra. 1952. Nouvelle détermination des intensités des groupes de structure fine de la transmutation ThC → (α,γ) AcC. *Comptes rendus des séances de l'Académie des Sciences*, 234, 1765.

Salgueiro, Lídia, Campos, M. A., and Ferreira, J. G. 1965. Le rendement de transition de Coster-Krönig LI→LIII du rhenium. *Portugaliae Physica*, 4(2), 131–134.

Salgueiro, Lídia and Carvalho, Luísa. 2001. Manuel Valadares (1904–1982). Facetas de uma personalidade: humana, científica e artística. In Simões, Ana (ed.), *Novas memórias de Professores Cientistas*. Lisbon: Faculty of Sciences of the University of Lisbon, pp. 70–77.

Scerri, Eric. 2013. Element 85. Astatin. In *A Tale of 7 Elements*. Oxford: Oxford University Press.

Schiebinger, Londa. 1987. The history and philosophy of women in science: a review essay. *Signs*, 12(2), 305–32. https://doi.org/10.1086/494323

Silveira, Marieta da. 1946. *Contribuição para o estudo das radiações do urânio X complexo*. PhD Thesis. Lisbon: University of Lisbon.

Simões, Ana, Carneiro, Ana, Diogo, Maria Paula, Carolino, Luís Miguel, and Mota, Teresa Salomé. 2013. *Uma história da Faculdade de Ciências de Lisboa (1911–1974)*. Lisbon: Faculty of Sciences of the University of Lisbon.

Simões, Ana and Diogo, Maria Paula. 2022. Introduction: science, technology and medicine in the making of Lisbon (1840–1940). Simões, Ana and Diogo, Maria P. (eds.), *Science, Technology and Medicine in the Making of Lisbon (1840–1940)*. Leiden: Brill, pp. 1–30.

Valadares, Manuel. 1933. *Contribution à la spectrographie, par diffraction cristalline, du rayonnement γ*. PhD Thesis, Paris: University of Paris Sorbonne. Paris: Masson.

Valadares, Manuel and Mendes, Francisco. 1938. Étude des satellites Lα, de l'élément 82 (Pb). *Comptes rendus de l'Académie des* sciences, 206, 744.

Valadares, Manuel. 1950. O Laboratório de Física da Faculdade de Ciências de Lisboa, sob a direcção do Prof. Dr. A. Cyrillo Soares (1930–1947) e a investigação científica. *Gazeta de Física*, 2, 93–106.

Valadares, Manuel and Salgueiro, Lídia. 1949. Les spectres L et gamma émis par la transmutation RaD → RaE. *Portugaliae Physica*, 3, 21–28.

Vieira, Glaphyra. 1989. O princípio de uma carreira. In *Jubileu de José Gomes Ferreira, Prof. Catedrático de Física da F.C.L.* Lisbon: Faculty of Sciences of the University of Lisbon, pp. 33–34.

13

Carolyn Parker's Electronic Frequencies

CHARNELL CHASTEN LONG

13.1 Introduction

As the first known African American woman to receive a postgraduate degree in physics, Carolyn Beatrice Parker's (1917–1966) inspiring story has been featured in many – albeit brief – online articles. These typically focus on her participation in a top-secret military research project (Mickens, 2021; Prescod-Weinstein, 2022).[1] According to Parker's sister, Juanita Parker Wynter, Parker's work was "so secret she couldn't discuss it, even with us, her family" (Warren, 1999, p. 216). The secret nature of Parker's work at Wright Field in Dayton, Ohio, has led many of her biographers to infer that Parker worked on the Dayton Project (Warren, 1999, p. 209; Mickens, 2021, pp. 8–9; Prescod-Weinstein, 2022), the branch of the Manhattan Project that processed the polonium used for the first atom bombs (Thomas, 2017). Scholars have further linked Parker's premature death to radioactive contamination from the Dayton Project (Mickens, 2021; Prescod-Weinstein, 2022). However, a finer analysis of declassified government documents reveals that Parker worked as a research physicist in the Aircraft Radio Laboratory at Wright Field (Stewart and Parker, 1944; Atlanta Daily World, 1945), and not at the Dayton Project as previously presumed. This discovery illuminates a less *heroic* narrative while also helping to unveil the compelling story of a Black woman physicist.

Thanks are due to North Carolina A&T State University Librarian, Harvey D. Long, who assisted me in locating Carolyn Parker in historic newspapers and archival collections. Also, I would like to thank the following for sending vital information via email: Sara Myers (Tennessee State Library and Archives), Michele Wilbanks (University of Florida), Karina Cooper, Corinne Mona, Chip Calhoun, Allison Rein (Niels Bohr Library & Archives), Lisa Moore (Amistad Research Center, Tulane University), Thera Webb (Massachusetts Institute of Technology), and Michelle Henderson (University of Michigan). Lastly, I would like to thank the hidden labor of those who remain nameless but assisted me by mailing me Parker's academic transcripts and answering important questions via automated chat systems at the National Archives and Records Administration and the Library of Congress.
[1] While some historical works use the term "top secret" to describe wartime research projects, the US National Security Information system in use from 1940 to 1946 classified them into three levels of increasing secrecy, "Restricted," "Confidential," and "Secret." The "Top Secret" level was not added until 1951 (Maus, 1996).

Parker's story is significant to the history of physics in part because it challenges traditional conceptualizations of what is considered a "scholarly" contribution and also because it illustrates the intersectional barriers that a young Black woman aspiring to become a quantum physicist encountered in the mid-twentieth-century US. Parker's scholarly work in physics was primarily experimental, applied, and pedagogical. She did not publish in scientific journals; instead, she contributed to applied research for the US military,[2] while often detouring to teach mathematics and physics in segregated schools and in historically Black colleges and universities (HBCUs) in the Jim Crow South. This non-traditional path does not diminish her contributions to the history of physics. As historian of science Rayvon Fouché argues in *Black Inventors in the Age of Segregation*, "by terminating our dependence on the over-inflated and *heroic* contemporary understandings of Black inventors and refocusing our analyses on their everyday experiences, we will allow black inventors' lives to speak" (Fouché, 2003, p. 8). In other words, we must remember Parker for who she was, not what we want her to be. In so doing, we can begin to hear Parker's true story and better understand her lived experience in physics as an African American woman.

13.2 A Note on Method

Archival silences and Parker's premature death at 48 years old limit historians' ability to hear her experiences from her own voice. Her voice was not captured in the oral history revolution of the 1970s, nor were her written contributions intentionally preserved in archival collections (Shopes, 2011, pp. 451–465). For this chapter, I have instead pieced together extant fragments to reconstruct her story. As Saidiya Hartman notes, scholars of Black history have to navigate silences and gaps in historical archives. Records on marginalized groups remain sparse and fragmented (Hartman, 2008). To find information about Parker, I have leaned into counter-archival traditions of Black print culture and HBCU library collections for clues. Black counter-archival practices grew out of the need to validate the Black experience that broader society disregarded, and that mainstream institutional archives failed to collect and preserve. The editors of Black pamphlets, newspapers, and periodicals aimed to capture Black life and, in turn, preserved the stories of families and communities that would have otherwise been lost to history (Nelsen, 2016; Greer, 2019). Black scientists proceeded similarly. They created scientific journals (i.e., *Beta Kappa Chi Bulletin*), shared their experiences in Black

---

[2] The National Archives and Records Administration will not release Parker's personnel files since some of them remain classified. At this time, only one (formerly) classified report that she co-authored is available (Stewart and Parker, 1944). However, since her co-author, Chandler Stewart Jr. acknowledged her on different publications (Stewart, 1945b, c), it is probably safe to assume that she contributed to multiple reports.

newspapers and periodicals (e.g., *Ebony Magazine, Chicago Defender, Baltimore Afro-American*), and, when possible, also shared their success stories with mainstream scientific journals (Long, 2023).

This chapter introduces many previously uncited materials found across various types of sources. It makes use of secondary sources to understand the context of materials such as census data, city directories, and oral histories in the Samuel L. Proctor Collection at the University of Florida to reconstruct Parker's childhood. I rely particularly heavily on the oral history of her sister, Julia Leslie Parker-Cosby (Parker-Cosby, 1985; Parker-Cosby, 2004), and I fill in gaps with oral histories that capture Black life in Gainesville, Florida, over the period of interest. I reconstruct Parker's undergraduate experiences with Fisk University catalogs, newspapers, and related secondary sources.

My treatment of Parker as a K–12 (kindergarten to twelfth grade) teacher introduces archival evidence from the A. J. Quinn Collection at the University of Florida and the Huntington High School yearbooks (conveniently captured by the Internet Archive). To understand her graduate school career at the University of Michigan, Ohio State University, and Massachusetts Institute of Technology (MIT), I rely heavily on transcripts, course catalogs, and her master's thesis. Parker's contributions to World War II have been recovered from scientific journals, declassified documents, and articles in Black newspapers. As needed, I rely on *The Journal of Negro Education* to understand the HBCU science curriculum and other secondary sources for context. In short, by relying on Black counter-archival traditions, I have reconstructed Parker's story to extend awareness of her accomplishments beyond her being historicized as the "first Black woman physicist," so as to better understand who she was as a physicist.

13.3 Raising a Scientist (1917–1933)

Parker was born on November 18, 1917, in Gainesville, Florida (US Census, 1920; Parker, 1953, p. 28). During the early twentieth century, African Americans in Florida experienced widespread racial violence and discrimination. Jim Crow, a set of racist laws that emerged in the US toward the late nineteenth century, structured the daily lives of most Americans, but its most harmful impact was reserved for the Black community. In particular, between 1905 and 1909, Florida passed a series of legislative measures that outlawed cohabitation, miscegenation, and racial integration of higher education, jails, public transportation, public waiting rooms, and ticket offices (Howard and Howard, 1994). Because of Jim Crow restrictions, the Parker family's two-story home also served as an early classroom and laboratory for the young Parker children. Parker's mother, Delia Ella Murrell Parker, was her first teacher.[3] Mrs. Parker, having earned a teaching certificate and worked as a teacher before Parker was born, could easily instruct Parker in basic arithmetic and literacy (Parker-Cosby, 1985, pp. 2–3; Parker-Cosby, 2004, p. 2).

Parker's father, Julius Augustus Parker, introduced Parker to science. As one of Gainesville's few Black physicians and pharmacists, Dr. Parker served the city's Black residents and delivered most of the children in the neighborhood (McPherson, 1976, p. 5; Parker-Cosby, 1985, p. 1; Vihlen, 1994, pp. 21–30). Dr. Parker could not work in a nearby hospital that served white patients (Gamble, 1995; Ward, 2003), so the family living room served as a reception area where Black patients waited to be called into their appointments (Parker-Cosby, 2004, p. 7). When surgery was required, Dr. Parker prepared one of the rooms in his home – or occasionally a room in his patient's home – for the procedure (Griffin, 1985, p. 18; Parker-Cosby, 1985, p. 3). When a patient required medication, he would retrieve the prescription from the backroom of his own house (Parker-Cosby, 1985, p. 33). Parker and her siblings learned at an early age to help out their father by answering calls and taking messages (Parker-Cosby, 1985, p. 3).

As the oldest daughter in the family, Parker took on a leadership role by helping care for her younger siblings: Mary, Juanita, Martha, Gloria, Julia, and Julius Jr. Before anyone went out to play, Parker made certain that the house was cleaned from top to bottom, and that all her siblings helped with cooking, cleaning, and washing dishes. Julius Jr. cleaned the yard since their father "felt his son should not work inside nor his girls outside" (Parker-Cosby, 1985, p. 3), his sister later recalled. While Dr. Parker subscribed to gendered ideas about the division of household labor, it is safe to assume he had a more open mind about work outside the home given that all of his daughters attended college and pursued professional careers (Parker-Cosby, 1985, p. 1).[4]

In 1925, eight-year-old Parker moved with her family to Tampa, Florida (Parker-Cosby, 2004, p. 1). There, Dr. Parker continued to practice medicine from home,[5] while Mrs. Parker attended Daytona Cookman Collegiate Institute (now Bethune-Cookman University) in preparation for her return to the classroom (Parker-Cosby,

---

[3] For more information about the US education system for African American children in the South during Jim Crow, see Anderson (2010).

[4] All of Parker's siblings went to college except Gloria, who died at a young age. Mary Parker Miller graduated from Talladega College with a BA in Mathematics and then obtained a MS in Mathematics from New York University. Juanita Parker Wynter graduated with a BS in Mathematics from Saint Augustine College (now Saint Augustine University) and then obtained a MS in Counseling from New York University. Martha Parker Anderson graduated with a BS in Social Science from Tennessee Agricultural and Industrial State College (now Tennessee State University) and then obtained a MS in Sociology from Temple University. Julia Leslie Parker-Cosby graduated with a BS in Chemistry from Fisk University and then obtained a MS in Medical Technology from Meharry Medical College. Parker's only brother, Julius Augustus Parker Jr., graduated with a BA in Chemistry from Fisk University (Parker-Cosby, 1985, p. 29; Warren, 1999, p. 209; Mickens, 2021, p. 7; Dillard, n.d.).

[5] The business directory address for Dr. Parker was a home address (US City Directories, 1932).

1985, p. 7). Tampa had a thriving Black community, complete with businesses, social clubs, civic societies, churches, and private and public Black schools (Howard and Howard, 1994). However, in 1930, during Parker's second year in high school, her father returned to Gainesville to care for his own ailing mother, Eliza, who had lost her eyesight. Mrs. Parker then took on work as a substitute teacher (Parker-Cosby, 1985, p. 3), possibly to support the family financially.

On April 27, 1933, tragedy struck the Parker household. Parker's younger sister, Gloria, died of appendicitis. According to her other sister Julia, Gloria might have survived had Dr. Parker still been living with the family in Tampa at that time and had he been available to treat his daughter (Parker-Cosby, 2004, pp. 5–6). Later that spring, Parker graduated from Middleton High School (Parker, 1953, p. 28), and the family reunited in Gainesville. Parker delayed enrolling in college to help the family and worked as a teacher in High Springs (Atlanta Daily World, 1945). Although her family valued the support, they held education in higher regard (Parker-Cosby, 2004, pp. 2–3), so the following year she did go off to college (Parker, 1953, p. 28).

13.4 Entering the World of Physics at Fisk University (1934–1938)

Family tradition led Parker to Fisk University, an historically Black university in Nashville, Tennessee, where a dedicated group of scientists ushered Parker into the world of physics. Parker's father had graduated from Fisk (Afro-American Advance, 1899, p. 2).[6] Founded in the late nineteenth century by members of the American Missionary Association to educate African Americans, Fisk was a cultural and intellectual powerhouse (Lovett, 2015, pp. 102–103). As the child of a proud Fiskite, Parker knew the importance of the institution and its rich history (Parker-Cosby, 1985, p. 8). By the time Parker enrolled in 1934, Fisk had earned accreditation, which rendered the university eligible for additional funding from the General Education Board, a private foundation, to build a modern science building and expand science materials in the library's holdings (Lovett, 2015, pp. 102–103). Fisk administrators had also recruited well-credentialed PhD faculty to teach lecture and laboratory courses in the newly built modern science building (Jones, 1930, pp. 29–37).[7] Because the faculty included the Black physicist Elmer Imes – the second African American to earn a PhD in physics (from the University of Michigan) in the US – Fisk was able to grant an undergraduate degree in physics (Mickens, 2018, p. 33).

---

[6] Parker's grandfather also worked as a physician, as noted in US City Directories (1887), but he learned medicine through the apprenticeship model (Vihlen, 1994, p. 20).

[7] St. Elmo Brady obtained a PhD in chemistry from the University of Illinois, Harry T. Folger a PhD in biology from Johns Hopkins University, and Clarence VanHorne a PhD in mathematics from the University of Chicago (Jones, 1930, p. 34).

By pursuing a physics degree at Fisk, Parker rejected rigid racial and gender norms of the early twentieth century, opting instead for a modern stance on womanhood. As historian Stephanie Shaw points out, during Parker's era, college-educated Black women were expected to become teachers, nurses, and home economists to help their communities (Shaw, 1996). The cultural movements of the 1920s created new possibilities, however. "New women" rejected gendered labor expectations and pursued careers in arenas that had traditionally been reserved for men (Beauboeuf-Lafontant, 2018).[8] Parker benefited from that new culture of Black womanhood. In a university publication, Fisk's Dean of Women, Mayme Upshaw Foster described a new type of Fisk woman that stood in stark contrast to "their predecessors of the 'age of correctness' because they are very real members of a new and modern social order, they necessarily express their ideals in a modern manner" (Foster, 1938, p. 8). In other words, Parker and her classmates were encouraged to embrace modern Black womanhood.[9]

At Fisk, Parker studied and trained directly with Imes, who complemented the standard undergraduate curriculum with graduate courses for students who aspired to attend graduate school elsewhere (Frazier, 1933, p. 337). He infused his own research interest in infrared (IR) spectroscopy into the undergraduate curriculum. His "Advanced Laboratory Technique" course, in particular, emphasized "experimental work under direction, affording students access to all the facilities of the laboratory" (Fisk University, 1935, pp. 88–89). By taking this course, Parker gained invaluable research experience in IR spectroscopy. This training eventually led her (see Section 13.5), like many of Imes' early students, to pursue graduate studies at the University of Michigan, supervised by Imes' former mentor, Harrison Randall (Mickens, 2018, p. 33).[10]

But first, on Wednesday, June 8, 1938, Parker graduated magna cum laude, the only student to receive a physics degree that day (Fisk University, 1938, p. 11). Fisk alumnus, sociologist, historian, and activist W. E. B. (William Edward Burghardt) Du Bois (1868–1963) delivered the commencement address titled "The Revelation of Saint Orgne the Damned," urging graduating seniors to fight for equality (Lewis, 2000, pp. 441–442). Parker walked off that commencement stage with honors, well prepared to enter the scientific workforce or to enroll in graduate school, but equality was elusive. Many industries had not opened their doors to Black women scientists at that time, and only a few Black graduates managed to secure

---

[8] The term "New women" is adapted from the conceptualization of the "New Negro Woman" and the "New Howard Woman" described in Beauboeuf-Lafontant (2018).

[9] Parker further expressed her modern womanhood through participation in a range of political, social, and intellectual organizations at Fisk. She notably joined the historically Black sorority Delta Sigma Theta, the homecoming planning committee, and Sigma Upsilon Pi, a Fisk honor society (Chicago Defender, 1935; Fleming and Burckel, 1950, p. 406).

[10] James Raymond Lawson and Lewis Clark also attended the University of Michigan (Mickens, 2018, p. 33).

employment, some by passing as white and others by working as the sole Black employee in their workplace, thus establishing a form of tokenism (Manning, 2007, pp. 54–57; Long, 2023).

13.5 From Teaching to Michigan and Back Again (1938–1942)

After graduation, Parker moved back with her parents (US Census Bureau, 1940) and took a teaching position at Rochelle Elementary, before joining the staff of her hometown's Lincoln High School the following year, in 1939 (Atlanta Daily World, 1945; Parker, 1953, p. 28). Lincoln High then served Black families in Gainesville and surrounding areas as one of only two accredited Black high schools in the whole state (Jones, 1976, p. 3). In the face of these challenges, as well as structural inequality and discrimination, many Black teachers during Jim Crow engaged in progressive pedagogical methods while affirming the humanity of Black children (Walker, 1996; Givens, 2021). Parker likely sought to humanize her students, and handwritten notes by her principal, A. Quinn Jones, attest to Parker's general abilities as an educator. She demanded excellence, engaged her students, and crafted well-organized lesson plans.[11] Even though leaving Lincoln High created a vacancy on the school staff, its principal always encouraged Lincoln teachers to attend graduate school (Jones and Jones, 1985, p. 21). After only a year at Lincoln High, in September 1940, Parker hence moved to the University of Michigan, in Ann Arbor, to pursue excellence herself, as a master's student.[12]

To earn an MA in physics, a Michigan student had to satisfy the general graduate school requirements and complete an additional 24 credit hours in physics (University of Michigan, 1942, p. 286).[13] Parker took three math courses, a series of physics courses – "Heat," "Mechanics of Solids," "Sound," "Light," and "Atomic & Molecular Structure" – and two laboratory courses, one in heat and the other in light.[14] Parker later reported having specialized in IR spectroscopy at University of Michigan (Dayton Journal, 1945c). Some of her courses, especially "Light," "Lab Work in Light," and "Atomic & Molecular Structures," were certainly relevant in this respect (University of Michigan, 1942, pp. 286–292). Although no record of a thesis was uncovered, she was likely mentored by Randall, Imes' advisor. Another Fisk alumnus, James Lawson graduated only a year earlier after completing a PhD in IR spectroscopy under his supervision (Dillard, n.d.). On February 10, 1941, Parker

---

[11] A. Quinn Jones, Miss Carolyn Parker- Pl. Demonstration Class, Friday, November 17, 1939, 10:30 am. Lincoln High School Correspondence 1928–1956. A. Quinn Jones Collection. University of Florida Archives.

[12] The University of Michigan registrar sent graduate verification and Parker's student transcript to the author. See also Dillard (n.d.).

[13] Based on her graduation application card, she stayed one year in a boarding house in Ann Arbor, Michigan.

[14] University of Michigan transcript released by the registrar's office to the author on February 16, 2024.

Figure 13.1 Carolyn Parker as a teacher at Huntington High School c. 1942 (Huntington High School, 1942, p. 17). Reproduced with permission.

applied for graduation with an MA in physics, and she graduated the following June (University of Michigan, 1941, p. 23).[15]

Twenty-three-year-old Parker returned to teaching the following fall, this time at Huntington High School in Newport News, Virginia (see Figure 13.1).[16] The community-centered curriculum emphasized democracy and student growth as the principal hoped to prepare its students to work in wartime industries (Kridel, 2015, pp. 48–51). Parker contributed to that mission by teaching eighth-grade students and by coaching girls' basketball (see Figure 13.2). She stayed, however, for only one academic year, before accepting a teaching position more commensurate with her graduate training at Bluefield State College (now Bluefield State University), in West Virginia.

At Bluefield State, Parker earned a modest living teaching math and physics (West Virginia State Board of Control, 1943, p. 415).[17] Like many HBCUs during World War II, Bluefield State centered their instruction on training students to

[15] Application date based on graduation application card received via email from the University of Michigan.
[16] The yearbook misspelled Carolyn's first name as "Caroly" (Huntington High School, 1942).
[17] According to a state report, Parker earned a salary of $1,296.77 for teaching one semester, which amounts to about $25,000 in 2024 dollars (West Virginia State Board of Control, 1943, p. 415).

Figure 13.2 Coaching the Girl's Basketball Team. Carolyn Parker is in the third row, first from left (Huntington High School, 1942, p. 80). Reproduced with permission.

contribute to the war effort. As illustrated in the 1943 short film, *Negro Colleges in the War Time*, the government enlisted Black scientists and HBCUs to train their students in the scientific and technical skills needed by the emerging war industries (US Office of War Information, 1943). To this end, Black scientists advocated for access to funding from the federal initiative for engineering, science, and management defense training (Branson, 1942a, 1942b, 1943).[18] Actual implementation varied by institution, but one key program that emerged from these funds was a training program in radio communication. This program educated and trained students to work as broadcasters, technicians, and engineers (Robinson, 1942). It's unclear whether Parker had anything to do with such a program at Bluefield State, but the zeitgeist might well have affected her subsequent career choices.

[18] For additional information about the status of physics at HBCUs during this period, see Woodson (1941).

13.6 Applying Physics to Electrical Engineering (1943–1947)

As the war progressed, some Black scientists, including Parker, indeed chose to leave HBCUs and join the military or governmental agencies in hopes of expanding opportunities for themselves (Plummer, 1996; Long, 2023). In March 1943, Parker became a research physicist in the Air Technical Service Command (ATSC) in Dayton, Ohio. She was one of the nearly 31,000 civilians then working at that military complex (US Air Force, 2015, p. 46), belonging specifically to the radio and radar subdivision of the engineering division located in the Aircraft Radio Laboratory (ARL) at Wright Field (Atlanta Daily World, 1945; Stewart and Parker, 1944; Carroll, 1945; Parker, 1953, p. 28; Nalty et al., 1994).[19]

At ARL, Parker focused on improving the technological capabilities of US Army Air Forces' planes and equipment alongside Chandler Stewart Jr., an electrical engineer and an Army Corps member (Stewart and Parker, 1944; Institute of Radio Engineers, 1945). Engineering divisions recruited physicists like Parker because their expertise could help advance electrical engineering innovations such as radio communication systems (Terman, 1976). As the electrical engineer on the project, Stewart trained all the aircraft inspectors and developed new methods for specialized electrical measurements of coaxial cables; as a physicist, Parker focused on theoretical modeling and computations of this system. Parker, who had two physics degrees and had taken courses on the physics of electrical circuits, was certainly up to the task.[20]

In partnership with Stewart, Parker co-authored a report on February 21, 1944, about their research on improving the testing of coaxial cables. The report was classified as "Restricted," the lowest of the three National Security Information classification levels then available (see Note 1), which meant that the research could be distributed internally amongst researchers and administrators but could not be communicated to the public (Maus, 1996).[21] According to the recently declassified report, Stewart and Parker focused on improving the radio antennae testing methods used by aircraft inspectors. Together, they determined the velocity of propagation of an electrical signal traveling through a cable, and the results helped reduce the time required to set up instruments for testing as well as measurement variability (Stewart and Parker, 1944).

---

[19] ARL was next door to – albeit distinct from – the Dayton Project. In particular, while the Dayton Project processed polonium that would be subsequently used in the atomic bomb, ARL did not (Thomas, 2017).

[20] According to the course description for electrical measurements, "the laboratory work is intended to give experiences and facility in handling electrical measuring instruments . . . covering modern methods of measuring current, resistance, electromotive force, and power, and the calibration of instruments employed, together with measurements to capacity in inductance and chromatism" (Fisk University, 1935, p. 89).

[21] Per the federal guidelines, the report therefore contained "engineering principles and design details, composition, method of processing or assembling, which are vital to the functioning or use of an article of material" (US War Department, 1939, p. 6).

Despite the restricted status of Parker and Stewart's findings, permission to publish was granted by the War Department, the results hence appearing in a series of publications: one in the *Journal of Applied Physics* (Stewart, 1945b) and two in the *Proceedings of the Institute of Radio Engineers* (Stewart, 1945a, 1945c). This work also eventually led to a patent (Stewart, 1949). Although Parker was acknowledged as having "assisted considerably in the theoretical work" (Stewart, 1945b, p. 614) and having participated in the "study of the problems of connector testing" (Stewart, 1945c, p. 609) she did not sign these publications. Parker's omission from the author list made her contributions to that research "less visible" in the acknowledgments section or altogether "invisible" on one publication (Stewart, 1945a) and the derived patent (Stewart, 1949), per the nomenclature of Cronin et al. (2004).

Parker's "less visible" and "invisible" scientific labor at ARL explains in part why Parker's scholarly contributions have been largely overlooked until now. Historian of science Margaret Rossiter has outlined how women's scientific labor has been undervalued throughout history and perception of scientific merit has been unequally distributed based on gender. This "Matilda effect" hides women's contributions as they are often excluded from scientific dictionaries, publications, and awards (Rossiter, 1993). In an analysis of the award structures, Willie Pearson found that Black scientists rarely received recognition for their scientific work or their merit (Pearson, 1989, pp. 71–86). Parker's scientific labor was hidden in the historical record since authorship serves as the primary form of social capital in science. It was further complicated by race and gendered expectations in the US. The only financial compensation Parker received was a five-dollar cash prize from the ATSC for her contributions to World War II (Dayton Journal, 1945a). This monetary gift certainly could neither make her contributions visible nor support her career in the same way as patents or scholarly publications could.

Although Parker's contributions to science were far from visible to the scientific community, her presence at ATSC was made amply visible in the Black community.[22] The Black press, which profiled Black scientists to illustrate Black people's intellectual capabilities and uplift them as symbols of racial progress in the US (Landrum, 2005), ran numerous stories about Parker's work at ATSC. "Woman physicist tests radio antennae at Wright Field, Ohio," proclaimed the *Atlanta Daily World* on July 13, 1945, subtitled "From teacher in Florida and Virginia only one in command." The article focused on Parker's position as one of two women scientists in the engineering division, touted her upbringing in Florida, and celebrated her success as a Black woman. It made a point to mention that Parker came from

---

[22] Articles about Carolyn Parker at the ATSC ran all across the US. See, for example, Atlanta Daily World (1945), Baltimore Afro-American (1945), Call (1945), Cleveland Call and Post (1945), Michigan Chronicle (1945), and New York Amsterdam News (1945).

a family of scientists and had earned a master's degree in physics (Atlanta Daily World, 1945). Another article explicitly positioned her as a race hero with the title "Women physicist to mark success" (New York Amsterdam News, 1945). The publicity validated Parker as a physicist against the backdrop of a wider world that did not adequately recognize her scholarly contributions.

Nearing the end of her time at ATSC in 1945, Parker joined a collective of Black scientists and inventors and served as the secretary and consultant of the grassroots movement to increase access to scientific industries (Dayton Journal, 1945c). The goal of the proposed Cosmopolitan Research Center (CRC) was to provide a space where all scientists and inventors could create "regardless of race, color, or creed" (Dayton Journal, 1945b). In addition to space, they sought financial support and training for young scientists to tinker with "aeronautical, electronic, photographic metallurgical, plastic, and other physical science fields" (Dayton Journal, 1945b). Many HBCU faculty and Black scientists served as CRC consultants, supporting the efforts of the budding movement. Although the center did not become a reality, Parker's participation further demonstrates her commitment to increasing access and training the next generation of scientists.

In January 1946, Parker enrolled at the Ohio State University Graduate Center at Wright Field, a joint operation between the university and the US Army Air Forces (Ohio State University, n.d.). Faculty members traveled "to Wright Field to advise and conduct courses" (Ohio State University, 1949). During her two semesters there, Parker took courses in electrical engineering and mathematics. However, in the summer of 1946, Parker returned to her original passion, physics, especially theoretical physics.[23]

13.7 Joining the Faculty at Fisk University (1947–1951)

In the fall of 1947, Parker returned to her alma mater, Fisk University, this time as an assistant professor (Nashville Tennessean, 1947). She joined a team of science faculty seeking to establish graduate studies in physics and build a scientific research infrastructure. Five years earlier, the first physics major at Fisk University, James Lawson, had replaced Imes on the faculty, following Imes' death. Lawson, who also studied IR spectroscopy with Randall (see Section 13.5), had convinced the University of Michigan to donate an IR spectrometer to Fisk University to help establish graduate studies in the field (deGregory, 2008). When Parker joined the Fisk faculty, Lawson was the only other physics professor (Fisk University, 1948). When, three years later, in 1950, Lawson left to join the physics department at another HBCU, Tennessee State University (deGregory, 2008), Nelson Fuson, a white

---

[23] Ohio State University transcript released to the author from the university registrar on March 5, 2024.

physicist and a Michigan graduate, took Lawson's place as chairperson and continued the department's effort to build a robust graduate program.

According to the 1951 issue of the *Fisk News*, the physics department had four full-time faculty members by then, and Parker was considered the "old-timer" (Fuson and Josien, 1951, p. 9). Apart from the departmental secretary, Parker was the only woman (Fisk University, 1950a). She also carried most of the undergraduate teaching load (Fuson and Josien, 1951, p. 9). Parker and another faculty taught the undergraduate courses, including survey courses for non-science majors and foundational physics classes for science majors (Fuson and Josien, 1951, pp. 8–9). The former aimed to teach social science and humanities students how physics and the scientific method impacted their everyday lives; the latter aimed to "give a sound technical foundation in physics to science majors, stressing through their training the importance of individual thinking and critical evaluation of the materials presented," according to the course catalog (Fisk University, 1950b, p. 11). The other faculty members taught graduate courses and oversaw the research of master's students. In addition to courses, the department hosted weekly seminars and the faculty could attend specialized courses on topics ranging from nuclear physics to glassblowing (Fuson and Josien, 1951, pp. 8–9).

During Parker's tenure at Fisk, nearly all the physics faculty conducted research on IR spectroscopy, utilizing the IR spectrometer that Lawson had helped secure. The department had also hired two students to fix up the machine shop and renovate the historic astronomical telescope. Fuson and a visiting professor, Marie-Louise Josien, further served as co-principal investigators on a research project supported by the Research Corporation for Science Advancement, US Public Health, and the General Education Board (Fuson and Josien, 1951, p. 9; Fisk University, 1953, pp. 13–14). The IR laboratory attracted many visitors, like Black chemist and director of Soya Products Division of Glidden Company Percy Julian as well as Ernest Bergmann of the Weizmann Institute in Israel. By 1953, the IR laboratory had given rise to 17 scientific papers, many of them co-authored with Fisk graduate students (Fisk University, 1953, p. 14). Parker was not listed as an author or acknowledged in any of these publications, so her level of involvement with research, if any, remains invisible. In all likelihood, Parker's heavy teaching load limited her ability to participate in the various departmental activities.

Although Fisk gave Black scientists space to conduct scientific research, Jim Crow laws continued to restrict the faculty's ability to engage in scholarly debates outside the university. For example, scientific societies often left membership decisions up to their local chapters, and this led to inconsistencies in the rules about membership eligibility (Long, 2023). While some chapters accepted membership applications from Black scientists, others did not. Additionally, some professional societies held meetings in segregated hotels, restaurants, or college campuses that banned African Americans (Manning, 2007, pp. 58–59). Parker had most likely joined the Institute of Radio Engineers when she worked for ASTC (Fleming and Burckel, 1950, p. 406), but belonging to the American Physical Society (APS) would be another matter. In other states, Black scientists belonged to APS, but archival evidence shows that Tennessee did not accept its first known Black member until 1948 during an APS membership vote in Oak Ridge.[24] As stated in the meeting minutes:

... the secretary presented the names of 118 people who had been nominated for membership. The names and qualifications were read, and all but seven were ruled eligible ... one of the nominees was recognized as being colored. This was discussed at some length by all present, and the unanimous opinion was expressed that any eligible person in the region would be elected if nominated.[25]

While the names of the Black scientists were not listed in the vote, the historical record reveals that race became a point of discussion, illuminating the way identity impacted membership decisions in the Jim Crow South. Parker may well have been the scientist in question. We cannot know for sure. However, we do know that Parker joined APS around that same time because she reported her membership to *Who's Who in Colored America* in 1950 (Fleming and Burckel, 1950, p. 406).

In response to segregation in science, HBCU faculty members created scientific meetings and conferences on Black college campuses (Long, 2023). During Parker's last academic year at Fisk, 1950–1951, the physics department even hosted a conference titled "The Infrared Spectroscopy Institute" (Fuson, 1954). Hosting the conference at Fisk guaranteed that Parker and other Black scientists could freely engage in scientific discussions, unlike, for example, hosting the meeting at Vanderbilt University (Manning, 2007, pp. 58–59; Long, 2023).[26] While Fisk scientists created their own spaces for scientific dialogue, they also continued to advocate for access to all scientific meetings so their faculty and peers could share their scientific research with the world. Fuson also stood up on behalf of his Black colleagues. He notably urged APS to host their 1952 meeting at Fisk; instead, they chose North Carolina State College (now North Carolina State University). Although Black scientists were then permitted to attend, they had to stay in a separate hotel and could not attend the

---

[24] It is important to note that in 1937, the Southeastern Section of the APS voted to grant membership to a Black physicist named Robert A. Horton (Talladega College). However, he is listed as being a member of the American Association of Physics Teachers, not APS. See, "Nomination for Membership," Box 1, Folder 15. American Physical Society Southeastern Section Records, 1934–2004. Niels Bohr Library & Archives.

[25] "Minutes of the Executive Committee at Oak Ridge, Tennessee, 8:45 P.M., April 8, 1948." Box 2, Folder 27. American Physical Society Southeastern Section Records, 1934–2004. Niels Bohr Library & Archives.

[26] In a 1951 open letter to *Science* magazine, Fisk University faculty objected to the decision of the Mathematical Association of America to host a conference at Vanderbilt. The choice of venue meant that African Americans would be denied banquet tickets since the dining hall did not serve Black people (Manning, 2007, pp. 58–59).

annual dinner.[27] Advocacy and protest would eventually push the Southeastern section of the APS to move their meetings to Fisk University in 1956.[28] Records are scarce, but during her tenure at Fisk from 1947 to 1951 Parker most likely attended the APS meetings at Fisk and at venues where Jim Crow prohibitions did not interfere. As a result of changes to Fisk's tenure and accreditation standards in the early 1950s,[29] however, Parker's situation became precarious. She now needed a doctorate to progress as a faculty member.

13.8 Comparing Nuclear Data with Theoretical Values (1951–1953)

In September 1951, Parker enrolled as a PhD student at MIT, where she was admitted based on her credentials from the University of Michigan. However, shortly after receiving a warning for her low grade in "Intro to Quantum Theory I" in January 1952, she switched to the MS program.[30] Parker then worked on a master's thesis under the supervision of David H. Frisch (Parker, 1953), a well-respected nuclear physicist who trained many graduate students, wrote textbooks, and even appeared in educational films (Weisskopf et al., 1992). Frisch's research agenda had grown out of his work on the Manhattan Project, for which he helped develop an electrostatic neutron detector for monitoring neutron-producing reactions (Frisch, 1946, 1950). From then on, Frisch devoted his career to experimental particle physics, specifically to understand nuclear binding energies and develop high-energy accelerators. When Parker entered Frisch's laboratory, a 300-MeV synchrotron had been in operation there for close to a year. His research was funded by the US Office of Naval Research and the US Atomic Energy Commission.[31]

Parker's research aimed to extend understanding of nuclear interactions, more specifically of positive and negative pions. The project was part of a research program that began in 1935 when Hideki Yukawa proposed a first theory that became known as the meson theory of nuclear forces (see Chapter 14). This theory predicted that a particle of mass intermediate between a proton and an electron – later named "meson" – would mediate the short-range interactions between nucleons (protons and neutrons, the particles that compose atomic nuclei). A few years later, a type of charged mesons, named "pi-mesons" and later "pions," were

---

[27] "Minutes of the Executive Committee Meeting University of Chattanooga," Box 3, Folder 20. American Physical Society Southeastern Section Records, 1934–2004. Niels Bohr Library & Archives; "Southeastern Section of the American Physical Society, Program of the Eighteenth Annual Meeting," Box 6, Folder 16. American Physical Society Southeastern Section Records, 1934–2004. Niels Bohr Library & Archives.

[28] Fisk University hosted the 26th annual meeting on March 29–31, 1956. See, "Program of the Twenty-Second Annual Meeting," Box 4, Folder 14. American Physical Society Southeastern Section Records, 1934–2004. Niels Bohr Library & Archives.

[29] "Permanent tenure and sabbatical leave at Fisk University," AMA Addendum, Box 17, Folder 9. Amistad Research Center, Tulane University.

[30] Parker's MIT transcript was released to the author on March 14, 2024.

[31] Based on the funding acknowledgement in publications, for example, in Cooper et al. (1954, p. 1209).

first detected in cosmic-ray experiments and then artificially produced at the 184-inch synchrocyclotron of the University of California, Berkeley. Cosmic-ray and accelerator experiments produced evidence that the pions interacted strongly with nuclei and could be identified with the nuclear quanta of meson theory. Parker's research aimed to extend these studies by exploring pion–nucleon interactions (Parker, 1953, pp. 1–5).

For her thesis Parker analyzed data collected in 1951 by "Dr. M.B. Scott, formerly of this laboratory," at the synchrocyclotron of the University of Chicago (Parker, 1953, p. 3), who had exposed photographic plates to beams of 122 MeV positive and negative pions. Parker compared the experimental data with theoretical range–energy values and sought information about the total cross-section of pions for nuclear interactions, which would be useful in "view of current studies being made of production of high energy mesons, nuclear interactions, etc" (Parker, 1953, p. 3). From her analysis, Parker concluded that the experimental observations of absorption ranges were in agreement with calculated values, and that the total cross-sections for positive and negative pions at 122 MeV were not noticeably different (Parker, 1953).

In parallel to this work, in 1952 she took a summer job as a physicist with the Geophysics Research Directorate at Air Force Cambridge Research Center (AFCRC) (Parker, 1953, p. 28). The US Air Force had established a close working relationship with MIT to develop a new air defense system for the continental US, and, as a MIT graduate student, Parker had access to the Cambridge facility.[32] The AFCRC site near MIT on Hanscom Field in Bedford, Massachusetts, conducted radar research in communication, weather, and computers (Liebowitz, 1985, p. 12). Parker's personnel files have not been declassified yet, but she may have worked in the weather radar laboratory in the geophysics division, which used IR spectroscopy to understand thermal radiation in the atmosphere. Certainly, as a physicist, she had expertise that would have been beneficial to the many research projects at Hanscom Field (Liebowitz, 1985, p. 12).

Thanks to her affiliation with MIT, in the spring of 1953, Parker earned membership in Sigma Xi, the scientific arm of the elite honor society, Phi Beta Kappa (Ellery, 1936; Sigma Xi, n.d.). At the time, not a single HBCU could charter a Sigma Xi chapter.[33] A combination of structural racism and chronic underfunding made it

---

[32] AFCRC contracted with various MIT professors, notably meteorologist Edward N. Lorenz (Lorenz, 1951).

[33] Sigma Xi is not listed as an HBCU honor society in a 1942 survey (Harris, 1942). HBCUs' infrastructure could not support a Sigma Xi chapter, which required, "apparatus and facilities available for research; there must be members of the faculty that have adequate training for research; there must have been a continuous output for research for a number of years; there must be appropriations for research" (Ellery, 1936). According to the *Beta Kappa Chi Bulletin* (Beta Kappa Chi Scientific Honor Society, 1947, p. 55) and the Howard University archives, faculty organized a Sigma Xi Club in 1947 but their petition for charter was not voted on and approved until December 27, 1956. "Sigma Xi Programs," Box 90, Moorland Spingarn Center, Howard University archives.

impossible for most HBCUs to meet the scientific society's written requirements regarding infrastructure and facilities, faculty training, continuous production of scientific research, and financial appropriations dedicated to research (Slaton, 2010, pp. 19–47; Lovett, 2015, pp. 98–137).[34] Being a member of Sigma Xi validated Parker's scientific abilities and, for the first time in her career, it afforded Parker a measure of prestige that all scientists needed to establish credibility.

Parker was awarded an MS in physics in September 1953, and the Hayden Library at MIT received Parker's master's thesis on December 18, 1953 (Parker, 1953). Her work, however, was not subsequently published in the scientific literature. From this point onward, the archival fragments of Parker's life also begin to disappear. According to family recollections, Parker continued working at the AFCRC, but illness prevented her from continuing her education (Mickens, 2021). Her academic transcript indicates that she registered as a "special graduate student" and "listener only" for one course during the fall of 1953, and then withdrew from MIT on February 1, 1954. She re-enrolled in the fall of 1954 but again withdrew for the last time on February 1, 1955. Clearly, she intended to pursue her graduate course in nuclear physics, but she was somehow unable to. Was she sick, or in some other way overwhelmed? It's hard to say. City directories then place Parker in Boston working as a laboratory technician (US City Directories, 1953, 1960). From 1953 to 1965, she lived in the Franklin Square House, a safe and affordable home for single, working women (City of Boston, 1953, 1965; Hinchliffe and Smith, 2012).[35] Other residents also worked as laboratory technicians, thus offering Parker a community. The (currently classified) AFRC archives, technical reports, and personnel files may one day offer more context on this period of her life, but what we now know is that she died on March 17, 1966, in Gainesville, Florida (Tampa Tribune, 1966).

13.9 Remembering Carolyn Parker

Parker's experimental research may not have been as heroic or glamorous as that of other scientists who worked on the atomic bomb, but Black women scientists' lives deserve to be remembered and included, repositioning their narratives in the broader history of science. Parker's significance to quantum physics goes beyond simply being the first Black woman in the US to obtain an advanced physics degree and to contribute through her MIT work to research in nuclear physics; her story illustrates the ways that race and gender biases influenced the production of

---

[34] For a history of how HBCUs have been underfunded, see Harris (2021).

[35] She is listed as living in Ohio in 1952. More research needs to be done to understand why this may be the case. In 1953–1954, she was listed as a student and in 1955–1965 as a laboratory technician in Boston ("List of Residents").

historical materials, leaving a lasting impact on how intellectual contributions to science are remembered or misremembered. In revisiting Parker's story, I have attempted to correct the record by including her intellectual contributions as a research physicist at the ARL. In doing so, we can see that even though her work was less visible in scientific records, she nonetheless made crucial contributions as evidenced by the acknowledgements section pertaining to her work, and her role on a "Restricted" government publication. In many ways, I was a bit saddened writing this chapter. Parker's career remained unstable; she often took a meandering path into research and moved around every few years. While life circumstances took Parker in many directions, the non-traditional pathways she pursued mirror that of other Black women scientists during the early twentieth century (Brown, 2002; Jordan, 2006). Without a doctorate, Parker had limited prospects of returning to Fisk University as an assistant professor due to the increased competition for employment and changing faculty credentialing requirements in the 1950s. Also, Parker does not appear to have received any acknowledgment from the scholarly community for her theoretical contributions, and the extant records suggest she was denied this privilege, in relevant part, by her own collaborator, Stewart.

Despite roadblocks, Parker's passion for scientific knowledge production remained intact throughout her life. This required a level of magisterial confidence rooted in a strong sense of cultural pride. Her confidence can be traced back to her baptism in HBCU culture. Historian of HBCUs, Jelani Favors' *Shelter in a Time of Storm*, argues that HBCUs baptize their students in what he calls the "second curriculum." Favors' defines the second curriculum as a "pedagogy of hope grounded in idealism, race consciousness, and cultural nationalism" (Favors, 2019, p. 5). In my view, Parker's dedication to science, despite the paucity of formal acknowledgment she received, illustrates how strongly her mentor Imes, and indeed Fisk as an institution, instilled a pedagogy of hope. This ethos helped Parker navigate from K–12 to graduate school and back again without giving up on her chosen career. Without a doubt, hope also came from her parents. Parker grew up learning from successful Black role models who pointed out the importance of HBCUs. Her parents raised children who all dared to dream and pursue professional careers in the US. The ability to do so despite external forces like racism and sexism represents the essence of the second curriculum. Hopefully, Parker's story, in turn, provides hope for others to witness and learn from as well.

References

Afro-American Advance. 1899. Fisk University. *The Afro-American Advance*, June 17.
Anderson, James D. 2010. *The Education of Blacks in the South, 1860–1935*. Chapel Hill, NC: University of North Carolina Press.

Atlanta Daily World. 1945. Women physicist tests radio antennae at Wright Field, Ohio. *Atlanta Daily World*, July 14, 1.

Baltimore Afro-American. 1945. Woman physicist first to test radio antennae at Wright Field. *Baltimore Afro-American*, July 14, 14.

Branson, Herman. 1942a. Physics training for the negro student. *American Journal of Physics*, 10, 201. https://doi.org/10.1119/1.1990377

Branson, Herman. 1942b. The role of the Negro College in the preparation of technical personnel for the war effort. *The Journal of Negro Education*, 11, 297–303. https://doi.org/doi:10.2307/2292666

Branson, Herman. 1943. The training of negroes for war industries in World War II. *The Journal of Negro Education*, 12, 376–385. https://doi.org/10.2307/2293057

Beauboeuf-Lafontant, Tamara. 2018. The New Howard Woman: Dean Lucy Diggs Slowe and the education of a modern black femininity. *Meridians*, 17, 25–48. https://doi.org/10.1215/15366936-6955065

Beta Kappa Chi Scientific Honor Society. 1947. Chapter news. *Beta Kappa Chi Bulletin*, 5(2), 55

Brown, Jeannette E. 2002. *African American Women Chemists*. New York: Oxford University Press.

Dillard, Angela. n.d. African American Student Project, Bentley Historical Library of the University of Michigan, https://africanamericanstudentproject.bentley.umich.edu

Call. 1945. Miss Carolyn Parker, woman physicist, tests radio antennae at ATSC, Wright Field. *The Call,* July, 29, A4.

Carroll, Franklin O. 1945. Research and development at Wright Field. *Journal of Applied Physics*, 16, 199–202. https://doi.org/10.1063/1.1707575

Chicago Defender. 1935. Plan Fisk homecoming. *Chicago Defender*, November 23, 10.

City of Boston. 1953. *List of Residents 20 Years of Ages and Over Boston Ward 8 Precinct 2 January 1 1953*. Boston, MA: City of Boston Printing Department, 21.

City of Boston. 1965. *List of Residents 20 Years of Ages and Over Boston Ward 8 Precinct 2 January 1 1965*. Boston, MA: City of Boston Printing Department, 5.

Cleveland Call and Post. 1945. Woman physicist pioneers in engineering division at Dayton Wright Field. *Cleveland Call and Post*, July 14, 1B.

Cooper, Daniel I., Frisch, David H., and Zimmerman, Robert L. 1954. Proton–proton scattering near the interference minimum. *Physical Review*, 94, 5, 1209. https://doi.org/10.1103/PhysRev.94.1209

Cronin, Blaise, Shaw, Debora, and La Barre, Kathryn 2004. Visible, less visible, and invisible work: patterns of collaboration in 20th century chemistry. *Journal of the American Society for Information Science and Technology*, 55(2), 160–168. https://doi.org/10.1002/asi.10353

Dayton Journal. 1945a. 48 field employees received merit awards cash for ideas. *Dayton Journal*, December 8, 7.

Dayton Journal. 1945b. Science workshop proposed for youth. *Dayton Journal*, October 25, 3

Dayton Journal. 1945c. Secretary appointed at research center. *Dayton Journal*, October 22, 3.

deGregory, Crystal A. 2008. Profile of African Americans in Tennessee: James Raymond Lawson, Nashville Conference on African American History and Culture. https://www.tnstate.edu/library/documents/James_Lawson.pdf

Ellery, Edward. 1936. Brief history of Sigma Xi, *Science*, 83, 610–611. https://doi.org/10.1126/science.83.2165.610

Favors, Jelani M. 2019. *Shelter in a Time of Storm: How Black Colleges Fostered Generations of Leadership and Activism*. Chapel Hill, NC: University of North Carolina Press.

Fisk University. 1935. *Fisk University Bulletin*. Nashville, TN: Fisk University.

Fisk University. 1938. *Anniversary Program Sixth-Fourth Annual Commencement, June 3–8, 1939*. Nashville, TN: Fisk University

Fisk University. 1948. *The Oval*. Nashville, TN: Fisk University.

Fisk University. 1950a. *The Oval*. Nashville, TN: Fisk University.

Fisk University. 1950b. *Fisk University Bulletin*. Nashville, TN: Fisk University.

Fisk University, 1953. Infrared spectroscopy. *Fisk News*, 26, 13–14.

Fleming, G. James and Burckel, Christian E. (eds.). 1950. *Who's Who in Colored America*. Yonkers-on-Hudson, NY: Christian E. Burckel and Associates.

Frazier, E. Franklin. 1933. Graduate education in negro colleges and universities. *Journal of Negro Education*, 329–341. https://doi.org/10.2307/2292203

Frisch, David H. 1946. The total cross sections of carbon and hydrogen for neutrons of energies from 35 to 490 keV. *Physical Review*, 70, 589. https://doi.org/10.1103/PhysRev.70.589

Frisch, David H. 1950. The uniform theory of nuclear binding energies. *Physical Review*, 84, 1169. https://doi.org/10.1103/PhysRev.84.1169

Foster, Mayme U. 1938. Fisk women today and yesterday. *Fisk News*, January–February, 8.

Fouché, Rayvon. 2003. *Black Inventors in the Age of Segregation: Granville T. Woods, Lewis H. Latimer, and Shelby J. Davidson*. Baltimore, MD: Johns Hopkins University Press.

Fuson, Nelson. 1954. Fisk Infrared Institute. *Physics Today*, 7(12), 28. https://doi.org/10.1063/1.3061476

Fuson, Nelson and Josien, Marie Louise. 1951. Physics. *Fisk News*, 24(3), 8–9.

Gamble, Vanessa. N. 1995. *Making a Place for Ourselves: The Black Hospital Movement, 1920–1945*. Oxford: Oxford University University.

Givens, Jarvis R. 2021. *Fugitive Pedagogy: Carter G. Woodson and the Art of Black Teaching*. Cambridge, MA: Harvard University Press.

Greer, Brenna W. 2019. *Represented: The Black Imagemakers Who Reimagined African American Citizenship*. Philadelphia, PN: University of Pennsylvania Press.

Griffin, Claranolle S. 1985. Transcript of an oral history interview of Claranolle Smith Griffin by Joel Buchanan on April 21, 1985, Samuel Proctor Oral History Program, University of Florida, Gainesville, FL.

Harris, Adam. 2021. *The State Must Provide: Why America's Colleges Have Always Been Unequal–and How to Set Them Right*. New York: Ecco, HarperCollins Publishers.

Harris, Nelson H. 1942. Honor societies in negro four-year colleges. *The Journal of Negro Education*, 11, 60–63. https://doi.org/10.2307/2292949

Hartman, Saidiya. 2008. Venus in two acts. *Small Axe*, 12, 1–14. https://doi.org/10.1215/-12-2-1

Hinchliffe, Beth and Smith, Bonnie Hurd. 2012. *The House That Love Built: The History of the Franklin Square House*. Belmont, MA: Franklin Square House Foundation.

Howard, Walter T. and Howard, Virginia M. 1994. Family, religion, and education: a profile of African–American life in Tampa, Florida, 1900–1930. *The Journal of Negro History*, 79, 1–17. https://doi.org/10.2307/2717663

Huntington High School. 1942. *The Huntingtonian Yearbook, 1942*. Newport, VA: Huntington High School.

Institute of Radio Engineers. 1945. Contributors. *Proceedings of the Institute of Radio Engineers*, 33, 70.

Jones, A. Quinn. 1976. Transcript of oral history interview of Allen Quinn Jones by Joyce Miller on October 27, Samuel Proctor Oral History Program, University of Florida, Gainesville, FL.

Jones, Frederica and Jones A. Quinn. 1985. Transcript of oral history interview of Frederica and A. Quinn Jones by Joyce Miller on July 30, 1985, Samuel Proctor Oral History Program, University of Florida, Gainesville, FL.

Jones, Thomas E. 1930. *Progress at Fisk University: A Summary of Recent Years*. Nashville, TN: Fisk University.

Jordan, Diann. 2006. *Sisters in Science: Conversations with Black Women Scientists about Race, Gender, and their Passion for Science*. West Lafayette, IN: Purdue University Press.

Kridel, Craig A. 2015. *Progressive Education in Black High Schools: The Secondary School Study, 1940–1946*. Columbia, SC: Museum of Education, University of South Carolina.

Landrum, Shane. 2005. *"In Los Alamos, I Feel Like I'm a Real Citizen": Black Atomic Scientists, Education, and Citizenship, 1945–1960*. MS Thesis. Waltham, MA: Brandeis University.

Lewis, David L. 2000. *WEB Du Bois, 1919–1963: The Fight for Equality and the American Century, Volume 2*. New York: MacMillan.

Liebowitz, Ruth P. 1985. *From the Cambridge Field Station to the Air Force Geophysics Laboratory, 1945–1985*. AFGL TR-85–0201. Hanscom AFB, MA: Air Force Geophysics Laboratory.

Long, Charnell C. 2023. *Into the Scientific Undercommons: Black Scientists and Fugitive Study, 1923–1961*. PhD Thesis. Madison, WI: University of Wisconsin – Madison.

Lorenz, Edward. 1951. Computations of the balance of angular momentum and the poleward transport of heat. Report No. 6, General Circulation Project No. AF 19–122-153. https://hdl.handle.net/2027/uc1.31822016274607

Lovett, Bobby L. 2015. *America's Historically Black Colleges & Universities: A Narrative History from the Nineteenth Century into the Twenty-First Century*. Macon, GA: Mercer University Press.

Manning, Kenneth R. 2007. African Americans in science. In Dodson, Howard and Palmer, Colin A. (eds.), *Ideology, Identity, and Assumptions*, New York: New York Public Library Schomburg Center for Research in Black Culture, pp. 49–95.

Maus, Cathy N. 1996. United States Office of Nuclear and National Security Information. History of classification and declassification. https://sgp.fas.org/othergov/doe/history .html

McPherson, Thomas B. 1976. Transcript of oral history interview of Thomas Benjamin McPherson by Joyce Miller on December 10, 1976, Samuel Proctor Oral History Program, University of Florida, Gainesville, FL.

Michigan Chronicle. 1945. Former school teacher in only negro woman physicist at ATSC. *Michigan Chronicle*, July 14, 13.

Mickens, Ronald E. 2021. Carolyn Beatrice Parker: a life in physics. *History and Philosophy of Physics Newsletter*, 15, 1–11.

Mickens, Ronald E. 2018. The life and work of Elmer Samuel Imes. *Physics Today*, 71, 10, 28–35. https://doi.org/10.1063/PT.3.4042

Nalty, Bernard C., Shiner, John F., and Watson, George M. 1994. *With Courage: The US Army Air Forces in World War II*. Maxwell AFB, AL: Air Force History & Museums Program.

Nashville Tennessean. 1947. Fisk appoints 7 to faculty. *Nashville Tennessean*, September, 28, 17.

Nelsen, R. Arvid. 2016. Race and computing: the problem of sources, the potential of prosopography, and the lesson of *Ebony* magazine. *IEEE Annals of the History of Computing*, 39, 29–51. https://doi.org/10.1109/MAHC.2016.11

New York Amsterdam News. 1945. Woman physicist to mark success. *New York Amsterdam News*, July 14, A8.

Ohio State University. 1949. 1500 Attend air school at Dayton. *The Lantern*, February 24, 2.

Ohio State University. n.d. Background and history of the division of continuing education. https://kb.osu.edu/server/api/core/bitstreams/b2d605fa-1c88-5b3c-bc3b-8c0230b17181/content

Parker, Carolyn B. 1953. *Range Distribution of 122 MeV π+ and π– Mesons in Brass*. MS Thesis. Boston, MA: Massachusetts Institute of Technology.

Parker-Cosby, J. Leslie. 1985. Transcript of an oral history interview of Julia Leslie Parker-Cosby by Joel Buchanan on July 23, 1985, Samuel Proctor Oral History Program, University of Florida, Gainesville, FL.

Parker-Cosby, J. Leslie. 2004. Transcript of oral history interview of Julia Leslie Cosby by Ruth Marston on March 3, 2004. Samuel Proctor Oral History Program, University of Florida, Gainesville, FL.

Pearson, Willie. 1989. *Black Scientists, White Society, and Colorless Science: A Study of Universalism in American Science*. Millwood, NY: Associated Faculty Press.

Plummer, Brenda G. 1996. *Rising Wind: Black Americans and US Foreign Affairs, 1935–1960*. Chapel Hill, NC: University of North Carolina Press.

Prescod-Weinstein, Chanda. 2022. Racial and economic barriers kept Carolyn Beatrice Parker from realizing her full potential. *Popular Science*, May 18. https://www.popsci.com/science/carolyn-beatrice-parker-profile/

Robinson, William H. 1942. Radio in negro colleges and universities. *American Journal of Physics*, 10, 319–321. https://doi.org/10.1119/1.1990408

Rossiter, Margaret W. 1993. The ~~Matthew~~ Matilda effect in science. *Social Studies of Science*, 23(2), 325–341.

Sigma Xi. n.d. Come up higher: the history of Sigma XI. Sigma Xi, the Scientific Research Honor Society. https://www.sigmaxi.org/about/history

Shaw, Stephanie J. 1996. *What a Woman Ought to Be and to Do: Black Professional Women Workers During the Jim Crow Era*. Chicago, IL: University of Chicago Press.

Shopes, Linda. 2011. Oral history. In Denzin, Norman K. and Lincoln, Yvonna S. (eds.), *The SAGE Handbook of Qualitative Research*, 4th ed. Thousand Oaks, CA: SAGE Publications, pp. 451–465.

Slaton, Amy E. 2010. *Race, Rigor, and Selectivity in US Engineering: The History of an Occupational Color Line*. Cambridge, MA: Harvard University Press.

Stewart, Chandler and Parker, Carolyn B. 1944. A method of measuring attenuation of short lengths coaxial cable. PB 6250. War Department Aircraft Radio Laboratory ARL Memorandum Report 159.

Stewart, Chandler. 1945a. A method of measuring attenuation of short lengths coaxial cable. *Proceedings of the Institute of Radio Engineers*, 33, 46–48, https://doi.org/10.1109/JRPROC.1945.230856

Stewart, Chandler. 1945b. A method of measuring the radiofrequency resistance of wires. *Journal of Applied Physics*, 16, 608–614. https://doi.org/10.1063/1.1707510

Stewart, Chandler. 1945c. Electrical testing of coaxial radio-frequency cable connectors. *Proceedings of the Institute of Radio Engineers*, 33, 609–619. https://doi.org/10.1109/JRPROC.1945.231197

Stewart, Chandler. 1949. US Patent No. 2,459,197. Washington, DC: US Patent and Trademark Office.

Tampa Tribune. 1966. Miss Carolyn Parker. *The Tampa Tribune*, March 19, 15.

Terman, Frederick E. 1976. A brief history of electrical engineering education. *Proceedings of the IEEE*, 64, 9, 1399–1407. https://doi.org/10.1109/PROC.1976.10333

Thomas, Linda C. 2017. *Polonium in the Playhouse: the Manhattan Project's Secret Chemistry Work in Dayton, Ohio*. Athens, OH: The Ohio University Press.

Walker, Vanessa S. 1996. *Their Highest Potential: An African American School Community in the Segregated South*. Chapel Hill, NC: University of North Carolina Press.

Ward, Thomas J. 2003. *Black Physicians in the Jim Crow South*. Fayetteville, NC: University of Arkansas Press.

Warren, Wini. 1999. *Black Women Scientists in the United States*. Bloomington, IN: Indiana University Press.

Weisskopf, Victor, Low, Francis, Osborne, Louis, and Frisch, David H. 1992. *Physics Today*, 45(7), 80–81. https://doi.org/10.1063/1.2809748

US Air Force. 2015. *Wright–Patterson Air Force Base: The First Century*. Dayton, OH: Wright-Patterson Air Force Base. https://www.wpafb.af.mil/Portals/60/documents/Index/History-of-WPAFB.pdf

US Census Bureau. 1920. Gainesville, Alachua, Florida T625_214 Enumeration District 10, 31A.

US Census Bureau. 1940. Gainesville, Alachua, Florida m-t0627-00573 Enumeration District 10, 6B.

US City Directories. 1887. *Webb's Jacksonville and Consolidated Directory of Representative Cities of East and South Florida 1887*. Jacksonville, TN: Booksellers and Stationers.

US City Directories. 1932. *Polk's Tampa City Directory 1932*. Jacksonville, TN: R.L. Polk and Co. Publishers.

US City Directories. 1953. *Polk's Boston City Directory in Two Volumes*. Boston, MA: R.L. Polk and Co. Publishers.

US City Directories. 1960. *Polk's Boston City Directory in Two Volumes*. Boston, MA: R.L. Polk and Co. Publishers.

US Office of War Information. 1943. Negro colleges in war time. United States Office of War Information, Domestic Branch Bureau of Motion Pictures, video. https://www.loc.gov/item/2020600704

US War Department. 1939. Army Regulations, Safeguarding Military Information, No. 380–5, June 10.

University of Michigan. 1941. *Ninety-Seventh Commencement, June 21, 1941*. Ann Arbor, MI: University of Michigan.

University of Michigan. 1942. *The Horace H. Rackham School of Graduate Studies Announcements 1941–1941*. Ann Arbor, MI: University of Michigan.

Vihlen, Sally P. 1994. *The Black Physician in Florida from 1900–1965: An Examination of the Desegregation Process*. MS Thesis. Tampa, FL: University of South Florida.

West Virginia State Board of Control. 1943. *Biennial Report of the State Board of Control of West Virginia 1939–1943*. Charleston, WV: Jarrett Printing Company.

Woodson, Harold W. 1941. The present status of physics in negro colleges. *American Journal of Physics*, 9, 180–183. https://doi.org/10.1119/1.1991664

14

The Chew–Low–Salzman Method and Freda Friedman Salzman: A Physicist Between Nuclear and Social Interactions

JENS SALOMON

14.1 Introduction

Freda Friedman Salzman was an American physicist, who in the 1950s was involved in deepening our understanding of nuclear physics. This chapter reviews one of her key research accomplishments, the Chew–Low–Salzman method (Salzman and Salzman, 1957), which she published in 1957 with her husband, George Salzman. That method was the first solution striving for numerical accuracy of the then salient Chew–Low model (Chew and Low, 1956a, b). The model, an early phenomenological attempt by Geoffrey Chew and Francis Low to describe nuclear interactions, was notable for the reasonableness of its theoretical basis (Cushing, 2005). This chapter elaborates on the Chew–Low model and its historical context in Section 14.3 and then focuses on the Chew–Low–Salzman method in Section 14.4. Because Friedman Salzman was not only a physicist, but also felt the "weight of being a woman," as she put it (Salzman, 1979a), Section 14.2 first reviews her life and academic struggles.

14.2 Biographical Notes

Freda Friedman was born in Brooklyn, New York, on May 12, 1927 to Jacob Friedman and Annie Mandelssohn Friedman, both originally from Russia.[1] Her childhood was ambivalent. On the one hand, science and mathematics were held in

I thank Friedman Salzman's daughter, Amy G. Parker, for sharing various materials and recollections. This article is a spin-off of research by Alexander Blum of the Max Planck Institute for the History of Science (MPIWG) and me. I am very grateful for many revealing discussions with Blum, especially on the Chew and the Chew–Low models. Furthermore, I want to express my warmest regards to the MPIWG for supporting this work and in particular for welcoming me during a research stay. I also want to thank the editors and two anonymous referees for many helpful proposals on how to improve this chapter. Let me also thank Andrea Reichenberger and Julia Franke-Reddig for giving me the opportunity to present an earlier version of this work at their "Women in Modern Physics" workshop in June 2023.
[1] Standard Certificate of Death, part of 83-M100–83-M173: T-155, 2, Papers of Freda Friedman Salzman, 1927–1981, Schlesinger Library, Harvard Radcliffe Institute.

Figure 14.1 Portrait of Freda Friedman Salzman during her undergraduate years. Schlesinger Library, Harvard Radcliffe Institute. Reproduced with permission.

high regard in her home environment,[2] and her family soon praised her abilities in this respect (Salzman, 1979a). On the other hand, the Great Depression and an authoritarian climate at home took their toll. Two of her siblings died young, and she herself suffered from rickets.[3] Their father could also be abusive.[4] Both aspects of her youth, her growing intellectual proficiency and the oppressiveness of her childhood, are somehow epitomized under her photograph in her high-school yearbook: "We found her in the woods, digging for square roots."[5]

After graduating from high school in 1944, Friedman (Figure 14.1) entered the evening session of Brooklyn College for one semester (Salzman, 1953a); in 1946, she switched to the day session while "tutoring, baby-sitting, and working summers, eager to continue her formal studies," her husband later recalled (Salzman, 1953a, 1971). During these undergraduate years, she met fellow

---

[2] Friedman Salzman's mother described her husband as a philosopher and teacher, who taught French, German, Russian, and maths in Russia ("in the old country") in an undated letter to Friedman Salzman's daughter Amy. See 83-M100–83-M173: T-155, 5, Papers of Freda Friedman Salzman, 1927–1981, Schlesinger Library, Harvard Radcliffe Institute.

[3] A. G. Parker, email to author, May 29, 2024.

[4] In a letter to Friedman Salzman, her sister Louise later made allegations of childhood sexual abuse against him. Letter from Louise Friedman Benett to Friedman Salzman, September 18, 1971. A. G. Parker personal collection.

[5] At least A. G. Parker recalled the caption to have read thus, email to author, June 6, 2024.

physics student George Salzman (Salzman, 1971), whom she married in 1948. A year later, in 1949, Friedman Salzman received a BSc with honors in physics (Salzman, 1953a).[6]

The couple pursued a graduate education alongside one another. At that time, the physics faculty of Brooklyn College included Melba Phillips, a distinguished nuclear physicist, who in 1935, in collaboration with J. Robert Oppenheimer, had provided the theoretical description of the type of nuclear reactions now known as the Oppenheimer–Phillips process (Oppenheimer and Phillips, 1935).[7] Phillips encouraged the couple to continue their education at the Graduate College of the University of Illinois (Salzman, 1971). There, Friedman Salzman worked with Geoffrey Chew (Salzman, 1953a) in nuclear physics, receiving a MSc in 1951 and a PhD in 1953 (Salzman, 1953a); her husband obtained his PhD the same year (Salzman, 1953b; Salzman and Taub, 1954), having worked with Abraham Taub (Mashhoon, 2001), the chief mathematician behind the first Illinois Automatic Computer (ILLIAC I).

The couple (Figure 14.2) strove to stay together through over a decade of postdoctoral positions, as can be retraced using a variety of sources.[8] Friedman Salzman was first a research associate at the University of Wisconsin at Madison in the summer of 1955 and then at the University of Rochester (December 1955–July 1956). From Salzman and Salzman (1957), it appears that she stayed at least one more year in Rochester albeit without holding a formal position there, while her husband did. From 1958 to 1959, and in the summer of 1962, she was a member of the CERN Theory Group.[9] From 1960 to 1964, she was a research associate at the University of Colorado at Boulder with a leave of absence at the University of Rome during the academic year 1961–1962 (Salzman and Salzman, 1962) as a National Science Foundation (NSF) Postdoctoral Fellow, and then as a NSF grantee.[10] In 1964–1965, she was a guest physicist at Northeastern University in Boston. During their postdoctoral years, their two daughters were born, Amy (born 1954) and Erica (1958–2003).[11] The daughters would move with

---

[6] Although Friedman Salzman mostly signed her publications as "Freda Salzman," omitting her birth name – the one exception which she signed "Freda F. Salzman" being Salzman and Snyder (1954) – her full surname is used in the following to more clearly distinguish her contributions from her husband's.

[7] Phillips was not the only prominent woman who then taught physics at Brooklyn College. Esther M. Conwell was also on the faculty, and possibly others.

[8] Unless indicated otherwise, the information about Friedman Salzman's career is based on two overviews of her professional positions in the Papers of Freda Friedman Salzman, 1927–1981, Schlesinger Library, Harvard Radcliffe Institute. They are part of the "Notification of Personnel Action," 1–49, 83-M100–83-M173: T-155, 55, and the "Curriculum Vitae of Freda Salzman" in 83-M100–83-M173: T-155, 2.

[9] In the acknowledgments of Salzman and Salzman (1960a), she and her husband "thank Professor Ferretti and Professor Fierz for the warm hospitality." Bruno Ferretti was the leader of the CERN (European Organization for Nuclear Research) Theory Group in those days, and Markus Fierz became his successor in 1959.

[10] In Salzman (1963), she thanks Professor Marcello Cini for his hospitality while in Rome.

[11] Biography, Papers of Freda Friedman Salzman, 1927–1981, Schlesinger Library, Harvard Radcliffe Institute. https://hollisarchives.lib.harvard.edu/repositories/8/resources/9171

Figure 14.2 Friedman Salzman with her husband on the University of Illinois campus in April 1954. A. G. Parker personal collection. Reproduced with permission.

the couple, and daycare issues were addressed by hiring elderly women as caregivers. To have time with her children, Friedman Salzman essentially gave up on a social life, as she later recalled. Still, Friedman Salzman did not consider these arrangements to be really satisfying for the children (Salzman, 1979a).[12]

Friedman Salzman and Salzman were then both offered professorships at two universities. They declined "an offer of dual tenure from a large urban campus of

---

[12] Nor did her daughter Amy, who later stated that her parents had "checked out" of childcare. A. G. Parker, video call with author, June 5, 2024.

a major midwestern state university," Salzman would later mention (Salzman, 1971), but they accepted offers at the newly established Boston campus of the University of Massachusetts in 1965. Although Friedman Salzman became an untenured associate professor (three-quarter-time) whereas her husband got a tenured professorship (Salzman, 1971), both were put in charge of establishing teaching and research programs (Lyons, 1982).[13] As Salzman later reported, he and Friedman Salzman got different kinds of positions because the "University policy [was] opposed to close relatives simultaneously holding tenure[d] positions within a department."[14] Salzman's understanding is confirmed by a letter from the university's president, John W. Lederle, in which he disapproves of tenured couples within departments, and by a letter from Arthur C. Gentile, who offered the positions to the couple.[15] Salzman (1971) stated that it was "invariably the wife" who was disadvantaged, thus explaining Friedman Salzman's untenured status and part-time appointment. Both aspects were linked. According to the 1940 "Statement of Principles on Academic Freedom and Tenure," untenured full-time positions were not permitted for more than seven years, but the restriction did not apply to part-time positions.[16] However, as Salzman then understood the nature of his wife's position, it was to be a "regular faculty post," which would be extended automatically on a yearly basis.[17] Friedman Salzman and Salzman were further informed that other couples working at the university had comparable arrangements and "that the arrangement[s] had proved entirely satisfactory for all concerned," he later recalled (Salzman, 1971).

Despite these reassurances, Friedman Salzman's position was terminated in 1969,[18] following a resolution of the university's chancellor – first announced in 1967 – to enforce the anti-nepotism policy of the university's trustees more strictly.[19] (Friedman Salzman maintained that she actually lost her position "due to internal politics" and that the anti-nepotism policy was only used as an excuse (Salzman, 1979a). Salzman (1971) added that the chancellor's actual motive was retribution as the couple had openly criticized "administrative failures and malpractices, a whole series of which led to his early resignation and

---

[13] Besides them, the physics faculty initially comprised of only one assistant professor and a visiting assistant professor (Salzman, 1971).

[14] Statement: Concerning Certain Initial Conditions of Employment at the University of Massachusetts at Boston (August 10, 1967), 1–42, 83-M100–83-M173: T-155, 55, Papers of Freda Friedman Salzman, 1927–1981, Schlesinger Library, Harvard Radcliffe Institute.

[15] Letters from Lederle to the Trustee Committee on Faculty and Program of Study, December 4, 1963, and from Gentile to Paul Gagnon, August 11, 1967, 83-M100–83-M173: T-155, 55, Papers of Freda Friedman Salzman, 1927–1981, Schlesinger Library, Harvard Radcliffe Institute.

[16] The corresponding excerpt reads: "Beginning with appointment to the rank of full-time instructor or a higher rank, the probationary period should not exceed seven years."

[17] See Note 14.   [18] Curriculum vitae of Freda Salzman, see Note 8.

[19] Memorandum of John W. Ryan, Chancellor, to Paul Gagnon (July 11, 1967), 1–35, 83-M100–83-M173: T-155, 55, Papers of Freda Friedman Salzman, 1927–1981, Schlesinger Library, Harvard Radcliffe Institute.

departure from the campus in the Fall of 1968.") In order to prevent Friedman Salzman from losing her position, the physics department "solicited letters of reference from a number of prominent physicists, among them several Nobel laureates" (Salzman, 1971).[20] The chancellor nevertheless acted against Friedman Salzman on his final day in office, and the decision was sustained by his successor (Salzman, 1971). Friedman Salzman resisted and, in December 1970, she "got unanimous decision from a tenure grievance committee recommending my reappointment," she later noted; still, the administration did not budge (Salzman, 1971, 1979a). She consulted a lawyer working at the Massachusetts Commission Against Discrimination (MCAD) who, according to Salzman (1979a), exclaimed: "My god, you have an open and shut case of discrimination!" The lawyer reported that the anti-nepotism rule was under attack on a more general basis at the time, and an investigation of the University of Massachusetts Amherst clearly documented that the anti-nepotism rule disadvantaged the wife (Salzman, 1979a). Friedman Salzman therefore filed a complaint with both the MCAD and the US Department of Health, Education and Welfare. But because the authorities stalled (Salzman, 1979a), she went public. She surmised that there was nobody in the Boston area who was not aware of her case at the time. As the situation became increasingly embarrassing, authorities eventually caved in. Friedman Salzman got her job back in 1972 (Salzman, 1979a) and was finally granted tenure in 1975.[21]

Despite all these hurdles that kept her from devoting time to physics and that had an impact on her health,[22] Friedman Salzman published two papers on the general theory of relativity during that period (Salzman and Salzman, 1969; Salzman, 1974). In parallel, she became involved in feminist movements. She was part of the Boston chapter of Science for the People and its Sociobiology Study Group as well as its Women's Issues Project Group.[23] In 1977 and in 1979, she notably published two articles in the *Science for the People* magazine (Salzman, 1977, 1979b), arguing against sociobiology, according to which evolutionary genetics determined social inequalities and a price had to be paid to overcome them. In the first article, titled "Are sex roles biologically determined?"

---

[20] As two lists of (possible) references and an (unsigned) letter from Friedman Salzman to "Robert J. [sic] Sachs" (May 13, 1968) suggest, among these physicists were Gerald Feinberg, Maurice Goldhaber, Robert Hofstadter, Tsung-Dao Lee, Robert G. Sachs, Abdus Salam, Victor Weisskopf, Arthur Wightman, Chen-Ning Yang – and many more. See 83-M100–83-M173: T-155, 55, Papers of Freda Friedman Salzman, 1927–1981, Schlesinger Library, Harvard Radcliffe Institute.

[21] Biography, see Note 11.

[22] The notification that her position would be terminated caused her "agitation and distress," and she asked "for a sick leave for an indefinite period." She added that already the first two years at the new university had been "difficult and trying" and had come at a "considerable cost to my mental and physical health." Friedman Salzman, letter to Marvin Antonoff, Acting Chairman of the Department of Physics, September 9, 1968, 83-M100–83-M173: T-155, 55, Papers of Freda Friedman Salzman, 1927–1981, Schlesinger Library, Harvard Radcliffe Institute.

[23] Biography, see Note 11.

Friedman Salzman remarked that this old idea had reemerged in the mid-1960s, and that there often was a hidden agenda behind research querying to what extent societal traits naturally emerge out of biological traits. She wrote (Salzman, 1977, p. 28):

The re-emergence of interest in sex-difference research is part of a general resurgence of biological determinist theories which try to demonstrate that our highly stratified society, based mainly on class, race and sex, is due to genetic differences between these groups and not to societal factors. Historically, biological determinism as a theory of the status quo has been widely publicized and encouraged during periods of considerable social unrest and questioning of societal institutions.

Supporting her points with succinct but well-researched and accessible lines of reasoning, she argued that theories of biological determinism benefited a wealthy elite, which could influence the direction of research: "The elite finds it far easier to promote theories which undermine demands for equality than to spend money to eliminate inequality" (Salzman, 1977, p. 28). This, however, contrasted with "there [being] at this time no credible scientific evidence that genetic basis exists for *any* socially significant human behavior trait" (Salzman, 1977, p. 31, emphasis in original).

In 1979, Friedman Salzman was diagnosed with breast cancer (Chen, 2020), to which she succumbed at the age of 53 on April 1, 1981.[24] As science writer Maggie Chen (2019) later put it: "In all respects, hers was an untimely death, with so many years lost through her academic struggle."

As Chen's biographical efforts have revealed (Chen, 2019, 2020), Friedman Salzman's legacy lives on in feminist circles nevertheless. By shedding light on Friedman Salzman's key professional accomplishments, the following sections take a first step toward accounting for an even more central aspect of her identity, her work as a nuclear physicist.

14.3 Coming to Terms with Nuclear Interactions

Nuclear interactions, for the purposes of this chapter, refer to our understanding of how the constituents of atomic nuclei, protons and neutrons, interact. In this context, a very fruitful conceptualization stems from the work of the Japanese theorist Hideki Yukawa, who in 1935 published a theory on treating nuclear interactions analogously to quantized electrodynamics and the Fermi theory of weak interactions (Yukawa, 1935), and who received the Nobel prize in 1949 for this breakthrough. His proposal indeed led to quantum field theories for nuclear interactions which contained an exchange particle mediating the interaction of nucleons. After a period of confused search for experimental validation, the nuclear

---

[24] Standard Certificate of Death, see Note 1.

exchange particles, termed "mesons,"[25] came to be identified with a new class of particles first discovered in cosmic radiation in 1947 (Lattes et al., 1947a, b), which were initially named "pi-mesons" and eventually became known simply as "pions." Three types of pions were discovered: two electrically charged ones (in 1947) and one neutral (in 1950). Extensive historical reviews of this key period have been produced (Mukherji, 1974; Darrigol, 1988; Pais, 1995; Kragh, 2023). For the sake of this chapter, it suffices to know that today, the pion–nucleon interaction is understood to be one among many processes of interest in the context of nuclear (and strong) interactions, but in the 1950s, understanding the pion–nucleon interaction was tantamount to understanding nuclear interactions as a whole.

Since protons and neutrons behave very similarly in nuclear interactions, they can be formally described as an "isospin doublet," that is, a single particle – called the "nucleon" – existing in two possible states defined by two values of a quantum variable named "isospin." Likewise, the three pions can be regarded as a single particle, the "pion," which corresponds to an isospin triplet. Other important properties of the nucleon and pion are their mass and spin (see Table 14.1 for a summary).

Thanks to its much heavier mass, the nucleon can be assumed, to first approximation, to be static and infinitely heavy in the context of low-energy interactions. The pion is by far the lightest particle when it comes to the understanding of nuclear interactions. The next lightest particles (the σ-meson and the kaons) have a mass of about 500 MeV (Workman et al., 2022) and hence are so heavy that they can be neglected in low-energy interactions involving nucleons.

Unsurprisingly, the discovery of the exchange particle of nuclear interactions opened up new possibilities. Experimentally, direct access to the pion led to innovative experiments that directly studied pion–nucleon scattering. In

Table 14.1 *Some properties of pion and nucleon (Workman et al., 2022). The first and second mass given correspond to neutral and charged particles, respectively.*

Property	Pion	Nucleon
Mass (MeV)	135/140	940/938
Spin	0	$\frac{1}{2}$
Isospin	Triplet	Doublet

---

[25] The meaning of the term "meson" has shifted over time. Nowadays, it refers to particles consisting of a quark and an antiquark. In this chapter, it refers to the nuclear exchange particle.

particular, early pion–nucleon scattering experiments discovered a prominent "resonance," that is, a steep increase and subsequent decrease in the number of scattering events with rising scattering energy. (Technically speaking, these experiments discovered a "p-wave resonance": a resonance with unit orbital angular momentum.) Nowadays, this peak is called the Delta resonance,[26] but historically, it was known as the (3,3) resonance, given that its spin and isospin are both 3/2.

On the theoretical front, the discovery of the pion, which enabled experimentalists to study phenomena involving only a single nucleon (such as pion–nucleon scattering), was particularly helpful. Relative to the nucleon–nucleon interactions that had previously been considered, single-nucleon phenomena were much easier to analyze. Friedman Salzman nicely describes the situation on the first page of her PhD thesis (Salzman, 1953a): "Although the original interest in the meson[27]– nucleon interaction was due to its role in the nuclear force problem, the single nucleon phenomena involving the meson field are more easily formulated and understood theoretically."

Even for this simpler system, it nevertheless took several years to reconcile theory and experiment. For quite some time, it was not even clear if the pion–nucleon coupling was weak or strong (Mukherji, 1974; Ruiz de Olano et al., 2022). In particular, theorists who opted for an interpretation that favored weak coupling struggled to reproduce the (3,3) resonance, one of the most salient phenomenological features of pion–nucleon scattering.

Enter Geoffrey Chew – Friedman Salzman's thesis advisor at the University of Illinois. Arguably, throughout the 1950s, Chew was the physicist most prolific at formulating models for nuclear interactions. Many models that are phenomenologically viable and exhibit novel theoretical ideas bear his name. Starting in 1953, Chew established the first phenomenologically adequate model of the pion–nucleon interaction, which became known as the Chew model. This model rested on a specific quantum field theory based on Yukawa's idea that nuclear forces are mediated by exchange particles. The first iteration of this model featured a draconian approach to high-energy contributions: cutting them off completely (Chew, 1953, 1954a). This truncation allowed Chew to evaluate important higher-order contributions.[28] He did so by establishing an integral equation,[29] a mathematical device which is also central to all the other models encountered in this chapter. By suitably adjusting the coupling constant (it turned out to be rather

---

[26] Short-lived Delta baryons are created at the resonance.

[27] According to the identification of the meson with the pion established at the beginning of this section, we can read "pion" for "meson" here and in the following.

[28] Technically speaking, standard renormalization methods do not render the quantum field theory used by Chew convergent. Rather, Chew (1954a, p. 1748) states that "convergence is achieved by inserting a cut-off factor." This peculiarity is related to the fact that the quantum field theory is non-local. Incidentally, the cutoff is applied to the Fourier transform of the source function of the pion–nucleon interaction.

[29] In an integral equation, an unknown function appears under an integral sign.

small) and the position of the cutoff, he could reproduce the (3,3) resonance while also predicting that no other resonances were present at low energies. All of this rendered the model phenomenologically viable.

In his first article on the Chew model, Chew focused on pion–nucleon scattering, but he also mentioned other processes (Chew, 1953). In particular, he referred to an ongoing analysis of "photopion production," the process through which a nucleon absorbs a photon and emits a pion.[30]

This was precisely the topic of Friedman Salzman's PhD thesis, *Photo-Meson Production from a Single Nucleon*, which she finished in 1953. (Her contribution to theoretical models first expounded by Chew would become a recurring pattern in the following years.) Implementing the ideas of the Chew model for this process required some additional thought. The thesis introduction presents photopion production as more complicated than mere pion–nucleon scattering. After all, the former involves electromagnetic interactions as well. In particular, this meant that Friedman Salzman had to reconsider which higher-order contributions to take into account. Using the values of the parameters that Chew had already determined for his model, Friedman Salzman was then able to directly compare her results with experimental data on photopion production. She found an improved agreement relative to former analyses based on weak coupling constant calculations. Ultimately, however, Friedman Salzman deemed that the results for photopion production were "somewhat disap-pointing" (Salzman, 1953a, p. 5) because they reproduced only some of the features of the experimental data, and hence she suggested a refinement of the "very crude single cut-off approach used by Chew." This proposal surely added pressure for Chew to refine his model.[31] At any rate, in 1954, he propounded a second iteration of the Chew model, which redefined the coupling constant[32] and enabled the application of perturbation theory (Chew, 1954b, c).

This advance, however, is less important for this chapter than the third model iteration, which Chew formulated in 1954 by incorporating further higher-order contributions (Chew, 1954d, e). Mathematically speaking, he did so by refining the integral equation given in Chew (1954c). Chew argued that, this time, the results were phenomenologically viable for other processes as well, such as photopion production, but he only solved the resulting integral equation by an approximate method proposed by John Gammel (1954). Friedman Salzman and James Snyder, a physicist and programmer who was on her PhD committee, were the ones who

---

[30] Photopion production had been experimentally studied since the end of the 1940s, when it was first discovered (McMillan et al., 1949).

[31] It is unclear how influential Friedman Salzman's PhD thesis was for Chew's upcoming work. However, only one year after her thesis publication, Chew explicitly stated his interest in Friedman Salzman's studies of photopion production. In his article (Chew, 1954e), which belongs to the third iteration of the Chew model, he announced the publication of further results by Friedman Salzman on photopion production, which were supposed to refine Chew's own treatment of that process.

[32] Technically speaking, this redefinition was a finite renormalization.

actually solved the integral equation numerically, using the ILLIAC I, thereby confirming the accuracy of Gammel's approximation (Salzman and Snyder, 1954). Admittedly, Friedman Salzman (Figure 14.3) and Snyder's task was straightforward. They used standard methods for the numerical solution of the integral equation, thus resulting in a half-page long article.

In 1956, the Chew–Low model was published by Chew and Francis Low, his colleague at the University of Illinois (Urbana). While it was related to the Chew model, it also involved novelties in quantum field theory, Low's specialty (Cushing, 2005). "After some sophisticated field-theory input from Low, Chew's fixed-extended-source model [i.e. the Chew model] became, in 1956, the Chew–Low

Figure 14.3 Friedman Salzman with her daughter Amy in February or March 1956. Amy is said to have babbled about pi-mesons when she was two years old (email from A. G. Parker, May 31, 2024). A. G. Parker personal collection. Reproduced with permission.

model" (Pickering, 1989, pp. 582–583), a historian of the field later stated. More specifically, the Chew–Low model introduced three theoretical novelties. The first was an equation, now known as the Low equation, that applied the so-called Heisenberg picture rather than the interaction picture (Low, 1955).[33] These "pictures" are formalisms of quantum mechanics which treat the time-evolution of physical processes in different fashions. The second novelty was their use of ideas of (the mathematical field of) complex analysis and concepts, which would later be associated with the so-called autonomous S-matrix program. At the time, the idea of replacing quantum field theory with the use of general principles such as unitarity, analyticity, and causality was first taking shape. Chew would be an active proponent of this approach in the 1960s (Cushing, 2005), and the article which introduced the Chew–Low model already hints at this development.

The third novelty was the so-called effective-range approach. The Chew model had one puzzling aspect. It worked relatively well phenomenologically even though it modeled high-energy contributions only crudely. As mentioned above, it just cut these contributions off beyond a certain energy. Why could high-energy contributions be so easily neglected? In the 1930s and 1940s, nuclear theorists had faced a similar situation when studying nucleon–nucleon interactions. As there was no evidence for pions yet, they treated the problem quantum mechanically and studied the shape of the potential of the nuclear interactions.[34] However, the available experimental data were compatible with rather different shapes. By suitably adjusting two parameters related to the width and depth of the potential, predictions based on the *relative potential shape* could be made compatible with experimental data.[35] Whereas it was expected that the shape of the potential would not leave a trace at very low energies because of the so-called centrifugal barrier, the experiments were performed at high enough energies that the shape of the potential was expected to leave its mark on the experimental data.

This situation prompted theoretical research and led to the effective-range approach of quantum mechanics, which explained why the details of the potential (such as its range) can only be determined approximately ("effectively"). In particular, the approach demonstrated that two (shape-insensitive) parameters indeed suffice to describe scattering data up to quite high energies. (One of the parameters is the so-called effective range, hence the name of the approach.) The

---

[33] The Low equation is related to the equations which were established by Lehmann, Symanzik, and Zimmermann in their 1955 landmark article (Lehmann et al., 1955). Low's work is also related to work of Gian Carlo Wick (1955).

[34] In simple terms, this shape determines how the force between two nucleons varies with the distance between them.

[35] In simple terms, this meant that scientists could only determine the general or "effective" range and strength of the nuclear interactions, but not how the strength of the nuclear interactions depends in detail on the distance between the nucleons.

upshot was that even a crude modeling of the potential led to phenomenologically viable results.

One of the scientists involved in the corresponding quantum mechanical proofs was – once more – Chew. (Chew and Goldberger (1949) contains more references to the experimental and theoretical situation.) It would hence have been only natural for him to wonder if the crude modeling of high-energy contributions in the original Chew model might be phenomenologically inconsequential up to quite high scattering energies for similar reasons. In any case, Chew and Low transferred the quantum-mechanical effective-range approach into quantum field theory. In so doing, they showed that a good part of the phenomenological consequences of the Chew–Low model was independent of the modeling of high-energy contributions.

The resulting article "Effective-range approach to the low-energy p-wave pion–nucleon interaction" could well have been titled "Effective-range approach to the Chew model." At the beginning of their article, Chew and Low (1956a) indeed explain that: "One of the main purposes of this paper is to show that the most important predictions of the model [i.e. the Chew model] are actually independent of its details."

Although no evidence has come to light to suggest that Friedman Salzman was involved in establishing the Chew–Low model, an understanding of that model provides context to what she and her husband subsequently accomplished.

14.4 The Chew–Low–Salzman Method

The Chew–Low–Salzman method was the first attempt at a numerically accurate solution of the Chew–Low model, and was published only one year later. In the meantime, the Chew–Low model had already successfully been applied to pion–nucleon scattering and photopion production. However, as Friedman Salzman and her husband noted in their article, these applications of the Chew–Low model were based on approximate methods.

Friedman Salzman and her husband instead solved the Chew–Low integral equation model numerically with the ILLIAC I. In some respects, the couple's task was similar to a challenge that Friedman Salzman and Snyder faced three years prior when solving the Chew model. Solving the Chew–Low model numerically, however, was significantly more challenging as, technically speaking, only the integral equations of the Chew model were "linear." To ease their task, the couple followed Chew and Low in using a clever transformation of these integral equations before solving them, but the solutions of the transformed equations did not need to correspond to solutions of the original ones. They therefore had to evaluate whether their numerical solutions truly solved the Chew–Low model. Although the couple found their solution to be problematic for a phenomenologically less important part of pion–nucleon scattering, for the most part it held true. While solving the Chew–Low model, Friedman Salzman and her husband had to overcome further obstacles. For example, when evaluating the integrals (of the integral equations), they found some significant high-energy contributions which had to be solved manually rather than through numerical methods. This is remarkable because the Chew–Low model suggested that high-energy contributions could be neglected entirely. In the end, the couple compared their solutions with experimental data, which they could reproduce quite well.

The article on the Chew–Low–Salzman method was Friedman Salzman's first widely received contribution in physics; it would be cited dozens of times over the subsequent decades.[36] Within these, of particular note are: (i) the highly influential article by Stanley Mandelstam (1958), which gave the Chew–Low–Salzman method its name only one year after its publication; (ii) Ken Wilson's PhD thesis (Wilson, 1961; he would later win the Nobel Prize in Physics), which makes use of Friedman Salzman and her husband's analysis in the context of a study of the Low equation (on which the Chew–Low model is based); and (iii) the article by Polish physicist Andrzej Kotański, which expands on Friedman Salzman and Salzman's work (Kotański, 1963). According to its introduction, this last work follows from "the realization of an idea suggested" to Kotański by the couple, thus evincing that they still had an active interest in discussing the Chew–Low–Salzman method well after its publication. The impact of the Chew–Low–Salzman method on nuclear physics might well have been even stronger had the Chew–Low model not quickly competed with other promising models for the nuclear interactions after its conception. (For more details, see Cushing (2005).)

Interestingly, Friedman Salzman and Salzman's article on the Chew–Low–Salzman method does not list its authors' names in alphabetical order, thus making Friedman Salzman its second author. This contrasts with all the other articles written by the couple. Why might that be? Perhaps Salzman first initiated this effort on his own. Chew and Low (1956a) mention that the Chew–Low model had already been solved numerically by him, a result that was to be discussed in a forthcoming paper. Salzman (1956) had also taken sole credit when presenting his preliminary findings at the Sixth Rochester Conference on High Energy Nuclear Physics, in 1956. However, he later co-authored on the same topic with his wife, after she had joined in the endeavor. Salzman (1957) indeed specifically mentioned the collaboration at the Seventh Rochester Conference the following year. A comparison of Salzman's first report (Salzman, 1956) and the couple's final

---

[36] According to a search of "Freda Salzman" on Google Scholar on April 5, 2024, Salzman and Salzman (1957) has 61 citations. By contrast, her most cited article before that publication, Salzman and Snyder (1954), has been cited only six times.

article reveals that further investigations took place in the meantime. It can be inferred that this gave Friedman Salzman ample opportunity to make a genuine contribution to their project. In particular, at the Sixth Rochester Conference, it was still unclear whether Salzman's solutions also satisfied "the original integral equation[s]" (Salzman, 1956) rather than only the transformed ones.

14.5 Epilogue

The successful collaboration of the couple continued after the publication of the Chew–Low–Salzman method. They even kept a six-foot-tall blackboard in their bedroom, which they used to discuss their research.[37] Their subsequent work on long-range interactions in collisions of nucleons with pions and nucleons, respectively, was particularly well received. In contrast, their work on general relativity was met with skepticism.[38] Altogether, the couple's collaboration led to at least six publications in the 1960s (Salzman and Salzman, 1960a, b, 1961, 1962, 1969; Salzman and Salzmann [sic], 1965).

As mentioned above, the Chew–Low model and the methods used for its solution, such as the Chew–Low–Salzman method, were later superseded by alternative approaches to nuclear (and strong) interactions, notably by quantum chromodynamics, which does not focus on the interactions of nucleons but rather on the interactions of their constituents. The Chew–Low model nevertheless played an influential role in the history of nuclear interactions as captured by Silvan S. Schweber's textbook, *An Introduction to Relativistic Quantum Field Theory* (Schweber, 1962). On page 391, he claims that "[t]he papers by Chew and Low are landmarks in quantum field theory. Every serious student of modern field theory is urged to study them, for no review of the articles can do full justice to the clarity and simplicity of the Wick–Chew–Low approach."[39] This brief quotation brings up two additional key features of Friedman Salzman and Salzman's article. First, it contains a concise summary of the Chew–Low model, thereby adding to its clarity. Second, to this day, the Chew–Low–Salzman article, as it captures the couple's struggle to solve the Chew–Low model, demonstrates in which sense that model is actually *not* that simple.

---

[37] This is at least true for their time at the University of Massachusetts Boston. A. G. Parker, emails to author, May 31, 2024 and June 8, 2024.

[38] An anonymous referee's report rejected the couple's draft "Invariance of the light cone and Synge-type transformations" submitted in 1969. The referee judged that "[t]he apparent paradox" presented in the article "is too easily resolved." A. G. Parker personal collection. The date of the draft is November 5, 1969, the date on the report reads December 12, 1969.

[39] As mentioned above, Wick (1955) found similar results to Low (1955) and hence Schweber mentions Wick here as well.

References

Chen, Maggie. 2019. The glass ceiling of nepotism. *Polycephalic Perceptions*, September 6. https://polycephalicperceptions.home.blog/2019/09/06/the-glass-ceiling-of-nepotism/.

Chen, Maggie. 2020. Wives, physics, and nepotism in academia. *Lady Science*, August 18. https://www.ladyscience.com/essays/wives-physics-nepotism-academia-freda-salzman-2020.

Chew, Geoffrey F. 1953. Pion–nucleon scattering when the coupling is weak and extended. *Physical Review*, 89(3), 591–593. https://doi.org/10.1103/PhysRev.89.591

Chew, Geoffrey F. 1954a. One of Schwinger's variational principles for scattering. *Physical Review*, 93(2), 341–343. https://doi.org/10.1103/PhysRev.93.341

Chew, Geoffrey F. 1954b. Renormalization of meson theory with a fixed extended source. *Physical Review*, 94(6), 1748–1754. https://doi.org/10.1103/PhysRev.94.1748

Chew, Geoffrey F. 1954c. Method of approximation for the meson–nucleon problem when the interaction is fixed and extended. *Physical Review*, 94(6), 1755–1759. https://doi.org/10.1103/PhysRev.94.1755

Chew, Geoffrey F. 1954d. Improved calculation of the p-wave pion–nucleon scattering phase shifts in the cut-off theory. *Physical Review*, 95(1), 285–286. https://doi.org/10.1103/PhysRev.95.285

Chew, Geoffrey F. 1954e. Comparison of the cut-off meson theory with experiment. *Physical Review*, 95(6), 1669–1675. https://doi.org/10.1103/PhysRev.95.1669

Chew, Geoffrey F. and Goldberger, Marvin L. 1949. On the analysis of nucleon–nucleon scattering experiments. *Physical Review*, 75(11), 1637–1644. https://doi.org/10.1103/PhysRev.75.1637

Chew, Geoffrey F. and Low, Francis E. 1956a. Effective-range approach to the low-energy p-wave pion–nucleon interaction. *Physical Review*, 101(5), 1570–1579. https://doi.org/10.1103/PhysRev.101.1570

Chew, Geoffrey F. and Low, Francis E. 1956b. Theory of photomeson production at low energies. *Physical Review*, 101(5), 1579–1587. https://doi.org/10.1103/PhysRev.101.1579

Cushing, James T. 2005. *Theory Construction and Selection in Modern Physics: The S Matrix*. Cambridge: Cambridge University Press.

Darrigol, Olivier. 1988. The quantum electrodynamical analogy in early nuclear theory or the roots of Yukawa's theory. *Revue d'histoire des sciences*, 41(3), 225–297. https://doi.org/10.3406/rhs.1988.4098

Gammel, John L. 1954. Elastic scattering of pions by nucleons and pion production in nucleon–nucleon collisions. *Physical Review*, 95(1), 209–216. https://doi.org/10.1103/PhysRev.95.209

Kotański, Andrzej. 1963. Consistent formulation of the Chew–Low theory for meson–nucleon scattering. *Acta Physica Polonica*, XXIV(1), 73–78.

Kragh, Helge. 2023. A terminological history of early elementary particle physics. *Archive for History of Exact Sciences*, 77(1), 73–120. https://doi.org/10.1007/s00407-022-00299-2

Lattes, Cesare M. G., Muirhead, Hugh, Occhialini, Giuseppe P. S., and Powell, Cecil F. 1947a. Processes involving charged mesons. *Nature*, 159(4047), 694–697. https://doi.org/10.1038/159694a0

Lattes, Cesare M. G., Occhialini, Giuseppe P. S., and Powell, Cecil F. 1947b. Observations on the tracks of slow mesons in photographic emulsions. *Nature*, 160(4066), 453–456. https://doi.org/10.1038/160453a0

Lehmann, Harry, Symanzik, Kurt, and Zimmermann, Wolfhart. 1955. Zur Formulierung quantisierter Feldtheorien. *Il Nuovo Cimento (1955–1965)*, 1(1), 205–225. https://doi .org/10.1007/BF02731765

Low, Francis E. 1955. Boson–fermion scattering in the Heisenberg representation. *Physical Review*, 97(5), 1392–1398. https://doi.org/10.1103/PhysRev.97.1392

Lyons, Donald M. 1982. Freda Salzman [Obituary]. *Physics Today*, 35(2), 89–90. https:// doi.org/10.1063/1.2914950

Mandelstam, Stanley. 1958. Determination of the pion–nucleon scattering amplitude from dispersion relations and unitarity. General Theory. *Physical Review*, 112(4), 1344–1360. https://doi.org/10.1103/PhysRev.112.1344

Mashhoon, Bahram. 2001. In memoriam: Abraham Haskel Taub. *SIAM News*, 34(7). https://archive.siam.org/news/news.php?id=570

McMillan, Edwin M., Peterson, Jack M., and White, R. Stephen. 1949. Production of mesons by X-rays. *Science*, 110(2866), 579–583. https://doi.org/10.1126/ science.110.2866.579

Mukherji, Viśvapriya. 1974. A history of the meson theory of nuclear forces from 1935 to 1952. *Archive for History of Exact Sciences*, 13(1), 27–102. https://doi.org/10.1007/ BF00327862

Oppenheimer, J. Robert and Phillips, Melba. 1935. Note on the transmutation function for deuterons. *Physical Review*, 48(6), 500–502. https://doi.org/10.1103/PhysRev.48.500

Pais, Abraham. 1995. *Inward Bound: Of Matter and Forces in the Physical World*. Oxford: Oxford University Press.

Pickering, Andy. 1989. From field theory to phenomenology: the history of dispersion relations. In Brown, Laurie M., Dresden, Max, and Hoddeson, Lillian (eds.), *Pions to Quarks: Particle Physics in the 1950s*. Cambridge: Cambridge University Press.

Ruiz de Olano, Pablo, Fraser, James D., Gaudenzi, Rocco, and Blum, Alexander S. 2022. Taking approximations seriously: the cases of the Chew and Nambu–Jona–Lasinio models. *Studies in History and Philosophy of Science*, 93(June), 82–95. https://doi .org/10.1016/j.shpsa.2022.02.013

Salzman, Freda. 1953a. *Photo-Meson Production from a Single Nucleon*. PhD Thesis, Urbana, IL: University of Illinois.

Salzman, Freda. 1963. Range of high-energy interactions. *Physical Review*, 131(3), 1345–1350. https://doi.org/10.1103/PhysRev.131.1345

Salzman, Freda. 1974. Gravitational field of a freely moving mass. *Il Nuovo Cimento B (1971–1996)*, 24(2), 157–188. https://doi.org/10.1007/BF02725955

Salzman, Freda. 1977. Are sex roles biologically determined? *Science for the People*, 9(4), 27–32, 43.

Salzman, Freda. 1979a. Physicist Freda Salzman. Interview by Eileen Zalisk as episode of series "Women in Science," radio station WBAI. Pacifica Radio Archive number: IZ1483. https://avplayer.lib.berkeley.edu/Pacifica/b23306718

Salzman, Freda. 1979b. Sociobiology: The controversy continues. *Science for the People*, 11(2), 20–27.

Salzman, Freda and Salzman, George. 1960a. Pion production from π^-–p collisions in the long-range interaction model. *Physical Review*, 120(2), 599–608. https://doi.org/ 10.1103/PhysRev.120.599

Salzman, Freda and Salzman, George. 1960b. Single-virtual-boson-exchange interaction in high-energy collisions. *Physical Review Letters*, 5(8), 377–379. https://doi.org/ 10.1103/PhysRevLett.5.377

Salzman, Freda and Salzman, George. 1961. High-energy nucleon–nucleon collisions. *Physical Review*, 121(5), 1541–1544. https://doi.org/10.1103/PhysRev.121.1541

Salzman, Freda and Salzman, George. 1962. Single virtual particle exchange model of high-energy inelastic glancing collisions. *Physical Review*, 125(5), 1703–1719. https://doi.org/10.1103/PhysRev.125.1703

Salzman, Freda and Salzman, George. 1969. Acceleration of material particles to the speed of light in general relativity. *Lettere al Nuovo Cimento (1969–1970)*, 1(17), 859–864. https://doi.org/10.1007/BF02911557

Salzman, Freda and Salzmann [sic], George. 1965. A theory of *T*-violation. *Physics Letters*, 15(1), 91–94. https://doi.org/10.1016/0031-9163(65)91142-X

Salzman, Freda F. and Snyder, James N. 1954. Numerical calculation of $\frac{3}{2} - \frac{3}{2}$ pion–nucleon reaction matrix and phase shifts. *Physical Review*, 95(1), 286–287. https://doi.org/10.1103/PhysRev.95.286

Salzman, G. and Taub, A. H. 1954. Born-type rigid motion in relativity. *Physical Review*, 95(6), 1659–1669. https://doi.org/10.1103/PhysRev.95.1659

Salzman, George. 1953b. *Born-Type Rigid Motion in Relativity*. PhD Thesis. Urbana, IL: University of Illinois.

Salzman, George. 1956. Numerical solution of the Chew–Low equation for meson–nucleon scattering. In *High Energy Nuclear Physics: Proceedings of the Sixth Annual Rochester Conference*. New York: Interscience Publishers, pp. III:14–III:18.

Salzman, George. 1957. Phase shifts predicted by the Chew–Low theory. In *High Energy Nuclear Physics: Proceedings of the Seventh Annual Rochester Conference*. New York: Interscience Publishers, pp. II:18–II:22.

Salzman, George. 1971. Discrimination at UMass: Woman scientist fights back. *Science for the People*, 3(2), 18–21.

Salzman, George and Salzman, Freda. 1957. Solutions of the static theory integral equations for pion–nucleon scattering in the one-meson approximation. *Physical Review*, 108(6), 1619–1628. https://doi.org/10.1103/PhysRev.108.1619

Schweber, Silvan S. 1962. *An Introduction to Relativistic Quantum Field Theory*. New York: Harper & Row.

Wick, Gian Carlo. 1955. Introduction to some recent work in meson theory. *Reviews of Modern Physics*, 27(4), 339–362. https://doi.org/10.1103/RevModPhys.27.339

Wilson, Kenneth G. 1961. *An Investigation of the Low Equation and the Chew–Mandelstam Equations*. PhD Thesis. Pasadena, CA: California Institute of Technology. https://doi.org/10.7907/VCDA-4Q31

Workman, Ronald L. et al. (Particle Data Group). 2022. Review of particle physics. *Progress of Theoretical and Experimental Physics*, 2022(8), 083C01. https://doi.org/10.1093/ptep/ptac097

Yukawa, Hideki. 1935. On the interaction of elementary particles. I. *Proceedings of the Physico-Mathematical Society of Japan. 3rd Series*, 17, 48–57. https://doi.org/10.11429/ppmsj1919.17.0_48

15

Out of the Ivory Tower: Maria Lluïsa Canut and X-Ray Crystallography

15.1 Introduction

Maria Lluïsa Canut Ruiz (1924–2005) was a physicist and one of the most prominent crystallographers in Spain during the Francoist regime. Together with the natural scientist José Luis Amorós Portolés (1920–2001), her partner in science and in life, she was awarded the Francisco Franco Prize in Sciences in 1963 for their joint contribution to crystallography, and she received the Research Recognition Award of the Southern Illinois University in 1968, among other accolades. Both also achieved international recognition.[1]

Canut spent her scientific career in Barcelona, Madrid, and Illinois. Over the course of approximately 20 years she published more than 60 scientific papers and co-authored three books, all exploring and steadily mastering a novel and original field of research: the relationship between X-ray diffuse scattering through crystals and their thermal dynamics. X-Ray crystallography traces its origins back to a ground-breaking discovery in quantum physics made in 1912: X-rays passing through crystals produce diffraction patterns that are directly related to the microscopic structure of the crystals (Eckert, 2012). Three decades later, diffuse zones observed in X-ray diffraction patterns began to be interpreted in terms of the dynamics and non-ideal features of crystal structures. This was Canut's and Amorós' field of interest.

Amorós has been recognized as a prominent crystallographer since the early 1960s.[2] Two of his former PhD students have also written about his scientific trajectory and

I thank Joan March, Agustí Ceba, and Guillem Pons for their generosity in sharing with me their archival findings and discussions about the topic, and Maria Victoria López-Acevedo and Sol López for their hospitality at UCM. I also thank Òscar Bagur for his support as in charge of the IME archives. Finally, I thank the editors for their admirable effort and patience in the revision of this chapter and the whole book.
[1] Many uncatalogued primary sources have been accessed to write the chapter, in particular, Canut's records at: Archivo General de la Administración (AGA, General Archive for the Administration), Centro Superior de Investigaciones Científicas (CSIC, Superior Council for Scientific Research), Universitat de Barcelona (UB, University of Barcelona), Southern Illinois University (SIU), Fundación Joan March (Joan March Foundation), and Institut Menorquí d'Estudis (IME, Menorcan Institute for Studies).
[2] See, for example, Ewald (1962).

contributions (López Andrés and López-Acevedo, 2002; López-Acevedo and López Andrés, 2004). Moreover, the library of the Faculty of Geology of the Complutense University of Madrid bears his name. Although Canut and Amorós worked as a scientific team, practically throughout their professional lives, Canut has not received similar attention. Hardly any secondary sources exist about her, other than a short note on Wikipedia and a blog entry.[3]

It is not for lack of interest in women of this period. The lives of other women scientists during Francoist Spain have indeed been studied, including those of Piedad de la Cierva and María Aránzazu Vigón, who contributed to the related fields of optics and nuclear physics, respectively (Alva Rodríguez, 2016; Romero de Pablos, 2017; Santesmases et al., 2017). Paloma Alcalá and Carmen Magallón also have offered an interesting overview of the integration of women into scientific careers in experimental sciences in Spain from the early twentieth to the early twenty-first century (Alcalá and Magallón, 2008), with special focus on the continuities and discontinuities due to the Spanish Civil War (1936–1939). A few examples emerge from their account, including de la Cierva, but not Canut. More recently, Fernando García Naharro (2021) analyzed the presence and profiles of women in applied sciences in the Centro Superior de Investigaciones Científicas (CSIC, Superior Council for Scientific Research) during Francoist Spain. García Naharro has shown that close links to men in the Spanish scientific field were especially important for women in applied sciences to prosper. The name of Canut, as the wife of Amorós, is mentioned just as an example of this phenomenon. Nowhere is Canut's trajectory explained.[4]

This chapter not only fills this gap but also reflects on the underlying causes for such an omission. In Section 15.2, I describe the beginnings of Canut's life and scientific career, and Section 15.3 analyzes the scientific and political contexts. In Sections 15.4–15.6, I examine Canut's scientific contributions in the three different stages of her career: Barcelona, Madrid, and Illinois. Next (Section 15.7), I briefly discuss her involvement in the second feminist wave in the US, which broke the continuity of her crystallography career. I then describe how Canut reinvented herself upon her return to Spain, and I finish with some concluding remarks.

15.2 Canut's Early Years: From Menorca to Barcelona (1924–1955)

Maria Lluïsa Canut Ruiz was born on May 26, 1924, in Maó, Menorca, one of the Balearic Islands in the Mediterranean Sea.[5] The daughter of Frederic Canut Clavé

---

[3] See Maria Lluïsa Canut, feminismo desde la física. Baleópolis, 159, June 11, 2012. https://baleopolis.blogspot .com/2012/06/maria-lluisa-canut-feminismo-desde-la.html

[4] About the contribution of women to science in Spain before the Civil War, see Magallón (2004). No similar work exists for scientific women after the Civil War, the generation to which Canut belonged.

[5] Canut's birth certificate, AGA records. See Note 1.

and Francisca Ruiz Ponsetí, she was a member of a renowned family of local entrepreneurs. Her mother was one of the 14 children of Pau Ruiz Verd, the founder of the most important metallurgical company in Menorca at the turn of the twentieth century. Among her 13 uncles and aunts, Estanislau Ruiz Ponsetí was a renowned mathematician and socialist politician, and Carles Ruiz Ponsetí one of the most important businessmen on the island from the 1940s to the 1970s (Méndez Vidal and Hernández Andreu, 2011). Menorca, however, was Canut's home only for a few years. Her family moved to Mallorca when she was still a child, and she graduated from high school there in 1942 with the highest qualification.[6]

At 18, Canut moved again, this time to Valencia, to pursue university studies. During the academic year 1942–1943, she enrolled in the Physics and Chemistry Section of the Science Faculty of the University of Valencia (UV). Her stay in Valencia lasted just one year, though. She continued her studies in physics at the University of Barcelona (UB) for the academic year 1943–1944 (Figure 15.1). For reasons that remain unclear, she then interrupted her university studies. It was not until 1949 that she resumed her education; she finally obtained her graduate degree in physical sciences three years later.[7]

Figure 15.1 Portrait of Canut in her 20s. Source: IME. Unknown author.

---

[6] Canut's academic certificate from the Colegio de la Pureza de María Santísima, 1943, UB record. See Note 1.
[7] Canut's academic records from UV and UB, between 1944 and 1952, and certificate of her 1952 degree examination, UB record. See Note 1.

Her personal situation changed markedly during this period. In 1950, she married José Ignacio Rived and gave birth to her only child.[8] As a mother, Canut pursued her professional aspirations. Immediately after graduating, she was appointed teacher of physics and chemistry at the secondary school Maragall in Barcelona, a job she kept for two years. In parallel, she started her doctoral studies at UB.[9]

Around that time, Canut's personal circumstances changed again, when Rived abandoned his family.[10] Being a single mother did not seem to intimidate Canut, and her graduate work stayed on schedule. In 1955, she started as an assistant at the Physics Section of CSIC in Barcelona,[11] and she obtained her doctoral degree for a thesis on experimental crystallography (Canut, 1955), developed at the Department of Crystallography and Mineralogy of CSIC, under the supervision of Mariano Velasco Durántez, chief of the Physics Section of CSIC in Barcelona, with the collaboration of Amorós. Amorós was the successor to Francisco Pardillo Vaquer as chief of the Department of Crystallography and Mineralogy of CSIC in Barcelona and had served in that role since 1954.

15.3 The Context

15.3.1 X-Ray Crystallography in Spain after the Civil War

Pardillo's research group in Barcelona was one of the three in Spain devoted to X-ray crystallography after the Civil War.[12] Pardillo had introduced the technique to the country already in 1913, having learnt it with Victor Goldschmidt at the University of Heidelberg, thanks to a fellowship of the progressive Junta de Ampliación de Estudios e Investigaciones Científicas (JAE, Committee for Extension of Studies and Scientific Research).[13] At the time of the Civil War, X-ray crystallography research in Pardillo's laboratory was still growing.

The Civil War caused a break at many levels in Spain, notably regarding scientific policy. Beyond old dichotomies that opposed science to Francoism,[14]

---

[8] Marriage certificate in Canut's AGA record. See Note 1.

[9] Canut's doctoral studies record, 1952 and 1953, UB record. Letters from the director of the secondary school Maragall, Juan Estevan, to José María Sánchez de Muniain, general director of high education of the National Ministry of Education, July 31, 1952, and August 25, 1953, AGA record. See Note 1.

[10] The moment at which she was left alone with the child is unclear. In a letter sent by Amorós on May 18, 1963, to the board of the Superior Council for Scientific Research (CSIC) in Spain to support Canut's research stays abroad, Amorós mentioned that Rived abandoned his duties as a father 11 years prior. This letter is contained in Canut's CSIC record. See Note 1.

[11] CSIC certificate of Canut's inauguration, February 11, 1955, CSIC record. See Note 1.

[12] See Mañes (2005) for further information about X-ray crystallography in Spain before and after the Civil War, in particular the three groups mentioned in this section.

[13] JAE was a public institution founded in 1907 to promote research and scientific education in Spain.

[14] Literature on Francoism and science is vast. For an overview, see Romero de Pablos and Santesmases (2008) and Janué i Miret and Presas i Puig (2021). The work of Roqué (2012) is specifically about physics during the Francoist dictatorship.

the new regime saw science development as a priority, but scientific practices changed in complex ways. To begin with, the regime erased previous scientific structures, such as JAE, and created new ones, such as CSIC, founded in 1939. CSIC took over JAE's functions to promote scientific research by making use of JAE's facilities and material resources, while implementing fundamental changes in its organization, such as the subordination of CSIC to the Ministry of Education. In practical terms, CSIC became the research superstructure of universities through the creation of centers, departments, and sections, whose researchers were university professors. In the late 1940s, new research categories not ascribed to universities were created: CSIC scientific collaborators and scientific researchers (Sánchez Ron, 2021). The budget and number of researchers considerably increased from the 1940s to the 1970s in this context, even though they remained insufficient.

Through the CSIC structure, the Francoist dictatorship could implement an interventionist scientific policy. The CSIC leadership determined the investment distribution among areas of research both at universities and at CSIC units, independently from specialized commissions evaluating research on purely academic criteria (Malet, 2008, p. 237). The research areas of applied sciences and technology were especially promoted, linked to the "spiritual and material needs of the Nation" (Malet, 2008, p. 227). Moreover, CSIC promoted the creation of numerous journals and publications, in which CSIC researchers published their results without peer review, thus leading to an endogamic system of scientific production of unequal quality and scarce international impact (Malet, 2008).

The recruitment of researchers also lacked objectivity, as it was not based solely on academic merits. The purge of the community of scientists was another draconian strategy used by the Francoist regime to reorganize science (Claret Miranda, 2006; Herran and Roqué, 2012). Almost half of the Spanish university professors who were in service before 1936 had died, or were exiled or retaliated against after the Civil War. In the 1940s, vacant and new positions were occupied by professors close to the new regime.

After having initially been purged, Pardillo was one of the professors rehabilitated at UB.[15] In 1943, he was assigned to the newly created Crystallography and Mineralogy Section (from 1950 onwards, referred to as Department) within the Center of Geological Studies Lucas Mellada of CSIC in Barcelona.

Amorós was one of the members of Pardillo's research group after the Civil War, then comprised five naturalists and one pharmacist. In the years 1953–1955, seven young collaborators were incorporated, two men and five women, among them

---

[15] Initially, he was purged for his collaboration with the previous government, but his relationship to power structures prior to the Civil War was ambiguous. In 1934, he had supported a public manifesto against the board of trustees of the UB. Therefore, he could have his professorship restored (Mañes, 2005, p. 68).

Canut. When Pardillo retired in 1954, Amorós became professor and led the department for one year before being appointed Professor of Crystallography and Mineralogy at the University of Madrid (UM) in 1955.

The department's unusual policy to promote research stays abroad was particularly remarkable. From 1948 to 1953, thanks to CSIC support, Amorós visited various British research centers, such as Kathleen Lonsdale's laboratory at the University College of London, the Massachusetts Institute of Technology and the Pennsylvania State University in the US, as well as several centers in France, Belgium, and the Netherlands. During these trips, Amorós learned of new methodologies for X-ray diffraction analysis and instrument building that were of interest for the Barcelona laboratory, since the university could not afford to buy all necessary instruments. As a result, the Barcelona group published a high number of papers about X-ray crystallography, X-ray experimental methods and instrument making, as well as optical mineralogy during the 1950s.

The second center for crystallography in Spain was in Madrid: the X-Ray Section of the newly created – in 1940 – Instituto de Física Alonso de Santa Cruz of CSIC, formerly Laboratorio de Investigaciones Físicas (LIF, Laboratory of Physical Research). Before the Civil War, LIF had stood out as a well-equipped and highly productive X-ray crystallography center. The head of the section before and after the conflict was Julio Palacios, one of the most significant Spanish physicists of the twentieth century, aligned with the new regime during the first years of the dictatorship.[16] After the Civil War, research on X-ray crystallography decreased, following Palacios' diversification of interests, but in the 1950s Palacios' disciple, the chemist Luis Rivoir, succeeded him as head of the X-Ray Section, and publications on the topic rose again, focusing on the X-ray determination of chemical structures. In 1958, two years after Amorós became Professor of Crystallography and Mineralogy at UM, Amorós started to lead a new Physical Crystallography Section of the institute, devoted to the study of diffuse scattering and the thermal dynamics of crystals (CSIC, 1958).

The third group devoted to X-ray crystallography after the Civil War, albeit significantly less well-equipped than the first two, was located in Sevilla, in the Section of Physics of the Institute of Physics-Chemistry Antonio de Gregorio Rocasolano (CSIC), led by a former disciple of Palacios in Madrid.

In short, by the mid-twentieth century X-ray crystallography was well established in Spain. Moreover, in 1950, the Spanish Society of Crystallography was created, motivated by the membership of some Spanish crystallographers, including Amorós, in the International Union of Crystallography.

---

[16] Due to later disagreements with the new regime, Palacios worked for some years in Portugal, see Chapter 12.

15.3.2 Women and Science in Spain after the Civil War

The new scientific policy also had consequences for women pursuing scientific careers. In the first decades of the twentieth century, women in Spain aligned with European movements to integrate women in educational and professional fields from which they had been traditionally excluded (Romero de Pablo and Santesmases, 2008). JAE supported coeducation and the scientific careers of women. This spirit of openness, however, was interrupted by the Francoist dictatorship. The new education system segregated boys and girls, and implemented curricular differences based on the different expected societal roles of men and women (Alcalá and Magallón, 2008).

After 1939, exile and purge forced a large number of both men and women researchers to suspend or abandon their careers. It was particularly difficult for women researchers. Due to new labor legislation, which purportedly aimed at "freeing women from factories and workshops" (Alcalá and Magallón, 2008, p. 159), marriage became another obstacle in women's professional paths. The director of the Balmes Institute of Sociology of CSIC specifically asserted that "women public servants were bad mothers, bad wives and bad servants" (Alcalá and Magallón, 2008, p. 160). In this context, it is not surprising that in its early years CSIC predominantly hired male university professors close to the new regime. Paloma Alcalá and Carmen Magallón even assert that "women were, in these initial moments of CSIC, nothing else than an accident" (Alcalá and Magallón, 2008, p. 160). The situation slightly improved with the call for new researchers in the 1940s and 1950s, but women researchers at CSIC remained a small minority, usually with lower status than men in the three research categories of scientific collaborators, scientific researchers, and research professors.

The minority of women who attended university and started a research career in these first decades of the Francoist regime usually, though not exclusively, came from elite families (Santesmases, 2000). Even for them, the career path was not easy. Intellectual capability was clearly not sufficient. Case studies show that supportive mentors were essential to women who pursued research, in particular in applied sciences (García Naharro, 2021). Close links to men developing scientific careers in Spain, especially members of the Francoist elite, allowed them to benefit from the opportunities offered by the evolving scientific policies concerning international cooperation programs and industrial investments, and by the emergence of new fields of knowledge of strategic interest (Romero de Pablos, 2017), such as X-ray crystallography.

15.4 Canut's Early Research in Barcelona

Canut succeeded in developing a scientific career despite this hostile context. She was a member of a local elite family, although not openly close to the new regime.

Even though she was married and a mother, practically alone with her child, she seized the opportunities she was given. Pardillo's laboratory group, devoted to the fertile area of X-ray crystallography and with international connections, welcomed a significant number of women collaborators and offered Canut a favorable context to start her scientific career. Importantly, Canut was able to count on the support of Amorós, with whom she formed a professional partnership from the very beginning, growing from a mentee to a collaborator.

Canut's dissertation was to form the foundation of a long-term research plan, which she thoroughly implemented, together with Amorós, during the following decade: the study of the thermal dynamics of crystals through the analysis of X-ray diffuse scattering (Canut, 1955). In theory, when ideal crystals are irradiated with X-rays, they give rise to diffraction patterns with definite spots that are regularly distributed. But in practice, this is typically not the case. Canut analyzed the diffuse patterns of X-ray scattering, which, according to her interpretation, were related to non-ideal and dynamical features, mainly the thermal vibration of atoms and molecules within the crystal.

This research direction was novel, based on the foundational works of Kathleen Lonsdale, one of the leading crystallographers of the time.[17] One of Lonsdale's works was especially relevant for Canut's research on the thermal dynamics of crystals. A 1941 publication co-authored by Lonsdale established the existence of diffuse spots in X-ray photographs, as well as radial "streaks," interpreted as evidence of qualitative relationships between the intensity, shape, size, position, and persistence of diffuse X-ray patterns of a series of organic and inorganic crystals to the nature, structure, perfection, orientation, and, notably, the temperature of crystals examined (Lonsdale and Smith, 1941).

Amorós, having worked with Lonsdale on X-ray and thermal motion in crystals in London from 1948 to 1950 (López-Acevedo and López Andrés, 2004, p. 85), no doubt brought these ideas back to Barcelona. Canut's first paper was a clear declaration of intent in this respect. Titled "Estudios acerca de la dinámica reticular en cristales moleculares: I, Difracción difusa del ácido oxálico dihidratado" (Velasco et al., 1954a) and part of her dissertation (Canut, 1955), this paper was the first of a series of 11 publications between 1954 and 1960, always co-authored with Amorós, and occasionally with others.

In these works, Canut tackled the problem from various perspectives. The first three papers, written in Barcelona, were mainly experimental (Velasco et al., 1954a, b; Amorós and Canut, 1955). X-Ray diffuse scattering patterns of two substances,

---

[17] Lonsdale was one of the first two women fellows of the Royal Society of London in 1945, the first woman professor at the University College London in 1946, and the first woman president of the International Union of Crystallography (Pycior et al., 1996). At University College London, Lonsdale created her own crystallographic research group.

dehydrated oxalic acid and adipic acid, were obtained and interpreted. Both acids had molecules united by hydrogen bridge bonds, with the difference that one molecule (dehydrated oxalic acid) was planar and the other (adipic acid) had a chain structure. According to the authors, the observed differences in diffuse scattering of the X-ray photographs of the two substances had to be due to differences in molecular morphology.

Canut also focused on methodological questions. The interpretation of X-ray photographs in terms of the real crystal structure and dynamics was a mathematical challenge. Moreover, each kind of experimental setup led to a different kind of imaging and called for a different interpretation method. For example, one could obtain X-ray photographs by using either a plane film or a cylindrical film. In Canut's experiments, X-rays reflected from a crystal impinged against a cylindrical photographic film that surrounded the crystal. By 1954, Canut proposed a graphical method to make a direct conversion between X-ray cylindrical photographs and reciprocal space[18] (Font-Altaba and Canut, 1954), in collaboration with Manuel Font-Altaba, doctor in pharmacy and natural sciences, and scientific collaborator of the Department of Crystallography and Mineralogy of UB.

Canut rarely published alone. Out of her six publications in Barcelona, only her dissertation was single-authored. Her publication list reflects the extent of her collaborations, in particular with Amorós. From 1956 to 1973, she published 61 papers, her dissertation, and three books. In only four papers was she the sole author, while 49 papers were co-authored with Amorós. In 24 of these 49, the only two authors were Amorós and Canut, and in nine she was the first author. Both scientists published about experiments, new experimental setups, theoretical questions, mathematical developments, and methodological issues concerning X-ray diffraction applied to the study of the thermal dynamics of crystals. From this list of publications between 1954 and 1973, which covers the whole period during which Canut was scientifically active, it is difficult to separate the tasks or interests of either Canut or Amorós.

To be sure, Amorós published more frequently than Canut. According to López-Acevedo and López Andrés (2004), he published 215 papers from 1945 to 1987. Papers concerning the thermal dynamics of crystals, however, were practically all co-authored with Canut. Amorós separately published in areas related to his natural science background, particularly geology, and dissemination papers and books.

X-Ray diffraction applied to crystallography was the research theme in which Canut's background as a physicist and Amorós' background in natural sciences were particularly complementary. At the same time, this collaborative dynamics reflects a more general story: that of the construction of a new interdisciplinary

---

[18] Reciprocal space is a mathematical space related to the real or direct space of the crystal via Fourier transforms.

field, namely, the study of solid materials in the context of the emerging quantum physics, which expanded greatly from the 1920s to the 1960s (Hoddeson et al., 1992; Martin, 2018).

15.5 The Consolidation of a Research Program in Madrid

Canut moved to Madrid when she was appointed temporary scientific collaborator of CSIC – officially from July 1, 1956 – at the Department of Crystallography. On January 1, 1960, her position became permanent. That same year, she also became head of the Crystal Thermodynamics Section of the Department of Crystallography at CSIC, the only woman head of a physics section. On September 1, 1966, she was promoted to scientific investigator of CSIC.[19] From 1956 on, Amorós also held an academic position in Madrid as Professor of Crystallography and Mineralogy of UM (1956–1966) and head of the Department of Crystallography at CSIC (López-Acevedo and López Andrés, 2004, p. 86).

It is unclear when Canut and Amorós became emotionally involved. What is clear is that during their stay in Madrid, Canut and Amorós not only published numerous scientific papers together, but they also succeeded in creating a productive school of crystallography, focused on the thermal dynamics of crystals, in the context of which several dissertations were written, and several international visiting scholars were hosted. In Madrid, seven scientific projects were supported: two by the Joan March Foundation, three by the US Air Force, and two by the US Army, all dealing with the thermal dynamics of crystals, as a continuation of the scientific plan initiated in Barcelona.[20] Amorós was the principal investigator of the first project, with Canut as the main collaborator, and Canut held one of the two Joan March Foundation fellowships. Both Amorós and Canut served as the principal investigators for the other projects.

The US-funded projects identified above must be understood in the context of a new Spanish policy more open to international relations after World War II (Santesmases, 2006). Nevertheless, obtaining such funds was not the norm. Given the uncertain economic situation in Spain at that time, these funds were

---

[19] CSIC's notification of Canut's appointment as a temporal scientific collaborator, April 8, 1958; CSIC's notification of Canut's appointment as a permanent scientific collaborator, May 4, 1960; Canut's service record at CSIC, April 1967, CSIC record. See Note 1. See also CSIC (1960, p. 267).

[20] The projects were "Física del estado sólido: propiedades térmicas en relación con la estructura" ("Solid state physics: thermal properties in relation to structure," Fundación Joan March), "Mecanismo de una transición polimórfica de tipo vibracional, con formación de superstructura" ("Mechanism of a vibrational-type polymorphic transition with superstructure formation," Fundación Joan March), "Research in the area of crystallography" (Contract: AF61(052)196), "Studies of thermal motion in crystals" (Contract: AF61(514)114), "Thermal expansion of solids" (Contracts: Grant AF-EOAR-62–92 and Grant AF-EOAR-596–64, European Office of Air Research), US Defense Army-91–591-EUC-1717, and US DA-91–591-EUC-27755.

probably key for Amorós and Canut's research to prosper. Of the 40 papers published by Canut in Madrid, 38 were part of externally funded projects.

Teamwork formed the basis of their research. For each project, Canut and Amorós built interdisciplinary teams that included graduates in natural sciences, chemical sciences, geology, mathematics, and engineering. Their research was assisted by technicians, human computers, a typist, and one secretary. Scientific collaborators included research fellows, PhD students (many of them women), and visiting scholars.

Their approaches to the problem were diverse. Canut's 40 papers published in Madrid can accordingly be divided into five groups.

The first group of papers dealt with the systematic exploration of other materials, such as more chain-like molecular crystals (Canut and Amorós, 1957a, 1957b; Amorós and Canut, 1958; Banerjee et al., 1961), planar molecules (Acha et al., 1958; Amorós et al., 1961a), planar molecules in layers linked through hydrogen bridges (Alonso et al., 1958b), planar molecules kept together by Van der Waals forces (Acha et al., 1958; Canut and Amorós, 1961b), spherical molecules (Canut and Amorós, 1958b), and molecules with tridimensional H-bonds (Canut and Amorós, 1960; Valdés and Canut, 1961) (Figure 15.2).

In a second group of papers, new experimental techniques and instruments were described, which improved measurement precision both at low and high temperatures, and increased the range of data obtained. Amorós and Canut developed an experimental method they dubbed "systematic Laue," which consisted of sending monochromatic X-rays onto a crystal at varying angles covering the whole reciprocal space (Canut, 1957; Amorós et al., 1960, 1961b, 1962; Madurga and Canut, 1962). Consequently, in a third group of papers, new methods to interpret X-ray photographs were introduced (Canut, 1959; Canut and Amorós, 1958a).

In a fourth group of papers, different phenomena were discussed in relation to the thermal dynamics of crystals: polymorphism, namely, changes in crystal forms that maintained chemical composition under varying thermal conditions (Alonso et al., 1958a; Amorós et al., 1958a, b; Amorós and Canut, 1962), the inversion of diffuse scattering, that is, the decreasing intensity of diffuse scattering for increasing temperatures above a certain maximum temperature (T_{max}) (Valdés and Canut, 1961; Canut and Amorós, 1961a), and thermal dilation (Neira and Canut, 1962; Amorós et al., 1964a, b; Felix et al., 1964; Amorós et al., 1965).

The fifth group of papers tackled theoretical questions. Canut and Amorós' most important theoretical contribution from Madrid was the difference Fourier transform (DFT) method (Acha et al., 1960; Amorós et al., 1961). Based on quantum models of crystal dynamics, they defined a theoretical function that accounted for the difference in intensity of X-ray diffraction between a crystal with thermal agitation and the same ideal non-agitated sample, for varying temperatures and

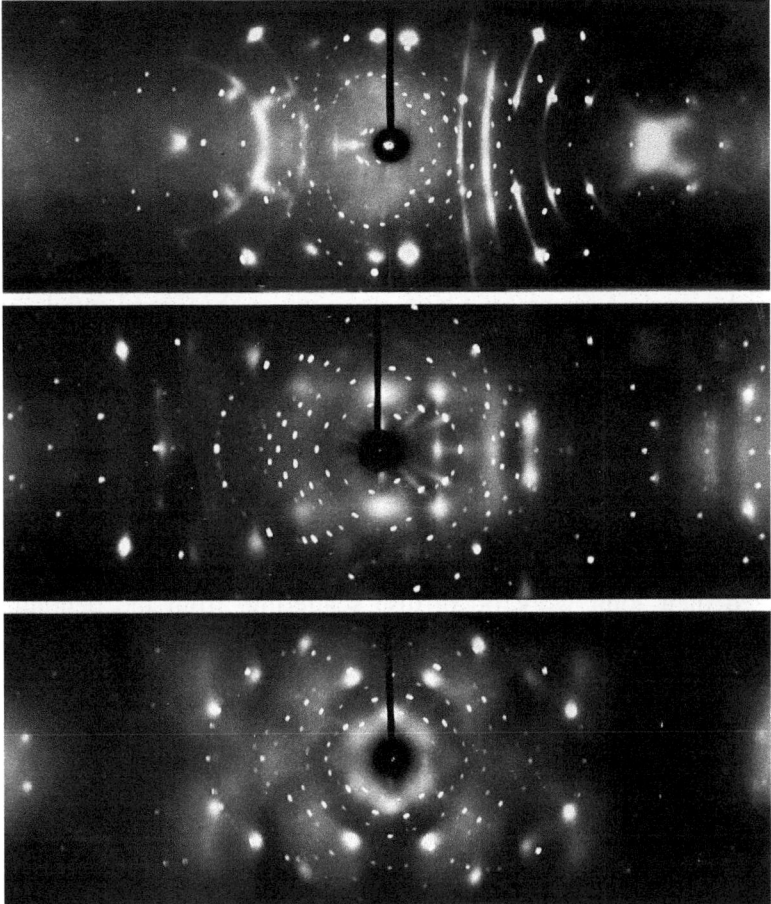

Figure 15.2 X-Ray diffuse scattering pattern through adipic acid (chain-like), pentaerythritol (planar), and hexamine (spherical). Source: Amorós. *Física del estado sólido: propiedades térmicas en relación con la estructura. Parte II*, research report for the Joan March Foundation, 1957–1958, p. 13.

orientations of the crystal. The method rested on the hypothesis that the continuous diffuse X-ray regions were due to the thermal agitation of molecules acting as independent rigid bodies vibrating at high frequencies.[21] Complex calculations led to a good agreement with experimental data. Another method was developed to overcome computational difficulties, the optical-DFT method. Macroscopic gratings emulating crystal planes of hypothetical agitated samples were constructed,

---

[21] Diffuse spots and streaks, in contrast to continuous diffuse X-ray regions, were not explained by DFTs. These other diffuse phenomena were thought to arise from the thermal-acoustic waves traveling through the crystal (Laval, 1954).

through which optical simulations of X-ray diffraction patterns were obtained (Canut et al., 1963).

A considerable number of PhD students from different disciplines (natural sciences, geology, physics), among them several women, co-authored Canut's and Amorós' publications during this period: Pilar Alonso, Antonio de Acha, Maria Luisa Valdés, M. Moreno (female), Félix Arrese, A. Carbonell (female), Eduardo Neira, and Moisés Gutiérrez. Carmen Madurga started as an undergraduate and Rajendra Banerjee was a visiting researcher from the University of Calcutta. Amorós, who worked for both CSIC and the university, could officially serve as their PhD advisor. Canut, who worked only for CSIC, could not. Nevertheless, no doubt she also contributed to their PhD research.

Canut and Amorós also published works in which they surveyed various topics related to their field of research. In these works, no additional author was involved. The most important of these is their 1965 book, *La difracción difusa de los cristales moleculares* (*The Diffuse Scattering of Molecular Crystals*), for which Amorós and Canut were awarded the 1963 Francisco Franco Prize in Sciences. This book summarized their joint work on diffuse scattering, from the beginning of the research program in Barcelona up to 1963 (Amorós and Canut, 1965).

Last but not least, from Madrid Canut and Amorós established a remarkable network of international collaborators, especially through Canut's research stays abroad. In 1958, she spent four months with Lonsdale working on the thermal dynamics of crystals.[22] In June–July 1962 as well as January–February 1963 she worked with Rolf Hosemann at the Fritz Haber Institute of the Max Planck Society, on molecular statistical disorder (Canut and Hosemann, 1964). In 1962, she stayed with Raymond Pepinsky at the Pennsylvania State University, learning about X-ray diffraction and order–disorder transitions in ferroelectric materials. During this time, she also attended international courses and conferences. From July 31 to August 18, 1961, she attended a course on fluctuations and irreversible processes in Newbattle (Scotland), and in 1960 Jean Laval, a crystallographer from the Collège de France, was invited to Madrid to give a course on lattice dynamics, in which Canut lectured about the DFT method.[23] During this period, she also participated in several international conferences.

Amorós, as head of the Department of Crystallography at CSIC, strongly supported Canut's stays abroad, as shown, for example, by a letter addressed to the President of the Division of Mathematical, Health and Natural Sciences of

---

[22] CSIC's grant notification for a research stay with Lonsdale, May 3, 1957, Canut's CSIC record. The CSIC record contains other official permissions to Canut to participate in international conferences and courses, and to carry out research stays abroad. See Note 1. Amorós summarizes Canut's research stays in a letter sent to the board of CSIC on May 18, 1963, Canut's CSIC record. See Note 1.

[23] Information about this course is found in Amorós' final report of the 1957–1958 project "Física del estado sólido: propiedades térmicas en relación con la estructura." Joan March Foundation's record. See Note 1.

CSIC, on May 18, 1963: "I am pleased to say that all of Dr. Canut's stays abroad have served to expand our work possibilities. . . . Experience has shown that only benefit for our department can be expected from her trip and stay."[24] This letter was written to support her nine-month research appointment at Southern Illinois University (SIU) as an associate professor, starting in September of 1963. In February 1964, Canut was appointed to a temporary full-time position at SIU, and in 1966 she became a permanent faculty member.[25]

Canut's and Amorós' international agendas were not independent. In 1962, when Canut went to Pennsylvania State University, Amorós was also there as guest professor at the Department of Physics and interim director of the Materials Science Research Center of the university. In 1965, one year before Canut's position at SIU became permanent, Amorós became Professor of Material Sciences in its Applied Science Unit (López-Acevedo and López Andrés, 2004, p. 86).

As Canut would later recall, she and Amorós were recruited together at SIU in 1963 because:

the School of Technology planned the creation of a new area of Materials Sciences in the Applied Science graduate program. For this purpose, the Dean of the School sought to attract to SIU a team capable of developing the new area. The team consisted of two scientists from UM. Dr. Canut-Amorós was one of them, and she was offered and accepted a lecturer position. Considering her publications and credentials, however, her appointment was changed from lecturer to associate professor by action of the Board of Trustees while she was still in Madrid.[26]

From 1965 onwards, Canut began to sign as "Maria Canut-Amorós," as Canut and Amorós married when they arrived at SIU in 1964.

15.6 Working in the Ivory Tower at Southern Illinois University

The "ivory tower" is the phrase Canut used to describe, in retrospect, her and Amorós' first five years with Amorós at SIU. In a lecture that she gave in 1999 at the conference in Barcelona, "*Seneca Falls: un siglo y medio de Movimiento Internacional de Mujeres*" ("Seneca Falls: A Century and a Half of the International Women's Movement"), she discussed her feminist activism in the US, and she recollected:

In 1964, when I arrived in the United States hired by Southern Illinois University (SIU), Carbondale campus, as an associate professor, my professional situation was enviable. The laboratories that my husband and I had established in the new building of the School of

---

[24] The translation is mine. See Note 1.
[25] Canut's CSIC service record, April 1967; CSIC's official permissions to Canut to stay at SIU, June 5, 1963, December 19, 1964, July 16, 1965, October 29, 1965, and December 5, 1966, CSIC's record. In 1966, CSIC granted Canut a leave of absence to continue with her SIU job permanently. See Note 1.
[26] "The Canut–Amoros Sex Discrimination Case Against SIU," 1972, p. 1, Canut's SIU record. See Note 1. For more information about the construction of the discipline of solid-state physics in the US, see Martin (2018).

Engineering and Technology, the calculation facilities provided in the Wham Computer Center, the instrumental equipment that we used for the first time, the new techniques that we were able to develop in those years, the contracts for fundamental research financed by the US Air Force Office of Scientific Research, the presentation of works at the meetings of the American Crystallographic Association, the publication in 1968 of the book *Molecular Crystals* by the accredited Wiley publishing house. It was all much more than I could have hoped to achieve in moving from the Higher Council for Scientific Research in Madrid to an American university. The SIU Board of Trustees, chaired by Delyte W. Morris, named me associate professor in March 1964, in 1966 I obtained a tenure, and in June 1970, I was promoted to professor. During the first five years, my husband and I felt so immersed in the scientific research we were carrying out, that we remained outside the social reality in which we lived, and without realizing it we were forging for ourselves the famous ivory tower, characteristic of scientists who live removed from the world and reality.[27]

From the Illinois ivory tower, Canut continued intensively with research on the thermal dynamics in crystals, analyzing novel effects and developing new approaches thanks to the new field of possibilities offered by the technological facilities at SIU. According to the university records, from 1965 to 1970, Canut was awarded three grants from the Air Force Office of Scientific Research to develop her research. In all of them, she was the principal investigator.[28] During the same period, Amorós was awarded three other grants as principal investigator.[29] In contrast to Madrid, in Illinois Canut and Amorós directed different – though clearly interrelated – projects.

Crystal dynamics continued as her main object of research. Teamwork also continued as the modus operandi. Amorós and Canut had different groups for each project, though some collaborators coincided. Because the projects were different yet interrelated, Amorós and Canut continued publishing and attending many conferences together. During this period, Canut directed nine master's theses and one PhD dissertation, published 20 scientific papers, was elected member of Sigma Xi (in 1967), and was granted, together with Amorós, the first Annual Research Recognition Award of the SIU Graduate Council for outstanding research. She also worked on two co-authored books (Amorós and Canut, 1968;

---

[27] Canut, "Una feminista en la Universidad norteamericana en los 70" (A feminist in the American university in the '70s), 1999. The translation is mine. The last corrected version of the lecture before printing is in the IME archive. See Note 1.

[28] "Disorder of Paraelectric and Antiferroelectric Phases of Ferroelectrics," AF-AFOSR-832–65, was the first grant. The second grant, AF-AFOSR-832–67 had several parts: "X-Ray Studies on Molecular Free or Hindered Rotation in Plastic Crystals," "A Set of Computing Programs Supporting the Selected Electron Shell Method," "Symmetry-Factor Maps: Computing Programs and Dichromatic Symmetry," and "STLPLT-Calcomp Plot of Crystallographic Projections of Laue Photographs." The third grant, AFOSR-832–67 was titled "X-Ray Diffraction Studies of Electron Polarization In Ferroelectrics." SIU record.

[29] "Study of Single Crystals under Thermal Gradients," AF-AFOSR-596–64 (1964–1966), AF-AFOSR-596–66 (1966–1968); "Temperature and Pressure Effects on Atomic Repulsion in Aerospace Electronic Materials," AF-AFOSR-68–1587 (1968–1970); "Study of Ferroelectrics," AF-AFOSR-69–0954 (1968). SIU record.

Amorós et al., 1975) and presented 10 papers to national and international scientific meetings.[30]

In Illinois, Canut's and Amorós' different specializations became apparent. The availability of the facilities of the Data Processing and Computing Center at SIU, in particular, marked a turning point in her research. Canut saw in high-speed computers and computer programming the seed of a major transformation in crystallography.

Since computer programming required an adequate mathematical approach to the physical problem being tackled, Canut also adopted new mathematical tools. She particularly aimed to calculate electron densities directly from the X-ray diffraction data of non-ideal crystals. To do so, Canut relied on Patterson functions, which are defined as the Fourier transforms of the experimental diffraction intensities and result in a set of interatomic vectors related to crystal atomic coordinates in real space.[31] Until then, Patterson functions had only been employed for crystals without any thermal agitation. Canut proposed two new Patterson functions, one describing the thermally agitated crystal, and the other the disordered crystal, in which ideal periodicity was disturbed. The calculation of these functions from experimental data was complex and required several interconnected computing programs, which she developed in the context of her first funded project (Canut, 1967; Amorós and Canut, 1967).

By relying on the calculation of Patterson functions, Canut, together with Amorós, developed the selected-electron-shell method to calculate electron densities from experimental X-ray data. The new mathematical approach allowed the separation of inner and outer electron contributions directly from experimental data in order to calculate electron densities from two different Patterson functions for inner and outer electron shells in non-ideal crystals. Canut's project, "X-Ray Studies on Molecular Free or Hindered Rotation in Plastic Crystals" implemented this idea in 24 FORTRAN programs – developed mainly by her – to be used in the laboratory to compare calculations with experimental data. The resulting papers were co-authored by both (Amorós and Canut-Amorós, 1968, 1969, 1970; Amorós et al., 1969; Canut-Amorós et al., 1970).

The development of computing programs became central to Canut's other Illinois projects: to determine electron densities and symmetry factors, to simulate ferroelectric phenomena, and to obtain crystallographic projections from experimental results.[32] Canut also paid attention to the creation of libraries of scientific

---

[30] See also "The Canut–Amoros Sex Discrimination Case Against SIU," 1972, pp. 1–2, Note 26.

[31] As a matter of fact, Patterson functions were what Rosalind Franklin used to derive the structure of DNA from her X-ray crystallographic photographs (unlike Watson and Crick, who were using modeling) (Osman Elkin, 2003).

[32] See Note 28 about Canut's Illinois funded projects.

computer programs for later reuse. The last part of Canut's last SIU project involved a systematic cataloging of these programs.

As computer programming grew increasingly central to her research, it became, itself, a new line of research. She decided to further develop her programming skills with the intention of offering courses on computer programming at SIU to attract new students and teach them new tools.[33] To that effect, she was granted a sabbatical leave from the end of 1970 to enroll in computer courses at Washington University in St. Louis. The leave was initially scheduled to last one and a half years, but various issues drew her back to SIU in 1971. To begin with, the arrival of a new dean, Thomas Jefferson, in 1969 had put the Applied Science Unit in limbo. She was informed that Jefferson had decided not to offer any of her proposed courses, which made her think that her new specialization was of no interest. In early 1971, Jefferson indeed eliminated the Applied Science Unit. Thus, in 1971, Canut started to come out of the ivory tower.

15.7 Out of the Ivory Tower: Canut as a Feminist in the US

Canut's exit from the ivory tower can be traced to two sets of interrelated circumstances. One of them was her disappointment with the new dean's decisions concerning her course proposals and the Applied Science Unit. Simultaneously, and in relation to that, Canut was becoming a feminist activist.

That activism did not come out of the blue. As she explained in her 1999 Barcelona conference lecture, her first contact with feminist movements occurred in 1969, when a group of SIU women students formed the Women's Liberation Front (WLF) to advocate for equal rights of men and women in using the campus library.[34] They requested that the opening hours of the library be the same for all, which had not been the case thus far. Canut and Amorós were the two SIU professors who, by their signatures, authorized this feminist group on the campus.

The creation and actions of this group must be understood in the context of the second feminist wave in the US (Rossiter, 1995). In 1963, Congress, under John F. Kennedy's presidency, approved the Equal Pay Act (EPA) to abolish wage discrimination on gender basis for equal jobs, except for administrative, executive, and professional positions. In 1967, President Lyndon B. Johnson signed Executive Order 11375, which prohibited discrimination of employment based on race, gender, color, religion, or national origin, this time without excluding teaching personnel.

---

[33] "The Canut–Amoros Sex Discrimination Case Against SIU," 1972. See Note 26. See also Note 1.
[34] Canut, "Una feminista en la Universidad norteamericana en los 70," 1999. See Note 27.

In parallel to these achievements, many feminist groups were created in defense of women's rights in the 1960s. The National Organization for Women (NOW), founded in 1966, inspired local feminist groups to mobilize throughout the country. The Women's Equity Action League (WEAL), funded in 1968, aimed at enforcing existing legislation against gender discrimination.

NOW and WEAL began to collaborate in 1970, making a call to all women university employees in the US to file claims against discriminatory practices by their colleges and universities, so that individual actions could become collective. The first claim, presented in 1970 to the Department of Health, Education, and Welfare (HEW), was brought against Harvard University. At the same time, a new action started for compliance with Executive Order 11375, promoted by the subcommittee on education of the Committee on Education and Labor: early hearings on education and employment of women were conducted with the goal of gathering information about their situation. The hearings were public.

Canut was not unaware of these developments. After Canut and Amorós' action in support of the constitution of WLF at SIU, Canut remained involved in feminist activities on campus. In 1969, she also participated in a roundtable on sexual politics at SIU, as a representative of the American Association of University Professors (AAUP), and she continued participating in WLF's actions at SIU.[35]

But it was in 1971 that she learned of the relevance of Executive Order 11375 and the above-mentioned hearings. She recalled reading an article in *Physics Today* about "Women in physics" (Brown et al., 1971). On April 19, 1971, Canut requested a copy of the hearings, which she received a month later. Upon reading the 1,261 pages from the subcommittee on education of the Committee on Education and Labor, she was surprised to learn that there was no report on the situation of women university employees at SIU. One of the reasons that SIU had not yet been included in the collective action against discrimination was, according to her, that local groups like WLF were not part of the national organizations, NOW and WEAL. Canut decided to become a member of both NOW and WEAL and hence contributed to breaking this isolation.

In May 1971, Canut filed a formal sex discrimination complaint with the Affirmative Action Office at SIU. This office had just opened in January 1971, and Canut became the first woman on the whole campus to file a claim. Among other things, she asserted that she had been denied the opportunity to teach summer courses, while her male colleagues were allowed to do so, even though she was more highly qualified. Moreover, she alleged gender-based salary differences.[36]

---

[35] Canut, "Una feminista en la Universidad norteamericana en los 70," 1999. See Note 27.

[36] Amorós supported Canut's claims. In a statement of Canut describing her complex relationship with the Affirmative Action Office at SIU, Canut declared, "I did this following the advice of my husband. He pointed out to me that the mere fact that Dean Jefferson had called him to offer two months research award instead of

The relationship with the Affirmative Action Office at SIU turned out to be difficult. To justify the university's decisions, the dean alleged to the Affirmative Action Office, orally, that Canut was not competent as a teacher. The Office accepted his arguments even though there was no written evidence for his assertion. Canut did not agree with the treatment. An atmosphere of hostility arose and, on May 14, she wrote a note to the dean expressing her intent to tender her resignation from his program, but not from the university as a whole. She expected to find a more suitable place elsewhere within SIU.[37]

Her resignation was interpreted by the dean, the vice chancellor, and the chancellor of SIU as Canut's resignation from the university in its entirety, despite her repeated clarifications to the contrary and the protests of the Academic Women for Equality association and of the AAUP against this interpretation by the administration and of the Board of Trustees.[38] The Faculty Council also asserted a lack of due process. Nonetheless, by June 1971, Canut was no longer a member of SIU. Without access to university facilities, she could no longer pursue her research. In response, she became more active as a feminist.

The integration of SIU into the national actions against sex discrimination accelerated in 1971. Information about salaries at SIU was made public in July 1971, and the wage inequality between men and women became plainly apparent; 170 women employed at SIU lodged a complaint with HEW against SIU, alleging gender discrimination, among them Canut. In August 1971, the Chicago Office of HEW sent a team to investigate the Canut–Amorós complaint. Meanwhile, new complaints from SIU flooded the Chicago office of HEW with new evidence of discrimination. In October 1971, a new chapter of NOW was organized at SIU, and Canut became Compliance Coordinator for Higher Education of NOW in Illinois. As a member of the federal network of NOW, she began to travel around the US giving talks and exchanging information about discrimination cases with other women professors and members of NOW.[39]

At the beginning of 1972, HEW personnel went to the SIU campus to continue the agency's investigation. In March 1972, HEW found evidence in favor of Canut in matters of disparate salaries, summer employment, sabbatical leave, and her supposed "resignation." Nevertheless, SIU did not follow HEW recommendations, and in May of the same year, Canut and Carolyn Weiss, a professor in the Department of Geography at SIU, brought their complaints to the State of Illinois

---

having offered one month to each of us, as he might have done even if he were short of funds, constituted a proof of sex discrimination." (Canut's statement on HEW visit at SIU, August 16, 1971, Canut's SIU record. See Note 1.)

[37] "The Canut–Amoros Sex Discrimination Case Against SIU," 1972. See Note 26.

[38] Canut statement on HEW visit at SIU on August 16, 1971, in David R. Derge, "A Summary of the Canut–Amoros Case," 1972, p. 14, SIU record. See Note 1.

[39] Canut, "Una feminista en la Universidad norteamericana en los 70," 1999. See Note 27.

for relief against the university. As early as 1971, the Illinois Fair Employment Practices Act had prohibited employment discrimination based on gender, and the Illinois Fair Employment Practices Commission (FEPC) enforced State law. Several other women professors filed similar actions. In December 1972, FEPC found prima facie evidence that sex discrimination existed at SIU in these two cases, and the two complainants asked for a public hearing. The hearings took place in Carbondale in 1973 and 1974, and the president, the deans, and other SIU executives were called to testify. After two years of investigations and negotiations, with no agreement between SIU and Canut, the case was reopened by HEW, which eventually ordered SIU to comply with the law. By that time, though, Canut was already living in Madrid.[40]

The dispute with SIU severely damaged Canut's scientific career. After 1971, Canut only published two scientific papers, both reviews of her previous work with Amorós (Amorós and Canut-Amorós, 1973a, 1973b). Canut seemed to feel that she had no other option than to reinvent herself. Her activism brought her face to face with matters of science policy. Before leaving the US for Spain in 1975, Canut served as program manager at the Washington office of the US–Spain and Latin America Cooperative Program, funded by the US National Science Foundation.[41] In 1975, Amorós returned to his former position as Professor of Crystallography and Mineralogy at UM (López-Acevedo and López, 2004), and Canut was reinstated to CSIC as a scientific investigator.[42] Franco also died in 1975 and the so-called Spanish Transition started, which lasted until the consolidation of the parliamentary system with the first democratic elections in 1982.

Immediately after Canut's return to CSIC, the Ministry of Science and Education appointed her to lead the Programa de Administración para el Desarrollo de Cooperación Educativa, Cultural y de Investigación entre España y los Estados Unidos (Administration Program for the Development of Education, Culture and Research Cooperation between Spain and the US), based on her previous experience in Washington. The program lasted until 1977,[43] after which Canut returned to CSIC as a scientific researcher, but not for long. Whereas Amorós combined research in mineralogy with teaching duties and supervision of PhD students, Canut reoriented her career toward data management through computation, in particular the informatization of libraries (Canut and Amorós, 1986). To some extent, she adapted the expertise she had acquired in computer programming

---

[40] Correspondence between HEW director and Canut, 1976, Canut's SIU record. See Note 1.
[41] Letter from the subsecretary of the Ministry of Education and Science to the CSIC's president, September 15, 1975, Canut's SIU record. See Note 1.
[42] CSIC's notification, October 16, 1975, CSIC record. See Note 1.
[43] See the epistolary exchange between the Ministry of Science and Education of Spain and the CSIC's secretary about Canut's appointment in the above-mentioned cooperation program, between 1976 and 1977, CSIC record. See Note 1.

acquired in Illinois to a new area of interest. One of the first projects in this direction was the compilation of the catalog of the Library of the Faculty of Geology of UM; its creation had been advocated by Amorós. Today, however, the library bears only his name.

From 1977 until Canut's retirement in 1989, she changed positions many times within CSIC, looking for a suitable place to cultivate her new skills. In 1980, she applied for a transfer to the Central Office of CSIC, to work on the organization of a central data bank of CSIC.[44] In January 1982, she was appointed as head of the cabinet of CSIC for almost two years. In 1984, she obtained a period of secondment at the Institut d'Estudis Baleàrics (Institute of Balearic Studies, a CSIC center) in Mallorca to reorganize bibliographic resources of all Balearic Islands using computer programs.[45] In 1985, she was appointed state employee at CSIC, according to the evolving administrative categories of the new democratic period.[46] In 1986, she enjoyed a research leave at the Institut Menorquí d'Estudis (Menorcan Institute for Studies), which she and Amorós had co-founded in 1985 with a group of other researchers. During this time, she carried out the informatization of the bibliographic sources of the academic *Revista de Menorca (Minorca Journal)*, which later led to the publication of *Anatomía de una cultura (Anatomy of a Culture)* (Canut and Amorós, 1989).

Upon Amorós' retirement in 1987, Canut applied again to CSIC for a transfer to the islands.[47] In 1988, she was finally appointed as the CSIC delegate to Universidad Nacional de Educación a Distancia (UNED, National University for Distance Education) in Menorca.[48] In UNED, one of her goals was the informatization of the UNED library and data. After a few months, she was dismissed from UNED for unclear reasons (Payeras, 1988), and in 1989 she finally retired.

From 1989 onwards, Canut's and Amorós' main intellectual engagement was with history; they published several historical books together (Figure 15.3). She never fully abandoned feminism. As mentioned before, she presented a paper in Barcelona on her experience as a US feminist activist at the 1999 conference, Seneca Falls: A Century and a Half of the International Women's Movement (see Section 15.6). Her last book was also on the history of a training school of female teachers in Mallorca, which pioneered the education of women teachers in the Balearic Islands (Canut and Amorós, 2000). Canut died in Mallorca in 2005, four years after Amorós.

---

[44] See CSIC's official notification of transfer, September 30, 1980, CSIC record. See Note 1.
[45] CSIC's notification of the secondment period, November 29, 1983, CSIC record. See Note 1.
[46] Title of state employee, December 17, 1985, CSIC record. See Note 1.
[47] CSIC's negative response to Canut's application for a transfer to the Institute for Advanced Studies in the Balearic Islands, February 16, 1987, CSIC record. See Note 1.
[48] CSIC's official notification to Canut, January 18, 1988, CSIC record. See Note 1.

Figure 15.3 Amorós and Canut together, probably in the 1980s. Source: IME. Unknown author.

15.8 Final Remarks

This story highlights one of the most prominent Spanish crystallographers, who started her scientific career at times when "women [at CSIC] (. . .) were nothing else than an accident" (Alcalá and Magallón, 2008, p. 160). Having started research in Barcelona, Canut and Amorós, partners in science and in life, built an X-ray crystallography research group in Madrid, which from 1956 to 1964 was extremely diverse in profile and productive in the number of publications, PhD students supervised, international collaborations and projects achieved, and research awards obtained. Given their international recognition, they were then recruited by SIU to further their research in material sciences. Both continued working in tandem until 1971, when the university penalized her feminist engagement and dismissed her.

Having different academic backgrounds – physics and natural sciences – Canut and Amorós contributed to a novel and interdisciplinary research field in the context of quantum theory and solid-state physics: the exploration of the thermal dynamics of crystals through X-ray diffraction. They did so from different perspectives: experimental, technical, methodological, and theoretical.

In 1962, Paul P. Ewald described Amorós as "the most active among the Spanish crystallographers." I would describe Canut in the same way, although she has by no means been remembered or recognized as fully as Amorós.

To be sure, Canut came from a local elite family in Menorca, yet her personal circumstances were not favorable for prospering in research in that period. As Ana Romero de Pablos and Fernando García Naharro have argued, having supportive mentors was very important for women during Francoist Spain to pursue research (Romero de Pablos, 2017; García Naharro, 2021). Canut was no exception. In Spain, Amorós always held higher positions than Canut.[49] Although he was therefore able to serve as a supportive mentor to her, such a relationship can also be a double-edged sword, since mentees have fewer chances to stand out than their mentors, especially in a society with gender bias that impacts the way scientists were and are viewed.

Another possible reason for the omission of Canut from the history of physics is the abrupt discontinuity in her scientific research due to her feminist activism when she was at the zenith of her career. The absence of a teaching position at the university in Spain – unlike Amorós – and the fact of holding only a CSIC research position, prevented her from officially supervising any PhD students. This difference became especially relevant upon her return to Spain in the 1970s. Whereas Amorós' PhD students after the 1970s have preserved his legacy, nothing similar has occurred with Canut's.

Altogether, this chapter seeks to highlight Canut's contribution to the development of X-ray crystallography, both in Spain and in the international context, and to understand her scientific contributions in relation to her personal circumstances and her struggle to prosper in research as a woman. My aim is also to vindicate her contribution as part of a collective effort, insofar as neither Amorós nor Canut would have obtained recognition without the other, and without the effort of their collaborators. Both Canut and Amorós were qualified scientists, working in tandem, and Canut's support was also fundamental for Amorós' scientific career. From this point of view, this case study is an invitation to further analyze the dynamics of scientific couples in the wake of quantum physics and specifically during Francoist Spain, when mentoring was so important for the minority of privileged women pursuing a scientific career.[50]

References

Alcalá, Paloma and Magallón, Carmen. 2008. Avances, ruptura y retrocesos: mujeres en las ciencias experimentales en España (1907–2005). In Romero de Pablos, Ana and

---

[49] In Illinois, she rapidly held the same university position as Amorós, although her salary did not catch up with his.

[50] The topic of scientific couples has been addressed in the historical literature, particularly in Pycior et al. (1996) and Lykknes et al. (2012). A systematic study in the case of Francoist Spain remains lacking.

Santesmases, Maria Jesús (eds.), *Cien años de política científica en España*. Bilbao: Fundación BBVA, pp. 141–169.

Acha, Antonio de, Canut, Maria Lluïsa, and Amorós, José Luis. 1958. Dinámica de redes en cristales moleculares: X, Difracción difusa térmica del naftaleno. *Boletín de la Real Sociedad Española de Historia Natural. Sección Geológica*, 56, 405–418.

Acha, Antonio de, Canut, Maria Lluïsa, and Amorós, José Luis. 1960. Interpretation of the extended continuous diffuse regions of X-ray thermal diffuse scattering of molecular crystals. *Zeitschrift für Kristallographie*, 114, 39–65. https://doi.org/10.1524/zkri.1960.114.16.39

Alonso, Pilar, Canut, Maria Lluïsa, and Amorós, José Luis. 1958a. Dinámica de redes en cristales iónicos: I, Difracción difusa de las formas polimorfas IV (entre −13° y 32°C) y III (entre 32° y 84°C) del nitrato amónico. *Boletín de la Real Sociedad Española de Historia Natural. Sección Geológica*, 56, 51–64.

Alonso, Pilar, Canut, Maria Lluïsa, and Amorós, José Luis. 1958b. Dinámica de redes en cristales moleculares: IX, Difracción difusa térmica en el pentaeritritol. *Boletín de la Real Sociedad Española de Historia Natural. Sección Geológica*, 56, 379–390.

Alva Rodríguez, Immaculada. 2016. Piedad de la Cierva: una sorprendente trayectoria profesional durante la Segunda República y el franquismo. *Arbor*, 192(779), a322. http://dx.doi.org/10.3989/arbor.2016.779n3012.

Amorós, José Luis and Canut, Maria Lluïsa. 1955. Desorden térmico. *Publicaciones del Departamento de Cristalografía y Mineralogía*, 2, 25–46.

Amorós, José Luis and Canut, Maria Lluïsa. 1958. Ondas térmicas en cristales en cadenas. *Boletín de la Real Sociedad Española de Historia Natural. Sección Geológica*, 56, 25–50.

Amorós, José Luis and Canut, Maria Lluïsa. 1962. El polimorfismo del nitrato amónico. *Boletín de la Real Sociedad Española de Historia Natural. Sección Geológica*, 60, 15–40.

Amorós, José Luis and Canut, Maria Lluïsa. 1965. *La difracción difusa de los cristales moleculares*. Madrid: CSIC.

Amorós, José Luis and Canut, Maria Lluïsa. 1968. *Molecular Crystals: Their Transforms and Diffuse Scattering*. New York: John Wiley and Sons.

Amorós, José Luis and Canut-Amorós, Maria Lluïsa. 1967. The difference Fourier-transform (DFT) method for direct crystal-structure determination. *Zeitschrift für Kristallographie*, 124, 262–274. https://doi.org/10.1524/zkri.1967.124.16.262

Amorós, José Luis and Canut-Amorós, Maria Lluïsa. 1968. On the effect of the electron shell structure of the atoms in X-ray diffraction. *Zeitschrift für Kristallographie*, 127, 5–20. https://doi.org/10.1524/zkri.1968.127.1-4.5

Amorós, José Luis and Canut-Amorós, Maria Lluïsa. 1969. Analysis of density distribution of the outer electrons in hexamine. *Zeitschrift für Kristallographie*, 129, 124–141. https://doi.org/10.1524/zkri.1969.129.16.124

Amorós, José Luis and Canut-Amorós, Maria Lluïsa. 1970. The selected-electron-shell method. Part I: theory. *Transactions of the Illinois State Academy of Science*, 63, 117–124.

Amorós, José Luis and Canut-Amorós, Maria Lluïsa. 1973a. Análisis de la densidad electrónica de los cristales. *Memorias de la Real Academia de Ciencias y Artes de Barcelona*, 42, 3–71.

Amorós, José Luis and Canut-Amorós, Maria Lluïsa. 1973b. Simetría dicromática en los mapas de factores de estructura. *Memorias de la Real Academia de Ciencias y Artes de Barcelona*, 42, 97–126.

Amorós, José Luis, Alonso, Pilar, and Canut, Maria Lluïsa. 1958a. Transformaciones polimórficas en monocristales. I. Formación de superestructura en la transición IV–V (−18°C) del nitrato amónico. *Boletín de la Real Sociedad Española de Historia Natural. Sección Geológica*, 56, 65–75.

Amorós, José Luis, Alonso, Pilar, and Canut, Maria Lluïsa. 1958b. Transformaciones polimórficas en monocristales. II. Transición IV–II (84°C) del nitrato amónico y forma metaestable II (55°C). *Boletín de la Real Sociedad Española de Historia Natural. Sección Geológica*, 56, 77–91.

Amorós, José Luis, Canut, Maria Lluïsa, Acha, Antonio de, Moreno, M., and Guibert, Miguel. 1960. Un difractómetro de rayos X para monocristal. *Revista de Ciencia Aplicada*, 14, 289–302.

Amorós, José Luis, Acha, Antonio de, and Canut, Maria Lluïsa. 1961a. L'agitation thermique dans les cristaux moléculaires: la diffusion des rayons X par l'acridine III. *Bulletin de la Société française de Minéralogie et de Cristallographie*, 84, 40–50. https://doi.org/10.3406/bulmi.1961.5448

Amorós, José Luis, Guibert, Miguel, Canut, Maria Lluïsa, and Arrese, Félix. 1961b. Un goniómetro Weissenberg vertical para altas y bajas temperaturas. *Revista de Ciencia Aplicada*, 15, 289–297.

Amorós, José Luis, Carbonell, A., and Canut, Maria Lluïsa. 1962. Un difractómetro de rayos X de dos limbos para altas y bajas temperaturas y su utilización en medidas absolutas de difracción difusa. *Revista de Ciencia Aplicada*, 16, 385–396.

Amorós, José Luis, Gutiérrez, Moisés, and Canut, Maria Lluïsa. 1964a. Sobre la dilatación térmica del nitrito sódico ferroeléctrico. *Boletín de la Real Sociedad Española de Historia Natural. Sección Geológica*, 62, 5–21.

Amorós, José Luis, Gutiérrez, Moisés, and Canut, Maria Lluïsa. 1964b. La dilatación térmica del nitro, NO_3K (sal de piedra). *Boletín de la Real Sociedad Española de Historia Natural. Sección Geológica*, 62, 23–39.

Amorós, José Luis, Canut, Maria Lluïsa, and Neira, Eduardo. 1965. Thermal expansion of β-succinic acid and α-adipic acid in relation to their crystal structures. *Proceedings of the Royal Society of London A*, 285, 370–381. https://doi.org/10.1098/rspa.1965.0110

Amorós, José Luis, Canut-Amorós, Maria Lluïsa, Montoto, Luis, and Singhabhandhu, Anamai. 1969. On the core and outer-electrons approximation to the X-ray scattering factor and electron density of atoms and ions. *Zeitschrift für Kristallographie*, 130, 241–253. https://doi.org/10.1524/zkri.1969.130.16.241

Amorós, José Luis, Buerger, Martin Julian, and Canut-Amorós, Maria Lluïsa. 1975. *The Laue Method*. New York: Academic Press.

Banerjee, Rajendra L., Canut, Maria Lluïsa, and Amorós, José Luis. 1961. X-Ray thermal diffuse scattering in azelaic and pimelic acids. *Indian Journal of Physics*, 35, 62–76.

Brown, Judith, Wolten, Gerard M., Daniel, Donald and Wood-Kyrala, Judith. 1971. Women in physics. *Physics Today*, 24, 9–11. https://doi.org/10.1063/1.3022869

Canut, Maria Lluïsa. 1955. *Dinámica de redes en cristales moleculares*. PhD Thesis. Barcelona: Universitat de Barcelona.

Canut, Maria Lluïsa. 1957. Equipo para el estudio experimental de la difusión de rayos X a baja temperatura. *Publicaciones del Departamento de Cristalografía y Mineralogía*, 3, 39–40.

Canut, Maria Lluïsa. 1959. Tables for conversion of cylindrical Laue patterns to stereographic projection. In Kasper, John S., and Lonsdale, Kathleen, *International Tables for X-Ray Crystallography. Vol II: Mathematical Tables*. Birmingham: Kynoch Press, pp. 168–174.

Canut, Maria Lluïsa and Amorós, José Luis. 1957a. Estudio acerca de la dinámica reticular en cristales moleculares: VI, Difracción difusa de los ácidos dicarboxílicos de la serie par: succínico y adípico. *Publicaciones del Departamento de Cristalografía y Mineralogía*, 3, 15–25.

Canut, Maria Lluïsa and Amorós, José Luis. 1957b. Estudio acerca de la dinámica reticular en cristales moleculares: VII, Difracción difusa de los ácidos dicarboxílicos de la serie impar: pimélico. *Publicaciones del Departamento de Cristalografía y Mineralogía*, 3, 27–31.

Canut, Maria Lluïsa and Amorós, José Luis. 1958a. Interpretación racional de los Lauediagramas. *Boletín de la Real Sociedad Española de Historia Natural. Sección Geológica*, 56, 15–24.

Canut, Maria Lluïsa and Amorós, José Luis. 1958b. Dinámica de redes en cristales moleculares: VIII, Difracción difusa térmica de la hexamina, $C_6H_{12}N_4$. *Boletín de la Real Sociedad Española de Historia Natural. Sección Geológica*, 56, 323–338.

Canut, Maria Lluïsa and Amorós, José Luis. 1960. Nota. Difracción difusa térmica del ácido oxálico dihidratado, $COOH-COOH,2H_2O$. *Boletín de la Real Sociedad Española de Historia Natural. Sección Geológica*, 58, 17–23.

Canut, Maria Lluïsa and Amorós, José Luis. 1961a. On the inversion temperature function of the first order (one phonon) scattering and the determination of Debye characteristic temperature. *Proceedings of the Physical Society of London*, 77, 712–720.

Canut, Maria Lluïsa and Amorós, José Luis. 1961b. Temperature dependence of the X-ray diffuse scattering of molecular crystals: naphthalene. *Journal of Physics and Chemistry of Solids*, 21, 146–155. https://doi.org/10.1016/0022-3697(61)90094-4

Canut, Maria Lluïsa and Amorós, José Luis. 1986. *Sistema de informatización de una biblioteca (SINBIB)*. Madrid: Editorial de la Universidad Complutense.

Canut, Maria Lluïsa and Amorós, José Luis. 1989. *Anatomía de una cultura: cien años de la Revista de Menorca*. Maó: Institut Menorquí d'Estudis.

Canut, Maria Lluïsa and Amorós, José Luis. 2000. *Maestras y libros, 1850–1912: la primera Normal femenina de Baleares*. Palma, Maó: Universitat de les Illes Baleares, Institut Menorquí d'Estudis.

Canut, Maria Lluïsa and Hosemann, Rolf. 1964. X-Ray analysis of ferroelectric domains in the paraelectric phase of $NaNO_2$. *Acta Crystallographica*, 17, 973–981. https://doi.org/10.1107/S0365110X64002523

Canut, Maria Lluïsa, Valdés, Luisa María, and Amorós, José Luis. 1963. Empleo de análogos ópticos en el estudio de la difracción difusa de rayos X por los cristales. *Revista de Ciencia Aplicada*, 17, 199–213.

Canut-Amorós, Maria Lluïsa. 1967. On the significance of two new Patterson functions for disordered crystals. *Zeitschrift für Kristallographie*, 124, 241–261. https://doi.org/10.1524/zkri.1967.124.16.241

Canut-Amorós, Maria Lluïsa, Casper, Tom, Walters, Craig, and Amorós, José Luis. 1970. The selected-electron-shell method. Part II: implementation. *Transactions of the Illinois State Academy of Science*, 63, 125–135.

Claret Miranda, Jaume. 2006. *El atroz desmoche. La destrucción de la Universidad española por el franquismo, 1936–1945*. Barcelona: Crítica.

CSIC. 1958. *Memoria de la Secretaria General del Consejo Superior de Investigaciones Científicas*. Madrid: Talleres Gráficos del CSIC.

CSIC. 1960. *Memoria de la Secretaria General del Consejo Superior de Investigaciones Científicas*. Madrid: Talleres Gráficos del CSIC.

Eckert, Michael. 2012. Disputed discovery: The beginnings of X-ray diffraction in crystals in 1912 and its repercussions. *Zeitschrift für Kristallographie*, 227, 27–35. https://doi .org/10.1107/s0108767311039985

Ewald, Paul Peter. 1962. *Fifty Years of X-Ray Diffraction. Dedicated to the International Union of Crystallography on the Occasion of the Commemoration Meeting in Munich July 1962*. Utrecht: International Union of Crystallography.

Félix, Ana, Maria Lluïsa, and Amorós, José Luis. 1964. Dilatación térmica y desorden de apilamiento en cristales moleculares: 2-2'-piridil. *Boletín de la Real Sociedad Española de Historia Natural. Sección Geológica*, 62, 187–197.

Font-Altaba, Manuel and Canut, Maria Lluïsa. 1954. Interpretación del Laue cilíndrico: falsilla de φ y ρ. *Publicaciones del Departamento de Cristalografía y Mineralogía*, 1, 151–156.

García Naharro, Fernando. 2021. In the land of men: Women in applied sciences at the CSIC. In Janué i Miret, Marició and Presas i Puig, Albert (eds.), *Science, Culture and National Identity in Fracoist Spain, 1939–1959*. Cham: Plagrave Macmillan, pp. 155–175.

Herran, Néstor and Roqué, Xavier. 2012. Los físicos en el primer franquismo: conocimiento, poder y memoria. In Roqué, Xavier (ed.), *La física en la dictadura. Físicos, cultura y poder en España (1939–1975)*. Bellaterra: Servei de Publicacions de la Universitat Autònoma de Barcelona, pp. 85–104.

Hoddeson, Lillian, Braun, Ernest, Teichmann, Jürgen, and Weart, Spencer (eds.). 1992. *Out of the Crystal Maze. Chapters from the History of Solid-State Physics*. New York: Oxford University Press.

Janué i Miret, Marició and Presas i Puig, Albert. (eds.). 2021. *Science, Culture and National Identity in Fracoist Spain, 1939–1959*. Cham: Plagrave Macmillan.

Laval, Jean. 1954. Théorie de la difusión des rayons X par les cristaux. I, II. *Journal de Physique et le Radium*, 15, 545–558, 657–666. https://doi.org/10.1051/jphys rad:01954001507-9054500 and https://doi.org/10.1051/jphysrad:01954001501065700

Lonsdale, Kathleen and Smith, Halley. 1941. An experimental study of diffuse X-ray reflexion by single crystals. *Proceedings of the Royal Society A*, 179(976), 8–50. https://doi.org/10.1098/rspa.1941.0075

López-Acevedo, Maria Victoria and López Andrés, Sol. 2004. José Luis Amorós Portolés: vida y obra. *Macla*, 1, 85–92. https://hdl.handle.net/20.500.14352/49837.

López Andrés, Sol and López-Acevedo, Maria Victoria. 2002. Recordando al Professor D. José Luis Amorós Portolés. 1920–2001. *Boletín de la Real Sociedad Española de Historia Natural. Sección Geológica*, 97(1–4), 139–158.

Lykknes, Annette, Opitz, Donald L., and van Tiggelen, Brigitte (eds.). 2012. *For Better or for Worse? Collaborative Couples in the Sciences*. Heidelberg: Birkhäuser.

Madurga Carmen, and Canut, Maria Lluïsa. 1962. Sobre un método nuevo de difracción de rayos X: el Laue girando. *Boletín de la Real Sociedad Española de Historia Natural. Sección Geológica*, 60, 41–46.

Magallón, Carmen. 2004. *Pioneras españolas en las ciencias. Las mujeres del Instituto Nacional de Física y Química*. Madrid: Consejo Superior de Investigaciones Científicas.

Malet, Antoni. 2008. Las primeras décadas del CSIC: investigación y ciencia para el franquismo. In Romero de Pablos, Ana and Santesmases, Maria Jesús (eds.), *Cien años de política científica en España*. Bilbao: Fundación BBVA, pp. 211–256.

Mañes, Xavier. 2005. *Determinación de estructuras cristalinas en España: inicios, desarrollo y consolidación (1912–1955)*. Master's Thesis. Barcelona: Universitat Autònoma de Barcelona.

Martin, Joseph D. 2018. *Solid State Insurrection. How the Science of Substance made American Physics Matter*. Pittsburgh, PA: University of Pittsburgh Press.

Méndez Vidal, Alfons and Hernández Andreu, Joan. 2011. *Trenta-cinc empresaris menorquins. Èxit individual i progres social*. Maó: Institut Menorquí d'Estudis.

Neira, Eduardo and Canut, Maria Lluïsa. 1962. Dilatación térmica del ácido oxálico dihidratado entre −185°C y 50°C y proceso de deshidratación. *Boletín de la Real Sociedad Española de Historia Natural. Sección Geológica*, 60, 195–212.

Osman Elkin, Lynne. 2003. Rosalind Franklin and the double helix. *Physics Today*, 56, 42–48. https://doi.org/10.1063/1.1570771

Payeras, Joan. 1988. Maria Lluïsa Canut. Una dama cibernética. *Menorca*, May 22.

Pycior, Helena M., Slack, Nancy G., and Abir-am, Pnina G. (eds.). 1996. *Creative Couples in the Sciences*. New Brunswick: Rutgers University Press.

Romero de Pablos, Ana. 2017. Mujeres científicas en la dictadura de Franco. Trayectorias investigadoras de Piedad de la Cierva y María Aránzazu Vigón. *Arenal*, 24(2), 319–348.

Romero de Pablos, Ana and Santesmases, Maria Jesús (eds.). 2008. *Cien años de política científica en España*. Bilbao: Fundación BBVA.

Roqué, Xavier (ed). 2012. *La física en la dictadura. Físicos, cultura y poder en España (1939–1975)*. Bellaterra: Servei de Publicacions de la Universitat Autònoma de Barcelona.

Rossiter, Margaret W. 1995. *Women Scientists in America. Before Affirmative Action 1940–1972*. Baltimore, MD: Johns Hopkins University Press.

Sánchez Ron, José Manuel. 2021. *El Consejo Superior de Investigaciones Científicas: una ventana al conocimiento (1939–2014)*. Madrid: Editorial CSIC.

Santesmases, María Jesús. 2000. *Mujeres científicas en España (1940–1970). Profesionalización y modernización social*. Madrid: Instituto de la Mujer.

Santesmases, María Jesús. 2006. Orígenes internacionales de la política científica. In Romero de Pablos, Ana and Santesmases, Maria Jesús (eds.), *Cien años de política científica en España*. Bilbao: Fundación BBVA, pp. 293–326.

Santesmases, Maria Jesús, Cabré i Pairet, Montserrat, and Ortiz Gómez, Teresa. 2017. Feminismos biográficos: aportaciones desde la historia de la ciencia. *Arenal*, 24(2), 379–404.

Valdés, Luisa Maria and Canut, Maria Lluïsa. 1961. El fenómeno de inversión en la difracción difusa del bencilo a baja temperatura. *Boletín de la Real Sociedad Española de Historia Natural. Sección Geológica*, 59, 41–48.

Velasco, Mariano, Canut, Maria Lluïsa, and Amorós, José Luis. 1954a. Estudios acerca de la dinámica reticular en cristales moleculares: I, Difracción difusa del ácido oxálico dihidratado. *Publicaciones del Departamento de Cristalografía y Mineralogía*, 1, 157–164.

Velasco, Mariano, Canut, Maria Lluïsa, and Amorós, José Luis. 1954b. Estudios acerca de la dinámica reticular en cristales moleculares: II, Difracción difusa del ácido adípico. *Publicaciones del Departamento de Cristalografía y Mineralogía*, 1, 165–171.

16

Ana María Cetto Kramis: Light in Quantum Mechanics and Open Science

MAR RIVERA COLOMER

16.1 Introduction

What better way to preserve the legacy of quantum physicists than through their own voices? This chapter, following a long tradition in the field, is based on an oral history interview with physicist Ana María Cetto Kramis, conducted in Mexico City in early September 2023 and now deposited in the Niels Bohr Archives of the American Institute of Physics (AIP) (Cetto, 2025). Her testimony contributes to completing, democratizing, and diversifying the record of the history of quantum physics, not only because she is a Mexican woman scientist but also because she is a living source for a minority interpretation of quantum mechanics, the stochastic approach. Sections 16.3 to 16.6 are based on her personal perspective on quantum mechanics and, more generally, on her understanding of science. Before diving into this material, however, Section 16.2 reviews the role that oral history has played and continues to play in the historiography of quantum mechanics.

16.2 Revisiting Oral History

At the outset of the 1960s, a few theoretical physicists and historians of science took note of the uniqueness of the moment in the history of physics. Thanks to the unprecedented disciplinary growth that physics was undergoing in the wake of the quantum revolution, the vast majority of physicists who had ever existed were alive at the time. But the quantum pioneers were quickly passing. As Gerald Holton then remarked, "in the sciences, we are now uniquely privileged to sit side-by-side with the giants on whose shoulders we stand" (Holton, 1961, p. 807). Yet, the literature in the history of physics still predominantly focused on the contributions of deceased male scientists (Wheeler, 1967). The novice historian of science,

Thanks to Ana María Cetto Kramis for her enthusiastic participation in four interview sessions. Special thanks to Adriana Minor for her close and valuable collaboration and warmth.

Thomas S. Kuhn (1922–1996), among others, recognized that although there were a growing number of deaths among the "giants," several key figures were still alive. Kuhn saw that interviewing the founders of quantum mechanics would allow historians to understand new aspects of the history of this theory, such as the personal scientific experiences of the work; it would also help to provide multiple perspectives on a single event, discern who built on the ideas of others, and clarify how new ideas related to older ones.[1] This vision would mark the inception of the project, Sources for History of Quantum Physics (SHQP).

SHQP began in August 1960, when a group of physicists and historians – including Kuhn – drafted a three-year project proposal. In early 1961, the initiative received funding from the National Science Foundation and was supervised by a specially appointed committee, the Joint Committee of the American Physical Society and the American Philosophical Society on the History of Theoretical Physics in the Twentieth Century. SHQP's goal was to find and preserve primary source materials about the history of quantum physics. To undertake this work, a nucleus staff was established, consisting of four young individuals with distinct roles: Kuhn, John L. Heilbron (1934−2023), Paul Forman, and Lini Allen. In total, counting additional staff members who handled routine research and editing tasks, 22 persons collaborated on the project, with women mostly serving as secretaries and research assistants. Karen Kivett, one of the longstanding research assistants of the project, distinguished herself for her skill in performing all types of archiving tasks and for taking on roles of responsibility, although little else is known about her (Kuhn et al., 1967).

Seizing the opportunity to consult *living sources*, SHQP collected primary source materials such as manuscripts, correspondence, lectures, seminars, personal notes, photographs, and other unpublished material, and created microfilms of some of these documents (te Heesen, 2020, p. 91). SHQP also adopted the oral history interview method, which, at the time, had not commonly been used in historiography and was a novelty in the quickly professionalizing field of history of science. In the early 1960s, the expansion of new media posed a threat to the collection of written sources, and oral history came to be seen as a new, alternative practice that mitigated this issue (te Heesen, 2020, p. 93). Oral history allowed the production of new sources, consisting of the memories of scientists, conveyed through their words, and the preservation of those memories as heritage. In that context, memory was perceived to be more efficient if elicited through oral processes. But, as the past president of the US Oral History Association, Linda Shopes, noted, "An interview for an oral history is inevitably an act of memory [. . .] molded

---

[1] Additionally, as noted by Anke te Heesen (2022), for Kuhn this project was an opportunity to test the ideas about the dynamics of scientific knowledge that he was expounding in his new book, *The Structure of Scientific Revolutions* (1962).

both by the moment of narration and by the history being told; [...] it is an expression of identity, consciousness, and culture" (Shopes, 2002, pp. 6–7). In oral history's procedure, the experience of the questioner – in this case, historians of science – also plays a significant role. SHQP further marks a shift in the historiography of science, being the first time historians could not only rely on existing sources but were actively generating them (te Heesen, 2021, p. 169). The outcome of the project was the Archive for the History of Quantum Physics, which originally was deposited at the Library of the American Philosophical Society in Philadelphia, the Library of the University of California, Berkeley, and the Universitets Institut for Teoretisk Fysik (University Institute for Theoretical Physics) – now Niels Bohr Institutet – in Copenhagen.

While oral history has been widely regarded as a valuable opportunity for sourcing knowledge that often eludes traditional historical records, such as everyday life, this potential was not fully leveraged within SHQP. SHQP instead aligned with the trend of the 1940s–1950s, when oral history primarily focused on elite interviews, involving conversations with individuals from the political and social elites to augment existing records (Chaelton et al., 2007; te Heesen, 2021, p. 163). Today, oral history also involves interviewing non-elite participants in past events to incorporate outlooks from marginalized groups who would otherwise be *hidden* from historical narratives (Perks and Thomson, 1998, p. ix). If we examine the oral history component of SHQP from this standpoint, we can see that it does not lend a voice to those who were silenced previously, such as women, individuals without a research position (students, assistants, partners ...), or participants from regions other than Europe or the US. It is a type of oral history that reaffirms the elitist image of scientific progress fashioned by traditional historiography of science.

A quantitative analysis of the SQHP oral histories highlights several interesting features concerning the participation of underrepresented groups in the history of quantum physics. For the contribution of women, the information is summarized in Table 16.1. The dearth of women among the over 100 interviewees reflects in part the historical context. In the early twentieth century, women faced significant obstacles to access universities, let alone to hold positions of responsibility within them; they were mostly considered *assistants*. As a result, far fewer women than men were formally involved in the development of quantum physics.

Part of the Archive for the History of Quantum Physics (AHQP), which was one of the outcomes of the SHQP project, is a catalog of authors who either participated in oral history interviews or contributed source material. Of the 279 physicists named in this catalog, only four are women (Irène Joliot-Curie, Maria Goeppert Mayer, Lise Meitner, and Hendrika Johanna van Leeuwen), accounting for just a little over 1% of the total number of physicists. Of these physicists, only Goeppert Mayer and Meitner were interviewed, even though Van Leeuwen was alive at that

Table 16.1 *Women interviewed for the SHQP project*

Name	Physicist	Oral history	Author catalog	Comments
Margrethe Nørlund Bohr (1890–1984)		X		Niels Bohr's wife
Irene Joliot-Curie (1897–1956)	X		X	
Elaine Charlton (1899–1983)		X		Douglas Hartree's wife *Unrecorded*
Hertha Sponer (1895–1968)	X	X		James Franck's wife
Hedwig Kohn (1887–1964)	X	X		*Unrecorded*
Maria Goeppert Mayer (1906–72)	X	X	X	
Lise Meitner (1878–1968)	X	X	X	
Betty Schultz (1898–1980)		X		Niels Bohr's secretary
Annemarie Bertel Schrödinger (1896–1965)		X		Erwin Schrödinger's wife
H. Johanna van Leeuwen (1887–1974)	X		X	

time (see also Chapter 2). If we focus specifically on the oral history archive, 102 individuals participated in interviews (some in group interviews) between 1961 and 1964. Among them, eight were women, representing less than 8% of the group. Of these eight women, four, as noted above, were physicists (Hertha Sponer, Hedwig Kohn, Goeppert Mayer, and Meitner), three were scientists' wives (Margrethe Nørlund Bohr, Elaine Charlton – the wife of Douglas Rayner Hartree – and Annemarie Bertel Schrödinger), and one was a secretary (Betty Schultz). Why do these women not appear in the catalog of authors? The wives of scientists, along with Sponer (a physicist and the wife of physicist James Franck, with whom she was interviewed jointly; see Chapter 3), are listed under their husbands' names.[2] The same applies to Schultz, Niels Bohr's secretary; her name does not appear in the catalog because her interview is subordinated to his. (One may wonder whether interviews with Nørlund Bohr and Schultz would have been conducted at all had Niels Bohr not passed away on November 18, 1962, in the midst of the series of interviews conducted by Kuhn.[3]) Additionally, seven interviews are listed as

[2] The interview of Sponer and Franck was conducted by Kuhn and Goeppert Mayer in July 1962.
[3] Schultz's interview took place more than a year later and was the only interview conducted by Forman, in which Kuhn did not participate. Margrethe Nørlund Bohr's interview was joint with her son Aage Bohr and his colleague, Léon Rosenfeld, in January 1963.

"unrecorded," two of which belong to women, Kohn and Charlton (see Table 16.1). These two interviews were conducted by Kuhn in March 1963 and June 1962, respectively, but no information is available as to why they were not recorded.

Thus, the project only considered Goeppert Mayer and Meitner, the only women who participated in an oral history interview and are listed as authors, as significant contributors to the legacy of quantum physics. Seven of the scientists featured in this volume belong to the SHQP generation, but only two, Van Leeuwen and Sponer, are referenced in AHQP, and none of the seven was interviewed individually.

Moreover, if we delve into the nationalities and the workplaces of the physicists listed in the catalog, out of 305 individuals (including those who are not in the author list), an overwhelming majority, approximately 75%, come from Europe, particularly from Central and Northern European countries, especially Germany, Denmark, and the Netherlands. Following this dominant group are US contributors, constituting more than 15% of the total participants. The remaining participants, who represent a minority of fewer than 30 physicists, amounting to less than 10% of the total, originated from countries beyond Europe and the US. These regions, ranked by participation, are the Soviet Union, Japan, Canada, and India. It's worth noting that these scientists significantly shaped their academic careers in Europe and the US. If we look closely at the interviews, a mere two physicists from non-European and non-US origins (although they worked at some point in these countries) took part in oral history interviews, Paul Sophus Epstein and Hideki Yukawa.[4] This minuscule representation accounts for less than 1% of the interviews. Since groundbreaking discoveries in modern physics in the time period covered by the project (1898–1933) predominantly unfolded in Europe and the US, it was inevitable that the SHQP project would exhibit a geographic bias. Yet, this bias also set the tone for the project that amplified and continued to amplify the voices of already-established male physicists and neglected the diverse threads that have woven the tapestry of quantum physics over time.

Nowadays, oral history has gained further methodological grounding. Since the 1990s, feminist historians have reignited the discussion about this type of historiography by suggesting that "the traditional oral history methodology overlooked the basic insight that grew out of the women's liberation movement" (Berger Gluch and Patai, 1991, p. 1). Such insight includes the understanding that personal experiences have political implications and the recognition of the intrinsic value of women's lived experiences, underscoring the need for their documentation. As gender historian Joan Sangster argues, "in using oral history as a means of exploring memory construction (where both researcher and narrator participate), then, it is

---

[4] The AHQP oral histories also include a recording of a 1932 lecture by Ernest Rutherford, who was born in New Zealand.

crucial that we ask how gender, race and class, as structural and ideological relations, have shaped the construction of historical memory" (Sangster, 1998, p. 92). This discussion has sparked a movement toward documenting *history from below*, valuing the standpoints of ordinary people as credible sources of knowledge.

If we analyze the current archive of the AIP, where the Niels Bohr Library & Archives is located, unfortunately, the situation has not changed much over the last 60 years. When searching for "quantum" and "oral history interviews," 193 items appear, corresponding to a total of 189 interviewed individuals (some were interviewed on multiple occasions by different people, and others participated in group interviews). Of these interviews, more than 50% are from the SHQP project. Overall there are 16 women, representing approximately 8% of the total. If one subtracts the eight women who participated in the Kuhn group's project, these eight women left are: Bertha Swirless Jefferys (1903–1999), Aida El-Kahdra, Janice Steckel, Laura H. Greene, Lene Hau, Mary Jo Nye, Myriam Sarachik (1933–2021), and Renata Wentzcovitch. All of them were interviewed in 2020, except Lady Jefferys, who was interviewed in 1997. A search for "women in physics" and "oral history interviews" turns up only 28 items, the vast majority also from 2020 onwards.[5] The same happens when we look at the origin of the physicists interviewed, as no country or language that was not already present in SHQP has been incorporated; mainly US voices have been added.

While oral history has the potential to recover voices that have been historically silenced in historiography, in the oral history of quantum physics this has not been the case thus far. Many historical narratives focus their attention on a small number of physicists whom they treat as representative figures and scientific heroes, with a near total disregard for the rest of the characters; this leaves a distorted and simplistic view of the field (de la Peña, 2003, p. 22). For years, the historiography of physics has tended to exclude many individuals, contributions, ideas, and context. As historian of science Naomi Oreskes has argued, the complementary images of objectivity and heroism in the history of science, by glorifying extraordinary individuals, contributed to the erasure of other actors (Oreskes, 1996, p. 103). The above data about the SHQP and the AIP oral history archives provides a clear picture of the lack of interest historians of physics have shown for this issue, at least until a few years ago.

By going beyond these limitations, oral history could help compose more inclusive pictures of the history of quantum physics. In particular, by capturing the voices of those who would otherwise have remained historiographically unheard, oral testimonies can provide distinctive and thought-provoking avenues

---

[5] The data shown were up to date as of October 2023. The list of oral histories has recently been extended, including new and more diverse scientists, as for example Janice Button-Shafer, Nergis Mavalvala, and Regina Maruyama (May 2024).

Figure 16.1 Ana María Cetto Kramis in her office at the Instituto de Física at UNAM (IFUNAM). Photograph taken by the author, September 2023.

for accessing essential insights into women scientists' diverse experiences and perspectives as well as into non-mainstream scientific developments. The interview of Ana María Cetto Kramis, photographed in her office in Figure 16.1, is a step in this direction.

16.3 Ana María Cetto Kramis: Her Beginnings

Ana María Cetto Kramis was born in 1946 in Mexico City, the second daughter of intellectual and open-minded refugees from the Nazi dictatorship. She grew up in a secluded yet unbound natural and social environment, among the lava fields and early inhabitants of El Pedregal, south of Mexico City.[6] Her mother, Gertrud Catarina Kramis, was a Swiss designer with a libertarian spirit, and her father, Max Ludwig Cetto Day, was a renowned German architect and intellectual. Her

---

[6] Jardines del Pedregal stands as a distinctive urban development within Mexico City, characterized by its avant-garde landscape architectural approach. During its early years, it was a meeting place for national and international intellectuals. Her parents were the founders of the development, building the first house in this privileged neighborhood.

parents showed her and her two sisters, Verónica and Bettina, how to reason, and they fostered a spirit of freedom and rebellion. In addition, having German as a second language provided her with a link with and access to Europe. Cetto has always expressed gratitude to her country, Mexico, for welcoming her family, and for allowing her to grow up in a healthy and privileged environment.

She completed her primary education at the German School Alexander von Humboldt of Mexico City, where she experienced an oppressive atmosphere. For her secondary school, she requested to transfer to a public high school, just like her older sister Verónica had done. She attended Secondary School N° 8, Tomás Garrigue Masaryk in San Pedro de los Pinos, where she found a more congenial environment, characterized by diversity, solidarity, and cooperation. However, at her father's insistence, she ultimately completed her two-year preparatory education at the German School once again. During this period, physics classes were conducted in German, which presented some challenges for her, as she felt it was not the same as thinking in Spanish. The linguistic breadth nevertheless allowed her to read the German magazines that were in her home, something quite unusual in Mexico at that time (Cetto, 2025).

Cetto's earliest memories of science were directly related to living in the secluded natural environment of El Pedregal. As she explained in the interview, her father used to make her think about the nature and behavior of materials. "These rocks have been here for centuries," he would say to her. "You perceive them as unchanging, but they are actually made up of tiny moving particles." Or "the materials seem solid, but are practically empty because the particles that move take up almost no space" (Cetto, 2025).[7] Although these ideas initially puzzled her, she had a strong inclination to get to the bottom of things. Her first contact with quantum science – as an 11-year-old – came from reading an article about Heisenberg's uncertainty principle in one of the German journals at home. But she didn't like it; something in her was already *protesting* (Cetto, 2025).

Her parents maintained close relationships with numerous intellectuals, both nationally and internationally, whom they would receive at home. Among them were the physicist and family friend, Manuel Sandoval Vallarta (1899–1977), and his wife, María Luisa Margáin Gleason. They encouraged Cetto to attend the seminars organized by Vallarta on Fridays at the Zócalo (in the city center), where the physicist gave lectures and regularly invited other physicists. Cetto did not understand all of the things that were said, but she nevertheless vividly remembers two lectures, one about cosmic rays and another given in 1962 by Robert Oppenheimer, who discussed the dangers of nuclear weapons and the

---

[7] The interview was conducted in Spanish, but for a better understanding the author of the chapter has translated all direct quotes as literally as possible.

need for greater understanding among cultures. She also remembers "Don Manuel" (as she refers to Vallarta) advising her to study physics even if she were to pursue a different career later on (Cetto, 2025).

In the early 1950s, the new Ciudad Universitaria (CU, University City) was designed, built, and inaugurated in the southern part of Mexico City (Minor García, 2021, p. 20). There, intellectual roots from the Enlightenment, liberalism, scientism, and positivism intertwined with the vibrant cultural milieu of the Mexican revolution, characterized by cultural nationalism and the expansion of new institutions to meet growing productive demands (Cepeda Flores, 2006). Cetto went there so often with her family that it felt like her second home.[8] For her, it was natural to pursue her studies at the Universidad Nacional Autónoma de México (UNAM, National Autonomous University of Mexico) in the new CU, but she wasn't quite certain which field to choose: geology, biology, and physics were all on her mind. Surprisingly, neither conversations with Vallarta nor attending his seminars proved as decisive as the information she found in the undergraduate program catalog (Cetto, 2025). It was the specific courses listed there that ultimately led her to theoretical physics.[9]

Cetto began her bachelor's degree in physics in 1963, when UNAM was not yet a high-enrollment university, and it did not segregate students by social class through the credit system or space allocation, as it does nowadays (Cetto, 2025). However, most students came from privileged backgrounds. Students took year-long courses as a cohort to which they would belong for their whole trajectory. Cetto (2025) recalls that, although there were initially a few women in her class, most of them unfortunately dropped out. Her strong sense of camaraderie instead developed within a subgroup of her class, a circle of physics enthusiasts who gathered almost every afternoon to do homework, engage in discussions, and more. A clear atmosphere of solidarity emerged among them.

At home, political topics were always openly discussed, and her sister Verónica actively participated in various social movements. As early as her secondary school years, Cetto also considered herself politicized (Cetto, 2025). She read left-wing newspapers and contributed to the school newspaper; she remembers writing an article about Sandinista ideology. Once she entered university, she joined *Nuevo Grupo* (New Group). The group primarily engaged in theoretical political discussions, but also occasionally participated in demonstrations and in drawing protest banners.[10]

She greatly appreciated her professors, all of whom were men, and some of whom later became her colleagues. Cetto (2025) highlights those who influenced her the most, including Juan Manuel Lozano Mejía, Francisco Medina Nicolau, and

---

[8] She even recalls helping to hang small stones in the iconic mural of the Central Library (Cetto, 2025).

[9] At that time, physics was still divided into two bachelor's degrees: theoretical physics and experimental physics.

[10] *Nuevo Grupo* represented a broad and pluralistic left. Other physics students participated in it, and it was notable for calling for the merging of the theoretical and experimental physics streams.

Fernando Alba Andrade (1919–2021), the first physics PhD recipient trained entirely in Mexico (Domínguez Martínez, 1999; Plasencia Gaspar et al., 2010). There was also Juan de Oyarzábal Orueta, who taught a first-year course titled "Selected topics in contemporary physics," her first formal encounter with quantum mechanics. She recalls that "the standard interpretation of this theory made [her] hair stand on end," and she voiced her dissent to her professor: "This isn't physics. It's merely an interpretation!" (Cetto, 2025). The feeling was similar when she took the quantum mechanics course with Fernando Prieto Calderón, whom she believed offered a very conventional perspective and to whom she insistently expressed her concerns. Despite her clear discontent and her complaints, she obtained top scores in both courses. At UNAM in the 1950s – unlike in the US at that time – being rebellious, outspoken, and critical was valued. Even in the Faculty of Science, these traits were common among students. In her pursuit of intellectual alternatives, she searched and read publications of Mir Publishers, from the Soviet Union. She anticipated a different viewpoint, but to her surprise, their content proved to be equally orthodox (Cetto, 2025).

Cetto completed her undergraduate thesis on non-radiative transitions in diatomic molecules under the supervision of Luis Estrada Martínez, a nuclear physicist, professor of electromagnetism, and a pioneer in science communication in Mexico. While writing her thesis on the tenth floor of the old Science Tower at UNAM, the young physicist enjoyed being treated as a colleague, even though she had no established rank and was a young woman (Cetto, 2025). In particular, she appreciated the support of Alfonso Mondragón Ballesteros and Tomás Brody Spitz. It was during this period that Luis Fernando de la Peña Auerbach, who had just returned to UNAM in 1965 after receiving a doctorate in particle physics and field theory at the M. V. Lomonosov Moscow State University in the Soviet Union, offered her a desk in his office. This was her first meeting with the man, 15 years her senior, who would become her collaborator and husband for over 60 years.

At the age of 21, she went to Harvard University to pursue graduate studies in biophysics, a program jointly administered by the physics department and the medical school. The institution's prestigious reputation was alluring. Motivated to find a complete biophysical explanation of photosynthesis starting from quantum mechanics, she enrolled in courses spanning biology, chemistry, crystallography, and beyond. She immersed herself in her studies full time, making only one exception to join a protest against the Vietnam War (Cetto, 2025). She didn't have any female classmates and had but one female professor, Carolyn Cohen (1929–2017), who became her thesis advisor. Cetto thought she was assigned to Cohen because they were both women (Cetto, 2025). Because Cohen felt discriminated against for being Jewish, however, she soon moved to Brandeis University and co-founded the Structural Biology Laboratory there. Both Cohen and her

Cornell University collaborator Roderic K. Clayton (1922–2011), a photosynthesis guru, offered Cetto the opportunity to continue her PhD studies with them. Cetto nevertheless opted to complete her exams and to defend a master's thesis titled *The Fate of Energy in the Light Phase of Photosynthesis* at Harvard before returning to Mexico, where her family and de la Peña, whom she had married the previous year, awaited her.

She was back in Mexico City in 1968, in time for the Olympic Games. It was also the year of the big *movimiento estudiantil* (Mexican student movement), in which the UNAM Faculty of Science played a significant role. Cetto and de la Peña participated in demonstrations and protests. But the protests were abruptly and violently repressed. The army and the police assaulted UNAM with a grenade explosion (known as *bazucazo*), and they detained, assassinated, and injured numerous individuals from the educational community (Rodríguez Kuri, 2003, p. 199). In the midst of these events, Cetto prepared and passed the comprehensive knowledge exams required for a second master's degree so as to qualify for a PhD program. Her intention was to continue in biophysics. In parallel to her own studies, she supervised undergraduate students doing research on photosynthesis physics. In the early 1970s, she even submitted a proposal to the Consejo Nacional de Ciencia y Tecnología (Conacyt, National Council of Science and Technology) to establish a biophysics laboratory. Mexico, however, did not yet have a research group on that theme; the proposal was rejected because it was not considered a "national priority" (Cetto, 2025).

Following this disappointment, Cetto refocused her attention on the foundations of quantum mechanics. She saw this research area as closely related to the topic of photosynthesis because both sought to answer questions regarding the interaction of light with matter, a topic that fascinated her. It was also a pursuit that she had harbored for some time, perhaps stemming from the discontent of her undergraduate years. Furthermore, it was a subject on which she had been working with de la Peña since 1966. Her doctoral thesis was conducted under his supervision. While their collaboration offered a great opportunity for learning and understanding, it wasn't always easy due to the hierarchical teacher–student dynamic – rather than fostering a collegial relationship – that de la Peña had established. De la Peña had always seen himself as the teacher and the top-notch mentor. This is a frequently observed phenomenon in academia, where women encounter explicit and concealed forms of sexism, like paternalism. Female students are frequently perceived by male mentors and colleagues as dependent on mentors for demonstrating proficiency, embodying the concept of "professional wives" (Grant et al., 1993, p. 83; Reskin, 1978, p. 1241). It is remarkable, however, that in this case, even though they went on to become husband and wife, Cetto perceived their relationship as more hierarchical than the one she had with her undergraduate professors. This suggests that the patriarchal gender norms were more persistent and prevalent in private life than in academic life, which affected their supervisor–student relationship. Cetto remembers how he would always say, "*Pasa al pizarrón!*" (go to the board!) until one day she rebelled and said "Enough! I won't do it anymore; now it's your turn" (Cetto, 2025). From that moment, which Cetto has mentioned on several occasions (in de la Peña's biography (Cetto, 2006), interviews, etc.), they began to work as colleagues, as a research team. In 1971, Cetto became the first woman to earn a PhD in physics in Mexico. She defended her thesis (with honors) titled *Investigaciones sobre una teoría estocástica de la mecánica cuántica* (Investigations on a stochastics theory of quantum mechanics). In it, she aimed not only to demonstrate that a coherent stochastic interpretation of quantum mechanics was possible but also that it was advantageous, both in terms of the conceptual benefits it offered and of the new possibilities it unveiled.

16.4 Collective and Dissident Quantum Mechanics?

Motivation for the stochastic theory of quantum mechanics stemmed from a very basic metaphysical principle, a realist philosophy that sparked an intuition – similar to that of Albert Einstein (1879–1955) – that quantum theory was incomplete, that there was something deeper beneath what the existing theory conveyed.

Formal education in quantum mechanics in Mexico began in the late 1930s with the return of Vallarta's students from the US (Minor García, 2019).[11] The 1950s–1970s were a period of rapid growth for the Faculty of Science, especially the Instituto de Física at UNAM (IFUNAM, Institute of Physics). However, it's important to note that although a course in quantum mechanics had been taught since 1945, only in 1956 did it become a mandatory part of the undergraduate physics program. It was during this period that the first notable research on this theory took place at UNAM, conducted by Mexican physicists who had gone to the US to pursue doctorates (de la Peña, 2003). In other words, systematic study and research in quantum mechanics in Mexico began about 20 years after the theory was established in Europe.

During that period, there also were significant advances in the establishment of public institutions, thus creating some opportunities for Cetto. For example, the Faculty of Science became the most important institution of science in the country in the 1960s (Ramos Lara, 2003).[12] As a result, in 1975, just four years after earning

---

[11] Vallarta was one of the first Mexicans to conduct research in quantum mechanics. He defended his doctoral thesis on Bohr's atomic model at the Massachusetts Institute of Technology (MIT) in 1924; then, he received a two-year scholarship to study in Germany, where his stay coincided with key figures of the time. After he obtained a professorship at MIT, younger Mexican physicists pursued their doctorates there under his supervision.

[12] Even today, science continues to be predominantly an academic and university-based activity in Mexico. This situation results in limited innovation and technological development and contributes to maintaining Mexico's technological dependence (Ramos Lara, 2003).

her doctorate, Cetto secured a professorship at IFUNAM. She was teaching and researching when, in 1973, her only daughter, Carolina de la Peña Cetto (1973–2018), was born. Cetto fondly remembers this time and doesn't recall work–family balance being an issue since they always had help from household workers, given their privileged socio-economic status. She does, however, recall her daughter complaining that they couldn't stop talking about the Schrödinger equation over breakfast (Cetto, 2025). Although Cetto does not explicitly acknowledge imbalance in the burden of household responsibilities, this does not mean that it did not exist. For example, if we consider the numerous interviews conducted with both spouses, we note that Cetto frequently talks about household management and how she handled tasks such as caring for the garden and occasionally cooking (Canal 22, 2019; Ramos, 2020). De la Peña, by contrast, offers no corresponding reference.

Interest in science as a collaborative practice has been part of a larger historical turn toward understanding science as a socially embedded activity: laboratories, networks, research schools, and scientific teams. Moreover, historians have also begun to view science as a family and couple practice. While scientific couples have been a subject of study, most notably in *Creative Couples in the Sciences* (Pycior et al., 1996) and in *For Better or For Worse? Collaborative Couples in the Sciences* (Lykknes et al., 2012), it is important to recognize that the majority of those studied were white heterosexual couples from Europe and the US. Cetto and de la Peña are a Mexican upper-class female–male scientific couple. In prior centuries, such relationships could in some cases facilitate women's access to science. In numerous scenarios, even collaborative couples have enacted conventional gender norms, featuring the husband as the prominent scientist–instructor and the wife as the assisting student. Owing to structural sexism, the wife is commonly viewed as subservient, and her contributions are often regarded as secondary (Pycior et al., 1996). Cetto began her married and physics life when she embarked on her doctoral journey under the supervision of her husband, a man who is 15 years her senior. Throughout history, one can observe various other scientific couples in which a woman married a fellow researcher, but marrying her mentor is less common.

Cetto (2025) emphasizes that their collaborative work is truly a team effort, and their individual contributions are not usually distinguished in publications. However, while the majority of her publications are co-authored, approximately 90% of the articles are collaborative efforts between her and her husband. She expresses that de la Peña has proven to be an excellent collaborator who knows how to respect and value his partner's ideas, enriching their work generously while also providing critical input when needed (Cetto, 2006, p. 247). If we examine Cetto's publications in the field of quantum mechanics, out of over 100 articles and chapters, fewer than 10 are authored solely by her. A similar pattern can be observed in their presentations at scientific conferences. As expressed by Cetto in her interview, the order of authorship is often arbitrary when citing their work (Cetto, 2025). In their collaborative efforts, they often divided topics to work on, as each had their own preferences, although they later reviewed each other's drafts. Her favorite topics revolve around the interactions between radiation and matter, while her weaker point is more abstract mathematics and philosophy. Like past scientific couples, such as Marie Skłodowska Curie and Pierre Curie, they use the inclusive "we" in their professional writing to try to ensure that Cetto receives equal recognition for their discoveries (Chiu and Wang, 2011, p. 21). Cetto (2025) mentions that the only work she ever published without de la Peña's inputs is his biography, *Navegantes sin fronteras. Homenaje a Luis de la Peña* (Navigator Without Borders. Tribute to Luis de la Peña) (Cetto, 2006). Nonetheless, it's worth noting that when they are interviewed separately, de la Peña tends not to discuss Cetto's role, while she frequently highlights – often above her own opinions or contributions – his role as husband and research partner.

Her specific contributions to the foundations of quantum mechanics have been especially numerous in recent years. They encompass elucidating the role of the vacuum field as a *quantizing* and matter-stabilizing agent, its role in causing electron spin, and its role as a source of major radiative corrections (Cetto, 2022). She has also contributed to deriving fundamental commutation relations, exploring the dilution of physical meaning in quantum mechanics operators and their implications for Bell-type inequalities (Cetto et al., 2021; Cetto and de la Peña, 2022).

One characteristic of the couple's careers is the various sabbaticals they undertook in Europe, during which they dedicated their time to researching and preparing publications, especially their books. Because it was challenging to find colleagues who shared their precise viewpoint, their goal was to network with those with reasonably similar interests. Cetto (2025) emphasizes their long walks to exchange ideas as well as to tap into her own intuition, as Cetto and de la Peña didn't always think alike. Their first sabbatical, in 1971–1972 at Birkbeck College in London came at the invitation of David Bohm (1917–1992), another quantum dissident with whom they had clear differences regarding quantum theory but also enriching discussions (Freire, 2019; Cetto, 2025). During this sabbatical, they also visited one of their main collaborators, Emilio Santos Corchero, from Universidad de Valladolid, whom they again visited in 1985. Others were at Université de Paris IV (1977–1978), at Università degli Studi di Roma "La Sapienza" (1984–1985), and at University College London (UCL: 1993–1994). While they no longer undertake sabbaticals, they have recently established contact with John Bush, the director of the Applied Mathematics Computational Laboratory at MIT, to study hydrodynamic analogies of quantum physics.

One collaborator to highlight in their legacy is Andrea Valdés-Hernández, a young physicist from UNAM. She began publishing with the scientific couple in 2008, as she was completing her PhD thesis under the supervision of de la Peña and Cetto was in Europe for reasons related to scientific diplomacy (see Section 16.6). Most of their 2008–2021 publications were co-authored by all three researchers. Valdés-Hernández's contributions to topics related to quantum entanglement are particularly noteworthy (Cetto, 2025).

Cetto's work stands out in the field of quantum mechanics because it provides a different way of thinking, a causal and transparent explanation, according to her, that independently and critically examines the fundamentals of the theory. Interpretations of quantum mechanics offer different ways of comprehending and making sense of the principles and equations of this theory. The standard one is usually called the Copenhagen interpretation (Howard, 2004). It emphasizes the significance of observers and measurements, and postulates that quantum particles exist in superposed states until they are measured, at which point their states collapse into one of the possible outcomes. There are other interpretations, such the pilot-wave interpretation (also known as Bohm's interpretation or causal interpretation), the many-worlds interpretation (MWI), and the stochastic interpretation. This last one is what Cetto and de la Peña defend, although for many researchers, "stochastic mechanics does not provide a clear intuitive model of behavior of quantum systems" (Santos, 2022, p. 1251). Cetto and de la Peña posit that the foundations of quantum mechanics are incomplete, and they view its stochastic interpretation as a logical consequence of this incompleteness. This belief was endorsed by a few of the *founding fathers* of quantum theory, Einstein being the best known of them. Cetto and de la Peña identify him as their *guía silencioso* (silent guide) (Santos, 2022, p. 1247; Cetto, 2025). From one of Einstein's famous quotes "God does not play dice," came the title of their first and most cited book, *The Quantum Dice: An Introduction to Stochastic Electrodynamics* (de la Peña and Cetto, 1996). In this monograph, they expounded the application of stochastic theory to electrodynamics and reviewed the history of stochastic approaches to quantum physics. They also showed that they understood stochastic theory differently from other theorists and, most importantly, they pointed to areas where further progress could be made. In this stream of work – which one can call "quantum mechanics as a stochastic theory" – they justified quantum mechanics as a theoretical description of processes involving stochasticity, an element of randomness (Cetto, 2025). Furthermore, they distinguished between classical stochastic processes (described by Einstein in 1905) and those described by quantum mechanics, which has distinct rules. In several respects, this first part of their work is similar to earlier work by Edward Nelson (1932–2014), since they arrived at the same result when deriving the Schrödinger equation (Santos, 2022, p. 1249; Cetto, 2025).

Around 1971, during their first sabbatical with Bohm, the Mexican couple began a second, longer (and still ongoing) stream of theoretical work, which aims to provide an answer to what causes stochasticity in quantum mechanics. For that, they have begun to discuss more in depth the perspective of stochastic electrodynamics (SED). Their deeper investigation into the origin of the quantum phenomenon is documented in the book they co-authored with Valdés-Hernández, *The Emerging Quantum: The Physics Behind Quantum Mechanics* (de la Peña et al., 2015) and in the articles they published prior to it (de la Peña et al., 1972; de la Peña and Cetto, 1975, 1977, 1982). SED is a theory that combines the laws of classical electrodynamics with the assumption that the vacuum is not empty, but there is a background zero-point field (ZPF) filling the whole space. Although the ZPF is a solution of the Maxwell equations, it deviates from classical physics by having a nonzero mean energy at zero temperature, resulting in a fluctuating field responsible for quantization. Cetto and de la Peña's fundamental hypothesis – developed in collaboration with Valdés-Hernández – is that every material system operates as an open system subject to continuous perturbations by this field. The distinguishing feature of the trio's hypothesis is that the perturbations produce a qualitative change, while for the other SED theorists the ZPF is nothing more than a classical perturbation. The Mexican physicists advance that, in essence, quantization does not stem from inherent attributes of matter or the radiation field but emerges from a deeper stochastic process due to an interaction with that field (de la Peña et al., 2015). Recently, they have been working on analogies with hydrodynamics to obtain insight from macroscopic examples, such as a droplet's trajectory that informs the behavior of particles under the effect of radiation (Cetto, 2025).

Cetto's collaborative work is an intellectual project that has spanned more than six decades. The results of this project are now being crystallized by Cetto and de la Peña into a textbook (in preparation). Their aim is to distill their understanding of the quantum phenomenon through SED, a perspective currently absent from the pedagogical literature. Cetto and de la Peña are aware that they are swimming against the mainstream current in physics, but they hope to pass on their legacy to young enthusiasts worldwide. Cetto is bringing to this new endeavor her prior experience authoring textbooks at the secondary- and high-school levels, including *Mecánica* (*Mechanics*, 1984) and *El mundo de la física* (*The World of Physics*, 1981–1993).[13] They want to present results that demonstrate that their theory does not violate quantum mechanics.

Cetto thinks that the usual theory, with its standard interpretations included, seems to tell us more about our knowledge and our way of thinking about nature

---

[13] *El mundo de la física* started as 10 booklets inspired by the Physical Science Study Committee and later became a three-volume edition. A bestseller among physics textbooks in Mexico, it offers a more informal and approachable style than other textbooks but does not include a gender perspective.

than about nature itself. What is more, she and de la Peña believe that "the scientific community is dogmatic, as reflected in the majority of textbooks, and it is essential to provide new generations an alternative that does not rely on 'spooky action'" (Cetto, 2025). She asserts that "quantum mechanics is currently presented as a mere set of postulates (including the Schrödinger equation, the Heisenberg equation, the uncertainty principle, etc.)" (Cetto, 2025). This approach has led to the view – dominant among the majority of scientists and students – that quantum mechanics is a complete theory, requiring no further exploration of its foundations. She believes that this perspective is closely tied to the contemporary pragmatism in science, a byproduct of industrialization (Cetto, 2025). She also finds that many colleagues show interest in her work and her husband's work when they hear about it, but they do not want to change their perspective on the subject, mainly due to the difficulty of publishing alternative viewpoints. This says a lot about the current scientific community and its relationship to or interaction with non-mainstream research and physicists who do not follow established trends. Cetto is a quantum dissident; she even claims (Cetto, 2025) to be "more dissident" than the quantum dissidents portrayed in Olival Freire Junior's book, *The Quantum Dissidents. Rebuilding the Foundations of Quantum Mechanics (1950–1990)* (Freire, 2015).

Although she hopes to gain more visibility for the work that she and her husband have conducted in the scientific community and mainstream journals, she emphasizes that she is not seeking recognition for herself, as she believes that collective efforts are what truly matter (Cetto, 2025). However, it's challenging to be an outlier in a scientific field. The popular image of science is that of a field in which the *brilliant genius* (typically a man) is celebrated even if his *modus operandi* isn't clear. Yet, this representation does not always align with Cetto and her partner's experiences in stochastics research. In conventional heterosexual marital collaborations, women's contributions to mutual scientific efforts have been frequently underrated, while men have often received greater professional acknowledgment (Lykknes et al., 2012, p. 1). Unfortunately, Cetto's case is no exception. Some articles referring to their work mention "de la Peña's theory," even though she is sometimes the first author (e.g., Nassar, 1985, p. 1; Santos, 2022, p. 1249). This subordinates her work to his, as he is the figure who receives full recognition. While de la Peña was granted the status of UNAM Emeritus Researcher in 1994 (at the age of 63) and a UNAM chair professor in 1997, Cetto remains a simple professor to this day (at the age of 78). This difference in recognition translates into her husband's salary being twice the amount of hers. Although seniority has probably been the main factor in their different academic ranks, Cetto and de la Peña feel that gender bias and the related "glass ceiling" have also played a role. Inequality is particularly apparent in terms of prizes. For example, the Premio Nacional de Ciencias y Artes (National Prize for Arts and Sciences) was awarded to him alone for work that they carried out together. Cetto (2025) explains that whenever they have inquired why she did not receive the same recognition, the response has consistently been silence. These differences in recognition, combined with de la Peña's devoted personality, have earned him the reputation of The Professor, and when students seek a thesis advisor, they primarily approach him, especially at the postgraduate level.

16.5 Darkness and Light in Open Science

Under-recognition from the scientific community can be imputed, in part, to the publishing process. According to Cetto (2025), very few journals accept work on the foundations of quantum mechanics, and fewer still accept works on non-mainstream interpretations, such as the stochastic interpretation. One journal that does is Springer's *Foundations of Physics*, founded in 1970. Other journals that have published Cetto and de la Peña's works on the stochastic interpretation include *Revista Mexicana de Física* (from the Mexican Physical Society), *Physical Review Letters* (by the American Physical Society), and *La Rivista del Nuovo Cimento* (by the Italian Physical Society). Cetto (2025) readily admits that she and de la Peña have faced challenges related to publishing. When they submit their work, the response often is that the manuscript is fine, but that the topic is not of interest to the journal, resulting in their articles being summarily rejected. She also believes that too few researchers work on this topic, a situation she attributes to the current pressure on publishing. Researchers tend to gravitate toward easier or more applicable topics, as a consequence of the industrialization of science. The focus is primarily on the products, which are the articles rather than the ideas, and on applied research at the expense of foundational work and dissenting viewpoints (Cetto, 2025). Her case is not isolated but serves as a good example to illustrate the character of the current academic system, or "fast science," in which the prevailing "publish or perish" pressures tend to reinforce existing disparities (Leite and Diele-Viegas, 2021). While it's challenging for Cetto and de la Peña to gain visibility, they are cited predominantly by authors from Europe and the US. This citation pattern suggests that in Mexico and in Latin America more generally, particularly among physicists, there is a tendency to align with the mainstream and to rely heavily on knowledge generated in northern contexts.

Like many other scientists who are also scientists' wives, Cetto has devised strategies to uphold her own *scientific identity* (Kohlstedt, 2012, p. vii). One such strategy has been to cultivate interests related to, yet distinct from, the foundations of quantum mechanics, such as the pursuit of open science and public outreach (presented in this section) and the development of international scientific cooperation (see Section 16.6).

Cetto's concern about scientific publication policies began in the early 1990s. A particularly important episode in this respect took place in 1993, during her second sabbatical in London, at UCL. At that time, Cetto was spending a lot of time at the UCL Main Library because she and her husband were writing their first book, *The Quantum Dice* (1996). During the previous two years (1990–1992), Cetto had been the director of the *Revista Mexicana de Física*, which had managed to get included in the Science Citation Index (SCI) of the Institute for Scientific Information. One day, she decided to look for this journal on the shelves of the UCL Main Library, which she knew was sent free of charge to that institution. To her surprise, she couldn't find it. Most Latin American scientific journals were also absent. This led her to realize that the problem was regional (Cetto, 2025).

In fact, there was a growing concern in Mexico and the rest of Latin America about this very issue, but it was being addressed piecemeal. Cetto, who was part of the Scientific Publishing Committee of the International Council for Science (ICSU), sought the support of Kai Inge Hillerud, the director of ICSU Press, to organize the first International Workshop on Scientific Publications in Latin America. This workshop took place at the end of 1994 as part of the renowned Feria Internacional del Libro (FIL, International Book Fair) in Guadalajara, Mexico. The objective was to invite journal editors, specialists in journal information systems, and creators of regional databases and libraries, among others, to understand the root of the problem and discuss it from a Latin American perspective. Cetto (2025) wanted to break free from a vicious cycle that led to underrepresentation or even outright neglect, lack of recognition, and low visibility of knowledge from the region. The recommendations that emerged from the workshop were published in a volume titled *Publicaciones científicas en América Latina/Scientific Publications in Latin America* (Cetto and Hillerud, 1995). From the collected proposals, Cetto chose to pursue one: the creation of a database of journals published in the region, which is nowadays known as Latindex (Regional Online Information System for Scientific Journals from Latin America, the Caribbean, Spain, and Portugal).

Latindex is a scientific–political project that reacts to the colonization of knowledge and the influence of the Global North (or what is often referred to as the international community), thus helping Latin America escape its state of dependency. It started with a comprehensive database of all academic journals in the region, including those focusing on science communication and those that are no longer extant. Subsequently, a set of quality criteria was established to collectively define what constitutes a good journal according to the idiosyncrasies and the historical, cultural, and scientific context of the region, without denying the value of international journals. This effort resulted in the creation of the catalog, which is not a ranking but rather a database of journals that meet a sufficient but flexible number of factors (Cetto et al., 1999, p. 249). These criteria have helped set certain standards in the region, thus supporting publishers to increase the visibility of their journals. The project also supports and promotes scientific publication in regional languages. (Nevertheless, Cetto has very often used English for her quantum mechanics publications to have an impact at other latitudes.) Latindex is a pioneer in open and free access.[14] Cetto thinks that Europe and the US should look to Latin America as an example of how to regain control of scientific publishing, de-commercialize it, eliminate predatory journals, and redefine the scientific evaluation system (Cetto, 2025).

Through her scientific cooperation, Cetto can exert a certain influence regarding this concern. Following her participation in ICSU committees, in the late 1990s, Federico Mayor Zaragoza, the then Director-General of UNESCO (1987–1999), invited her to join an International Scientific Advisory Board, in which she actively participated for several years. There, Cetto emphasized the necessity for scientists and science to engage with society. As a result, she was entrusted with the task of developing the scientific program for the World Conference on Science for the 21st Century in Budapest in 1999. Cetto was viewed as someone who broadened the predominantly northern perspectives that had often characterized such *international* conferences. She also took on the responsibility of gathering, analyzing, and filtering recommendations and conclusions that came from the regional meetings, held at UNESCO's regional offices prior to the conference. With this material, she crafted a declaration and an action plan. She later edited and published it (Cetto, 2000). According to her, this book set the basis for future discussion on science and open access. In a way, it depicted the situation and functioning of Latin America, recognizing an unequal distribution of knowledge, and advocating for the free circulation of knowledge and scientific communication. However, Cetto (2025) also recognizes, based on her experience as a working scientist, that things are not as straightforward as one might hope from the field of action and do not change by decree. Cetto is also part of the board of the Directory of Open Access Journals and of UNESCO's directive Open Science Committee. The latter published in 2021 a rather advanced set of non-binding recommendations (UNESCO, 2021). However, according to Cetto (2025), there is still too much room for maneuvering. This puts open science at risk of falling into the hands of those who interpret open access as a gateway for commercial and profit-driven purposes, as is the case with the majority of Northern-based publishing houses.

---

[14] The term "diamond open access" is avoided intentionally, as it is a term from the Global North that appropriated practices that were already in use in other regions. In fact, Cetto points out that these terms originated in the Global North but were common practices in Latin America and had always been in operation in Latin America (Cetto, 2025).

In addition to her concerns about scientific publication, partly due to the Latin American situation and the challenges of publishing on the fundamentals of quantum mechanics in high-impact journals, Cetto has always considered science communication to be an essential aspect of her work (via textbooks, popular publications, museums, and more). Her efforts in popularizing physics have predominantly centered around the phenomena of light. According to her, "light has always pursued me! It envelops and fills our surroundings, yet it also possesses something intangible and guards its mysteries. Is there any evidence that light is composed of photons?" (Cetto, 2025). One notable work is her book, *La luz. En la naturaleza y en el laboratorio* (*Light. In Nature and in the Laboratory*) (Cetto, 1986) published under the guidance and with the cooperation of the actress María del Carmen Farías Roman and the physicist Alejandra Jáidar Matalobos (1937–1988). Following its success, in 1994, the rector of UNAM, José Sarukhán Kermez, proposed the creation of the Museo de la Luz (Museum of Light) in the historic building of the former San Pedro and San Pablo temple.[15] Cetto (2025) taught herself museology and drew inspiration primarily from two museums, the Exploratorium in San Francisco and the exploratory hands-on science centre in Bristol, UK, now called "We The Curious."

These experiences related to light and her diplomacy throughout her career (see Section 16.6) prepared Cetto to become the promoter of the International Year of Light and Light-based Technologies 2015, after two years of intense work alongside UNESCO colleagues. Additionally, she played a significant role in the inauguration of the "Lights and Shadow" exhibition at the Miró Hall of UNESCO's headquarters in Paris in 2018. The invitation to participate in the latter event arose almost by chance when she visited her former colleagues for a matter related to the Museo de la Luz's renovation. Cetto (2025) underscores the usefulness and impact of such occasions, as they involve numerous local activities that can shed light on environmental issues and other relevant matters. For instance, in Mexico, this led to the emergence of the project "Luces sobre la ciudad" (Lights above the City), which she coordinates through the Environmental Coalition on Standards network, an interdisciplinary organization focused on education, science, technology, and innovation in Mexico City.

---

[15] To design, build, and install the Museo de la Luz, Cetto formed a team composed of Luis de la Peña, Salvador Cuevas, Carlos Vázquez-Yánez, Emma Orozco, Glinda Irazoque, Manuel Martín, Pablo Pacheco, Giovanna Recchia, and Humberto Ricalde. The museum was inaugurated on November 18, 1996. In 2011, it was relocated to the Patio Chico of San Ildefonso, also in Mexico City. Unfortunately, despite Cetto's determined efforts, the museum is currently closed to the public. For more information about the museum https://patrimonio-cyt-cdmx.colmex.mx.

16.6 In Between Two Worlds. Regionality in Global Science

Cetto's first international academic experience occurred in 1966, during her final undergraduate year, when she attended the International Physics Congress in Beijing, China, alongside Virgilio Beltrán López and de la Peña. This rare opportunity for a student was made possible because there were surplus invitations available (Cetto, 2006, p. 239). During the Cold War, it was difficult to obtain a visa for the US if a scholar had visited China or another communist country, so many fellow Mexican researchers stayed away. She had another cross-cultural experience during her graduate studies at Harvard University. This institution provided her with significant resources in terms of facilities and materials. She also had the opportunity to hone her English skills, and thus achieved proficiency in three languages, Spanish, German, and English; this proved to be a significant advantage for her trajectory. But she also noted clear disparities in the student environment compared to that of UNAM. Despite the generous resources, Harvard's environment was characterized by competition and a notable absence of solidarity (Cetto, 2025).

Back in Mexico, Cetto's research in quantum mechanics became closely intertwined with a commitment to a particular approach to scientific inquiry that promotes collaborative and open work. Before turning 30 and with one small child, she began to assume roles related to administration at UNAM. Her initial position, elected by the faculty and some students, was that of Coordinator of the Department of Physics during the academic years 1975–1977. This period proved challenging, as UNAM's administrative structures were undergoing significant changes. Subsequently, she held the position of Director of the Faculty of Science (1978–1982). In this position, she encountered, for the first time, explicit differential treatment due to her sex, such as questioning of her leadership and comments about her youth and experience, and hence she needed to demonstrate her capabilities and qualifications as Director (Cetto, 2025).

Her interest in international scientific cooperation began to grow in the 1970s and 1980s, when she and her husband meticulously studied statistical yearbooks published by the United Nations Statistics Division. Their objective was to assess the scientific and technological development of countries, particularly in the aftermath of international conflicts. It was at this time that Cetto keenly perceived the glaring disparity in scientific participation among nations, a concern that deeply troubled her.

As director of the Faculty of Science, she was invited to attend the General Assembly of the International Union of Pure and Applied Physics (IUPAP) in 1981. Previously, Mexico had participated only sporadically in the organization, primarily because of its high membership fee. Cetto continued to be a part of the IUPAP, although the atmosphere was not as international as one might have hoped. As is often the case, the assembly was composed of all white, European, North American, and Soviet male physicists. This made her feel quite uncomfortable since, being short and soft-spoken, she often went unnoticed, as if she did not exist (Cetto, 2025). Her goal then became to expand the participation of physicists from other countries. To this effect, she listened attentively for a long time until she acquired the language and communication style of physicists from hegemonic countries. She spoke in a way that the members could understand, drawing their attention by discussing topics to which they were not accustomed (Cetto, 2025). Her efforts bore fruit in 1988, when she was invited to represent Latin America at the Third World Academy of Science (TWAS, now the World Academy of Science for the Advancement of Science in Developing Countries). TWAS was founded by the Pakistani physicist Abu Ahmad Muhammad Abdus Salam (1926–1996) with the support of UNESCO and the International Atomic Energy Agency (IAEA) to promote the exchange of knowledge among scientists. As the most prominent scientists from each country were typically male, TWAS had minimal female representation. After concerns were raised about this gender gap, Salam and the ICSU Executive Director, Julia Marton-Lefèvre, organized a gathering exclusively for female scientists in 1989. To do so they reached out to several academies of sciences and requested two or three names of their most distinguished women scientists. This process led to the selection of Cetto and Ruth Sonabend Moszkiewicz (also known as Ruth Gall), a Mexican physicist of Polish origin who specialized in the study of cosmic rays (Gall Sonabend, 2005). At this meeting, the Third World Organization for Women in Science (TWOWS) was established, with Cetto serving as one of the inaugural vice presidents; she represented Latin America and the Caribbean within the organization.

Cetto (2025) acknowledges that before being invited to TWOWS (now the Organization for Women in Science for the Developing World, OWSD), she did not consider gender bias in science a priority. After her designation, TWOWS decided to conduct a nationwide survey in Mexico to analyze the issues and obstacles women scientists faced. According to Cetto (2023), the poll revealed that while the glass ceiling was evident, the problem of undervaluation and segregation of women's work was much less significant in Mexico than in Europe and the US. She believes this difference may be attributed to the fact that the scientific field in Mexico was not as competitive as in other regions, given its relatively lower maturity level in scientific production. And this environment, she argues, was more open, flexible, and accommodating, which allowed more women to do science. Even though she does not consider herself a feminist, she now believes that this issue needs to be addressed. Moreover, she believes it also affects men, and therefore they must also participate in solving it (Cetto, 2025). She has written several articles, chapters, and a bulletin called *Supercuerdas* (with the collaboration of Hortensia González Gómez in the 1990s) on the status of women in science in Mexico and Latin America more generally. Apart from the academy, Cetto also become part of Mujeres por la soberanía nacional y la unidad latinoamericana (Musial; Women for the Sovereignty and Integration of Latin America), through which she engaged in discussions with Mexican intellectuals like Ifigenia Martínez y Hernández, Sol Arguedas Urbina (1928–2015), Ruth Gall (1920–2003), Elena Urrutia (1932 –2015), and others, on topics of common interest, particularly sovereignty.[16] In her role as representative from the marginality or periphery, she has also served as a member of the ICSU Committee on Science and Technology in Developing Countries (COSTED), both as vice president and co-secretary during the period 1990–2001.

From the 1990s onward, her involvement in international cooperation, as she calls it (Cetto, 2025), surged. Not only has she been active in organizations focused on science in peripheral countries and on academic publications, but she has also consistently been regarded as a *spokesperson* for the vision of science in Latin America and the Caribbean. She began as a member of the Executive Council of the International Network of Engineers and Scientists, which allowed her to become a member of the Pugwash Conferences in 1992. These international meetings bring together scientists and policymakers to address global security challenges, particularly those related to nuclear weapons proliferation and disarmament. They are exclusive and closed events, which can only be attended by personal invitation (Kraft and Sachse, 2019).[17] The presence of Mexicans had been very limited. Although members did not formally represent specific countries, Cetto took on a role as the Latin American representative, and she brought important issues to the table, such as scientific, technical, and economic dependence. She was a member when the organization was awarded the Nobel Peace Prize in 1995. She then served as President of the Executive Committee from 1997 to 2002, becoming only the second woman – after Dorothy Mary Crowfoot Hodgkin (1910–1994) – to hold this position. She noted the scarcity of women in these conferences, suggesting that there must have been bias in access (Cetto, 2025). Nevertheless, she learned to manage conflicts and establish networks with prominent figures during her involvement in Pugwash.

Cetto continued participating in Pugwash until she was appointed Deputy Director General (2003–2011) of the IAEA, which meant she had to resign from this previous position due to the potential conflict of interest. Pugwash was not the only collaboration which she had to let go, but she was clear that she would not abandon quantum mechanics or the Latindex project. Although Pugwash and the IAEA

---

[16] Musial notably published press releases in *La Jornada*, a Mexican newspaper founded in 1984.
[17] Cetto was invited to join by Mexican physicist Octavio Miramontes Vidal, who although being a member was not very active within the organization.

could be viewed as pursuing similar goals, Cetto found moving from a non-governmental organization to an intergovernmental institution challenging (Cetto, 2025). During this period, Cetto was in charge of the IAEA's technical cooperation program, for which she had to exercise mediation skills and adapt to the institution's remarkably rigid hierarchy. Her work promoting peaceful applications of nuclear energy led to the IAEA receiving the Nobel Peace Prize in 2005. This position was the most enduring, and she perceived more fully the essence of scientific diplomacy as she witnessed how diplomats represented their countries and personal agendas, and how challenging it could be to pursue common interests while serving everyone, without favoring any individual country (Cetto, 2025).

A consistent aspect of her diplomatic career in the 1990s was her engagement with UNESCO. She felt a deep connection to this institution due to their shared values. From this relationship, it is worth highlighting her role in the scientific program of the World Conference on Science (1999), as well as her involvement in drafting and editing its documents, particularly the Declaration on a New Social Contract for Science (Cetto, 2000). This earned her an active role on the Governing Board of the United Nations University (UNU) in Tokyo, Japan, and a position as UNESCO–UNAM Chair on Science Diplomacy and Scientific Heritage recently created in Mexico. In addition, she has been an active participant in the Open Science Committee, as discussed in Section 16.5.

These experiences have led Cetto to live outside of Mexico for more than a decade, particularly in Vienna and in other parts of Europe. It was her decision to stay abroad for such a long period. She was highly motivated by the work, which was a continuous learning experience, as each day brought something new. Despite promises to join her, her husband remained in Mexico City. This meant that Cetto lived abroad alone or, during some periods, with her daughter Carolina. Despite the separation, she continued collaborating with de la Peña. Their publication rate decreased somewhat, but she never completely abandoned her research on the foundations of quantum mechanics.[18]

After all her international experiences, Cetto refuses to use the term "scientific diplomacy," opting instead for *scientific cooperation* (Cetto, 2025). She believes that diplomacy is not science, and science is not diplomacy; moreover, bridging the gap between science and the diplomatic sphere is far from trivial. As a scientist, being diplomatic is challenging, as it entails advocating for governments' positions, which may go against one's own principles. Ultimately, although scientists play a relevant advisory role, they do not have final decision-making authority. Cetto considers that working in regional capacities has helped her to develop a multilateral

---

[18] During Cetto's absence, de la Peña took on another collaborator, Valdés-Hernández (Cetto, 2023) (see Section 16.4).

perspective and the ability to assist countries in acquiring independence in scientific and technical fields. However, Cetto has become a representative of Latin America in each and every one of her experiences, even if this was not the initial purpose of these positions. This also clearly illustrates how science and diplomacy operate through personal connections, insofar as individuals are selected and invited to assume positions of responsibility. Additionally, from her positions in international organizations such as the IUPAP and the ICSU, she noticed consistent collaboration between institutions and strong interrelationships among their respective members. These practices leave little room for other participants, and the same voices tend to dominate. Membership by invitation through personal connections appears to her more like a way of fulfilling the quota of members from a marginalized region – in this case, Latin America – rather than of showing a genuine interest in addressing underrepresentation. Whether intentionally or not, such practices can perpetuate a cycle of familiar faces and networks, potentially hindering diversity and inclusion efforts. It's essential to reflect on whether Cetto's positions, had they been occupied by someone from Europe or the US, would have been mainly identified with those regions, and pigeonholed as voices for the Global North. Furthermore, speaking of Latin America as a single entity overlooks the significant social, economic, and other differences within the region. The same concern arises with appointments of *token women* in scientific and diplomatic institutions (Rossiter, 1997, p. 171; Ferguson, 2015, p. 386; Prügl, 2015, p. 623).

16.7 Conclusions. All-In-One: International Year of Quantum Science and Technology

Coming from a family of European origin, growing up in a Mexican environment, and witnessing the trajectories of close friends all influenced Cetto's career in scientific cooperation. On the one hand, her goal has been to develop an alternative way to conduct diplomacy in science, one different from the northern strategies. On the other hand, her scientific realism and collaborative work with her partner, de la Peña, have led her to adopt a minority, dissenting view on the foundations of quantum mechanics, even when that has meant challenging the scientific community and struggling with the hurdles of publication. Her commitment to open science and scientific publications further demonstrates how Latin America can serve as a role model for achieving honest open access.

Cetto's career paths converge at a common culminating point, the 2025 International Year of Quantum Science and Technology (IYQ). This initiative was born in the IUPAP scientific community and was led by Mexico, with Cetto being one of its key promoters. Her prior experience with international years, her valuable connections in the diplomatic sphere, and her understanding of bureaucratic processes contributed to her taking on a leadership role. Moreover, until 2023, she served as the President of the Sociedad Mexicana de Física (SMF, Mexican Physics Society). It might be surprising that the proposal for the IYQ was not led by a Central European country, given that quantum mechanics originated there, or by the US, given its significant role in the development of quantum technologies. However, the US left UNESCO from 2018 until 2023, and countries like Germany have historically declined to lead similar initiatives in the past. In contrast, Mexico is respected by many nations for its foreign policy and possesses a sufficient level of scientific development without being labeled a developed country, which is advantageous for promoting quantum technologies as a global asset.[19]

Quantum theory, far from remaining solely an academic pursuit confined to the scientific domain, has been enriched by its applications, similar to other major branches of physics in the past. It has evolved into a vibrant source of research programs running the gamut from foundational principles to basic and applied projects. The common objective of the IYQ is to showcase quantum mechanics applications and their substantial economic impact (International Year of Quantum Science and Technology Partners, 2024). This aim addresses the growing public mistrust of science and scientists. For Cetto, this year aligns perfectly with her over-50-year legacy of defending a different understanding of quantum phenomena. She sees a significant opportunity to rectify historical and physical misconceptions and misinterpretations of physical and quantum formalism. She believes that, like much of physics, quantum theory has been pragmatically and instrumentally developed, often focusing solely on applications and outcomes while neglecting the fundamental roots and formalisms themselves (Cetto, 2025). To this end, the draft proposals for IYQ also discussed the significance of international collaboration and the free sharing of scientific information, which is now known as open science.

Cetto has consistently demonstrated a critical or alternative perspective in her experiences, actively seeking spaces that had not yet been widely explored or discussed. However, she has also benefited from friendships and privilege, especially when she began to participate actively in diplomacy, although her role and voice have been mostly confined to the Latin America domain. Throughout her work, especially in quantum mechanics, her recognition has been very limited. This is partly due to stochastic mechanics remaining a minority perspective, and one that is undervalued in the physics community, and partly due to the associated struggle of publishing in high-impact journals. Moreover, relative to her husband, she has received less recognition because of gender discrimination, which remains an issue even today. Given these structural asymmetries it is especially important to recognize the power of oral history as a historiographic source that allows us to hear her important voice.

---

[19] It is worth noting, however, that while Mexico leads the initiative, the American Physical Society (APS) and the Deutsche Physikalische Gesellschaft (DPG, German Physical Society) have created the official website and will organize most of the activities on the agenda.

References

Berger Gluch, Sherna and Patai, Daphne (eds.). 1991. *Women's Words: The Feminist Practice of Oral History*. New York: Routledge.

Canal 22. 2019. Voz de mujer. Ana María Cetto. Online videoclip, April 14. https://youtu .be/6ZpzNJQy3Y8?si=f52gE8Q5b3cNeq2z

Cepeda Flores, Francisco Javier. 2006. *El Prometeo en México: Raíces Sociales y Desarrollo de la Facultad de Ciencias, UNAM 1867–1980*. Ciudad de México: Facultad de Ciencias Físico Matemáticas.

Cetto, Ana María. 1986. *La Luz. En la naturaleza y en el laboratorio (Ciencia para todos)*. Ciudad de México: Fondo Cultura Económica.

Cetto, Ana María and Hillerud, Kai-Inge (eds.). 1995. *Publicaciones científicas en América Latina/Scientific Publications in Latin America*. Ciudad de México: Fondo Cultura Económica.

Cetto, Ana María, Alonso, Octavio, and Rovalo, Lourdes. 1999. Latindex, a dos años de su concepción. In Cetto Kramis, Ana María, and Alsonso, Octavio (eds.), *Revistas Científicas en América Latina/Scientific Journals in Latin America*. Ciudad de México: Fondo de Cultura Económica, pp. 247–254.

Cetto, Ana María (ed.). 2000. *World Conference on Science: Science for the Twenty-first Century; A New Commitment*. Paris: UNESCO.

Cetto, Ana María. 2006. Apuntes para una biografía ilustrada. In Cetto, Ana María (ed.), *Navegantes sin fronteras. Homenaje a Luis de la Peña*. Ciudad de México: Instituto de Física, pp. 233–258.

Cetto, Ana María, de la Peña, Luis, and Valdés-Hernández, Andrea. 2021. On the physical origin of the quantum operator formalism. *Quantum Studies Mathematics and Foundations*, 8(2), 229–236. https://doi.org/10.1007/s40509-020-00241-7

Cetto, Ana María. 2022. Electron spin correlations: probabilistic description and geometric representation. *Entropy*, 24(10), 1439. https://doi.org/10.3390/e24101439

Cetto, Ana María and de la Peña, Luis. 2022. Role of the electromagnetic vacuum in the transition from classical to quantum mechanics. *Foundations of Physics*, 52, 84. https://doi.org/10.1007/s10701-022-00605-6

Cetto, Ana María. 2025. Oral history interview of Ana María Cetto Kramis by Mar Rivera Colomer on September 1 to 7, 2023. Niels Bohr Library & Archives, American Institute of Physics, College Park, MD USA.

Chaelton, Thomas L., Myers, Lois E., and Sharpless, Rebecca (eds.). 2007. *The History of Oral History: Foundations and Methodology*. Lanham: Altamira Press.

Chiu, Mei-Hung and Wang, Nadia Y. 2011. Marie Curie and science education. In Chiu, Mei-Hung, Glimer, Penny J., and Treagust, David F. (eds.), *Celebrating the 100th Anniversary of Madame Marie Sklodowska Curie's Nobel Prize in Chemistry*. Rotterdam: Sense Publishers, pp. 9–39.

de la Peña, Luis, Cetto, Ana María, and Brody, Tomas A. 1972. On hidden-variable theories and Bell's inequality. *Lettere al Nuovo Cimento*, 5, 177–181. https://doi.org/10.1007/BF02815921

de la Peña, Luis and Cetto, Ana María. 1975. Stochastic theory for classical mechanical systems. *Foundations of Physics*, 5(2), 355–370. https://doi.org/10.1007/BF00717450

de la Peña, Luis and Cetto, Ana María. 1977. Derivation of quantum mechanics from stochastic electrodynamics. *Journal of Mathematical Physics*, 8, 1612–1622. https://doi.org/10.1063/1.523448

de la Peña, Luis and Cetto, Ana María. 1982. Does quantum mechanics accept a stochastic support? *Foundations of Physics*, 12, 1017–1037. https://doi.org/10.1007/BF01889274

de la Peña, Luis and Cetto, Ana María. 1996. *The Quantum Dice: An Introduction to Stochastic Electrodynamics*. London: Springer.

de la Peña, Luis. 2003. La mecánica cuántica en México. La visión desde la física. In Ramos Lara, María de la Paz (ed.), *La Mecánica cuántica en México: una visión interdisciplinaria a cien años de su nacimiento*. Ciudad de México: Siglo XXI de España Editores, pp. 21–44.

de la Peña, Luis, Cetto, Ana María, and Valdés-Hernández, Andrea. 2015. *The Emerging Quantum: The Physics Behind Quantum Mechanics*. London: Springer.

Domínguez Martínez, Raúl. 1999. *Historia de la física nuclear en México*. Ciudad de México: Plaza y Valdés Editores.

Ferguson, Lucy. 2015. This is our gender person: the messy business of working as a gender expert in international development. *International Feminist Journal of Politics*, 17(3), 380–397. https://doi.org/10.1080/14616742.2014.918787

Freire, Olival Jr. 2015. *The Quantum Dissidents. Rebuilding the Foundations of Quantum Mechanics (1950–1990)*. Heidelberg: Springer.

Freire, Olival Jr. 2019. On the legacy of a notable quantum dissident: David Bohm (1917–1992). In Matthews, Michael R. (ed.), *Mario Bunge: A Centenary Festschrift*. Cham: Springer, pp. 349–360.

Gall Sonabend, Olivia. 2005. Ruth Gall. In Viesca López, Georgina (ed.), *Biografías de personajes ilustres. Vol. 4, Ciencia y tecnología en México en el siglo XXI*. Ciudad de México: Academia Mexicana de Ciencias, pp. 93–110.

Grant, Linda, Ward, Kathryn B., and Forshner, Carrie. 1993. Mentoring experiences of women and men in academic physics and astronomy. In Urry, Meg C., Danly, Laura, Sherbert, Lisa.E., and Gonzaga, Shireen (eds.), *Proceedings of Women in Astronomy Meeting*. Baltimore: Space Telescope Science Group, pp. 81–86.

Holton, Gerald. 1961. On the recent past of physics. *American Journal of Physics*, 29, 805–810. https://doi.org/10.1119/1.1937623

Howard, Don. 2004. Who invented the "Copenhagen interpretation"? A study in mythology. *Philosophy of Science*, 71(5), 669–682. https://doi.org/10.1086/425941

International Year of Quantum Science and Technology Partners. 2024. Home. 100 years of quantum is just the beginning . . ., February 28. https://quantum2025.org/en

Kohlstedt, Sally G. 2012. Foreword: the material and personal value of care. In Lykknes, Annette, Optiz, Donald L., and van Tiggelen, Brigitte (eds.), *For Better or for Worse? Collaborative Couples in the Sciences*. Basel: Springer-Birhäuser, pp. v–viii.

Kraft, Alison and Sachse, Carola. 2019. *Science, (Anti-)Communism and Diplomacy. The Pugwash Conferences on Science and World Affairs in the Early Cold War*. Leiden-Boston: Brill.

Kuhn, Thomas S., Heilbron, John L., Forman, Paul, and Allen, Lini. 1967. *Sources for History of Quantum Physics. An Inventory and Report*. Philadelphia, PN: The American Philosophical Society. https://www.amphilsoc.org/guides/ahqp/index.htm

Leite, Luciana and Diele-Viegas, Luisa M. 2021. Juggling slow and fast science. *Nature Human Behavior*, 5(2), 409. https://doi.org/10.1038/s41562-021-01080-1

Lykknes, Annette, Optiz, Donald L., and van Tiggelen, Brigitte (eds.). 2012. *For Better or for Worse? Collaborative Couples in the Sciences*. Basel: Springer-Birhäuser.

Minor García, Adriana. 2019. *Cruzar fronteras: Movilizaciones científicas y relaciones interamericanas en la trayectoria de Manuel Sandoval Vallarta (1917–1942)*. Ciudad de México: CISAN-UNAM & El Colegio de Michoacán.

Minor García, Adriana. 2021. Atoms in the campus: Van der Graaff accelerators and the making of two major Latin American universities in the 1950s Brazil and Mexico. *Annals of Science*, 78(4), 504–530. https://doi.org/10.1080/00033790.2021.1949493

Nassar, Antônio B. 1985. Derivation of a generalized nonlinear Schrödinger–Langevin equation. *Physics Letters A*, 109(1–2), 1–3. https://doi.org/10.1016/0375-9601(85)90377-9

Oreskes, Naomi. 1996. Objectivity or heroism? On the invisibility of women in science. *Osiris*, 11, 87–113. https://doi.org/10.1086/368756

Perks, Robert, and Thomson, Alistair (eds.). 1998. *The Oral History Reader*. New York: Routledge.

Plasencia Gaspar, Leticia, Ramos Lara, María de la Paz, and Lozano Mejía, Juan Manuel. 2010. La formación profesional del físico en la UNAM. Trayectoria de su plan de estudios. *Perfiles Educativos XXXIII*, 131, 155–175.

Prügl, Elisabeth. 2015. Neoliberalising feminism. *New Political Economy*, 20(4), 614–631. https://doi.org/10.1080/13563467.2014.951614

Pycior, Helena M., Slack, Nancy G., and Abir-Am, Pnina G. (eds.). 1996. *Creative Couples in the Science*. New Brunswick: Rutgers University Press.

Ramos, Lenina. 2020. Ana María Cetto: "Admiro a Einstein, pero en la cocina sigo mis propias leyes." Milenio, February 18, Lado B. https://www.milenio.com/ciencia-y-salud/ana-maria-cetto-admiro-einstein-cocina-sigo

Ramos Lara, María de la Paz. 2003. *La Mecánica cuántica en México: Una visión interdisciplinaria a cien años de su nacimiento*. Ciudad de México: Siglo XXI de España Editores.

Reskin, Barbara F. 1978. Scientific productivity, sex, and the location in the institution of science. *American Journal of Sociology*, 83(5), 1235–1243. https://doi.org/10.1086/226681

Rodríguez Kuri, Ariel. 2003. Los primeros días. Una explicación de los orígenes inmediatos del movimiento estudiantil de 1968. *Historia Mexicana*, 53(1), 179–228.

Rossiter, Margaret W. 1997. Which science? Which women? *Osiris*, 12, 169–185. https://doi.org/10.1086/649272

Sangster, Joan. 1998. Telling our stories: feminist debates and the use of oral history. In Perks, Robert and Thomson, Alistair (eds.), *The Oral History Reader*. New York: Routledge, pp. 87–100.

Santos, Emilio. 2022. Stochastic interpretations of quantum mechanics. In Freire, Olival Jr, Bacciagaluppi, Guido, Darrigol, Olivier, et al. (eds.), *The Oxford Handbook of the History of Quantum Interpretations*. Oxford: Oxford University Press, pp, 1247–1263.

Shopes, Linda. 2002. What is oral history? Making sense of evidence. History Matters: the US Survey on the Web. http://historymatters.gmu.edu

te Heesen, Anke. 2020. Thomas S. Kuhn, earwitness: Interviewing and the making of a new history of science. *Isis*, 111(1), 86–97. https://doi.org/10.1086/708277

te Heesen, Anke. 2021. Spoken words, written memories: early oral history and elite interviews. *History of Humanities*, 6(1), 163–178. https://doi.org/10.1086/713261

te Heesen, Anke. 2022. *Revolutionäre im Interview: Thomas Kuhn, Quantenphysik und Oral History*. Berlin: Wagenbach Verlag.

UNESCO. 2021. *Recommendation on Open Science*. Paris: UNESCO.

Wheeler, John A. 1967. Preface. *Sources for History of Quantum Physics*. Philadelphia, PN: The American Philosophical Society. https://www.amphilsoc.org/guides/ahqp/index.htm

Index

For EU product safety concerns, contact us at Calle de José Abascal, 56–1°,
28003 Madrid, Spain or eugpsr@cambridge.org.

www.ingramcontent.com/pod-product-compliance
Ingram Content Group UK Ltd.
Pitfield, Milton Keynes, MK11 3LW, UK
UKHW051855101225
465912UK00009B/519